Lecture Notes in Computer Science 3560

Commenced Publication in 1973
Founding and Former Series Editors:
Gerhard Goos, Juris Hartmanis, and Jan van Leeuwen

T0223727

Viktor K. Prasanna   Sitharama Iyengar
Paul Spirakis   Matt Welsh (Eds.)

Distributed Computing
in Sensor Systems

First IEEE International Conference, DCOSS 2005
Marina del Rey, CA, USA, June 30 – July 1, 2005
Proceedings

 Springer

Volume Editors

Viktor K. Prasanna
University of Southern California
Department of Electrical Engineering, Los Angeles, California, 90089-2562, USA
E-mail: prasanna@usc.edu

Sitharama Iyengar
Louisiana State University
Department of Computer Science, Baton Rouge, LA 70802, USA
E-mail: iyengar@bit.csc.lsu.edu

Paul Spirakis
Research Academic Computer Technology Institute (RACTI) and Patras University
Riga Fereou 61, 26221 Patras, Greece
E-mail: spirakis@cti.gr

Matt Welsh
Harvard University
Division of Engineering and Applied Sciences, Cambridge MA 02138, USA
E-mail: mdw@eecs.harvard.edu

Library of Congress Control Number: 2005927485

CR Subject Classification (1998): C.2.4, C.2, D.4.4, E.1, F.2.2, G.2.2, H.4

ISSN   0302-9743
ISBN-10   3-540-26422-1 Springer Berlin Heidelberg New York
ISBN-13   978-3-540-26422-4 Springer Berlin Heidelberg New York

Springer is a part of Springer Science+Business Media

springeronline.com

© Springer-Verlag Berlin Heidelberg 2005
Printed in Germany

Typesetting: Camera-ready by author, data conversion by Scientific Publishing Services, Chennai, India
Printed on acid-free paper   SPIN: 11502593   06/3142   5 4 3 2 1 0

Message from the Program Chair

This volume contains the papers presented at the 1st *IEEE International Conference on Distributed Computing in Sensor Systems* (DCOSS 2005), which took place in Marina del Rey, California, from June 30 to July 1, 2005. DCOSS focuses on distributed computing issues in large-scale networked sensor systems, including systematic design techniques and tools, algorithms, and applications.

The volume contains 26 contributed papers, selected by the Program Committee from 85 submissions received in response to the call for papers. In addition, it also contains the abstracts of the invited poster on session entitled "distributed sensor systems in the real world" and abstracts from the poster session based on the response to call for posters.

I would like to thank all of the authors who submitted papers, our invited speakers, the panelists, the external referees we consulted and the members of the Program Committee. I am indebted to the Program Vice Chairs Sitharama Iyengar, Pavlos Spirakis, and Matt Welsh for their efforts in handling the review process and for their recommendations.

I would like to thank Mani Chandy for organizing a panel entitled "From Sensor Networks to Intelligence."

Several volunteers assisted us in putting together the meeting. I would like to thank Bhaskar Krishnamachari for his inputs in deciding the meeting focus and format, Loren Schwiebert for handling the student scholarships, Wendi Heinzelman for publicizing the event, Amol Bakshi for many helpful suggestions on the meeting focus and publicity, and Pierre Leone for his assistance in putting together these proceedings.

It was a pleasure working with José Rolim, General Chair, and Sotiris Nikoletseas, Vice General Chair. Their invaluable input in putting together the meeting program and in shaping the meeting series is gratefully acknowledged.

I would like to acknowledge support from the US National Science Foundation, Microsoft Research, and from the Centre Universitaire d'Informatique of the University of Geneva.

Finally, I would like to thank my student Yang Yu for his assistance in administering the manuscript management system.

The field of networked sensor systems is rapidly evolving. It is my hope that this meeting series serves as a forum for researchers from various aspects of this interdisciplinary field; particularly I hope it offers opportunities for those working in the algorithmic, theoretical and high-level aspects to interact with those addressing challenging issues in complementary areas such as wireless networks, signal processing, communications, and systems composed of these underlying technologies.

July 2005   Viktor K. Prasanna, Program Chair

Organization

General Chair

Jose Rolim — University of Geneva, Switzerland

Vice General Chair

Sotiris Nikoletseas — University of Patras and CTI, Greece

Program Chair

Viktor K. Prasanna — University of Southern California, USA

Program Vice Chairs

Algorithms:

Paul Spirakis — CTI and University of Patras, Greece

Applications:

Sitharama Iyengar — Louisiana State University, USA

Systems:

Matt Welsh — Harvard University, USA

Steering Committee Chairs

Jose Rolim — University of Geneva, Switzerland

Steering Committee

Josep Diaz — UPC Barcelona, Spain
Deborah Estrin — University of California, Los Angeles, USA
Phil Gibbons — Intel Research, Pittsburgh, USA
Sotiris Nikoletseas — University of Patras and CTI, Greece
Christos Papadimitriou — University of California, Berkeley, USA

Kris Pister   University of California, Berkeley,
                                  and Dust, Inc., USA
Viktor Prasanna   University of Southern California,
                                  Los Angeles, USA

Posters/Presentations Chair

Bhaskar Krishnamachari   University of Southern California, USA

Proceedings Chair

Pierre Leone   University of Geneva, Switzerland

Publicity Chair

Wendi Heinzelman   University of Rochester, USA

Publicity Co-chair

Erdal Cayirci   Yeditepe University and Istanbul Technical
                                  University, Turkey
Sanjay Jha   Univeristy of New South Wales, Australia

Student Scholarships Chair

Loren Schwiebert   Wayne State University, USA

Finance Chair

Germaine Gusthiot   University of Geneva, Switzerland

Sponsoring Institutions

- IEEE Computer Society Technical Committee on Parallel
  Processing (TCPP)
- IEEE Computer Society Technical Committee on Distributed
  Processing (TCDP)

- University of Geneva
- Research Academic Computer Technology Institute (RACTI)

Supporting Organizations

- US National Science Foundation
- Microsoft Research

Program Committee

Tarek Abdelzaher	University of Virginia, USA
Micah Adler	University of Massachussetts, Amherst, USA
Prathima Agrawal	Auburn University, USA
James Aspnes	Yale University, USA
N. Balakrishnan	Indian Institute of Science, India
Azzedine Boukerche	University of Ottawa, Canada
Richard Brooks	Clemson University, USA
John Byers	Boston University, USA
Krishnendu Chakrabarty	Duke University, USA
David Culler	University of California, Berkeley, USA
Kevin A. Delin	NASA/JPL, USA
Josep Diaz	Technical University of Catalonia, Spain
Jeremy Elson	Microsoft Research, USA
Deborah Estrin	University of California, Los Angeles, USA
Deepak Ganesan	University of Massachussetts, Amherst, USA
Johannes Gehrke	Cornell University, USA
Phil Gibbons	Intel Research, Pittsburgh, USA
Ashish Goel	Stanford University, USA
Wendi Heinzelman	University of Rochester, USA
Jennifer Hou	University of Illinois at Urbana-Champaign, USA
R. Kannan	Louisiana State University, USA
Elias Koutsoupias	University of Athens, Greece
Evangelos Kranakis	Carleton University, Canada
P.R. Kumar	University of Illinois at Urbana-Champaign, USA
Margaret Martonosi	Princeton University, USA
Rajeev Motwani	Stanford University, USA
Badri Nath	Rutgers University, USA
Stephan Olariu	Old Dominion University, USA
David Peleg	Weizmann Institute of Science, Israel
S. Phoha	Pennsylvania State University, USA

Cristina Pinotti	Università degli Studi di Perugia, Italy
Kris Pister	University of California, Berkeley, and Dust, Inc., USA
S.V.N. Rao	Oak Ridge National Lab, USA
Satish Rao	University of California, Berkeley, USA
Jim Reich	Palo Alto Research Center, USA
Shivakumar Sastry	University of Akron, USA
Christian Scheideler	Johns Hopkins University, USA
John A. Stankovic	University of Virginia, USA
Janos Sztipanovits	Vanderbilt University, USA
Bhavani Thuraisingham	National Science Foundation, USA
Jan van Leeuwen	Utrecht University, The Netherlands
Jay Warrior	Agilent Labs, USA
Peter Widmayer	ETH Zurich, Switzerland
Jie Wu	Florida Atlantic University, USA
Feng Zhao	Microsoft Research, USA

Referees

Tarek Abdelzaher	Josep Diaz	Xin Liu
Zoe Abrams	Prabal Dutta	Konrad Lorincz
Micah Adler	Jeremy Elson	Zvi Lotker
Luzi Anderegg	Mihaela Enachescu	Samuel Madden
James Aspnes	Deborah Estrin	Geoff Mainland
Amol Bakshi	Deepak Ganesan	Amit Manjhi
Boulat Bash	Phil Gibbons	Pedro Marron
Alan Albert Bertossi	Ashish Goel	Richard Martin
Pratik Biswas	Benjamin Greenstein	Margaret Martonosi
Paul Boone	Christopher Griffin	Myche McAuley
Azzedine Boukerche	Tian He	Rajeev Motwani
Richard Brooks	Wendi Heinzelman	Philip Moynihan
Scott Burleigh	Bo Hong	Michael Murphy
John Byers	Jennifer Hou	Badri Nath
Jason Campbell	Sitharama Iyengar	James Newsome
Krishnendu Chakrabarty	Shaili Jain	Ioanis Nikolaidis
Bor-rong Chen	Rajgopal Kannan	Mirela Notare
Stefano Chessa	Elias Koutsoupias	Regina O'Dell
Tai-Lin Chin	Evangelos Kranakis	Santosh Pandey
Maurice Chu	P.R. Kumar	David Peleg
Mark Corner	Vatsalya Kunchakarra	Shashi Phoha
Razvan Cristescu	Ioannis Lambadaris	Cristina M. Pinotti
Jun-Hong Cui	Philip Levis	Kristofer Pister
David Culler	Kathy Liszka	Rajmohan Rajaraman
Kevin Delin	Ting Liu	Nageswara Rao

Satish Rao
S.S. Ravi
Jim Reich
Jose Rolim
Shivakumar Sastry
Anna Scaglione
Christian Scheideler
Prashant Shenoy
Victor Shnayder
Mitali Singh
Priyanka Sinha

Sang Son
Pavlos Spirakis
John Stankovic
Ivan Stojmenovic
Janos Sztipanovits
Bob Veillette
Yong Wang
Jay Warrior
Mirjam Wattenhofer
Michele Weigle
Jennifer Welch

Matt Welsh
Cameron Whitehouse
Peter Widmayer
Alec Woo
Anthony Wood
Jie Wu
Qishi Wu
Yang Yu
Feng Zhao
Yi Zou

Table of Contents

Short Papers

Invited Posters

Contributed Posters

Algorithmic Problems in Ad Hoc Networks*

Christos H. Papadimitriou

UC Berkeley
christos@cs.berkeley.edu

The ad hoc networks research agenda aligns well with the methodology and point of view of theoretical computer science, for a variety of reasons: The resource frugality that permeates the theory of algorithms and copmplexity is a perfect match for the regime of resource scarcity that is prevalent in ad hoc networks, imposed by small, power-limited processors. The potentially large scale makes principled approaches more attractive, and necessitates analysis – occasionally even asymptotic analysis. Also, the interplay of geometry, graph theory, and randomness that is a characteristic of problems arising in ad hoc networks creates many challenging problems amenable to mathematical treatment. And the field's young age and lack of established paradigms further foment and reward theoretical exploration. This talk will review recent work by the author and collaborators on geographic routing and clock synchroniuzation.

---

* Research supported by NSF ITR grant CCR-0121555, and by a grant from Microsoft Research.

Five Challenges in Wide-Area Sensor Systems

Phillip B. Gibbons

Intel Research Pittsburgh
phillip.b.gibbons@intel.com

Most sensor network research to date has targeted ad hoc wireless networks of closely co-located, resource-constrained, scalar sensor motes. Such work has tended to overlook the potential importance of *wide-area sensor systems*–heterogeneous collections of Internet-connected sensor devices. A wide-area sensor system enables a rich collection of sensors such as cameras, microphones, infrared detectors, RFID readers, vibration sensors, accelerometers, etc. to be incorporated into a unified system capable of accessing, filtering, processing, querying, and reacting to sensed data. The sensor system can be programmed to provide useful *sensing services* that combine traditional data sources with tens to millions of live sensor feeds.

While the unique and many challenges of resource-constrained sensor networks are apparent, the limited research into wide-area sensor systems may indicate that researchers have not seen any interesting challenges in such systems. After all, where are the research challenges in sensor systems blessed with powerful, powered, Internet-connected computers with significant memory and storage?

In this talk, I will describe five research challenges in wide-area sensor systems that I believe are worthy of further pursuit. The challenges are: (1) scalability, (2) ease of service authorship, (3) robustness/manageability, (4) actuation, and (5) privacy. Although certainly not an exhaustive list, I believe these five make the case that wide-area sensor systems are an interesting area with many challenges and open questions. In discussing the research challenges, I will highlight techniques used to address some of the challenges within IrisNet [1, 2, 3, 4, 5], a wide-area sensor system we have developed over the past three years.

References

1. IrisNet (Internet-scale Resource-Intensive Sensor Network Service). http://www.intel-iris.net/.
2. S. Chen, P. B. Gibbons, and S. Nath. Database-centric programming for wide-area sensor systems. In *Proceedings of the First IEEE International Conference on Distributed Computing in Sensor Systems*, 2005.
3. A. Deshpande, S. K. Nath, P. B. Gibbons, and S. Seshan. Cache-and-query for wide area sensor databases. In *Proceedings of the 2003 ACM SIGMOD International Conference on Management of Data*, pages 503–514, 2003.
4. P. B. Gibbons, B. Karp, Y. Ke, S. Nath, and S. Seshan. Irisnet: An architecture for a worldwide sensor web. *IEEE Pervasive Computing*, 2(4):22–33, 2003.
5. S. Nath, Y. Ke, P. B. Gibbons, B. Karp, and S. Seshan. A distributed filtering architecture for multimedia sensors. In *Proceedings of the First IEEE Workshop on Broadband Advanced Sensor Networks*, 2004.

V. Prasanna et al. (Eds.): DCOSS 2005, LNCS 3560, p. 2, 2005.
© Springer-Verlag Berlin Heidelberg 2005

Challenges in Programming Sensor Networks

Feng Zhao
Senior Researcher

Microsoft Research
zhao@microsoft.com

The proliferation of networked embedded devices such as wireless sensors ushers in an entirely new class of computing platforms. We need new ways to organize and program them. Unlike existing platforms, systems such as sensor networks are decentralized, embedded in physical world, and interact with people. In addition to computing, energy and bandwidth resources are constrained and must be negotiated. Uncertainty, both in systems and about the environment, is a given. Many tasks require collaboration among devices, and the entire network may have to be regarded as a processor.

We argue that the existing node-centric programming of embedded devices is inadequate and unable to scale up. We need new service architectures, inter-operation protocols, programming models that are resource-aware and resource-efficient across heterogeneous devices that can range from extremely limited sensor motes to more powerful servers. I will supplement these discussions with concrete examples arising from our own work and the work of others.

Biography

Feng Zhao (http://research.microsoft.com/ zhao) is a Senior Researcher at Microsoft, where he manages the Networked Embedded Computing Group. He received his PhD in Electrical Engineering and Computer Science from MIT and has taught at Stanford University and Ohio State University. Dr. Zhao was a Principal Scientist at Xerox PARC and founded PARC's sensor network research effort. He serves as the founding Editor-In-Chief of ACM Transactions on Sensor Networks, and has authored or co-authored more than 100 technical papers and books, including a recent book published by Morgan Kaufmann - Wireless Sensor Networks: An information processing approach. He has received a number of awards, and his work has been featured in news media such as BBC World News, BusinessWeek, and Technology Review.

V. Prasanna et al. (Eds.): DCOSS 2005, LNCS 3560, p. 3, 2005.

Distributed Proximity Maintenance in Ad Hoc Mobile Networks

Jie Gao[1,**], Leonidas J. Guibas[2], and An Nguyen[2]

[1,**] Center for the Mathematics of Information, California Institute of Technology,
Pasadena, CA 91125, USA
jgao@ist.caltech.edu
[2] Department of Computer Science, Stanford University, Stanford, CA 94305, USA
{guibas, anguyen}@cs.stanford.edu

Abstract. We present an efficient distributed data structure, called the
D-SPANNER, for maintaining proximity information among communicating mobile nodes. The D-SPANNER is a kinetic sparse graph spanner on the nodes that
allows each node to quickly determine which other nodes are within a given distance of itself, to estimate an approximate nearest neighbor, and to perform a
variety of other proximity related tasks. A lightweight and fully distributed implementation is possible, in that maintenance of the proximity information only
requires each node to exchange a modest number of messages with a small number of mostly neighboring nodes. The structure is based on distance information
between communicating nodes that can be derived using ranging or localization
methods and requires no additional shared infrastructure other than an underlying
communication network. Its modest requirements make it scalable to networks
with large numbers of nodes.

1 Introduction

Collaborating intelligent mobile node scenarios appear in a wide variety of applications. Consider aircraft flying in formation: each plane must be aware of the locations
and motions of its neighbors in order to properly plan its trajectory and avoid collisions. If some aircraft are fuel tankers, each plane may need to determine the nearest
tanker when its fuel is low. Or take the case of a search team in a rescue operation
where collaboration among team members is essential to guarantee coordinated search
and exhaustive coverage of the rescue area. Current research provides many other similar multi-node collaboration examples, from the deployment of sensors in a complex
environment by a set of mobile robots [15] to the intelligent monitoring of forests by
suspended mobile sensors [16]. In all these scenarios a loosely structured collaborative
group of nodes must communicate in order to engage in joint spatial reasoning, towards
a global objective. The spatial reasoning required almost always involves proximity information — knowing which pairs of nodes are near each other. Proximity information
plays a crucial role in these scenarios because nearby nodes can interact, collaborate,
and influence each other in ways that far away nodes cannot.

** Work was done when the author was with computer science department, Stanford University.

V. Prasanna et al. (Eds.): DCOSS 2005, LNCS 3560, pp. 4–19, 2005.

Motivated by such examples, we consider in this paper the task of maintaining proximity information among mobile nodes in an ad hoc mobile communication network. We provide a data structure that allows each node to quickly determine which other nodes are within a given distance of itself, to estimate an approximate nearest neighbor, and to perform a variety of other proximity related tasks. As a matter of fact, proximity is important not only for the tasks the network has to perform, such as those illustrated above, but for building the network infrastructure itself. Mobile nodes typically use wireless transmitters whose range is limited. Proximity information can be essential for topology maintenance, as well as for the formation of node clusters and other hierarchical structures that may aid in the operation of the network. For example, it is sometimes desirable to perform network deformation so as to achieve better topology with lower delay [6].

There has not been much work in the ad hoc networking community on maintaining proximity information. A closely related problem is to keep track of the 1-hop neighbors, i.e. the nodes within communication range, of each node. This is a fundamental issue for many routing protocols and the overall organization of the network. But even this simple problem is not easy. If nodes know their own positions and those of their neighbors, then a node can be alerted by its neighbor if that neighbor is going to move out of the communication range. But knowing when new nodes come within the communication range is more challenging. A commonly used protocol for topology discovery is for all the nodes to send out "hello" beacons periodically. The nodes who receive the beacons respond and thus neighbors are discovered. However, a critical issue in this method is how to choose the rate at which "hello" messages are issued. A high beacon rate relative to the node motion will result in unnecessary communication costs, as the same topology will be rediscovered many times. A low rate, on the other hand, may miss important topology changes that are critical for the connectivity of the network. Unfortunately, as in physical simulations, the maximum speed of any node usually gates this rate for the entire system. Recently Amir *et al.* proposed a protocol for maintaining the proximity between 'buddies' in an ad hoc mobile network so that a node will be alerted if its buddies enter a range with radius R [2]. The basic idea is that each pair of buddies maintains a strip of width R around their bisector. The bisector is updated according to the new node locations when one of the nodes enters the strip. Although the nodes move continuously, the bisector is only updated at discrete times. This scheme, however, does not scale well, when the number of buddies is large — as would be the case when we care about potential proximities among all pairs of nodes.

The general problem of maintaining proximity information among moving objects has been a topic of study in various domains, from robot dynamics and motion planning to physical simulations across a range of scales from the molecular to the astrophysical. It would take us too far afield to summarize these background developments in other fields on distance computations, collision detection, etc. One relatively recent development in proximity algorithms that provides the basic conceptual setup for the current work is the framework of *kinetic data structures* (or KDSs for short) [4, 14]. The central idea in kinetic data structures is that although objects are moving continuously, an underlying combinatorial structure supporting the specific attribute(s) of interest changes only at discrete times. These critical events can be detected by maintaining a cached set of assertions about the state of the system, the so-called *certificates* of the KDS, and exploiting knowledge or predictions about the node motions. A certificate failure invokes the KDS repair mechanism that reestablishes the desired combinatorial structure and incrementally updates the certificate set. Many KDSs related to proximity have appeared in recent years [9, 5, 12, 13, 3, 1].

All current KDS implementations are centralized, requiring a shared or global event queue where events are detected and processed. In our setting, a centralized event queue would require that location/trajectory updates from all the nodes be sent to a central location, leading to unacceptably high communication costs. These centralized approaches are thus not directly applicable to ad hoc mobile networks, where a distributed implementation is always desirable and often necessary for the additional reasons of fault tolerance and load balancing.

A first contribution of this paper is to extend traditional centralized kinetic data structures to the domain of ad hoc mobile networks. We introduce the notion of *distributed kinetic data structures* (dKDSs for short) that demand no shared infrastructure other than a communication network among the mobile nodes and are therefore ideally suited to ad hoc mobile network settings. In a dKDS each node holds a small number of the centralized KDS certificates that are relevant to its portion of the global state. The KDS is maintained globally by exchanging messages among the nodes and updating the local certificate sets. As with any KDS implementation, an important issue to address is the interface between the KDS and the node motions — this directly determines the complexity of predicting or detecting KDS certificate failures. We describe two possible interfaces between motion information and the kinetic data structure: the *shared flight-plan* and the *distance threshold* motion models, each with its own advantages.

Our second contribution is the development of a lightweight and efficient distributed kinetic data structure for proximity maintenance among mobile nodes. The structure is based on our earlier work [10] on proximity maintenance in a centralized setting. In [10] we introduced the DEFSPANNER data structure that forms a sparse *graph spanner* [8] of the complete graph on the nodes, when edges are weighted via the Euclidean distance. A spanner has the property that for every pair of nodes there is a path in the spanner whose total length is a $(1 + \epsilon)$ approximation of the Euclidean distance between the nodes. Thus the spanner implicitly and compactly encodes all distance information. Here ϵ is a parameter that trades-off the approximation quality against the spanner size (number of edges needed). The D-SPANNER structure introduced in this paper, a dKDS for proximity maintenance among mobile nodes, represents a major reworking of our DEFSPANNER structure so as to make it decentralized. This is a highly non-trivial task, as classical KDSs gain much of their efficiency by assuming that failed certificates are processed in exact chronological order, one at a time. Instead, in the distributed setting, certificates may fail asynchronously at different nodes and the corresponding repair processes will run in parallel and must be coordinated. To our knowledge, this is the first distributed implementation of any KDS, attesting to the difficulty of the task.

Once we have a spanner data structure, we can efficiently answer many types of proximity queries. Effectively, the spanner replaces an expensive continuous search over the 2-D plane with a lightweight combinatorial search over the spanner subgraph. For instance, the spanner can be used to detect and avoid collisions between moving vehicles, especially unmanned vehicles, a task that is very difficult for any kind of on-demand discovery scheme. The spanner can do this, because only pairs of nodes with a spanner edges between them can possibly collide [10]. The continuous maintenance of the D-SPANNER guarantees that every possible potential collision is captured and predicted. As another example, a node u can locate all nodes within a distance d of itself by just initiating a restricted broadcast along the spanner edges that stops when the total distance traversed is $(1+\epsilon)d$. The set of nodes thus discovered is guaranteed to include all nodes within distance d of u; a further filtering step can remove all false positives, of which there will typically be few. The set of nodes within distance d can also be maintained through time so that u gets alerted only when nodes join the set or drop off from the set. Additionally, the D-SPANNER keeps track of a $(1+\varepsilon)$-nearest neighbor for *every* node in the network at *any* time. The neighbor of p in the spanner with shortest edge length is guaranteed to be within distance $(1+\varepsilon)$ the distance between p and q^*, the true nearest neighbor of p. As shown in [10], the D-SPANNER gives us a nice hierarchical structure over the nodes across all scales at any time, since it implicitly defines approximate k-centers of the nodes for any given k. The D-SPANNER also gives a well-separated pair decomposition [7], which provides an N-body type position-sensitive data aggregation scheme over node pairs. For further discussion of all these properties of the spanner, see [10].

In this paper we focus on the maintenance of the D-SPANNER in a distributed environment and analyze its maintenance and query costs. The spanner is overlayed on a communication network among the nodes. We assume the existence of a routing protocol that enables efficient communication between a pair of nodes, so that the communication cost is roughly proportional to the distance between the communicating nodes. The spanner is stored distributedly, having each node keep only its incident edges in the spanner, and is maintained by relevant pairs of nodes exchanging update packets. A query of the graph structure is performed by communicating with other mobile nodes following the edges of the D-SPANNER structure. We show that the D-SPANNER has the following attractive features:

- The D-SPANNER can be efficiently maintained in a distributed fashion under both the *shared flight-plan* and the *distance threshold* models (these will be formally defined below).
- The total communication cost for communicating the flight plans or location information for the initial setup of the D-SPANNER is almost linear in the weight of the minimum spanning tree of the network.
- Even though fixing a failed certificate may involve $O(\log n)$ nodes, the D-SPANNER can be repaired in such a way that at any moment each node updates the D-SPANNER locally using at most $O(1)$ computation and communication steps. Furthermore, multiple certificate failures can be fixed concurrently, without interference.
- The spanner structure is a hierarchy that scales well with the network size and the geometry of the nodes. Due to the hierarchical structure of the spanner, far away node pairs are updated less frequently than close-by pairs. Under reasonable motion assumptions, the communication cost related to the D-SPANNER maintenance incurred by a node p is $O(\log n)$ for each unit distance that p moves.

These properties show that the spanner is a lightweight data structure that gracefully scales to large network sizes. We validate these theoretical results with simulations of D-SPANNERs on two data sets, one with generated vehicle motions and one with real airline flight data. Our simulations show that, for both artificial and real-world data, the D-SPANNER can be maintained efficiently so that, at any time step, only a tiny part of the spanner needs to be updated. We observe that in practice, for reasonable data sets, the spanner has a much smaller spanning ratio compared to the theoretical worst-case bound discussed below. We also present some trade-offs in maintaining the D-SPANNER on different data sets.

2   Distributed Kinetic Data Structures

In the classical KDS setting it is assumed that, in the short run, moving objects follow motions that can described by explicit *flight plans*, which are communicated to the data structure. Objects are allowed to modify their flight plans at any time, however, as long as they appropriately notify the KDS. These flight plans form the basis for predicting when the KDS certificates fail — a typical certificate is a simple algebraic inequality on the positions of a small number of the objects. These predicted failure times become events in a global event queue. Upon certificate failure, the KDS repair mechanism is invoked to remove the failed certificate and update the structure as necessary.

In the distributed setting appropriate for ad hoc mobile communication networks, we must distribute, and possibly duplicate the certificates among the mobile nodes themselves. Furthermore, we must give up the notion that we process certificate failures in a strict chronological order, as nodes will process certificate failures independently of each other, as they detect them. Although it is not so readily apparent, the classical KDS setting depends quite heavily on the assumption that the KDS repair mechanism is invoked after exactly one certificate failure has occurred. This will no longer be true in the distributed case and thus the dKDS repair mechanism must be considerably more sophisticated. As we already remarked, globally broadcast flight plans are not appropriate in a distributed setting. While in some applications, like a dKDS for aircraft, it makes perfect sense to allow communicating aircraft to exchange flight plans, in others, as in our rescue team example, only something much weaker can be assumed. In the following we propose two motion models that are appropriate for ad hoc mobile networks. They are *shared flight plan model* and *distance threshold model*.

The Shared Flight Plan Model: In this motion model we assume that nodes have flight plans, but these are known only to the nodes themselves — unless they are explicitly communicated to others. A node must incur a communication cost to transmit its flight plan to another node. Furthermore, a node receiving a flight plan will assume that it is valid until the plan's owner communicates that the plan has changed. Thus each node has the responsibility of informing all other nodes who hold its flight plan of any changes to it.

The Distance Threshold Model: In this weaker model we only assume that a node knows its own position. Its prediction for future motion is either not available or too inaccurate to be useful. A node u may communicate its current position to node v, as needed by the dKDS. Associated with the (u, v) communication is a distance threshold $\delta(u, v)$; node u undertakes to inform node v if it ever moves more than $\delta(u, v)$ from the position previously communicated. Note that δ is a function of the pair of nodes — different v's may need updates about the changing position of u at different rates.

The evaluation and analysis of a dKDS is also somewhat different from the evaluation of a traditional KDS. A KDS has four desired performance properties: efficiency, responsiveness, locality and compactness [4]. In an evaluation of the properties of a KDS, we usually assume that the nodes follow pseudo-algebraic motions[1]. Efficiency captures how many events a KDS processes, as compared to the number of changes in the attribute of interest. The responsiveness of a KDS measures the worst-case amount of time needed to update its certificate cache after an event happens. Locality measures the maximum number of certificates in which one object ever appears. Finally, compactness measures the total number of certificates ever present in the certificate cache. Low values on these measures are still desirable properties for a dKDS. In the distributed setting, however, we have to include communication costs. First, we want to bound the total communication cost for exchanging flight plans or position information. We compare this with the cost of the optimal 1-median of the communication network[2], which is a lower bound on the communication cost of sending all the flight plans to a central server, assuming that no aggregation is done. The second difference is that, when a certificate kept at node u fails, some certificates held at other nodes of the structure may need to be updated. The cost of communication to the nodes keeping these certificates must be taken into account in the processing cost of the event. Finally, locality is an especially important property for a dKDS, since it determines how evenly one can distribute the set of certificates among the mobile nodes. If a node u is involved in certificates with too many other nodes, not only is u heavily loaded by holding many certificates, but also the update cost for u's flight plan changes is high, since u's new flight plan has to be delivered to many other nodes through costly communication.

There are also practical issues in maintaining a dKDS, because of the involvement of an underlying communication network. Ideally we put the dKDS on top of a TCP-like network layer, so that communication between nodes can be assumed reliable, without packet loss. Otherwise when reliable communication is not available, packet delay and loss must be considered in the process of the repair of the structure. We also assume that the communication cost is proportional to the Euclidean distance between the two communicating nodes — this is a reasonable assumption in dense networks. Finally, given that the ad hoc network environment is inherently parallel, special care is needed to make sure that KDS updates can handle race conditions and avoid deadlocks. Later we use the D-SPANNER as an example to show how these issues are handled.

---

[1] A motion is called algebraic with degree s if each coordinate of the motion is an algebraic function of degree s or less. For a definition of pseudo-algebraic, see [4].

[2] A 1-median of a graph $G = (V, E)$ is defined as $\min_{v \in V} \sum_{u \in V} \tau(u, v)$, where $\tau(u, v)$ is the shortest path length of u, v.

3   The D-SPANNER and Proximity Maintenance

For an ad hoc mobile network, a spanner is a sparse graph on the nodes that approximates the all pairs of distances. Specifically, a $(1 + \varepsilon)$-spanner is a graph on the nodes such that the shortest path distance between p and q in the spanner is at most $(1 + \varepsilon)$ times the Euclidean distance of p and q — $(1 + \epsilon)$ is called the stretch factor. In [10] we proposed a $(1 + \varepsilon)$-spanner, called DEFSPANNER that can be maintained under motion. In this paper, we show that the DEFSPANNER can also be maintained in a distributed fashion, where the nodes have no information about the global state and obtain information only through communication steps whose cost must be taken into account. This distributed version is denoted as a D-SPANNER. We emphasize that while there are many maintainable proximity structures under the KDS setting, D-SPANNER is the first kinetic structure of any kind that can be maintained distributedly. We start by reviewing the centralized DEFSPANNER in [10].

3.1   D-SPANNER

The definition of DEFSPANNER [10] requires the notion of discrete centers. A set of *discrete centers* with radius r for a point set S is defined as a maximal subset $S' \subseteq S$ such that any two centers are of a distance at least r away, and such that the balls with radius r centered at the discrete centers cover all the points of S. Notice that the set of discrete centers needs not be unique.

The DEFSPANNER G on S is constructed as follows. Given a set S of points in the d-dimensional Euclidean space, we construct a hierarchy of discrete centers so that S_0 is the original point set S and S_i is a set of discrete centers of S_{i-1} with radius 2^i, for $i > 0$. Intuitively, the hierarchical discrete centers are samplings of the point set at exponentially different spatial scales. Then we add edges to the graph G between all pairs of points in S_i whose distance is no more than $c \cdot 2^i$, where $c = 4 + 16/\varepsilon$. These edges connect each center to other centers in the same level whose distance is comparable to the radius at that level. Since the set of discrete centers is not unique, the DEFSPANNER is non-canonical. In fact, this is the main reason why DEFSPANNER admits an efficient maintenance scheme.

The *aspect ratio* of S, denoted by α, is defined by the ratio of the maximum pairwise distance and minimum pairwise distance. In this paper we focus on point sets with aspect ratio bounded by a polynomial of n, the number of nodes. This is a natural assumption on ad hoc mobile network because the maximum separation between nodes is bounded due to the connectivity requirement of the network and the nodes are physical objects so they usually have a minimum separation. Under this assumption, $\log \alpha = O(\log n)$. When convenient, we assume that the closest pair of points has distance 1, so the furthest pair of S has distance α.

We use the following notations throughout the paper. Since a point p may appear in many levels in the hierarchy, when the implied level is not clear, we use $p^{(i)}$ to denote the point p as a node in level S_i. A center $q^{(i)}$ is said to *cover* a node $p^{(i-1)}$ if $|pq| \le 2^i$. A node $p^{(i-1)}$ may be covered by many centers in S_i. We denote $P(p)$ one of those centers and call it the *parent* of $p^{(i-1)}$. The choice of $P(p)$ is arbitrary but fixed. We also call p a *child* of $P(p)$. A node p is called a nephew of a node q if $P(p)$ and q are neighbors. Two nodes p and q are *cousins* if $P(p)$ and $P(q)$ are neighbors. For a point p in level S_i, we recursively define $P^{j-i}(p)$ as the ancestor in level S_j of $p^{(i)}$ by $P^0(p) = p^{(i)}$, $P^{j-i}(p) = P(P^{j-i-1}(p))$. We note that if p is in level i, $p^{(j)} = P(p^{(j-1)})$ for each $j \leq i$, i.e. p is the parent of itself in all levels below i. For notational simplicity, we consider p a neighbor of itself in all levels in which it participates.

In [10], we show that a DEFSPANNER (and of course now its distributed version, the D-SPANNER), has a spanning ratio of $1 + \varepsilon$ and a total of $O(n)$ edges. A detailed list of results from [10] is shown below.

Theorem 1 (in [10]). *For a* DEFSPANNER *on a set of points S with aspect ratio α, the following hold.*

1. *If $q^{(i)}$ is a child of $p^{(i+1)}$, then $q^{(i)}$ and $p^{(i)}$ are neighbors, i.e. there is an edge from each point q to its parent.*
2. *If p and q are neighbors, then the parents $P(p)$ and $P(q)$ are neighbors.*
3. *The distance between any point $p \in S_0$ and its ancestor $P^i(p) \in S_i$ is at most 2^{i+1}.*
4. *The hierarchy has at most $\lceil \log_2 \alpha \rceil$ levels.*
5. *Each node p has $O(1)$ neighbors in any given level, and thus it has at most $O(\log_2 \alpha)$ neighbors totally.*
6. *G is a $(1 + \varepsilon)$-spanner when $c = 4 + 16/\varepsilon$.*

When the nodes move around, the discrete center hierarchy and the set of edges in the DEFSPANNER may change. Nodes may lose neighbors or gain neighbors in the spanner. Just as in our earlier discussion of maintenance of the 1-neighbors of a node, discovering the loss of a neighbor is easy while detecting new neighbors is hard. What makes the spanner work is the fact that property (2) above implies that before two nodes can become neighbors at a given level, their parents must already be neighbors at the next level up. Thus the search for new neighbor pairs can be confined to cousin pairs only.

The D-SPANNER has the same structure as the DEFSPANNER, though the internal constants involved in the structure are larger. The key difference between the D-SPANNER and the DEFSPANNER is in the way the structure is stored and maintained. In the centralized case, the spanner is fully repaired after every event, corresponding to a single certificate failure. In the distributed setting, D-SPANNER is computed and stored distributedly. In particular, each node only stores its own presence in the discrete centers hierarchy and its edges on each level. Certificates are handled locally. When a certificate fails, we communicate with other relevant nodes to have the D-SPANNER repaired. Since the network is inherently parallel and communication takes finite time, other certificates may fail concurrently and multiple repair processes may be active in parallel. We introduce the notion of a relaxed D-SPANNER to enable multiple certificates failures to be handled simultaneously in the network.

3.2   Relaxed D-SPANNER

To make the maintenance manageable, we make use of a variant, the *relaxed* D-SPANNER. The intuition is that whenever a certificate in a D-SPANNER fails it may take $O(\log n)$ communications for the D-SPANNER to be repaired, and up to $O(\log n)$

new edges may be established. Doing all that at once is not possible in a distributed setting. Instead, the repair is done in stages; between stages we relax our constraints on the spanner through the concept of a relaxed D-SPANNER. Relaxed D-SPANNERs make it possible to deal with multiple certificate failures by encoding the simultaneous failures with *relaxed parents*, the removal of which can be done in parallel.

Fix a constant $\gamma > 2$; we call a node $q^{(i)}$ a γ-*relaxed parent*, or simply a *relaxed parent*, of $p^{(i-1)}$ if $|pq| \leq \gamma \cdot 2^i$. A *relaxed* D-SPANNER is similar to a regular D-SPANNER except that if a node $p^{(i)}$ is not covered by any node in S_{i+1}, we do not require p itself to be in S_{i+1} but only require p to have a relaxed parent in S_{i+1}.

From (3) in Theorem 1, it is easy to see that for any node p, $|pP^i(p)| \leq 2^{i+1} < \gamma \cdot 2^i$, and thus we can intuitively think of the i-th level ancestors of a node p as its potential relaxed-parent, if p is in level S_{i-1}. When all nodes in a relaxed D-SPANNER have parents, the relaxed D-SPANNER is a D-SPANNER. Analogous to (8) in Theorem 1, we can prove that a relaxed D-SPANNER is by itself a $(1+\varepsilon)$-spanner when $c = \gamma \cdot (4+16/\varepsilon)$.

Theorem 2. *A relaxed* D-SPANNER *is a* $(1+\varepsilon)$-*spanner when* $c = \gamma \cdot (4+16/\varepsilon)$.

When nodes move and some certificates fail, we first make the D-SPANNER into a relaxed D-SPANNER which will be repaired after certain communications made to other nodes. The notion of relaxed parents guarantees that the structure is not far away from a real D-SPANNER so that local communications can fix it up, as the following lemma suggests (with proof omitted).

Lemma 1. *Let* q *be a relaxed parent of* p *in* S_i, *and* $c > 4 + \gamma$. *If* r *is a parent of* p, *then* q *and* r *are neighbors in* S_{i+1}. *If* p *is inserted to* S_{i+1}, $p^{(i+1)}$'*s neighbors must be cousins of* q.

In the above description we always double the radius when we construct the discrete centers hierarchy. In fact, we chose a factor of 2 just for the simplicity of explanation. We could have chosen any factor $\beta > 1$ and construct a hierarchy of discrete centers such that in level i, the nodes are more than β^i apart, and the edges connect nodes that are closer than $c \cdot \beta^i$. We would then call a node q in S_{i+1} a relaxed parent of a node p in S_i if $|pq| \leq \gamma \cdot \beta^i$, where $\gamma > \beta/(\beta-1)$. We choose $c > (2\beta+\gamma)/(\beta-1) > \beta(2\beta-1)/(\beta-1)^2$ so that D-SPANNER can be maintained.

3.3   The D-SPANNER Under the Shared Flight-Plan Model

The maintenance of the D-SPANNER in the centralized KDS framework was analyzed in [10]. In the dKDS setting, the absence of a centralized event queue and the lack of strict order in certificate failure processing invalidates much of the approach.

The basic idea for maintaining a D-SPANNER in a dKDS framework is as follows. Two nodes always communicate their flight plans if they are involved in a certificate. When the points move, the D-SPANNER is updated through a series of *relaxed* D-SPANNERs. When a certificate in a D-SPANNER fails, one or more relaxed parents may appear in the D-SPANNER, making it a relaxed D-SPANNER. Nodes in a relaxed D-SPANNER communicates with each other to restore the structure to a D-SPANNER. During the restoration, other certificates may fail. It is guaranteed, however, that we will eventually get a D-SPANNER after a period of no additional certificate failures, provided that the update time is small enough with respect to the velocity of the nodes and that $c > 4 + \gamma$.

We show that it only requires a constant number of communications to repair the structure locally when a certificate fails or to transform one relaxed D-SPANNER to another. We note that in a KDS setting c must be larger than 4 in order for the D-SPANNER to be maintainable. In the dKDS setting, we require that $c > 4 + \gamma > 6$. It worths pointing out that under the assumption that $c > 4 + \gamma$, a generic relaxed D-SPANNER cannot be maintained in either the KDS framework or the dKDS framework. We use a very specific series of relaxed D-SPANNERs during the maintenance of the D-SPANNER in the dKDS framework to make things work.

The certificates in a dKDS are similar to the certificates in KDS. They are all distance certificates, asserting that the distances between given pairs of nodes are above or below a certain threshold. A certificate about the distance between two nodes u and v is stored in both the event queues of u and v. Since u and v have the flight plans of each other, both of them can evaluate the first time when the certificate fails. Specifically, a node p as a point in S_i maintains four kinds of certificates:

1. *Parent certificate*: asserts that p is covered by its parent $P(p)$ if p has one, i.e., $|pP(p)| \leq 2^{i+1}$;
2. *Short edge certificates*: assert that $|pq| \geq 2^i$ for each neighbor $q \in S_i$ of p;
3. *Long edge certificates*: assert that $|pq| \leq c \cdot 2^i$ for each neighbor $q \in S_i$ of p;
4. *Potential edge certificates*: assert that $|pq| > c \cdot 2^i$ for each non-neighbor cousin q of p.[3]

When a certificate fails, each node involved in the certificate updates its event queue and performs the updates to the D-SPANNER. The updates for the certificates are as follows:

1. *Parent events*: When a parent certificate fails, we make the former parent a relaxed parent.
2. *Short edge events*: When $|pq| < 2^i$, the node with the lower maximum level, say p, removes itself from S_i. The children of p in S_{i-1} now have q as a relaxed parent, and $p^{(i-1)}$ has q as its parent.
3. *Long edge events*: When $|pq| > c \cdot 2^i$, the edge is simply dropped. The long edge certificate on pq is deleted from the certificate lists of both p and q. Accordingly some potential edge certificates between the cousins of p, q are dropped also.
4. *Potential edge events*: When a potential edge fails, a new edge is simply added. p and q communicate with each other about their children. A long edge certificate on pq and potential edge certificates between the cousins of p, q will be added to the certificate lists of both p and q.

For the first two types of certificate failure, the update introduces nodes with relaxed parents. We can find the true parents for these nodes by the following procedure. For a node $p^{(i)}$ with a relaxed parent $q^{(i+1)}$, we find a parent for $p^{(i)}$ using Lemma 1, namely, looking among all neighbors of $q^{(i+1)}$ for one that covers $p^{(i)}$. If such a node is found,

---

[3] Note that by (2) in Theorem 1, a pair of nodes cannot be neighbors before they first become cousins. Thus the potential edge certificates capture all possible edges that may appear in the near future.

$p^{(i)}$ has a parent. If not, p is at least 2^{i+1} away from every node in S_i. Thus p must be promoted to level S_{i+1}. There are two possible cases. In the case when $q^{(i+1)}$ has a parent r, it is easy to see that $p^{(i+1)}$ has r as a relaxed parent, and by Lemma 1, the neighbors of $p^{(i+1)}$ can be found among the cousins of $q^{(i+1)}$. In the case when q only has a relaxed parent, we have to wait until $q^{(i+1)}$ has a parent before p can be promoted to level S_{i+1}. Note that this is the only time when a node has to wait for another node. When a node has to wait, it always waits for a node in one level higher. This monotonic order guarantees that deadlock cannot occur. We note that the number of communications to search for a parent or to promote a node up one level is $O(1)$. In the worst case, a parent certificate failure may move a node all the way up the hierarchy. In the absence of certificate failures that generate nodes with relaxed parents, all the nodes with relaxed parents will eventually have parents, and thus we eventually obtain a D-SPANNER.

We currently do not handle the case when a relaxed parent of a node moves too far from the node. We could avoid dealing with this situation by noting that when a node $p^{(i)}$ first has a relaxed parent, the distance from it to the relaxed parent is always less than $2^{i+2} < \gamma \cdot 2^{i+1}$. If p and its relaxed parent move slowly enough or γ is large enough, p finds a true parent before its relaxed parent moves far away from it. If p loses its relaxed parent before it can find a parent, we can treat p as a newly inserted node and use the D-SPANNER dynamic update as explained in [11].

3.4   The D-SPANNER Under the Distance Threshold Model

The distance threshold motion model allows even less information, when compared with the shared flight plans motion model — a node only knows its current location. For each level S_i, we take a distance threshold $d_i = \mu \cdot 2^i$, $\mu = (c - 4 - \gamma)/8$, such that for each certificate in S_i involving a pair of nodes (u, v), we let u and v to inform each other their locations whenever any of them moves a distance of d_i from the last exchanged location.

Under this model, each node u predicts a certificate failure based on its current location and the last communicated location of its partner v, denoted by v_0. Node v may have moved since the last time it posted its location. However $|vv_0| \leq d_i$. Since u and v do not have the most updated locations of each other, they may not agree on when a certificate on u, v should fail. For example, in u's view of the world, a long edge certificate on u, v fails, i.e., $|uv_0| \geq c \cdot 2^i$. Then u informs v of appropriate updates and v changes its spanner edges. Although according to v it is possible that the long edge certificate has not failed yet, we can still guarantee that $|uv| \geq (c - \mu) \cdot 2^i$. In summary, under the distance threshold model, we maintain a μ-approximation to an exact D-SPANNER: If a node $p \in S_i$ covers a node q, $|pq| \leq (1 + \mu) \cdot 2^i$; Two nodes $p, q \in S_i$ have $|pq| \geq (1 - \mu) \cdot 2^i$; Two nodes $p, q \in S_i$ with an edge have $|pq| \leq (c + \mu) \cdot 2^i$; Two nodes $p, q \in S_i$ without an edge have $|pq| \geq (c - \mu) \cdot 2^i$. Notice that when two nodes $p, q \in S_i$ have a distance $(c - \mu) \cdot 2^i \leq |pq| \leq (c + \mu) \cdot 2^i$, p, q may or may not have an edge.

In the global view, we maintain a μ-approximate D-SPANNER through a series of γ-relaxed D-SPANNERs, with $\gamma \geq 2(1 + \mu)$. A μ-approximate D-SPANNER is a $(1 + \varepsilon)$ spanner for appropriately chosen parameters c, μ. The updates on certificates failures are the same as in the shared flight plan model. The following lemma (with proof omitted) is a more general version of Lemma 1, which shows that a node with a relaxed parent can either find a parent or promote itself to one level higher under the distance threshold model.

Lemma 2. *Let $c > 4 + \gamma$, and q be a relaxed parent of p_1 in S_i at time 1. Assume that all points involved move less than $d_i = \mu \cdot 2^i$, $\mu = (c - 4 - \gamma)/8$, between time 1 and time 2. At time 2, if r is a parent of p, then q and r are neighbors in S_{i+1}; if p is inserted to S_{i+1}, $p^{(i+1)}$'s neighbors must be cousins of q.*

Since we have less information in the distance threshold model, communications may be required for internal verification of the D-SPANNER, without resulting in any combinatorial changes in its structure. The cost of maintaining the D-SPANNER in the distance threshold model is the same as the cost of maintaining the D-SPANNER in the shared flight-plan model plus the cost of regular position updates. In the next section we will show a concrete bound on the maintenance cost.

3.5   Quality of Distributed Maintenance

In this subsection we study the memory, computation and communication cost of maintaining a D-SPANNER.

Memory Requirement for Each Node: The total number of certificates at any time is always linear in the number of nodes. Each node is involved in $O(1)$ certificates for each level it participates in. Since there are at most $\log \alpha$ levels, each node only has $O(\log \alpha)$ certificates in its queue. The D-SPANNER thus scales well when the network size increases.

The Startup Communication Cost for Exchanging Flight Plans: In order for the nodes in a D-SPANNER to build their event queues, the nodes that are involved in a certificate will have to inform each other of their flight plans. Note however, that communications between two nodes are not of equal costs, with communications between far away nodes costing more than those between close-by nodes. Under our assumption that the cost of a multi-hop communication between two nodes is proportional to the Euclidean distance between them, we show that the cost of exchanging flight plans between the nodes in a D-SPANNER is low by the following theorem, whose proof is omitted.

Theorem 3. *The total length of the edges in a D-SPANNER is at most $O(\log \alpha)$ times the total length of the minimum spanning tree (MST) of the underlying points. The summation of the distances between all cousin pairs in a D-SPANNER is $O(\log \alpha)$ times the total length of the MST of the underlying points.*

Notice that if we use the centralized KDS, the communication cost of every node sending their flight plans (or locations) to a central server is at least the weight of the optimal 1-median of the communication network, which is at least the weight of the minimum spanning tree. On the other hand, it is not hard to construct an example[4] such that the optimal 1-median of the network has weight $\Omega(n)$ times the total length of the

---

[4] Assume a list of nodes are staying on the x-axis with distance 1 between adjacent nodes. The communication network is just the chain.

MST. With this in mind, Theorem 3 implies that, in both the shared flight plans and distance threshold motion model, the startup communication cost of the D-SPANNER, i.e., the cost of communicating the shared flight plans or locations between the pairs of nodes in the certificates, is $O(\log \alpha)$ times the weight of the MST. Thus by using the distributed spanner we save substantially on the communication cost.

In the distance threshold motion model we update the locations of the nodes even if no certificate failures happen. Since the distance threshold, d_i, depends exponentially on i, the level number, nodes on higher levels update their positions less frequently but their communication costs are higher. More precisely, assume that the highest level that a node p appears is i. When p moves a distance d, the total cost of communicating its locations to its neighbors is bounded by $\sum_{j=1}^{i} O(c \cdot 2^j) \cdot d/d_j = O(di) \leq O(d \log \alpha)$. Thus the amortized communication cost of location updates for each node p is bounded by $O(\log \alpha)$ per unit distance that p moves.

The Total Number of Events Handled in a D-SPANNER: In [10] we showed that the number of combinatorial changes of a $(1 + \varepsilon)$-spanner can be $\Omega(n^2)$ and the number of certificate failures of a D-SPANNER is at most $O(n^2 \log n)$ under pseudo-algebraic motion, where each certificate changes from TRUE to FALSE at most a constant number of times. This claim is still true for the distributed environments. Thus the number of events we process in a D-SPANNER is close to the optimum.

The Communication Cost of Certificate Updates in a D-SPANNER: In a D-SPANNER, when a certificate fails, a node can repair the spanner locally with a constant number of communications, though the repair may introduce relaxed parents which have to be dealt with later. Each time a search for a true parent from a relaxed parent takes a constant number of communications, and a node may actually get promoted all the way up the hierarchy. In the worst case the node is promoted in $O(\log \alpha)$ levels. We first note that the cost of promoting a node up a level, introduced by resolving relaxed parents, can be amortized on the short edge events where a node is demoted down a level, since a node can not be demoted without first being promoted. Thus in the following study we neglect the cost of fixing relaxed parents and only consider the communication cost of repairing the certificates.

We bound the communication cost of certificates updates under pseudo-algebraic motions. We examine the update costs for the certificates defined on a particular pair of nodes p, q with p on level i. Nodes p, q may be involved in five different kinds of certificates, as shown in section 3.3. In particular, we characterize the certificate failures into three categories by the distance between p, q when the certificate fails: event A includes the parent event with p being the parent of q and the short edge event; Event B includes the long edge event and the potential edge event; Event C includes the parent event with q being the parent of p. Notice that when event A happens, the distance between p, q is exactly 2^i. When event B happens, the distance between p, q is exactly $c \cdot 2^i$. When event C happens, the distance between p, q is 2^{i+1}. Since p, q follow pseudo-algebraic motion, the number of times that events A (or B, C) happens is only a constant. Furthermore, it is not hard to see that between events in different categories, the distance between p, q must change by at least 2^i. Therefore, suppose the distance between p, q changes monotonically by d, the update cost for certificates failures with p, q is bounded by $O(c \cdot 2^i) \cdot d/2^i = O(d)$. Without loss of generality assume p moves faster than q, then p moves at least a distance $d/2$. Thus we can charge the update cost to p so that p is charged of communication cost $O(1)$ for each unit distance p moves. Since p has a constant number of neighbors in level i and p may appear in at most $O(\log \alpha)$ levels, we combine the costs in all levels together such that on average a node p with highest level h incurs $O(h)$ communication cost for each unit distance it moves. Therefore we have the following theorem.

Theorem 4. *Under pseudo-algebraic motion, the communication cost incurred by a node p related to the* D-SPANNER *maintenance is at most $O(\log \alpha)$ for each unit distance that p moves.*

Several other networking issues need to be addressed regarding the D-SPANNER in an ad hoc mobile network setting. These include load balancing, dynamic updates, and trade-offs between maintenance and query costs. We have investigated these issues but due to lack of space in these proceedings, we omit the discussion here. The reader is referred to the full version of the paper [11] for the particulars.

4   Simulations

To show that the D-SPANNER is well behaved in practice, we computed and maintained the D-SPANNER in two sets of simulations. The simulations validate our theory that the D-SPANNER is a sparse structure that can be maintain efficiently under motion.

In the first simulation, we considered a set of moving cars in an 11 by 11 block region of a city downtown. There were 20 one-lane road, 5 roads in each of the North, South, East, and West directions. Cars entered the region on one side and left the region on the other side. Each car moved at a random but constant speed between 0.2 and 0.3 block per second. Each car stopped if necessary to avoid collisions with other cars then moved again at its assigned speed. We allowed cars to disappear (say to stop in parking structures) within the city blocks. The number of cars n in the downtown area was a parameter of the simulation. In our simulations, n took one of the following values: 30, 50, 100, 150, 200, and 250. We kept the number of cars in a simulation constant by introducing a new car whenever some other car left the scene. We constructed and maintained a D-SPANNER on the cars. We used $\beta = 3$ and $c = 4.1$, i.e. the nodes in level S_i were at least 3^i apart, and edges in level S_i were at most $c \cdot 3^i$ long.

The degree of a node is proportional to its memory requirement and is also a measure of the work it does in maintaining the D-SPANNER. From Figure 1 (i), the average degree of the nodes in the D-SPANNER was around 9. The D-SPANNER was thus reasonably sparse, say comparing to planar graphs which have an average node degree of 6. The maximum degree of a node grew slowly when the number of cars increased. The result agrees with the theory, which predicts that the size of the spanner is linear, and the maximum degree grows as $O(\log n)$.

When $\beta = 3$ and $c = 4.1$, the theory predicts that the D-SPANNER has a spanning ratio at most 16.36. The spanning ratio in our experiments was much smaller. In all experiments, the spanning ratio fluctuated between 2 and 4 most of the time, and the average spanning ratio over time was about 2.7.

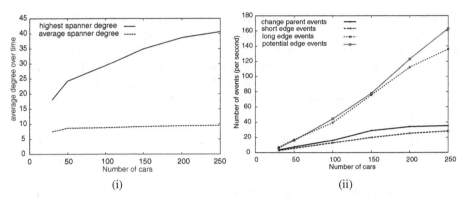

Fig. 1. (i) Average and maximum degree; (ii) Update frequency

Figure 1 (ii) shows the average number of events processed each second in the simulations. Even though the cars moved fast, the rate at which the events happened was low, suggesting that the D-SPANNER was stable. We noted that most certificate failures were among the long edge and the potential edge certificates. These two types of failures were cheap, as we only had to remove or to add an edge to the D-SPANNER to fix them. There were much fewer failures of the other two types which gave rise to relaxed parents and the involved procedure to remove them.

In the second set of experiment, we simulated a D-SPANNER on the flights in the US. The flight trajectories were from real flight data on July 27, 2002. There were from 4000 to 5000 air planes during the period of simulation. Each plane obtained its location about once every 60 seconds, and the planes often moved at a speed from 300 to 500 miles per hour, a fairly high speed compared to the distance separating the air planes. Under this condition, we traded in the query cost for lower maintenance cost by only maintaining some higher levels of D-SPANNER. We artificially increased the unit distance in the D-SPANNER and implicitly maintain all bottom edges. For each node in the bottom level, we only maintained its parent in S_1. The unit distance in the modified D-SPANNER we maintained was 16 miles. We essentially fully maintained a D-SPANNER on a dynamically selected set S_1 of air planes that were at least 48 miles from each other, and for each of those air planes, maintained a collection of air planes it covered, i.e. air planes within its 48 miles radius.

As in the simulation for cars, we constructed a spanner with $\beta = 3$ and $c = 4.1$. We found that on average there were about 14,000 edges in the modified D-SPANNER. The spanning ratio of the D-SPANNER restricted to the nodes in level S_1 was 3.15 on average. Note that since there were many implicit edges, if we considered the entire D-SPANNER with all those implicit edges, the spanning ratio would be much lower. Because we took a large value as the unit radius, the maintenance of the structure was relatively cheap. The entire structure processed 3.98, 1.63, 3.03, and 2.88 events per second for parent events, short edge events, long edge events, and potential edge events respectively. We noted that from time to time, due to missing data, some planes changed their positions too much between their consecutive position updates. When a node moved more than a certain distance threshold, we removed the node from the D-SPANNER, updated its position, then inserted it back to the D-SPANNER.

Acknowledgements. The authors gratefully acknowledge the support of NSF grants CCR-0204486 and CNS-0435111, as well as the DoD Multidisciplinary University Research Initiative (MURI) program administered by the Office of Naval Research under Grant N00014-00-1-0637.

References

1. P. Agarwal, J. Basch, L. Guibas, J. Hershberger, and L. Zhang. Deformable free space tilings for kinetic collision detection. *International Journal of Robotics Research*, 21(3):179–197, 2002.
2. A. Amir, A. Efrat, J. Myllymaki, L. Palaniappan, and K. Wampler. Buddy tracking - efficient proximity detection among mobile friends. In *Proc. of the 23rd Conference of the IEEE Communications Society (INFOCOM)*, volume 23, pages 298–309, March 2004.
3. J. Basch, J. Erickson, L. J. Guibas, J. Hershberger, and L. Zhang. Kinetic collision detection between two simple polygons. In *Proc. of the 10th ACM-SIAM symposium on Discrete algorithms*, pages 102–111, 1999.
4. J. Basch, L. Guibas, and J. Hershberger. Data structures for mobile data. *J. Alg.*, 31(1):1–28, 1999.
5. J. Basch, L. J. Guibas, and L. Zhang. Proximity problems on moving points. In *Proc. 13th Annu. ACM Sympos. Comput. Geom.*, pages 344–351, 1997.
6. A. Basu, B. Boshes, S. Mukherjee, and S. Ramanathan. Network deformation: traffic-aware algorithms for dynamically reducing end-to-end delay in multi-hop wireless networks. In *Proceedings of the 10th annual international conference on Mobile computing and networking*, pages 100–113. ACM Press, 2004.
7. P. B. Callahan and S. R. Kosaraju. A decomposition of multidimensional point sets with applications to k-nearest-neighbors and n-body potential fields. *J. ACM*, 42:67–90, 1995.
8. D. Eppstein. Spanning trees and spanners. In J.-R. Sack and J. Urrutia, editors, *Handbook of Computational Geometry*, pages 425–461. Elsevier Science Publishers B.V. North-Holland, Amsterdam, 2000.
9. J. Gao, L. Guibas, J. Hershberger, L. Zhang, and A. Zhu. Discrete mobile centers. *Discrete and Computational Geometry*, 30(1):45–63, 2003.
10. J. Gao, L. Guibas, and A. Nguyen. Deformable spanners and applications. In *Proc. ACM Symposium on Computational Geometry*, pages 190–199, June 2004.
11. J. Gao, L. J. Guibas, and A. Nguyen. Distributed proximity maintenance in ad hoc mobile networks. http://graphics.stanford.edu/~jgao/spanner-dcoss-full.pdf, 2005.
12. L. Guibas, J. Hershberger, S. Suri, and L. Zhang. Kinetic connectivity for unit disks. In *Proc. 16th Annu. ACM Sympos. Comput. Geom.*, pages 331–340, 2000.
13. L. Guibas, A. Nguyen, D. Russel, and L. Zhang. Collision detection for deforming necklaces. In *Proc. 18th ACM Symposium on Computational Geometry*, pages 33–42, 2002.
14. L. J. Guibas. Kinetic data structures — a state of the art report. In P. K. Agarwal, L. E. Kavraki, and M. Mason, editors, *Proc. Workshop Algorithmic Found. Robot.*, pages 191–209. A. K. Peters, Wellesley, MA, 1998.
15. A. Howard, M. J. Matarić, and G. S. Sukhatme. An incremental self-deployment algorithm for mobile sensor networks. *Autonomous Robots Special Issue on Intelligent Embedded Systems*, 13(2):113–126, 2002.
16. W. J. Kaiser, G. J. Pottie, M. Srivastava, G. S. Sukhatme, J. Villasenor, and D. Estrin. Networked infomechanical systems (NIMS) for ambient intelligence. CENS Technical Report 31, UCLA, December 2003.

Adaptive Triangular Deployment Algorithm for Unattended Mobile Sensor Networks*

Ming Ma and Yuanyuan Yang

Dept. of Electrical and Computer Engineering,
State University of New York, Stony Brook, NY 11794, USA

Abstract. In this paper, we present a novel sensor deployment algorithm, called *adaptive triangular deployment (ATRI)* algorithm, for large scale unattended mobile sensor networks. ATRI algorithm aims at maximizing coverage area and minimizing coverage gaps and overlaps, by adjusting the deployment layout of nodes close to equilateral triangulations, which is proved to be the optimal layout to provide the maximum no-gap coverage. The algorithm only needs location information of nearby nodes, thereby avoiding communication cost for exchanging global information. By dividing the transmission range into six sectors, each node adjusts the relative distance to its one-hop neighbors in each sector separately. *Distance threshold strategy* and *movement state diagram strategy* are adopted to avoid the oscillation of nodes. The simulation results show that ATRI algorithm achieves much larger coverage area and less average moving distance of nodes than existing algorithms. We also show that ATRI algorithm is applicable to practical environments and tasks, such as working in both bounded and unbounded areas, and avoiding irregularly-shaped obstacles.

1 Introduction and Background

In recent years, wireless sensor networks are playing an increasingly important role in a wide-range of applications, such as medical treatment, outer-space exploration, battlefield surveillance, emergency response, etc. [4, 5, 6]. A wireless sensor network is generally composed of hundreds or thousands of distributed mobile sensor nodes, with each node having limited and similar communication, computing and sensing capacity [3, 7]. The resource-limited sensor nodes are usually thrown into an unknown environment without a pre-configured infrastructure. Before monitoring the environment, sensor nodes must be able to deploy themselves to the working area. Although sensor nodes are designed with low power consumption in mind, they can survive only very limited lifetime with current technologies [7, 8, 9]. Furthermore, low computing capacity, limited memory and bandwidth of sensor nodes prohibit the use of high complexity algorithms and protocols. All these special characteristics of sensor networks bring a lot of challenges to developing reliable and efficient algorithms and protocols for such networks.

---

* The research was supported in part by NSF grant numbers CCR-0207999 and ECS-0427345 and ARO grant number W911NF-04-1-0439.

V. Prasanna et al. (Eds.): DCOSS 2005, LNCS 3560, pp. 20–34, 2005.

Although most of existing work on sensor networks currently is still at the laboratory level, see, for example, UCB-smart dusts [1], MIT-μAMPS [2], and UCLA-WINS [7], it is expected that sensor network technologies will be applied widely in various areas in the near future [4, 5, 6]. One of the main functionalities of a sensor network is to take place of human-being to fulfill the sensing or monitoring work in dangerous or human-unreachable environments, such as battlefield, outer-space, volcano, desert, seabed, etc. [4, 5, 6]. These environments usually have the following similarities: large scale, unknown, and full of unpredicted happening events which may cause the sudden failure of sensors. Unlike in a well-known environment, it is impossible to throw sensor nodes to their expected targets in an unknown working area or provide a map of the working area to sensor nodes before the placement. However, most of previous work in this area assumed that all nodes are well-deployed or a global map is pre-stored in the memory of all sensor nodes. In fact, in some situations the environment may be completely unknown to the newly coming sensor nodes. Without the control of the human being, sensor nodes must be "smart" enough to learn the working area by themselves, and deploy themselves to the expected working targets.

2 Related Work

There has been some previous work on the maximum coverage problem for sensor networks in the literature. *Potential-field-based deployment algorithm* [14] assumes that the motion of each node can be affected by virtual force from other nodes and obstacles. All nodes will explore from a compact region and fill the maximum working area in a way similar to the particles in the micro-world. Although this approach can maximize the coverage area, since the main idea of this algorithm is obtained from the micro-world, the nodes in the network may oscillate for a long time before they reach the static equilibrium state, like the particles in micro world. The oscillation of nodes consumes much more energy than moving to the desired location directly. Moreover, this algorithm can only be used in a bounded area, since nodes must be restricted within the boundary by the virtual force from boundary. Without the boundary, each node will not stop expelling others until there is no other nodes within its transmission range. *The Virtual Force algorithm (VFA)* [17] divides a sensor network into clusters. Each cluster head is responsible for collecting the location information of the nodes and determining their targets. The cluster architecture may leads to the unbalanced lifetime of the nodes and is not suitable for the networks that do not have powerful central nodes. *Constrained Coverage algorithm* [15] guarantees that each node has at least K neighbors by introducing two virtual forces. However, it still does not have any mechanism to limit oscillation of nodes. *Movement-Assisted Sensor Deployment algorithms* [16], which consist of three independent algorithms *VEC*, *VOR* and *MiniMax*, use Voronoi diagrams to discover the coverage holes and maximize the coverage area by pushing or pulling nodes to cover the coverage gaps based on the virtual forces. In the VEC algorithm, the nodes which have covered their corresponding Voronoi cells do not need to move, while other nodes are pushed to fill the coverage gaps. In VOR, nodes will move toward the farthest Voronoi vertices. The MiniMax algorithm moves nodes more gently than VOR, thereby avoiding the generation of new holes in some cases. Compared to potential-field-based deployment algorithm, movement-assisted sensor deployment algorithms reduce the oscillation and save the energy consumed by node movement. All three algorithms assume that each node knows its Voronoi neighbors and vertices. However, Voronoi diagram is a global structure, and all Voronoi vertices and cells can only be obtained when the global location information of all nodes in the network is known [13], which means that each node must exchange its current location information with all other nodes to acquire its corresponding Voronoi vertices and cell. For each location update message, it may take $O(n)$ one-hop communications to reach all other nodes, where n is the total number of nodes. In the case when the GPS system is unavailable, the error in one-hop relative locations of the nodes may be accumulated. Thus, the error for two far away nodes may be significant. In addition, so far most existing algorithms are only concerned with deploying nodes within a bounded area. VOR and Minimax algorithms are based on Voronoi diagrams and require every Voronoi cell to be bounded. However, by the definition of Voronoi graph [13], each periphery Voronoi cell is unbounded, since it contains a Voronoi vertex at infinity. Thus, VOR and Minimax algorithms cannot work correctly in this case.

Finally, though all the algorithms discussed above intended to maximize the node coverage, minimize the coverage overlap and gap, and deploy nodes uniformly, they did not answer a fundamental question in the deployment: what type of node layout can provide the maximum coverage with the smallest overlap and gap? We will address this issue in the next section.

3   Adaptive Triangular Deployment Algorithm

In this section, we first find out the optimum node layout for sensor deployment, and then present the deployment algorithm based on this layout. We now give the assumptions used in this paper. In order to make the proposed algorithm work in the environment where the GPS system is unavailable, such as outer-space, seabed and in-door, global location information should not be required. In the area without the GPS system, we assume that each sensor can estimate the relative location information to nearby nodes by detecting the relative distances and angles [12],[18]. In addition, we assume that each node has a unique reference direction, which can be easily obtained from electronic compass. Each node is able to monitor a unit-disk-shape area centered at itself, though our algorithm can work for other regular or irregular shapes of sensing area with only a minor modification.

3.1   The Ideal Node Layout for Maximum Coverage

Similar to the deployment algorithms discussed in the previous section, one of the important goals of our algorithm is to maximize the coverage area. However, before we design a maximum coverage algorithm, we need to know what type of node layout can provide the maximum coverage for a given number of nodes. We assume that each node can sense only a very limited disk-shaped range with radius r, called sensing range. In order to find the ideal node layout for maximum coverage, we introduce the Delaunay triangulation [13] to describe the layout of the network. Let N be the set of n nodes, which are randomly thrown in the plane, and T be a Delaunay triangulation of N, such that no nodes in N are inside the circumcircle of any triangle in T. Suppose that a large number of sensor nodes are randomly thrown in a two-dimensional field. The entire sensing area can be partitioned into some Delaunay triangles, whose vertices represent sensor nodes. We assume that the number of nodes is so large that the entire working area consists of a large number of Delaunay triangles. We have the following theorem regarding the optimum node layout.

Theorem 1. *If all Delaunay triangles are equilateral triangles with edge length $\sqrt{3}r$, the coverage area of n nodes is maximum without coverage gap.*

Proof. Since the entire working area can be decomposed into a large number of Delaunay triangles, if we can prove that the no-gap coverage area in any Delaunay triangle is maximized when the lengths of all its edges equal to $\sqrt{3}r$, then the maximum coverage area of n nodes can be obtained. Let C_{n_0}, C_{n_1} and C_{n_2} be the circles centered at the points n_0, n_1 and n_2, respectively, which denote the sensing range of corresponding nodes. Without loss of generality, we assume that circle C_{n_0} and circle C_{n_1} cross at point O, where O and n_2 locate on the same side of edge $\overline{n_0 n_1}$ as shown in Figure 1. Let $\phi_0 = \angle n_2 n_0 O$, $\phi_1 = \angle n_0 n_1 O$ and $\phi_2 = \angle n_1 n_2 O$. From Figure 1, since C_{n_0} and C_{n_1} cross at O, we can obtain $|\overline{n_1 O}| = |\overline{n_0 O}| = r$. In order to maximize the area of the triangle without coverage gap, $|\overline{n_2 O}|$ should equal r. The area of triangle $\triangle n_0 n_1 n_2$, denoted as $A(\triangle n_0 n_1 n_2)$, can be calculated as the summation of the area of the following three triangles $\triangle n_0 n_1 O$, $\triangle n_0 n_2 O$, and $\triangle n_2 n_1 O$.

$$A(\triangle n_0 n_1 n_2) = A(\triangle n_0 n_1 O) + A(\triangle n_0 n_2 O) + A(\triangle n_2 n_1 O)$$
$$= r^2(\sin \phi_0 \cos \phi_0 + \sin \phi_1 \cos \phi_1 + \sin \phi_2 \cos \phi_2)$$
$$= \frac{r^2}{2} \times (\sin(2\phi_0) + \sin(2\phi_1) + \sin(2\phi_2))$$

Since $|\overline{n_1 O}| = |\overline{n_0 O}| = |\overline{n_2 O}| = r$, we have $\phi_0 + \phi_1 + \phi_2 = \frac{\pi}{2}$. It can be shown that when $\phi_0 = \phi_1 = \phi_2 = \frac{\pi}{6}$ and the lengths of all three edges $\overline{n_0 n_1}$, $\overline{n_1 n_2}$ and $\overline{n_2 n_0}$ equal $\sqrt{3}r$, the area of triangle $\triangle n_0 n_1 n_2$ is maximized. Therefore, if all Delaunay triangles are equilateral triangles with edge length $\sqrt{3}r$, the no-gap coverage area in a plane is maximized. ∎

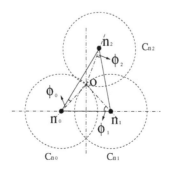

Fig. 1. The maximum no-gap coverage area in a triangle can be obtained, if and only if the lengths of all three edges of the Delaunay Triangle equal $\sqrt{3}r$

We have considered the maximum coverage problem. Now we derive the minimum average moving distance of the nodes under a uniform deployment density. We assume that initially all sensor nodes are in a compact area near the origin of the polar coordinate system and eventually will be deployed to a disk-shaped area S with radius D, that is, the sensors are uniformly distributed over area $S = \pi D^2$. We have the following theorem regarding the minimum average moving distance.

Theorem 2. *When sensor nodes are deployed from a compact area to a disk-shaped area S with radius D, the minimum average moving distance of the nodes is $\frac{2D}{3}$.*

Proof. When nodes are uniformly distributed over area S, the minimum average moving distance D_{avg} can be computed as the average distance from the origin of the polar coordinate system. Let (ρ, θ) denotes the polar coordinate of the node. Thus,

$$D_{avg} = E(\rho) = \int_0^{2\pi} \int_0^D \frac{\rho}{\pi D^2} \rho d\rho d\theta = \frac{2D}{3} \tag{1}$$

∎

By plugging $S = \pi D^2$ into (1), we have

$$D_{avg} = \frac{2}{3} \sqrt{\frac{S}{\pi}} \tag{2}$$

It should be pointed out that D_{avg} can be achieved only when every node directly moves to its final position during the deployment. Thus it represents the minimum average moving distance in the ideal situation. However, this optimum value is difficult to achieve when the final position of each node is unknown before the deployment and there are obstacles that may block the movement of the nodes. Nevertheless, D_{avg} can serve as a lower bound on the average moving distance of the nodes for any deployment algorithm. We will compare our algorithm with D_{avg} in the simulation section.

3.2   The Triangular Deployment Algorithm

In this subsection, we introduce a simpler version of our adaptive triangular deployment algorithm. We have known what type of node layout can maximize the coverage in a plane. Now the problem is how to deploy nodes from a compact area or an irregular layout to a perfect one. A large number of sensor nodes are usually randomly thrown into the working area or placed in a bunch. As discussed earlier, exchanging global location/topology information during such a dynamical deployment period would put a heavy traffic burden to the network. In fact, each unicast communication between two nodes in the network may need $O(n)$ one-hop communications, where n is the network size. Furthermore, when a bunch of nodes locate within a compact area and most of them need to communicate with others at the same time, the communication will be very inefficient due to the collision at the MAC layer [10]. Thus, the node movement decision should be based on local information in the deployment process. Since the location information is updated periodically, as a result, each node can only decide its movement periodically.

We now present an algorithm to deploy the sensor nodes close to a perfect equilateral triangular layout with the maximum coverage. The basic idea of the algorithm is to adjust the distance between two Delaunay neighbors to $\sqrt{3}r$ in three different coordinate systems, namely, XY, $X'Y'$ and $X''Y''$, where the angles between X'-axis, X''-axis and X-axis are $\frac{\pi}{3}$ and $\frac{2\pi}{3}$, respectively. In Figure 2(a), the transmission range of a sensor node is divided into six sector areas, called sectors 1 to 6 anticlockwise. X-axis, X'-axis and X''-axis symmetrically partition sectors 1 and 4, sectors 2 and 5, and sectors 3 and 6. The radius of each sector equals the transmission range of the node. The location of the nodes in sectors 1 and 4, sectors 2 and 5, and sectors 3 and 6, are expressed by XY, $X'Y'$ and $X''Y''$ coordinates, respectively.

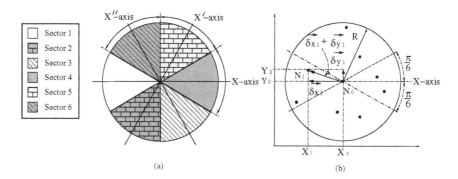

Fig. 2. Local movement strategies based on the location of one-hop neighbors

In each sector, the node adjusts its location along the corresponding axis based on the location of its Delaunay neighbors. Since global location of all sensor nodes is needed to determine Delaunay triangulations and Delaunay neighbors, in practice, we use the nearest neighbors to the node in each sector instead of Delaunay neighbors. For example, as shown in Figure 2(b), node N_1 is the nearest neighbor of node N_0 in sector 1, where their coordinates in XY coordinate system are (X_1, Y_1) and (X_0, Y_0). Let the location vector $\overrightarrow{\delta x_1}$ and $\overrightarrow{\delta y_1}$ denote vectors $[(X_1 - X_0), 0]$ and $[0, (Y_1 - Y_0)]$, respectively. If $|\overrightarrow{\delta x_1} + \overrightarrow{\delta y_1}| < \sqrt{3}r$, it means there is too much coverage overlap between node N_0 and node N_1. Thus, the movement of N_0 should be opposite to N_1 for $\frac{|\overrightarrow{\delta x_1} - \sqrt{3}r\frac{\overrightarrow{\delta x_1}}{|\delta x_1|}|}{2}$ to reduce the coverage overlap. On the contrary, if $|\overrightarrow{\delta x_1} + \overrightarrow{\delta y_1}| > \sqrt{3}r$, a coverage gap may exist between node N_0 and node N_1. We let N_0 move towards N_1 for $\frac{|\overrightarrow{\delta x_2} - \sqrt{3}r\frac{\overrightarrow{\delta x_2}}{|\delta x_2|}|}{2}$ to fill the coverage gap. Besides the movement on X-coordinate, the movement vector of N_0 projected on Y-coordinate equals $\overrightarrow{\delta y_1}/2$. Thus, the movement vector of N_0, $\overrightarrow{\delta v_1} = \frac{\overrightarrow{\delta x_1} - \sqrt{3}r\frac{\overrightarrow{\delta x_1}}{|\delta x_1|} + \overrightarrow{\delta y_1}}{2}$.

In general, for sector s, each node searches the nearest neighbor within the sector and calculates the relative horizontal and vertical location vectors $\overrightarrow{\delta x_s}$ and $\overrightarrow{\delta y_s}$ along its corresponding axis. Here, $\overrightarrow{\delta x_s}$ and $\overrightarrow{\delta y_s}$ are expressed by relative coordinates correspond-ing to sector s. The movement vector of a node in sector s, $\overrightarrow{\delta v_s}$, can be expressed as

$$\overrightarrow{\delta v_s} = \frac{\overrightarrow{\delta x_s} - \sqrt{3}r\frac{\overrightarrow{\delta x_s}}{|\overrightarrow{\delta x_s}|} + \overrightarrow{\delta y_s}}{2} \tag{3}$$

After the movement vectors in all six sectors are obtained, they will be transferred into uniform coordinates and added in order to obtain the total movement vector for the current round. As will be seen in the simulation results, after several rounds of such adjustments, the layout of the network will be close to the ideal equilateral triangle layout. As a result, the coverage area of the network will be maximized.

3.3   Minimizing Oscillation

We have proved that equilateral triangular layout can maximize the coverage, and also proposed a simple algorithm to adjust the network from an irregular layout to the ideal equilateral triangle layout. However, since the global location information of the net-work is difficult to obtain in the deployment process, it is impossible for each node to move to its desired target directly. Thus, sensor nodes may move back and forth fre-quently before it reaches its desired target. To make the algorithm suitable to real-world applications, another important issue is to reduce the total moving distance of the nodes in the deployment.

Recall that the moving strategy in our triangular deployment algorithm is that if the horizontal distance between two neighbors are longer than $\sqrt{3}r$, the sensors will move towards each other to shorten the gap between them. On the contrary, they will move away from each other to reduce the coverage overlap. According to this strategy, nodes move all the time, unless the network reaches the perfect layout or its maximum rounds. In order to reduce the oscillation, we adopt a threshold strategy by using two distance thresholds, T_1 and T_2, instead of $\sqrt{3}r$ for making moving decisions, where $T_1 = \sqrt{3}r + \varepsilon$ and $T_2 = \sqrt{3}r - \varepsilon$, where ε is a small constant. As described in Figure 3(a), the Y-coordinate d denotes the distance between the node and its nearest neighbor. When two far away nodes move towards each other and the distance between them decreases to T_1, two nodes stop moving. On the other hand, when two close nodes move apart and the distance between them increases to T_2, they will stop and keep the current distance between them. This moving strategy guarantees that the node will not move if it locates between T_1 and T_2 away from its neighbors, so that the node is affected less when its neighbors move slightly. If the adjustment granularity is too small, which is given by $\Delta d = T_1 - T_2 = 2\epsilon$, T_1 and T_2 are close to $\sqrt{3}r$ at the beginning of the deployment process.

However, if Δd is too large, it is impossible to adjust the network to the perfect equilateral triangular layout. In order to solve this problem, we can let the thresholds T_1 and T_2 be the function of time t. As shown in Figure 3(b), the adjustment granu-larity decreases as time t increases. That is, $T_1 = \sqrt{3}r + \epsilon(t)$ and $T_2 = \sqrt{3}r - \epsilon(t)$, respectively, where $\epsilon(t)$ is called threshold function. In practice, since the algorithm is executed round by round, the threshold can be changed to a function of the number of rounds. $\epsilon(Rd_{cur}) = \frac{\sqrt{3}}{4}r \times e^{-Rd_{cur}/Rd_{total}}$ is an example of the threshold function and will be used in the simulation in Section 4, where Rd_{cur} and Rd_{total} are the numbers of the current round and the total round, respectively.

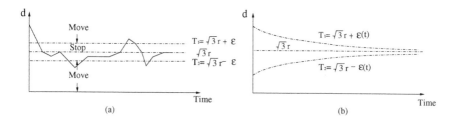

Fig. 3. Threshold strategy for reducing node oscillation

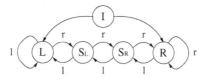

Fig. 4. Movement state diagram for reducing node oscillation. L: Move left; R: Move Right; S_L and S_R: Stay; I: Initial state; l: Expected moving direction is left; r: Expected moving direction is right

The second strategy, called movement state diagram, is to use a state diagram to reduce the node oscillation. Each movement can be considered as a vector and be decomposed into the projection vectors on X and Y coordinates. For X coordinate, nodes can only move left or right. Oscillation exists when a node moves towards the opposite direction of the previous movement. In order to avoid oscillation, nodes are not allowed to move backwards immediately. Two state diagrams are applied in movement vectors projected on X and Y coordinates separately. Figure 4 shows an example of movement state diagram for X coordinate, which contains 5 states and is used in our simulation. The diagram has 5 states: L, R, S_L, S_R and I, and two transitions l and r. L and R denote the movement to the left and the right respectively. If the state of a node is S_L or S_R, it has to stay where it is till the next round. I is the initial state. l and r represent the moving decision to the left and the right made by the triangular deployment algorithm. For example, a node plans to move left after running the triangular algorithm, which means that the current transition is l. Then, it needs to check its current state on its state diagram. If its current state is L, I or S_L, the next state will go to L after the transition l. A node can move left only when its next state is L. If the current state of the node is S_R or R, the next state will transit to S_L or S_R upon the transition l, respectively. Thus, the node can not move until next round. The movement control on Y direction follows a similar procedure. Simulation results in Section 4 show that the distance threshold strategy and movement state diagram strategy can reduce a significant amount of movements during the deployment.

3.4   Avoiding Obstacles and Boundaries

We have discussed how to deploy nodes in an open area. However, in most real-world applications, the working area is more likely bounded partially or entirely. Also, some irregularly-shaped obstacles may locate within the working area. In order to make the deployment algorithm more practical, sensors must be able to avoid obstacles and boundaries. Because an accurate map of the sensing area may not be always available before the deployment, sensor nodes can detect obstacles only when they move close enough to the obstacles. As discussed above, the triangular deployment algorithm can only adjust the relative position of two sensor nodes. However, unlike sensor nodes, obstacles and boundaries usually have irregular shapes and continuous outlines. In order to enable the triangular deployment algorithm to adjust the relative positions between sensor nodes and obstacles and boundaries with only a minor modification, the outlines of obstacles and boundaries are abstracted as many virtual nodes, which surround obstacles and boundaries closely. As shown in Figure 5, each small dotted circle around obstacles and boundaries denotes a virtual node. In practice, after each sensor node detects the outlines of obstacles or boundaries within its sensing range, from a sensor node's point of view, the outlines of obstacles or boundaries can be considered as a lot of virtual nodes. Like real sensor nodes, these virtual nodes also "push" real nodes away when real nodes locate too close to them, or "pull" real nodes close to them when real nodes locate too far away from them. Similar to the basic triangular deployment algorithm described earlier, after a real sensor node divides it sensing range into six sectors, it takes account of both real nodes and virtual nodes in each sector of its sensing range. When the real node finds other real or virtual nodes locate too close to itself in each sector, it moves away from them. On the contrary, when it locates too far away from other real or virtual nodes, it moves towards them to fill the coverage gap. One difference between the virtual nodes and the real nodes is that virtual nodes can not move. Another difference between real nodes and virtual nodes is that virtual nodes can cover a zero-size area. Thus, in order to avoid the coverage gap or overlap between real nodes and virtual nodes, the distance between real nodes and virtual nodes should be adjusted to $\sqrt{3}r/2$ instead of $\sqrt{3}r$, by decreasing the deployment radius r_d from r to $\frac{r}{2}$, where the deployment radius r_d is defined as the radius of the range the node should cover. We can revise our algorithm as follows for deploying nodes in an area with obstacles. In each sector, if the nearest neighbor of the node is a virtual node, the node sets its deployment radius r_d to $\frac{r}{2}$. If its nearest neighbor is a real node, its r_d still equals its sensing radius r. And then, the node runs the triangular deployment algorithm by replacing r with r_d. Figure 5 shows an example of the layout after nodes are deployed in a partially bounded area with irregularly-shaped obstacles. We can see that the distance between real nodes is still $\sqrt{3}r$, while the distance between real nodes and virtual nodes (obstacles or boundaries) is $\sqrt{3}r/2$.

By applying distance threshold, movement state diagram and adaptive adjustment strategies, we obtain the adaptive triangular deployment algorithm (ATRI) that is summarized in Table 1. As will be seen in our simulation results, by incorporating these strategies, our adaptive triangular deployment algorithm can drive nodes to avoid obstacles in the area and deploy them with different densities based on the requirements of tasks.

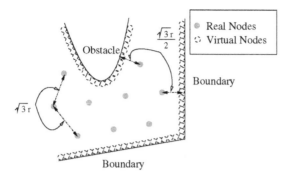

Fig. 5. Examples of the adaptive deployment in a bounded area with obstacles

Table 1. Adaptive Triangular Deployment Algorithm

Adaptive Triangular Deployment Algorithm
for each round
for each node $i = 1$ to n
Broadcast "Hello" message containing its location information to its one-hop neighbors;
Receive "Hello" message from nearby nodes and obtain their location information;
Detect obstacles or boundaries within its sensing range and obtain location of virtual nodes;
Divide its transmission range into 6 sectors;
for each sector $s = 1$ to 6
Adjust its sensing radius r_d adaptively based on location of virtual nodes;
Calculate the threshold value THR;
Calculate location vector $\vec{\delta x}$ and $\vec{\delta y}$ to its nearest neighbor/virtual node in the coordinate system;
if $(0 < \|\delta x_s\| \leq \|\sqrt{3}r_d - THR\|)$ **or** $(\|\delta x_s\| \geq \|\sqrt{3}r_d + THR\|)$
Calculate and store the movement vector for sector s, $\vec{\delta v_s} = \dfrac{\vec{\delta x} - \sqrt{3}r_d \frac{\vec{\delta x}}{\|\delta x\|} + \vec{\delta y}}{2}$;
else
$\vec{\delta v_s} = \vec{0}$;
end if
Transfer movement vectors of all sectors into uniform coordinates;
Add them up to obtain the total movement vector for node i;
Check the state diagram to decide if move or not and make transitions on the state diagram;
Move;

4 Simulation Results

This section presents a set of experiments designed to evaluate the performance and cost of the proposed algorithm. Besides the ideal flat open area, the simulation is also run in the more practical environments, where irregularly-shaped obstacles may block the movement of nodes. All sensor nodes are equipped with omnidirectional antennas, which can reach as far as 8 meters away. Each sensor node can sense the occurring of events within a radius of 3 meters away from itself. At the beginning of the experiments, a large number of nodes are randomly placed within a $1m \times 1m$ compact square which is centered at point $(25m, 25m)$. Then the nodes will explode to a large, evenly deployed layout. In order to limit the oscillation of nodes, the same movement state diagram depicted in Figure 4 is used in all scenarios. The distance threshold function $\epsilon(Rd_{cur}) = \frac{\sqrt{3}}{4}r_d \times e^{-Rd_{cur}/Rd_{total}}$, where Rd_{cur} and Rd_{total} are the numbers of current round and total round. We measure the total coverage area and the average moving distance per node and compare them with existing algorithms.

4.1   Performance and Cost Evaluation

In this subsection, we compare the performance and cost of VEC and ATRI algorithms. VEC algorithm has similar performance as VOR and Minimax algorithm in a bounded area. In addition, like ATRI algorithm, VEC algorithm can be used in both unbounded and bounded areas, while VOR and Minimax algorithms have to know the boundary information. For the sake of simplicity, we did not take account of the communication cost for exchanging location information and assume that location information is error-free, though VEC algorithm needs global location information and is more vulnerable to inaccurate location information than ATRI algorithm. In order to see the effects of various node densities, 100 nodes are randomly placed into a $1m \times 1m$ square around point $(25m, 25m)$ at the beginning, then they explode from a compact area to a large area. Both algorithms run for 100 rounds. In order to evaluate the performance and cost of the two algorithms, two metrics are measured for each simulation round: total coverage area and average moving distance, which are defined as the coverage area of 100 nodes and accumulated moving distance per node from the beginning of the simulation. Figure 6(a) shows the total coverage area of both algorithms as the simulation rounds increase. We can see that the total coverage of both algorithms increase rapidly to as high as $2200m^2$ before round 40, and then go smoothly after round 40. At round 100, ATRI stops at around $2600m^2$, while VEC is close to $2500m^2$. From round 10

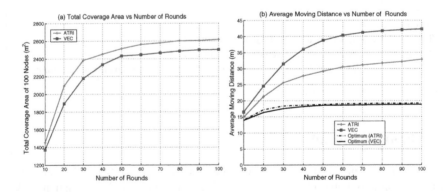

Fig. 6. The total coverage area and average moving distance of ATRI and VEC algorithms for 100 runs of simulations

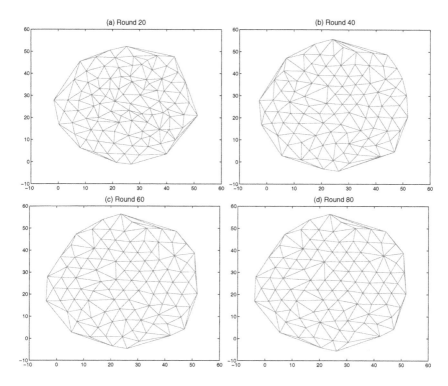

Fig. 7. Triangular layout of 100 nodes at round 20, 40, 60 and 80

to 100, ATRI always leads VEC for about $50m^2$ to $100m^2$. Figure 6(b) describes the average moving distance when simulation rounds range from 10 to 100. Average moving distance has the similar increasing trend as the total coverage area, as the simulation rounds increase. In both algorithms, after round 60, nodes are deployed evenly and their average distance is close to $\sqrt{3}r$. Most nodes do not need to move except minor adjustments. As shown in Figure 7, after round 60, most nodes form the equilateral triangle layout. There are no obvious changes between the layout of round 60 and that of round 80, because the layout of nodes is already very close to the ideal equilateral triangular layout after round 60. In addition, from total coverage area of both algorithms, we also calculate optimum average moving distances and plot them in Figure 6(b). As discussed earlier, given a fixed total coverage area, the optimum average moving distance can be calculated by (2). Recall that optimum average moving distances can only be obtained when the working environment is well known and each node knows its expected target before the deployment. Without the map of the environment, nodes have to move in a zigzag manner, which leads to average moving distances of both algorithms are more than optimum values. However, compared to VEC algorithm, ATRI still saves up to 50% of the optimum average moving distance from round 10 to 100.

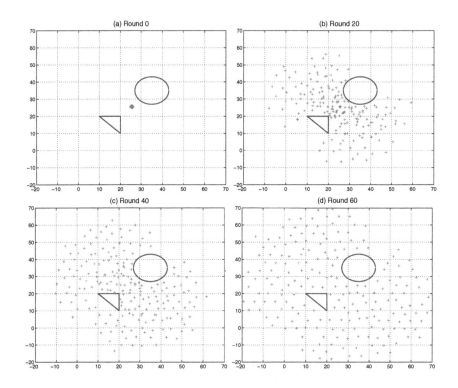

Fig. 8. Simulation results for non-uniform deployment: (a) Initial snapshot; (b) Round 20; (c) Round 40; (d) Round 60

Table 2. Exploring sensing area with obstacles : Moving distance vs coverage area

Number of Rounds	20	40	60
Total Coverage Area (m^2)	2032	3504	4415
Average Moving Distance (m)	17.45	33.33	46.13
Optimum Average Moving Distance (m)	16.95	22.26	25.00

4.2   Exploring the Area with Obstacles

In order to show that ATRI algorithm works well in the sensing area with obstacles, a circular and a triangular obstacles are placed in the sensing area. As shown in Figure 8, the circular obstacle is centered at point $(35m, 35m)$ with radius $8m$. Vertices of the triangular obstacle are $(10m, 20m)$, $(20m, 10m)$ and $(20m, 20m)$, respectively. Figure 8(a),(b),(c) and (d) show the deployment layouts at rounds 0, 20, 40 and 60, respectively, where each "+" symbol denotes the position of its corresponding node. At the beginning, similar to the previous scenario, 200 sensor nodes are randomly thrown into a $1m \times 1m$ square around point $(25m, 25m)$. As the simulation runs, we can see from Figure 8(b),(c) and (d) that, the total coverage enlarges round by round. At round

60, nodes are evenly deployed and no nodes enter the triangular or the circular region during the deployment. Though obstacles block the movement of some nodes, ATRI algorithm still performs very efficiently. During the simulation, we measure total coverage area and average moving distance at round 20, 40 and 60. In addition, we also calculate optimum average moving distance by plugging total coverage area into (2). All three metrics are shown in Table 2. In practice, optimum average moving distance is difficult to reach unless the desired optimum position of each node is known before the deployment and no obstacles block the movement of the nodes. In the situation that the map of the environment is unavailable and the movement of some nodes is blocked by obstacles, we can see that the movement cost of ATRI algorithm is very reasonable compared to the optimum value.

5   Conclusions

In this paper, we have proposed a new adaptive deployment algorithm for unattended mobile sensor networks, namely, adaptive triangular deployment (ATRI) algorithm. We have introduced an equilateral triangle deployment layout of sensor nodes and proved that it can produce the maximum no-gap coverage area in a plane. Although only the location information of one-hop neighbors is used for the adjustment of each node, the algorithm can make the overall deployment layout close to equilateral triangulations. In order to reduce the back-and-forth movement of nodes, the distance threshold strategy and movement state diagram strategy are proposed, which limit the oscillation and reduce the total movement distance of nodes. ATRI algorithm can be used in both bounded and unbounded areas. It also supports adaptive deployment. Without the map and information of the environment, nodes can avoid obstacles. In addition, ATRI algorithm is run in a completely distributed fashion by each node and based only on the location information of nearby nodes.

References

1. http://robotics.eecs.berkeley.edu/ pister/SmartDust/, 2004.
2. http://www-mtl.mit.edu/research/icsystems/uamps/, 2004.
3. I. F. Akyildiz, W. Su, Y. Sankarasubramaniam, and E. Cayirci, "A survey on sensor networks," *IEEE Communications Magazine*, Aug. 2002, pp. 102-114.
4. A. Cerpa, J. Elson, D. Estrin, L. Girod, M. Hamilton, and J. Zhao. "Habitat monitoring: Application driver for wireless communications technology," *Proc. 2001 ACM SIGCOMM Workshop on Data Communications in Latin America and the Caribbean*, April 2001.
5. S. Chessa and P. Santi, "Crash faults identification in wireless sensor networks", *Computer Communications*, Vol. 25, No. 14, pp. 1273-1282, Sept. 2002
6. L. Schwiebert, S. K. S. Gupta, and J. Weinmann, "Research challenges in wireless networks of biomedical sensors," *Proc. ACM/IEEE MobiCom*, pp. 151-165, 2001.
7. G. Asada, T. Dong, F. Lin, G. Pottie, W. Kaiser, and H. Marcy, "Wireless integrated network sensors: low power systems on a chip," *European Solid State Circuits Conference*, The Hague, Netherlands, Oct. 1998.

8. R. Min, M. Bhardwaj, N. Ickes, A. Wang, A. Chandrakasan, "The hardware and the network: total-system strategies for power aware wireless microsensors," *IEEE CAS Workshop on Wireless Communications and Networking*, Sept. 2002.

9. The Ultra Low Power Wireless Sensor Project, http://www-mtl.mit.edu/jimg/project_top.html, 2004.

10. IEEE standard for Wireless LAN Medium Access Control (MAC) and Physical Layer (PHY) Specifications, *IEEE Standard 802.11*, June 1999.

11. E. D. Kaplan, ed., *Understanding GPS – Principles and Applications*, Norwood MA: Artech House, 1996.

12. S. Capkun, M. Hamdi, J.P. Hubaux, "GPS-free positioning in mobile ad-hoc networks," *Proc. Hawaii Int. Conf. on System Sciences*, Jan. 2001.

13. M. de Berg, M. van Kreveld, M. Overmars and O. Schwarzkopf, Computational Geometry Algorithms and Applications, Springer-Verlag, 1997.

14. A. Howard, Maja J. Mataric, and G. S. Sukhatme, "Mobile sensor network deployment using potential fields: A distributed scalable solution to the area coverage problem," *Proc. DARS02*, Fukuoka, Japan, 2002, pp. 299-308.

15. S Poduri, and G. S. Sukhatme, "Constrained Coverage for Mobile Sensor Networks," *IEEE Int'l Conference on Robotics and Automation*, pp.165-172, 2004.

16. G. Wang, G. Cao, and T. La Porta, " Movement-Assisted Sensor Deployment," *Proc. IEEE Infocom*, March 2004.

17. Y. Zou and K. Chakrabarty, "Sensor deployment and target localizations based on virtual forces," *Proc. IEEE Infocom*, 2003.

18. N. Bulusu, J. Heidemann and D. Estrin, "Adaptive beacon placement," *Proc. ICDCS-21*, Phoenix, Arizona, April 2001.

An Adaptive Blind Algorithm for Energy Balanced Data Propagation in Wireless Sensors Networks[*]

Pierre Leone[1], Sotiris Nikoletseas[2], and José Rolim[1]

[1] Computer Science Department, University of Geneva,
1211 Geneva 4, Switzerland
[2] Computer Technology Institute (CTI) and Patras University,
P.O. Box 1122, 261 10, Patras, Greece

Abstract. In this paper, we consider the problem of energy balanced data propagation in wireless sensor networks and we generalise previous works by allowing realistic energy assignment. A new modelisation of the process of energy consumption as a random walk along with a new analysis are proposed. Two new algorithms are presented and analysed. The first one is easy to implement and fast to execute. However, it needs a priori assumptions on the process generating data to be propagated. The second algorithm overcomes this need by inferring information from the observation of the process. Furthermore, this algorithm is based on stochastic estimation methods and is adaptive to environmental changes. This represents an important contribution for propagating energy balanced data in wireless sensor netwoks due to their highly dynamic nature.

1 Introduction

Load balancing is a common important problem in many areas of distributed systems. A typical example is that of shared resources such as a set of processors, where it is of interest to assign tasks to resources without overusing any of them. A related but different aspect of load balancing appears in the context of sensor networks, where tiny smart sensors are usually battery powered: an important goal of data processing is to balance the total energy consumed among the entire set of sensors. However, limited local knowledge of the network, frequent changes in the topology of the network and the specifications of sensors, among others, make load balancing in sensors nets significantly different of classical load balancing in distributed systems.

To our knowledge, these considerations were first pointed out in the field of sensor networks in [9]. In this paper the authors deal with the problem of devising energy balanced sorting algorithms. In a subsequent paper [6] the authors

---

[*] This work has been partially supported by the IST Programme of the European Union under contract numbers IST-2001-33135 (CRESCCO) and 001907 (DELIS).

V. Prasanna et al. (Eds.): DCOSS 2005, LNCS 3560, pp. 35–48, 2005.

deal with the problem of energy balanced data propagation in sensor networks. They propose a randomised data propagation protocol and provide recursive and closed form solutions for the appropriate data propagation probabilities.

Before describing our contributions we present the problem previously stated in [6]. Formal definitions are deferred to the next section. An important area of application of sensor networks is the monitoring of a given region. Tiny smart sensors are scattered in a given region in order to detect and monitor some phenomena. Once a sensor detects the occurrence of an event it is responsible to inform (through wireless transmissions) a particular station (representing the end users of the network), called sink, about the occurrence of this event. Since the energy necessary to transmit a data through radio waves is proportional to some power of the distance of transmission (usually square power), sensors located far away from the sink are prone (if they would transmit directly to the sink) to run out of their available energy before sensors located closer to the sink. This leads to the idea that the data traffic has to be handled by the network with multiple hops to the sink, allowing only short distance communication. However, this strategy tends to overuse sensors located close to the sink since these sensors have to handle the entire set of events. There is then a trade-off between long and short distance communication to forward a data to the sink in order to make the life time of the whole network longer. A possible probabilistic protocol divides the set of sensors into slices or ring sectors [6]. The first slice is composed of sensors at *unit* distance from the sink, the second slice of sensors at distance 2, and so on. Here the distance is the maximal number of hops necessary to send a data to the sink. Sensors may communicate directly to the sink with probability $(1 - p_i)$. In this case the consumed energy is proportional to i^2 with i is the slice number the sensor belongs to. In order to save energy, a sensor can probabilistically decide with probability p_i to transmit its data to a sensor belonging to the next slice to the sink (slice $i \rightarrow$ slice $i - 1$). In this case the amount of consumed energy is proportional to a constant which is assumed to be 1 for convenience. This is illustrated in Figure 1. The problem is to determine with which probability p_i a sensor located in slice i has to transmit to the next slice (or to directly transmit to the sink with probability $1 - p_i$) in order to balance the consumed energy among all the sensors.

In the literature, probabilistic data propagation protocols are proposed, in [5] (LTP, a local optimisation protocol) and [4] (PFR, a limited flooding protocol). Both protocols are energy efficient and fault-tolerant, but tend to strain close to the sink sensors. [1] proposes a variable transmission range protocol (VTRP) that tends to bypass strained sensors lying close to the sink.

In [2] the authors propose a novel and efficient energy-aware distributed heuristic (EAD), to build a special rooted tree with many leaves that is used to facilitate data-centric routing. In [3] a new protocol (PEQ) is proposed that uses a *driven delivery of events* mechanism that selects the fastest path and reduces latency. [7] inspired our work here since it proposes some stochastic models (Markov chains, dynamic systems) for dynamic sensor networks.

In this section, we generalise the energy balanced data propagation problem by allowing unrestricted realistic energy assignment and we propose two new probabilistic protocols, one of which is adaptive. Our analysis is based on the modelisation of the process of energy consumption as a random walk in $\mathcal{R}^n$. The first algorithm we propose is relatively similar to the one suggested in [6] and corresponds to offline computation of the probabilities p_i of transmission to the next slice. Although very easy to implement and fast in execution it suffers from an important **weakness**; namely the probability of occurrence of the events per slice, i.e. the probability λ_i have to be known. This particularity allows very efficient computations of the probabilities p_i. However, this property is not realistic or at least we gain in flexibility and adaptability to devise an algorithm able to solve the problem without any assumption concerning these probabilities. The analysis of the problem is new and leads to a formal definition of the problem of energy balanced data propagation.

The second algorithm is adaptive and based on stochastic approximation methods. The algorithm does not assume that the probabilities of occurrence of the events are known and infers their values from the observation of the events. We refer to such an algorithm as **blind** algorithm for energy balanced data propagation to stress the fact that there is no *a priori* knowledge on the statistics concerning the localisation of the events. The algorithm can be accordingly implemented on any given network and run on the fly, allowing online adaptation of the parameters of the network. This characteristic is important if the parameters of the network are prone to change (this appears frequently in sensor networks). This algorithm is an important contribution of this section. Generally, adaptive algorithms, like the one proposed here, are most appropriate for wireless sensor networks because of their evolving nature due to dynamic properties of the networks such as sensors failures, obstacles, etc., leading to topology changes. We also formally define in a broader sense the problem of energy balanced data propagation and show formally under which conditions the problem is well formulated.

2   Framework and Formal Definition of the Problem

In this section we state formally the framework and notations and state the problem of energy balanced data propagation in wireless sensor networks. Notice that as a result of the analysis of the problem, that is presented in the next section, we show that the problem as stated in this section is well formulated.

The number of slices is denoted by n. The main assumption we need in this section is that the energy consumed per sensor to handle the data to the sink is the same among sensors belonging to a particular slice. This means that sensors belonging to the same slice exhaust their available energy more or less simultaneously. Both following assumptions give sufficient conditions validating this assumption. Notice that these assumptions are based on a probabilistic selection of a sensor belonging to a slice for data transmission. Different protocols can then be proposed.

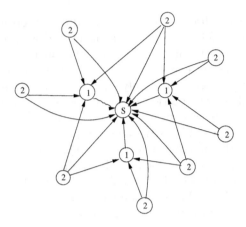

Fig. 1. The sink (S) with the first two slices of sensors

We assume that the probability that an event is detected by a given sensor depends uniquely on the slice the sensor belongs to. This means that we can define and estimate $\lambda_1, \lambda_2, \ldots, \lambda_n$ ($\sum_i \lambda_i = 1$) where λ_i is the probability that an event occurs in slice number i. For example, this property is satisfied if the events are uniformly randomly distributed on the monitored region. Indeed, in this particular situation, the probabilities λ_i are proportional to the area covered by the i-th slice. Moreover, when a data is transmitted from slice i to slice $i-1$ the selected sensor belonging to the slice $i-1$ is uniformly selected among the whole set.

The probability p_i, $i = 1, \ldots, n$ denotes the probability that a sensor belonging to the slice i sends a data to a sensor belonging to the "next" slice $i-1$. The complementary probability $1 - p_i$ denotes the probability that the sensor sends the data directly to the sink. Then, when a data is handled by a sensor belonging to the i-th slice the amount of consumed energy is a constant (assumed to be 1 for convenience) with probability p_i and i^2 with probability $1 - p_i$. By definition we have $p_1 = 1$ because sensors belonging to the first slice can do nothing else than transmitting to the sink.

The number of sensors belonging to the i-th slice is denoted by S_i. It might be the case that there is a strong relationship between S_i, λ_i but this is not essential.

The total energy available at the i-th slice is denoted by E_i, thus $e_i = E_i/S_i$ is the available energy per sensor. The energy can be seen as a given amount of energy available at the start or as a rate of consumable energy.

An important aspect of our analysis is to model the energy consumption for handling a given event as a random walk in R^n. We group the available scaled energy of each slice as a vector

$$\left(E_n/S_n, E_{n-1}/S_{n-1}, \ldots, E_1/S_1\right)^T. \tag{1}$$

To start, consider an event generated in the n−th slice and consider the complete process of handling the data generated by the event to the sink. We have different possibilities for the scaled energy consumed in the different slices corresponding to the different paths of the data. The data is directly transmitted to the sink with probability $(1-p_n)$. The corresponding consumed energy vector per sensor is $(n^2/S_n, 0, \ldots, 0)^T$. The data can alternatively be transmitted to the next $(n-1)$−th slice from which it is directly transmitted to the sink with probability $p_n(1 - p_{n-1})$. The corresponding consumed energy vector in this situation is $(1/S_n, (n-1)^2/S_{n-1}, 0, \ldots, 0)^T$. Repeating this enumerative process, we describe all possible events with their probability and the corresponding vector of energy consumed.

Formally, we denote the $U = \{U_1, U_2, \ldots\}$ the set of vectors describing the energy consumption for handling an event. Denoting by Ω the set of possible events we obtain a random variable $\Omega \to U$ which describes the energy consumed for handling an event. If we associate to each event its probability we have our probability space $(\Omega, \mathcal{P}(\Omega), P)$.

The process of energy consumption is described as a random walk in R^n with the energy consumed for handling m events in the form $X_1 + X_2 + X_3 + \ldots + X_m$, where X_i are independent random realisations of the random variable $\Omega \to U$. The law of large numbers implies that $X_1 + X_2 + \ldots + X_m \to mE(X)$ thus, to ensure energy balanced data propagation we must have

$$E(X) = \lambda\big(E_n/S_n, E_{n-1}/S_{n-1}, \ldots, E_1/S_1\big)^T. \qquad (2)$$

Intuitively, if the expected consumed energy does not satisfy (2) then there is a slice in which sensors will run out the available energy, described by (1), before the sensors belonging to others slices. The network stops working prematurely. Moreover, if (1) describes the rate of consumable energy requirement (2) amounts to preserving the ratio of consumed energy per slice. An energy assignment vector is a vector of the form (1) meaning that the ratio of energy consumed in slice i with respect to slice j should be E_i/E_j.

We later prove that that the set of admissible energy assignment vectors is $\{v \in R^n : v_i \geq 0, \| v \| = \text{constant}\}$ and to each such vector there is a unique assignment of the probabilities p_i. Besides this existential result, we propose two new protocols for calculating the optimal probabilities in an efficient manner. The first protocol assumes a certain amount of local knowledge, while the second one implicitly estimates the statistics of the events and is able to appropriately adapt to changes in the network parameters.

3   An Aware Strategy for Balanced Energy Dissipation

To ensure energy balance we have to determine for each slice i the probability of transmitting a given data to the next slice p_i, the data being transmitted directly to the sink with probability $(1-p_i)$. This section deals with this problem assuming an a-priori knowledge of the probabilities λ_i of the distribution of occurrences of the events among different slices. The first slice, located just before the sink, has only to transmit the data to the sink directly ($p_1 = 1$). Hence, if n is the total number of slices we have $n - 1$ unknown probabilities $p_2, \ldots, p_n$. The other free parameter to be determined is the factor λ appearing in (2).

Consider a node in the i-th slice which has to transmit a data. The data has to be transmitted because of an event occurring in the i-th slice with probability λ_i. The data can also be transmitted because it was previously generated by the preceding $(i + 1)$-th slice. This occurs with probability $\lambda_{i+1} \cdot p_{i+1}$. The event can also be transmitted due to an event generated in the $(i + 2)$-th slice, this occurs with probability $\lambda_{i+2} \cdot p_{i+2} \cdot p_{i+1}$ and so on up to the n-th slice. Then, a data is transmitted from the i-th slice with probability

$$\lambda_i + \lambda_{i+1}p_{i+1} + \lambda_{i+2}p_{i+2}p_{i+1} + \ldots + \lambda_n p_n p_{n-1} \cdot \ldots \cdot p_{i+1}. \tag{3}$$

The mean dissipated energy per sensor on the i-th slice is of the form

$$p_i \frac{1}{S_i} + (1 - p_i)\frac{i^2}{S_i}. \tag{4}$$

Then the mean energy dissipated in the i-th slice is of the form

$$\left(\lambda_i + \lambda_{i+1}p_{i+1} + \ldots + \lambda_n p_n p_{n-1} \cdots p_{i+1}\right)\left(p_i \frac{1}{S_i} + (1 - p_i)\frac{i^2}{S_i}\right) = \lambda e_i, \tag{5}$$

where the equality is imposed to ensure energy balanced data propagation through the network. With $p_{n+1} = \lambda_{n+1} = 0$ we define the x_i value as

$$x_i = \lambda_i + \lambda_{i+1}p_{i+1} + \ldots + \lambda_n p_n p_{n-1} \cdots p_{i+1}, \tag{6}$$

which satisfies the recurrence relation

$$x_i = p_{i+1}x_{i+1} + \lambda_i, \quad i = n, \ldots, 1, \tag{7}$$

with the convention $p_{n+1} = 0$. Solving (5) for p_i, $i = n, \ldots, 2$ we get

$$p_i = \frac{i^2 x_i - S_i e_i \lambda}{(i^2 - 1)x_i} = \frac{i^2}{i^2 - 1} - \frac{S_i e_i}{(i^2 - 1)x_i}\lambda, \quad i = n, \ldots, 2. \tag{8}$$

Since $p_1 = 1$ we solve (5) with $i = 1$ for λ and get

$$\lambda = \frac{x_1}{S_1 e_1}. \tag{9}$$

This last equation is actually a constraint on λ. To investigate properties of this constraint, we define a function $(\lambda, E_1, \ldots, E_n) \to \lambda' = f(\lambda)$ by substituting λ by λ' in (9). A fixed point of this function, i.e. $\lambda = f(\lambda)$, determines through (6) and (8) a solution of our problem. The recursive scheme described in Figure 2 calculates the value of the fixed point of this function. This algorithm is executed by the sink based on the knowledge of the energy assignment vector, the probabilities of occurrence of the events on the different slices and the number of sensors per slices. This information may come directly from the sensors (both at set-up or during protocol evolution) which know the slice they belong to.

Actually, there is a more efficient way of solving this problem, based on the result of Proposition 1, as illustrated in Figure 3.

Proposition 1. *The function $f(\lambda)$ defined through equations (6), (8) and (9) is linear. Then, we can write*

$$f(\lambda) = a + b\lambda, \tag{10}$$

with a and b real constants defined by

$$S_1 e_1 a = \lambda_1 + \lambda_2 C_2 + \lambda_3 C_3 C_2 + \ldots + \lambda_n C_n C_{n-1} \ldots C_2, \tag{11}$$

$$S_1 e_1 b = -D_2 - D_3 C_2 - D_4 C_3 C_2 - \ldots - D_n C_{n-1} \ldots C_2, \tag{12}$$

with

$$C_i = \frac{i^2}{i^2 - 1}, \quad D_i = \frac{S_i e_i}{i^2 - 1}. \tag{13}$$

Proof. Let us fix a value of λ. Computing $f(\lambda)$ amounts to recursively computing the parameters x_i for $i = n, n-1, \ldots, 1$ and then computing $f(\lambda)$ with (9). We prove by induction that the x_i values are all linear in λ. For $i = n$, we get $x_n = \lambda_n$. So, the assertion is true for $i = n$ Let us now assume it is true for $i = n, n-1, \ldots, k+1$. Using (7) and (8) we get $x_k = p_{k+1} x_{k+1} + \lambda_k = C_{k+1} x_{k+1} - D_{k+1} \lambda + \lambda_k$, which is linear in λ establishing the first part of the assertion. Since the x_i are linear functions in λ we introduce the notation

$$x_i = a_i + b_i \lambda, \quad i = n, \ldots, 1. \tag{14}$$

The coefficients a_i and b_i are recursively determined by

$$a_i = \lambda_i + a_{i+1} C_{i+1}, \tag{15}$$

$$b_i = b_{i+1} C_{i+1} - D_{i+1}, \tag{16}$$

with inital conditions $a_n = \lambda_n$ and $b_n = 0$ (i.e. $x_n = \lambda_n$). These equations prove (11) (12). To get (15) and (16), we write again (8) as $p_i = C_i - \frac{D_i}{x_i}\lambda$, which can be inserted into formula (7) to obtain $x_i = \lambda_i + x_{i+1} C_{i+1} - D_{i+1}\lambda$. Inserting (14) into this last equation leads to the recursive expression for a_i and b_i.

In the preceding computations, the dependence of the function $f(\lambda)$ on the total energy per slice (the E_i parameters) was not stated clearly, neither was the physical interpretation of this vector. The next results show that the entries of this vector are not important by themselves. What is important is the ratio E_i/E_j between the total mean energy available at the i-th slice with respect to the j−th slice.

Initialise $p_2, \ldots, p_n$ and λ
Initialise NbrLoop=1
while not convergence
   $x \leftarrow 0$
   for $counter = n$ **to** 2
      $x \leftarrow x + \lambda_{counter}$
      $p_{counter} \leftarrow p_{counter}(\lambda, p_{counter+1}, \ldots, p_n)$ with (8)
      $x \leftarrow x p_{counter}$
   end for
   $x \leftarrow x + \lambda_1$
   Compute λ_{inter} with (9)
   $\lambda \leftarrow \lambda + (\lambda_{inter} - \lambda)/nbrLoop$
   $nbrLoop \leftarrow NbrLoop + 1$
end while

Fig. 2. Pseudo-code for iterative solution of (2)

1. The Sink compute the fixed point of $f(\lambda)$ defined in Proposition 1
2. The Sink sends to every sensor the relevant λ value
3. Each sensor computes its probability p_i

Fig. 3. High-level description of the energy balanced data propagation protocol

Corollary 1. *Consider two total energy assignment vectors which are linearly dependent,*

$$\begin{pmatrix} E_n \\ E_{n-1} \\ \vdots \\ E_1 \end{pmatrix} = \mu \begin{pmatrix} E'_n \\ E'_{n-1} \\ \vdots \\ E'_1 \end{pmatrix},$$

with μ a real non-zero constant. Then, the fixed points, λ and λ' of the functions

$$f(\lambda, E_1, \ldots, E_n) = \lambda, \ \ and \ f(\lambda', E'_1, \ldots, e'_n) = \lambda'$$

are related by

$$\lambda' = \mu\lambda. \tag{17}$$

Moreover, the corresponding probabilities p_i and p'_i of transmitting directly to the next slice are equal.

Proof. This follows from the result of theorem 1. The value of the fixed point are given respectively by

$$\lambda = \frac{a}{1 - b}, \ \ and \ \ \lambda' = \frac{a'}{1 - b'}.$$

Direct inspection of the expressions for the coefficients of both functions, using $S_i e_i = E_i$ and $S'_i e'_i = E'_i$ shows that $\mu a = a'$ and $b = b'$ which implies

(17). We know that the probabilities p_i of transmitting directly to the next slice depend only on the λ and λ' values which are fixed points of the functions $f(\lambda, E_1, \ldots, E_n) = \lambda$ and $f(\lambda', E_1', \ldots, e_n') = \lambda'$. By hypothesis, we know that $\mu\lambda = \lambda'$. However, in the equations determining the $p_i's$ values only the products λE_i and $\lambda' E_i'$ are involved. Since these products are equal the result is proved. For the sake of completeness, we mention that we can also prove by induction that the values x_i and x_i' values are the same.

To conclude this section we finally state the result which shows how the energy balanced data propagation problem can be well formulated.

Proposition 2. *Given an energy assignment vector belonging to the set $\{v \in R^n : v_i \geq 0, \| v \| = constant\}$ then there exist unique probabilities p_i, $i = 2, \ldots, p_n$ to solve the energy balanced data propagation problem.*

Proof. Probabilities p_i are determined by the fixed point of the function defined by proposition 1. This fixed point is unique because the function is linear. Moreover, with our hypothesis the parameters a and b, see (11) and (12), satisfy $a > 0$ and $b < 0$ implying the existence of a fixed point.

This result shows formally that the problem is well formulated and possesses a unique solution if the energy assignment vector is restricted to belongs to the set $\{v \in R^n : v_i \geq 0, \| v \| = constant\}$.

4   A Blind Strategy for Balanced Energy Dissipation

In this section, we deal with the problem of the estimation of the probabilities p_i of transmitting a data directly to a sensor which belongs to the next slice being blind to the probabilities λ_i of occurrence of events in a given slice. The blindness assumption is more general and realistic and allows the design of adaptive algorithms that appropriately adjust to the network parameters. However, we do not estimate directly the λ_i probability but directly the values of x_i (6). One reason for this is that the x_i values have probabilistic interpretation in terms of the path of the data through the different slices of the networks.

Proposition 3. *Consider an event occurring in slice $i = 1, \ldots, n$ with probability λ_i. The event is handled by the network which conveys it to the sink. Define A_i the event : "The data passes through slice number i", and $\mathbf{1}_i$ the indicator function of event A_i. Then*

$$Prob(\mathbf{1}_i = 1) = x_i. \tag{18}$$

Proof. To compute the probability that the data passes through slice i we can pass in review the different scenarios leading to the realisation of the event A_i. A necessary condition is that the event is generated in slice i, $i + 1$, $\ldots$, n. If we denote G_i the event : " The event is generated in slice i", we have

Initialise $\tilde{x}_0 = \lambda, \ldots, \tilde{x}_n$
Initialise NbrLoop=1
repeat forever
    Send $\tilde{x}_i$ values to the stations which compute their p_i probability
    wait for a data
    process the data
    for i=0 **to** n
      if the data passed through slice i **then**
        $X \leftarrow 1$
      else
        $X \leftarrow 0$
      end if
      Generate R a $\tilde{x}_i$-bernoulli random variable
      $\tilde{x}_i \leftarrow \tilde{x}_i + \frac{1}{NbrLoop}(X - R)$
      Increment $NbrLoop$ by one.
    end for
end repeat

Fig. 4. Pseudo-code for estimation of the x_i value by the sink

$$\mathrm{Prob}(\mathbf{1}_i = 1) = \sum_{j=i}^{n} \mathrm{Prob}(\mathbf{1}_i = 1 | G_j).$$

Because $\mathrm{Prob}(\mathbf{1}_i = 1 | G_j) = \lambda_j p_j p_{j-1} \ldots p_{i+1}$ if $j > i$ and $\mathrm{Prob}(\mathbf{1}_i = 1 | G_i) = \lambda_i$ the last equation leads to (18).

This result is useful for devising a blind strategy for balanced energy. Indeed, from the sink point of view the realisation of the events A_i can be observed if we assume that each sensor handling an event appends to the data associated to it the slice number the stations belongs to.

    We describe the blind algorithm for energy data propagation. The algorithm does not know about the probability λ_i of occurrences of the events in the slices and indirectly estimates them. The algorithm is illustrated in Figure 4 in pseudo-code like form. The sink starts to assign values $\tilde{x}_i$ for the estimation of the x_i values and λ. For convenience, and since there are not intrinsic differences between λ and x_i we introduce the notation $x_0 = \lambda$. Each sensor is assigned a $\tilde{x}_i$ value depending on the slice number it belongs to and then computes the probability p_i of transmitting directly to the next slice using formula (8). As already mentioned, sensors add information to the propagated data to make possible for the sink to determine the slices a given data passed through. Based on these observations the sink recursively estimates the probability that the data passes through a given slice i. This probability is given by (see equation (6))

$$x_i = \lambda_i + \lambda_{i+1} p_{i+1} + \ldots + \lambda_n p_n p_{n-1} \cdots p_{i+1}.$$

Here, we used x_i without tilde to refer to the real probability of the event A_i which is the observable event. Moreover, we have seen that they can be written as (see equations (10)(14))

$$x_i = a_i + b_i\lambda, \quad i = 1,\ldots,n, \text{ and, } \lambda = x_0 = a + b\lambda.$$

This means that from the point of view of the sink an event A_i occurs with probability x_i given above.

Proposition 4. *The algorithm illustrated in Figure 4 converges in probability to the solution of the energy balanced data propagation. Precisely this means that*

$$Prob\big((\lambda_n - \frac{a}{1-b})^2 > \epsilon\big) \to 0, \quad \text{when} \quad n \to \infty,$$

with a and b defined in Theorem 1 and λ_n defined recursively by

$$\lambda_{n+1} = \lambda_n + \frac{1}{n}(X_n - R_n),$$

with X_n a Bernoulli random variable with parameter $a + b\lambda_n$ (the observable event A_1) and R_n a Bernoulli random variable with parameter λ_n (generated internally by the sink).

Proof. Since the values $\tilde{x}_i$ for $i = 1,\ldots,n$ depend on the $\tilde{x}_0 = \lambda$ value it is enough that the algorithm converges for λ_n. First notice that

$$E(X_n - R_n|\lambda_n) = a + (b-1)\lambda_n.$$

So, we get

$$E\big((\lambda_{n+1} - \frac{a}{1-b})^2|\lambda_n\big) = (\lambda_n - \frac{a}{1-b})^2 + 2\frac{b-1}{n}(\lambda_n - \frac{a}{1-b})^2 + \frac{1}{n^2}E(X_n - R_n|\lambda_n).$$

Taking expectation in both sides of this equality we get

$$E\big((\lambda_{n+1} - \frac{a}{1-b})^2\big) = E\big((\lambda_n - \frac{a}{1-b})^2\big) + 2\frac{b-1}{n}E\big((\lambda_n - \frac{a}{1-b})^2\big)$$

$$+ \frac{1}{n^2}E(X_n - R_n|\lambda_n)$$

$$= \ldots$$

$$= E\big((\lambda_1 - \frac{a}{1-b})^2\big) + 2(b-1)\sum_{j=1}^{n}\frac{1}{j}E\big((\lambda_j - \frac{a}{1-b})^2\big)$$

$$+ \underbrace{\sum_{j=1}^{n}\frac{1}{j^2}\underbrace{E(X_n - R_n|\lambda_j)}_{\text{bounded}}}_{\text{convergent}}.$$

Since $b < 0$ and $E\big((\lambda_{n+1} - \frac{a}{1-b})^2\big) > 0$ the first sum on the right side of this equation converges. Hence,

$$E\big((\lambda_j - \frac{a}{1-b})^2\big) \to 0 \quad \text{when} \quad j \to \infty,$$

which implies the convergence in probability.

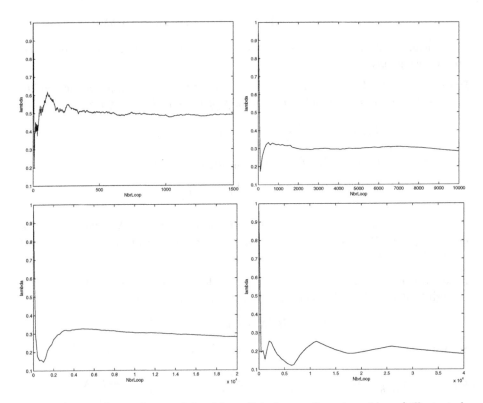

Fig. 5. Numerical experiment of algorithms (λ in terms of number of loops) illustrated in Figure 4 with $3, 10, 20, 30$ slices from left to right and top to bottom

5   Numerical Experiments and Conclusion

Recursive stochastic estimation procedures are very useful for solving practical problems and leads to adaptive protocols such that the one presented in Figure 4. The main drawback of these methods is the slow rate of convergence, typically of order $\mathcal{O}(1/\sqrt{n})$ and the lack of a robust stopping criterion. Intuitively, this is due to the fact that the procedure tracks the true value of the parameter to be estimated with correction of order $\mathcal{O}(\frac{1}{n})$. This implies that the procedure is robust in the sense that every possible value of the parameter is reached, but makes the estimate of the number of steps necessary very difficult.

Numerical experiments are then presented in order to validate the efficiency of the blind protocol introduced in this section. The framework we choose for our experiment is the same as the particular one described in [6]. This framework is realistic and allows us to compare our numerical results. The probability that an event occurs in slice number $i = 1, \ldots, n$ is proportional to the area of this slice and is given by $\lambda_i = \frac{2i-1}{n^2}$. We choose to deal with energy balanced data propagation and we have for the energy assignment vector

$$E(X) = \lambda\big(E_n/S_n, E_{n-1}/S_{n-1}, \ldots, E_1/S_1\big)^T = \lambda\big(1, 1, \ldots, 1\big).$$

We simulate the algorithm executed by the sink and illustrated in Figure 4. We start by arbitrarily fixing $\tilde{x}_1 = 0.5$ for $i = 2, \ldots, n$ and $\lambda = 1$. This last choice corresponding to the worst a priori estimation possible. We simulate the occurrence of the events with respect to the known probability λ_i. Notice that these probabilities are known from the simulation but are undirectly estimated by the algorithm (the sink). The path of the data generated by the event are simulated using the successive values of the probabilities p_i for $i = 2, \ldots, n$ which are computed on the basis of the $\tilde{x}_i$ values using formula (8). Once the path is simulated the sink updates the values of $\tilde{x}_i$ and a new event is generated. We proceed the simulations for 3, 10, 20 and 30 slices. These experiments are reported in figure 5. We observe from the experiments that, as expected, we quickly get a good estimation of the value of λ but need many more iterations to get high precision estimate due to the convergence of order $\mathcal{O}(1/\sqrt{n})$.

We discussed in this section the problem of balancing the consumed energy per sensors for the process of data propagation thourgh wireless sensor networks. The main novelty contained in this section is related to the very realistic hypothesis that no a priori knowledge on the probability of the region in which the events occur is known. We show that stochastic approximation methods can be applied and lead to protocols able to estimate these probabilities. Although high precision estimate needs many iterations of the estimation process, good estimations are provided by the algorithm prior to convergence.

An other important point to develop is to estimate the energy consumed before convergence of the algorithm and whether energy are wasted during this period of time or not.

A possible extension of this problem is not to consider slices of sensors but to consider all the sensors individually. In this situation, all the sensors have their own energy restriction and their own probability of observing an event. In this situation, it seems important that the algorithm can be devised in a distributed way. Indeed, if sensors are considered individually the process of broadcasting the x_i values from the sink to the sensors leads to an important traffic of data and it is likely that in this situation the impact of collisions cannot be longer ignored. Moreover, in the situation described in this section we ignore problems related to the size of the data sent to the sink. Indeed, when sensors add some information to the propagated data, such as the slice number, the size of the data can become prohibitive with respect to small memory capacity of smart sensors. In our situation it seems likely that the number of slice is of reasonable order, but if the sensors are individualised, the situation can prove to be very different.

References

1. T. Antoniou, A. Boukerche, I. Chatzigiannakis, S. Nikoletseas and G. Mylonas,
   • • •• ••••••• •• ••••• ••• ••••••••••••• • •••••••• ••• • •••• •••••••••••• ••
   •• ••• • •••• • ••• ••••, in Proc. 37th Annual ACM/IEEE Simulation Symposium
   (ANSS'04), IEEE Computer Society Press, pp. 43 52, 2004.

2. A. Boukerche, X. Cheng, and J. Linus, • •••••• • ••• • ••••• •• •••• • ••••• • ••
   • ••••••••• • ••• ••••, in Proc. of ACM Modeling, Analysis and Simulation of
   Wireless and Mobile Systems (MSWiM), pp. 42-49, Sept 2003.
3. A. Boukerche, R. Werner, N. Pazzi and R.B. Araujo, • • ••••• ••••• • ••••••• •
   ••• • •••••••• • ••• • •••• • •••••••• ••• • • ••••••• •••••• • ••• ••••, First International Workshop on Algorithmic Aspects of Wireless Sensor Networks (ALGO-
   SENSORS), Turku, Finland, Lecture Notes in Computer Sciences 3121, Springer-
   Verlag, July (2004).
4. I. Chatzigiannakis, T. Dimitriou, S. Nikoletseas, and P. Spirakis, • • ••••••••••
   • ••••••• ••• • • •••• ••• ••••••• • ••• ••••••••••• •• •• ••• • • ••• • ••• ••••,in
   the Proceedings of the 5th European Wireless Conference on Mobile and Wireless
   Systems beyond 3G (EW 2004), pp. 344-350, 2004.
5. I. Chatzigiannakis, S. Nikoletseas and P. Spirakis, •• ••• • ••• • •••••• •• ••• • ••••
   • ••••••• • ••• • •••••••••, in the Proceedings of the 2nd ACM Workshop on
   Principles of Mobile Computing (POMC), ACM Press, pp. 9-16, 2002.
6. C. Efthymiou, S. Nikoletseas and J. Rolim, • ••••• • ••• ••• • ••• • •••••••••••
   •• • ••••• ••••••• • ••• ••••, invited paper in the Wireless Networks (WINET,
   Kluwer Academic Publishers) Journal, Special Issue on "Best papers of the 4th
   Workshop on Algorithms for Wireless, Mobile, Ad Hoc and Sensor Networks
   (WMAN 2004)", to appear in 2005.
7. P. Leone, J. Rolim, • •• •••• • • ••••• •••• • •••• ••• • ••••• ••••• • ••• •••,
   First International Workshop on Algorithmic Aspects of Wireless Sensor Networks
   (ALGOSENSORS), Turku, Finland, Lecture Notes in Computer Sciences 3121,
   Springer-Verlag, July (2004).
8. H. Robbins, S. Monro, • •••••••••• • •••••• •••• • •••••, Ann. Math. Stat. 22,
   400-407, 1951.
9. M. Singh, V. Prasanna, • •••••• •••• ••••• • •••••• •••••• •••••• •• • • ••• •••
   • ••• • •••••• •••••• • ••• •••, In Proc. First IEEE International Conference on
   Pervasive Computing and Communications - PERCOM, 2003.

Sensor Localization in an Obstructed Environment

Wang Chen, Xiao Li, and Jin Rong

Department of Computer Science and Engineering,
Michigan State University, East Lansing, MI 48824
{wangchen, lxiao, rongjin}@cse.msu.edu

Abstract. Sensor localization can be divided into two categories: range-based approaches, and rang-free approaches. Although range-based approaches tend to be more accurate than range-free approaches, they are more sensitive to errors in distance measurement. Despite of the efforts on recovering sensors' Euclidean coordinates from erroneous distance measurements, as will be illustrated in this paper, they are still prone to distance errors, particularly in an obstruction abundant environment. In this paper, we propose a new algorithm for sensor localization based on Multiscale Radio Transmission Power(MRTP). It gradually increases the scale level of transmission power, and the distance is determined by the minimal scale of received signals. Unlike the range-based approaches, which treat each measured distance as an approximation of the true one, in our new approach, the measured distance serves as a constraint that limits the feasible location of sensors. Our simulations have shown that the MRTP-based approach is able to provide accurate and robust estimation of location, especially in an area abundant in obstructions, where most current approaches fail to perform well.

1 Introduction

One of the critical issues in sensor network research is to determine the physical positions of sensors. This is because: (1) sensed data are meaningful to most applications only when they are labeled with geographical position information; (2) position information is essential to many location-aware sensor network communication protocols, such as packet routing and sensing coverage. It has been a challenging task to design a practical algorithm for sensor localization given the constraints that are imposed on sensors, including limited power, low lost and small dimension. Furthermore, algorithms may have to be applied to inaccessible areas where sensors are randomly deployed. The objective of this paper is to design a localization algorithm, which can adapt to a complex terrain given a very limited support in hardware infrastructure.

To reduce the cost and dimension of sensors, most work suggests reusing radio signals for measuring the relative distance between adjacent sensors. However, since the radio signals can be strongly affected by the environment, distances

V. Prasanna et al. (Eds.): DCOSS 2005, LNCS 3560, pp. 49–62, 2005.
© Springer-Verlag Berlin Heidelberg 2005

inferred from radio signals tend to be inaccurate and unreliable. In the context of this paper, we categorize distance measurement errors into two groups: *micro-error* refers to the case where the distance measurement is only slightly affected by environments, such as atmospheric conditions, and the error is within a tolerable extent; *macro-error* refers to the case when radio signals are severely distorted by environments, such as obstructions, and measured distances are significantly different from their actual values.

Many efforts have been made to recover sensors' geographical coordinates from the error-prone distance measurements using radio signals. The previous work on sensor localization algorithms can be divided into two categories: the range-based approaches, and the range-free approaches. Both types of approaches employ a few GPS-aided sensors as reference beacons, Physical positions of the rest sensors are estimated based on the knowledge of their relative positions to the beacons. They differ from each other in that the range-based approaches utilize the measured distances between adjacent nodes for location estimation, while the range-free approaches only consider the connectivity information.

To alleviate the errors in distance measurement, several algorithms have been proposed in the range-based approaches. The main idea is to treat each measured distance as an approximation of the true distance between two nodes, and the optimal location of a sensor node is found by minimizing the difference between the measured distance and the distance computed from the assumed location of the sensor. In a measurement friendly environment, given multiple distance measurements are available for each sensor node, errors from individual distance measurement will be averaged out through the optimization procedure. As a result, accurate estimation of locations for sensor nodes can still be achieved with micro-error distance measurements. However, as will be illustrated later, the results of range-based approaches can be dramatically deteriorated, if a macro-error happens to one distance measurement, which is a common case in an obstructed environment. In summary, the current state-of-the-art range-based approaches can only handle micro-error distance measurements. Its accuracy can be significantly degraded given distance measurements are macro-erroneous. The range-free approaches can be viewed as binary versions of the range-based approaches: they only determine whether or not one sensor node is within the transmission range of the other. They are less dependable on the accuracy of distance measurement, and therefore are more robust to errors in the distance measurement. In fact, our empirical studies have shown that the range-free approaches can outperform the range-based approaches in a deployed area that is abundant in obstructions and distance measurements tend to be macro-erroneous.

Because range-based approaches are sensitive to the reliability of measurement, some sophisticated hardware is being developed to achieve reliable distance measurement. On the other hand, the range-free approach reuses the wireless communication radio to determine the connectivity between adjacent sensors and requires no extra hardware support. Due to its simplicity, the range-free approach is suitable for the applications where location accuracy is less critical while the cost and dimension are the main concern.

In order to achieve the high accuracy as the range-based approaches and maintain the measurement fault-tolerance as the range-free approaches, we propose a new algorithm for sensor localization based on the Multiscale Radio Transmission Power (MRTP) of beacons. Unlike the range-based approaches, which translate a distance from the Received Signal Strength Indicator (RSSI)[1], the MRTP approach gradually increases a beacon's transmission range by increasing the scales of its transmission power, and the upper bound of the distances between a beacon and other sensors are determined by the minimal audible radio signals received by the sensors. By using the discrete scale levels of transmission ranges to estimate distance between sensors, we reduce estimation errors caused by the fluctuation in radio signals. Furthermore, unlike the range-based approaches, where measured distances are used as approximation for the true distances, In MRTP approach, measured distances only serve as upper bounds for the true distances that define a set of feasible locations for sensors. By avoiding approximating true distances with the measured distances, the MRTP approach will be resilient to the macro-error distance measurement, which usually is significantly different with the true distance.

The rest of the paper is organized as follows. The previous work is reviewed in Section 2. The proposed MRTP-based approach is detailed in Section 3. The performance evaluation is conducted in Section 4. We conclude the work in Section 5.

2   Related Work

As the first range-based approach, the Received Signal Strength Indicator(RSSI) [1] based technique reuses the wireless communication infrastructure for measuring distance between adjacent sensors. However, the distance estimated by RSSI is usually inaccurate and unreliable[2][3]. This is because radio signals can be severely affected by the layout of complex terrains such as obstructions and shadow effects. To achieve better accuracy in distance measurement, Time of Arrival (ToA) is proposed to estimate the transmission distance of radio or sound signals based on their duration of flights. Since high precision clocks that measure the flying time of radio frequency(RF) signals are not available yet, Ultrasonic Time of Arrival (UToA) is adopted in the previous work[2][3][4]. Despite of its high accuracy, the UToA requires extra hardware infrastructure, which usually consumes more energy than other approaches. Furthermore, the UTOA-based approaches suffer from a poor range and lack of support to omnidirectional measurement.

To tolerate measurement errors in range-based approaches, statistic techniques such as least squares are proposed to find sensors' locations that can best fit distance measurements. It has been pointed out in robust statistic[5] that results of lease squares analysis can be completely corrupted by a single outlier. In this paper, we specialize this general idea in sensor localization and investigate how macro-errors impact final localization results. We also propose how to filter out macro-errors in the specific context of sensor networks.

Compared to the range-based approaches, the range-free approaches are usually low cost and can achieve adequate accuracy of localization for certain sensor network applications. Similar to the RSSI based technique, the communication RF is reused in the range-free approaches. Instead of measuring the strength of received signals, the range-free approaches determine the connectivity between sensors by measuring the percentage of radio packets that are successfully received. The determined connectivity information between different sensor nodes is then used to estimate the relative positions for sensor nodes.

Obviously, the connectivity is a coarse description of the distance status between sensors. To improve the accuracy for the range-free approaches, the principles of compensating low capacity of individual nodes with high global redundancy of the whole network is widely adopted in Centroid[6], DV-hop[7] and APIT[8] localization algorithms. According to the Centroid algorithm, the position of a sensor node is roughly estimated as the centroid of a polygon whose vertices are the beacons "visible" to the node. The accuracy of the Centroid algorithm is inversely proportional to the granularity of polygons formed by beacons. Simulation shows that sufficient accuracy can be achieved only when beacons are densely deployed. To have more beacons available to a sensor node, the DV-hop algorithm computes the shortest multi-hop path between beacons and the sensor node by flooding a radio signal initiated from the beacons. The distances between beacons and the sensor node is then approximated as the minimal number of hops multiplied with the average distance per hop. Assisted by the DV-hop algorithm, distances from a sensor node to beacons can be obtained even when they are far beyond the reach of radio signals. When quite a few beacons are within the radio range of a sensor node, the sensor node can be located within a certain number of triangles by the APIT algorithm. The ultimate position can be further "pinpointed" within the smaller intersection of those overlapped triangles.

The above discussion is limited to distances and connectivity between beacons and sensor nodes. Other approaches, such as Convex optimization[9] and Multidimensional Scaling (MDS)[10][11][12], have been proposed to take into account the distance and connectivity information between non-beacon nodes to further improve the accuracy in location estimation.

3   MRTP Algorithm for Sensor Localization

In order to motivate the new algorithm for sensor localization, we will first briefly review the Maximum Likelihood Estimation (MLE) approach and the Centroid approach, which are representatives of the range-based approaches and range-free approaches, respectively. The detailed description of the MRTP-based approach is presented after the overview of the existing algorithms.

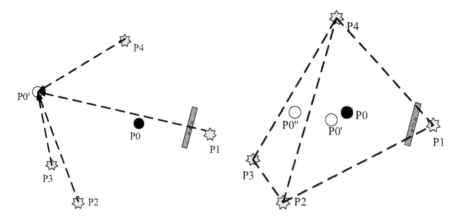

Fig. 1. MLE approach   **Fig. 2.** Centroid approach

3.1   MLE Approach

The Received Signal Strength Indicator (RSSI) has been widely used for the scenario when the radio signal is the only available resource for measuring distance between sensor nodes. It utilizes the property that a radio signal attenuates exponentially during its transmission[13], and the distance between a sender and receiver can be inferred from the attenuation of radio signals.

Due to the irregularity of radio propagation patterns, the distance converted from the RSSI value is usually inaccurate. To reduce the impact of the inaccurate distance measurement, several optimization algorithms have been proposed. One example is the Maximum Likelihood Estimation (MLE) approach. It assumes that each measured distance is an approximation of the corresponding true distance. Thus, the optimal position for a sensor node is found by minimizing the differences between the measured distances and the distances estimated using the assumed position of the node. More formally, this idea can be described as follows:

Let $\{P_i, 1 \leq i \leq n\}$ be the set of beacon nodes. Let P_0 be the node whose position is unkown, and $\widehat{P}_0$ be its estimated position. For each beacon P_i, let d_i be its measured distance to P_0, and $|P_0P_i|$ be its estimated Euclidean distance to node P_0. To minimize the difference between the measured distances and the estimated ones, we can refer to the following optimization problem:

$$\widehat{P}_0 = \arg\min_{P_0} \sum_{i=1}^{n} (|P_0P_i| - d_i)^2$$

Since the above objective function computes the aggregated difference between measured distances and the estimated ones, small errors in individual distance measurement will be averaged out, and as a result will not affect the accuracy of final estimated position significantly. Thus, the MLE approach tends to be reasonably accurate given that the measured distances are micro-erroneous.

However, given an unfriendly measurement environment where distance measurements are macro-erroneous, the accuracy of estimated positions based on the MLE approach can be dramatically deteriorated. This idea is illustrated in Figure 1, where nodes P_1, P_2, P_3 and P_4 are beacons, and P_0 is the sensor with unknown position. Because P_1 is hidden behind an obstruction, its radio signals will be attenuated considerably. As a result, the measured distance between P_1 and P_0 based on the RSSI will be much larger than its actual value. As a consequence, the estimated position for P_0, represented by the empty circle P_0' , is pushed far away from its true position. This simple example indicates that the MLE approach is particularly fragile to the large errors caused by macro-error distance measurements. In the above example, a single large error in distance measurements can significantly distort the estimation of position, even all the rest measurements are perfectly correct. The key reason for such a failure is because the MLE approach treats every measured distance as an approximation of the true distance, which is no longer true in macro-error distance measurement. Similar to the MLE approach, the estimation results of Convex Optimization and MDS may be corrupted by macro-error distance measurements between sensor nodes.

3.2   Centroid Approach

Different from range-based approaches, range-free approaches do not seek the accurate distance measurement. Instead, they find out the connectivity information between sensors by checking the percentage of radio packets that are successfully received. A receiving sensor learns that it is within the radio transmission range of a sender when the percentage exceeds a certain threshold. The connectivity information is then aggregated to determine the location for sensor nodes. For example, in the Centroid approach, the position of a sensor is estimated as the centroid of the polygon that is formed by the surrounding "visible" beacons. Since the range-free approaches are less dependable on the accuracy of measured distances, they tend to be more resilient to the large errors in macro-error distance measurements. To illustrate this point, a scenario similar to Figure 1 is shown in Figure 2. Considering two possibilities: 1) beacon P_1 is visible to sensor P_0. Therefore P_0', i.e., the centroid of the quadrangle $P_1P_2P_3P_4$, becomes the estimated position. 2) beacon P_1 is "invisible" to P_0. Then, the estimated position will be P_0'', the centroid of triangle $P_2P_3P_4$. In either case, the location estimated by the Centroid approach is closer to the true location than the MLE approach. Clearly, the key reason for the robust performance of the range-free approaches is because no attempt is made to measure the distance that can approximate the true distance.

3.3   MRTP Approach

In order to achieve the high accuracy as the range-based approaches and meanwhile maintain the measurement fault-tolerance as most range-free approaches, we propose a new localization algorithm, which uses distances as upper bound constraints to confine a small area containing the estimated sensor. Two possible

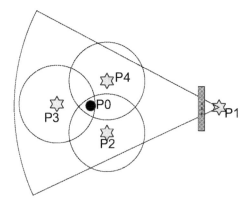

Fig. 3. MRTP approach

techniques can be used to obtain the distance upper bound constraints between beacons and a sensor. The first approach is to quantize the RSS into multiple scales as illustrated in the work[14]. Each RSS scale is mapped to a distance, which can be obtained through empirical test. When a sensor receives radio signals from a beacon, it compares measured RSS with empirical data. The distance upper bound constraints can be obtained by finding out the smallest RSS scale which is larger than measured RSS. The second approach is proposed when sensor nodes are incapable of measuring RSS. The main idea is to gradually increase the radio transmission power of sensors, until the radio packets can be "heard" by some receivers or the maximum transmission power is reached. Since each scale of transmission power corresponds to certain transmission range, given that a receiver receives a radio signal from a sender, it is able to infer, based on the scale of received signal, within which range it is from the sender. To be more precise, let P_0 be the receiver with unknown position, and P_1 be the sender. When P_0 receives a signal from P_1 at power scale level of m_1, the following inequality will hold for receiver P_0, i.e., $|P_0P_1| \leq f(m_1)$, where $|P_0P_1|$ stands for the Euclidean distance from receiver P_0 to sender P_1, and $f(m_1)$ is a function that maps the transmission power scale level of signals into its transmission range, which can be determined using empirical data. The above case can be generalized to multiple beacons, in which a different inequality constraint is introduced for every beacon P_i. Let $\{m_1, m_2, \ldots m_r\}$ be the set of signal levels that P_0 has received from beacons $\{P_1, P_2, \ldots P_r\}$. Thus, the following set of inequalities will hold for receiver P_0, i.e. $|P_0P_i| \leq f(m_i)$ $(1 \leq i \leq r)$.

However, the above set of inequality only determines the area that receiver will stay in. In order to pin down the exact location, we need to further decide which position among the confined area is more likely to be correct than other positions. To solve this problem, we follow the Centroid approach, i.e., when a receiver detected a signal from a sender, it must be close to the sender. Thus, within the determined area, we believe the position that is closest to all 'audible' senders is most desirable one. This simple idea can be formulated into the following optimization problem, i.e.,

$$\widehat{P}_0 = \arg\min_{P_0} \sum_{i=1}^{r} |P_0 P_i|^2 \tag{1}$$

$$\text{subject to } |P_0 P_i| \leq f(m_i) \ \forall 1 \leq i \leq r$$

The above optimization problem is a typical quadratic constraint and quadratic programming (QCQP) problem and can be solved efficiently using second order cone programming algorithms[15].

Compared to the range-free approach with uniform transmission power, the MRTP approach is more accurate, especially when the receiver is close to sensors. It is interesting to note that the Centroid approach can be viewed as a special case of Equation (1) when all the upper bounds $f(m_i) = \infty$. This result indicates that the centroid approach can be viewed as a MRTP approach under the assumption that the distance information is extremely inaccurate. However, in practice, by scanning the transmission power from low to maximum, distance measurement with reasonable accuracy can be achieved, at least for the beacon nodes that are close to the receivers. Under that circumstance, we will expect the proposed MRTP algorithm to provide more accurate estimation of location than the range-free approach.

Compared to the MLE approach, the MRTP approach is more tolerant to large errors in macro-error distance measurement. This is because the MLE approach treats each estimated distance as an approximation of the true distance and applies the maximum likelihood estimation to find a position such that all estimated distances are fitted well. In the case of obstructions that block between beacons and receivers, the estimated distances can be significantly far away from the true distance. As a result, fitting location with regard to the corrupted distance measures can significantly degrade the accuracy in location estimation. An example of such a scenario is illustrated in Figure 1.

In contrast, in the proposed MRTP approach for location recover, we never treat the estimated distance (i.e., $f(m_i)$) as a good approximation of the true distance. Instead, an estimated distance only provides the upper bound as to which range the receiver will be away from the sender. Due to the competition between different constraints and the objective function, only certain constraints are effective to determine the receiver's position and the rest inequalities are ignored. To better illustrate this point, an example is shown in Figure 3, in which an obstruction blocks between P_4 and the receiver P_0. As a result, more transmission power is required to send the radio packets from P_4 to P_0, which results that the measured upper bound is much larger than the actual distance between P_0 and P_4. For the other three beacons P_1 P_2 and P_3, since there is no block of obstructions, more accurate upper bound estimation between them and the receiver can be achieved. Since the distance between the receiver P_0 and each sender is less than the estimated upper bound, for each sender, we draw a circle to represent the feasible area that receiver P_0 can stay. The final admissible area for P_0 is the intersection of all circles. As indicated in Figure 3, because of the loose upper bound estimation for the distance between beacon P_4 and P_0, the circle for P_4 has no impact in determining the final intersection. This analysis indicates that the proposed algorithm is able to filter out certain erroneous upper bound estimation, thus is more robust than the range-based approaches in an environment where multiple obstructions exist and correct distance estimations are difficult to obtain. The MRTP approach is based the assumption that RSS is always attenuated by obstructions. However, radio signals may be strengthened due to the multipathing phenomena, thus estimated distances may be smaller than their true values. Distances with mixed errors make the sensor localization more complicated. Further investigation is required to seek a complete solution in such a complicated case.

4   Comparison of MLE, Centroid and MRTP in Large Scale Simulations

In the simulations below, we compare the performance of three methods, including MLE, Centroid algorithm and the MRTP approach. Radio signal is assumed to be the only method for determining the relative positions of sensor nodes. Nodes are randomly deployed in a 100m by 100m square area. The simulation is repeated multiple times with different number of nodes and beacons. The metrics defined below are used in our simulation:

- **Estimation Error:** the Euclidean distance between the estimated location of a sensor node and its actual position. For the ith unknown node, it is defined as $\sigma_i^2 = (\widehat{x}_i - x_i)^2 + (\widehat{y}_i - y_i)^2$, where $(\widehat{x}_i, \widehat{y}_i)$ is the estimated location and (x_i, y_i) is the real one.
- **Average Estimation Error:** the overall accuracy of a localization algorithm. It is defined as $\sigma^2 = \sum_{i=1}^{N} \sigma_i^2 / N$, where N is the total number of unknown nodes.

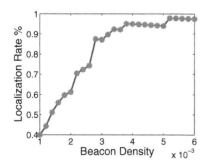
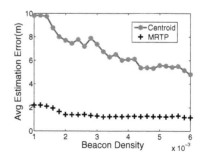

Fig. 4. Impact of beacon density on localization rate of the Centroid approach

Fig. 5. Comparison of average estimation error of the Centroid and MRTP approach

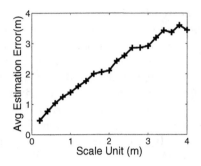
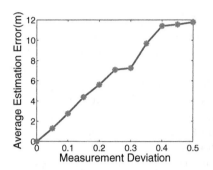

Fig. 6. Estimation error vs. scale unit of MRTP

Fig. 7. Average estimation error vs. measurement deviation of MLE

- **Median Estimation Error:** the median value of ordered Estimation Errors of all estimated sensors.
- **Localization Rate:** the ratio of successfully located nodes to the total number of nodes.
- **Radio Transmission Range:** the maximum distance the radio signal can be propagated from a beacon to any nodes.
- **Beacon Density:** the number of beacons per unit area.

4.1   Performance in Open Flat Area

We compare the performance of centroid, RSSI and MRTP algorithms in an open flat area, where the radio signal propagation is slightly affected by the atmospheric conditions such as temperature. This is related to the case where estimated distance can be micro-erroneous, but not macro-erroneous.

As discussed before, the Centroid approach requires that at lest three beacons as immediate neighbors for a sensor to locate itself. When the number of beacons are fixed, we can increase the radio transmission range of each beacon to have more beacons visible to sensor nodes, thus to locate more sensors. However, a larger radio transmission range will increase the localization error. This is because the measurement granularity of the Centroid approach is proportional to the radio transmission range. Thus, a larger radio transmission range will result in a coarser estimation of positions for sensor nodes. The above dilemma is due to the uniform transmission range used by the Centroid algorithm. In a random deployment, some sensor nodes are close to beacons, while others stay far away from the beacons. For the nodes with multiple beacons in their vicinity, a shorter radio transmission range is more preferable for better accuracy. For other nodes far away from beacons, a longer radio transmission range is required for them to be located. The only way to improve both Localization Rate and Estimation Error in Centroid approach is to increase the Beacon Density, as proposed in [16]. Figure 4 and Figure 5 shows that both the Localization Rate and the Average Estimation Error are improved as the number of beacons increases. By using the non-uniform radio transmission range in each beacon, our MRTP algorithm

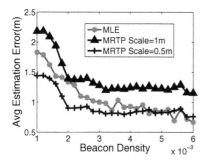

Fig. 8. Impact of beacon density on average estimation error

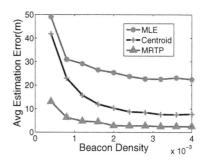

Fig. 9. Performance comparison in an obstructed environment

successfully conciliates the contradiction between the Localization Rate and the estimation accuracy.

As illustrated before, the estimated location for an unknown node is confined to the intersection of multiple circles. When a sensor node is close to multiple beacons, an accurate estimation can be achieved since the intersection area is tightly bound. By increasing the range of radio transmission of beacons, we are able to reach the sensor nodes that are far away from all beacons, thus their locations can still be computed. Figure 5 shows that the MRTP approach achieves significantly improved accuracy than the Centroid algorithm in estimating locations. It also indicates that the location accuracy tends to saturate when a certain threshold of beacon density is reached. Our simulation also shows that all unknown nodes can be located when the maximum transmission range of each beacon covers the entire area. The MRTP algorithm outperforms the Centroid algorithm because the multiscale radio transmission range provides more distance knowledge than the fixed radio transmission range. When comparing the performance of MRTP and MLE approaches, the accuracy of distance measurement is also critical to the accuracy for location estimation. We define the metrics below to represent the accuracy of distance measurement of MRTP and MLE respectively.

- **Scale Unit:** the Scale Unit determines the length of increment in the transmission range when a beacon's transmission power scale is escalated to the next adjacent level. Here, we assume that the increment in transmission range is uniformly distributed.
- **Distance Measurement Standard Deviation (DMSD):** To simulate the error in distance measurement of the RSSI that is caused by the radio's irregularity, a normal distribution is used to model the noise in measured distance, with a mean at the true distance m and a standard deviation $\sigma \bullet m$, where coefficient σ is the noise factor.

Figure 6 and Figure 7 shows that the Average Estimation Error increases when the Scale Unit and the coefficient σ of the DMSD increase. The simulation shows

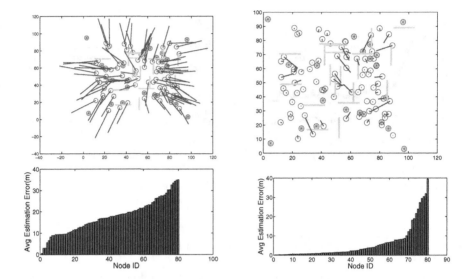

Fig. 10. Performance of the MLE approach

Fig. 11. Performance of the MRTP approach

MRTP and MLE achieve the same accuracy when the Scale Unit is 1m and the coefficient is 0.1, which means the MRTP approach with Scale Unit of 1m has the same localization accuracy as the MLE approach whose Standard Deviation is 10% of the measured distance.

Figure 8 shows that the Average Estimation Error decreases, for both MRTP and MLE, as the number of beacons increases. Note that according to Figure 8, a steady decrease in the Average Estimation Error is observed for MLE, while the performance of MRTP appears to saturate when the number of beacons exceeds a certain threshold. This is because the MRTP approach treats the estimated distance as an upper bound of the true distance. This assumption is no longer true when the true distance is within the range of the random noise in distance measurement. In contrast, since the MLE approach treats the measured distance as approximation of true distance, the noise in distance measurement is averaged out through solving the optimization problem in Equation (1). However, as we will see later in this paper, we trade the accuracy of location estimation with its fault-tolerance. From the analysis above, we can conclude that: in the open flat area with fixed number beacons, MRTP approach can achieve more accurate estimation result than Centroid approach, and the estimation accuracy of both MRTP and MLE approaches are determined by the basic distance measurement accuracy.

4.2 Performance in an Obstruction Abundant Environment

The simulation shows that in the flat open area, where the distance measurement is slightly affected by some random factors, the location estimation accuracy of both

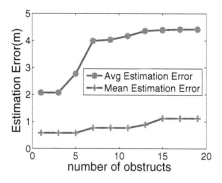

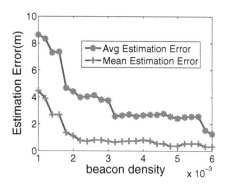

Fig. 12. Impact of obstruction on estimation error

Fig. 13. Impact of beacon density on estimation error

MRTP and MLE approaches is determined by the basic distance measurement accuracy. In such a case, the MRTP approach can be viewed as a multilevel quantized MLE approach. It appears that MRTP approach is different from the MLE approach only in the format of distance measurement. In this section, we consider a realistic setup, in which multiple obstructions are placed to severely deteriorate the accuracy of distance measurements. In other words, the distance measurements can be macro-erroneous under this environment. The performance comparison of MLE, Centroid and MRTP approaches is shown in Figure 9, which illustrates that MRTP outperforms MLE in an obstruction abundant environment.

Figure 10 and Figure 11 show an example to compare MLE and MRTP, where each gray bar represents an obstruction, and each solid line represents the Estimation Error. The distribution of Estimation Errors is also shown at the bottom of corresponding graph, which plots the Estimation Errors of all nodes in the increasing order of the errors. The comparison demonstrates that almost all the estimation results in MLE are severely corrupted, while more than half of the sensor nodes are located with high accuracy in MRTP. Some of the nodes in MRTP cannot estimate their position accurately because almost all the beacons are invisible to those nodes; therefore the estimated node is not tightly constrained in a small area. Figure 12 shows that both the Average Estimation Error and the Median Estimation Error increase as the number of obstructions increases. This is because more beacons are hidden behind the obstructions. As shown in Figure 13, by increasing the Beacon Density in the deployed area, the estimation accuracy can be improved because more beacons are seen by the unknown nodes.

5 Conclusion

In this paper, we propose the MRTP sensor network localization algorithm, which uses distance upper bound constraints to achieve accurate localization results in a complex terrain, especially in an obstructed environment, where most current approaches fail to perform well. The MRTP approach outperforms previous approach in that the impact of macro-errors caused by obstructions can be automatically filtered out in the localization algorithm, thus the accuracy of the final localization results is achievable. The simulation shows that the MRTP approach can achieve better localization results than both Centroid and MLE approaches.

References

1. J. Hightower, C. Vakili, G. Borriello, and R. Want, "Design and calibration of the spoton ad-hoc location sensing system," • • •   • • • • • • • •• • • • • • • • • •• • • • •• • • • • • • • • • • • , August 2001.
2. A. Savvides, C. Han, and M. B. Strivastava, "Dynamic fine-grained localization in ad-hoc networks of sensors," in Proc. • • • •• • • , 2001.
3. K. Whitehouse, "The design of calamari: an ad-hoc localization system for sensor networks," • • •• •• • • • • •• • • • •• • • • • • • • • •• • •• • • • • • • • • •• •, 2002.
4. D. Moore, J. Leonard, D. Rus, and S. Teller, "Robust Distributed Network Localization with Noisy Range Measurements," in Proc. • • • • • •, 2004.
5. P. Huber, • • • • • • • •• • • • • •. Wiley, New York, 1981.
6. N. Bulusu, J. Heidemann, and D. Estrin, "GPS-less Low Cost Outdoor Localization for Very Small Devices," •• • • • • • • • • • • • • • • • • • •• •• • • • • • • • • • • •, 2000.
7. D. Niculescu and B. Nath, "Ad hoc Positioning System (APS)," in Proc. • • • • • • • • • , 2001.
8. T. He, C. Huang, B. M. Blum, J. A. Stankovic, and T. F. Abdelzaher, "Range-Free Localization Schemes in Large Scale Sensor Networks," in Proc. • • •• • •• , 2003.
9. L. Doherty, K. S. J. Pister, and L. E. Ghaoui, "Convex Position Estimation in Wireless Sensor Networks," in Proc. •• • • • • • , 2001.
10. Y. Shang, W. Ruml, Y. Zhang, and M. P. J. Fromherz, "Localization from mere connectivity," in Proc. • • •• • • •, 2003.
11. Y. Shang and W. Ruml, "Improved MDS-Based Localization," in Proc. •• • • • • • • , 2004.
12. X. Ji and H. Zha, "Sensor Positioning in Wireless Ad-hoc Sensor Networks with Multidimensional Scaling," in Proc. •• • • • • • , 2004.
13. S. Y. Seidel and T. S. Rappaport, "914 MHz Path Loss Prediction Models for Indoor Wireless Communications in Multifloored Buildings," •• • • • • • • • • • •• • • •• • • • •• • • • • •• • • • • • • • • • •, vol. 40, pp. 209–217, 1992.
14. N. Patwari and A. O. Hero, "Using proximity and quantized rss for sensor localization in wireless networks," in Proc. • • • • • , 2003.
15. M. S. Lobo, L. Vandenberghe, S. Boyd, and H. Lebret, "Applications of Second-order Cone Programming," • •• •• • • •• •• • • •• • •• • • •• ••• •• • , 1998.
16. N. Bulusu, J. Heidemann, and D. Estrin, "Adaptive Beacon Placement," in Proc. •• • • • , 2001.

Stably Computable Properties of Network Graphs

Dana Angluin, James Aspnes, Melody Chan, Michael J. Fischer,
Hong Jiang, and René Peralta

Yale University

Abstract. We consider a scenario in which anonymous, finite-state sensing devices are deployed in an ad-hoc communication network of arbitrary size and unknown topology, and explore what properties of the network graph can be stably computed by the devices. We show that they can detect whether the network has degree bounded by a constant d, and, if so, organize a computation that achieves asymptotically optimal linear memory use. We define a model of stabilizing inputs to such devices and show that a large class of predicates of the multiset of final input values are stably computable in any weakly-connected network. We also show that nondeterminism in the transition function does not increase the class of stably computable predicates.

1 Introduction

In some applications, a large number of sensors will be deployed without fine control of their locations and communication patterns in the target environment. To enable the distributed gathering and processing of information, the sensors must constitute themselves into an ad-hoc network and use it effectively. A fundamental question in this context is whether there are protocols that determine enough about the topological properties of this network to exploit its full potential for distributed computation, when sensors are severely limited in their computational power.

We consider a model introduced in [1] in which communication is represented by pairwise interactions of anonymous finite-state devices in networks of finite but unbounded size. These systems correspond to sensor networks with intermittent two-way communications between nearby nodes. A **communication graph** describes which pairs of nodes are close enough to eventually interact. Our goal is to explore what graph-theoretic properties of the communication graph are **stably computable**, where a property is stably computable if all sensors eventually converge to the correct answer. The model assumes that the devices have no identifiers and only a few bits of memory each. Because each device is so limited, their collective ability to achieve nontrivial computation must be based on their capacity to organize a distributed computation in the network.

V. Prasanna et al. (Eds.): DCOSS 2005, LNCS 3560, pp. 63–74, 2005.
© Springer-Verlag Berlin Heidelberg 2005

In this setting, the structure of the network has a profound influence on its computational potential. If n is the number of vertices in the communication graph, previous results show that $O(\log n)$ bits of memory are sufficient for a nondeterministic Turing machine to simulate any protocol in the all-pairs communication graph, in which every pair of vertices is joined by an edge [1]. In contrast, we give a protocol that can determine whether the communication graph is a directed cycle and if so, use it as a linear memory of $O(n)$ bits, which is asymptotically optimal in terms of memory capacity. More generally, we show that for every d there is a protocol that can organize any communication graph of maximum degree d into a linear memory of $O(n)$ bits, also asymptotically optimal.

For general communication graphs, we show that any property that is determined by the existence of a fixed finite subgraph is stably computable, as are Boolean combinations of such properties. In addition, there are protocols to compute the following graph properties: whether the communication graph G is a directed star, whether G is a directed arborescence, whether G contains a directed cycle, and whether G contains a directed cycle of odd length. Furthermore, for any positive integer d, there is a protocol that stabilizes to a proper d-coloring of any d-colorable graph, but does not stabilize if the graph is not d-colorable.

In the model of [1], the sensor readings were all assumed to be available as inputs at the start of the computation. In this paper, we extend the model to allow for a more realistic scenario of stabilizing inputs, that may change finitely many times before attaining a final value. In addition to allowing fluctuations in the inputs, these results allow composition of stabilizing protocols: a protocol that works with stabilizing inputs can use the stabilizing outputs of another protocol. We generalize two fundamental theorems to the case of stabilizing inputs: all the **Presburger-definable** predicates are stably computable in the all-pairs graph, and any predicate stably computable in the all-pairs graph is stably computable in any weakly connected graph with the same number of nodes.

Another powerful tool for the design of protocols is to permit a nondeterministic transition function; we give a simulation to show that this does not increase the class of stably computable predicates.

1.1   Other Related Work

Population protocols and similar models as defined in [1–3] have connections to a wide range of theoretical models and problems involving automata of various kinds, including Petri nets [4–7], semilinear sets and Presburger expressions [8], vector addition systems [9], the Chemical Abstract Machine [10, 11] and other models of communicating automata [12, 13]. See [1] for a more complete discussion of these connections.

The potential for distributed computation during aggregation of sensor data is studied in [14, 15], and distributed computation strategies for conserving resources in tracking targets in [16, 17]. Issues of random mobility in a wireless packet network are considered in [18].

Passively mobile sensor networks have been studied in several practical application contexts. The **smart dust** project [19,20] designed a cloud of tiny wireless microelectromechanical sensors (MEMS) with wireless communication capacity, where each sensor or "mote" contains sensing units, computing circuits, bidirectional wireless capacity, and a power supply, while being inexpensive enough to deploy massively.

2   The Model of Stable Computation

We represent a network communication graph by a directed graph $G = (V, E)$ with n vertices numbered 1 through n and no multi-edges or self-loops. Each vertex represents a finite-state sensing device, and an edge (u, v) indicates the possibility of a communication between u and v in which u is the initiator and v is the responder.[1] We assume G is *weakly connected*, that is, between any pair of nodes there is a path (disregarding the direction of the edges in the path). The **all-pairs graph** contains all edges (u, v) such that $u \neq v$.

We first define protocols without inputs, which is sufficient for our initial results on graph properties, and extend the definition to allow stabilizing inputs in Section 4.

A **protocol** consists of a finite set of **states** Q, an **initial state** $q_0 \in Q$, an **output function** $O : Q \to Y$, where Y is a finite set of output symbols, and a **transition function** δ that maps every pair of states (p, q) to a nonempty set of pairs of states. If $(p', q') \in \delta(p, q)$, we call $(p, q) \mapsto (p', q')$ a **transition**.

The transition function, and the protocol, is **deterministic** if $\delta(p, q)$ always contains just one pair of states. In this case we write $\delta(p, q) = (p', q')$ and define $\delta_1(p, q) = p'$ and $\delta_2(p, q) = q'$.

A **configuration** is a mapping $C : V \to Q$ specifying the state of each device in the network. We assume that there is a global start signal transmitted simultaneously to all the devices, e.g., from a base station, that puts them all in the initial state and starts the computation. The **initial configuration** assigns the initial state q_0 to every device.

Let C and C' be configurations, and let u, v be distinct nodes. We say that C goes to C' via **pair** $e = (u, v)$, denoted $C \xrightarrow{e} C'$, if the pair $(C'(u), C'(v))$ is in $\delta(C(u), C(v))$ and for all $w \in V - \{u, v\}$ we have $C'(w) = C(w)$. We say that C can go to C' in one step, denoted $C \to C'$, if $C \xrightarrow{e} C'$ for some edge $e \in E$. We write $C \xrightarrow{*} C'$ if there is a sequence of configurations $C = C_0, C_1, \ldots, C_k = C'$, such that $C_i \to C_{i+1}$ for all i, $0 \leq i < k$, in which case we say that C' is **reachable** from C.

A **computation** is a finite or infinite sequence of population configurations $C_0, C_1, \ldots$ such that for each i, $C_i \to C_{i+1}$. An infinite computation is **fair** if for every pair of population configurations C and C' such that $C \to C'$, if C

---

[1] The distinct roles of the two devices in an interaction is a fundamental assumption of asymmetry in our model; symmetry-breaking therefore does not arise as a problem within the model.

occurs infinitely often in the computation, then C' also occurs infinitely often in the computation. We remark that by induction, if $C \xrightarrow{*} C'$ and C occurs infinitely often in a fair computation, then C' also occurs infinitely often in a fair computation.

2.1   Leader Election

As an example, we define a simple deterministic leader election protocol that succeeds in any network communication graph. The states of the protocol are $\{1, 0\}$ where 1 is the initial state. The transitions are defined by

$$(1) \ (1, 1) \mapsto (0, 1)$$
$$(2) \ (1, 0) \mapsto (0, 1)$$
$$(3) \ (0, 1) \mapsto (1, 0)$$
$$(4) \ (0, 0) \mapsto (0, 0)$$

We think of 1 as the leader mark. In every infinite fair computation of this protocol starting with the initial configuration in any communication graph, after some finite initial segment of the computation, every configuration has just one vertex labeled 1 (the leader), and every vertex has label 1 in infinitely many different configurations of the computation. Thus, eventually there is one "leader" mark that hops incessantly around the graph, somewhat like an ancient English king visiting the castles of his lords. Note that in general the devices have no way of knowing whether a configuration with just one leader mark has been reached yet.

2.2   The Output of a Computation

Our protocols are not designed to halt, so there is no obvious fixed time at which to view the output of the computation. Rather, we say that the output of the computation stabilizes if it reaches a point after which no device can subsequently change its output value, no matter how the computation proceeds thereafter. Stability is a global property of the graph configuration, so individual devices in general do not know when stability has been reached.[2]

We define an **output assignment** y as a mapping from V to the output symbols Y. We extend the output map O to take a configuration C and produce an output assignment $O(C)$ defined by $O(C)(v) = O(C(v))$. A configuration C is said to be **output-stable** if $O(C') = O(C)$ for all C' reachable from C. Note that we do not require that $C = C'$, only that their output assignments be equal. An infinite computation **output-stabilizes** if it contains an output-stable configuration C, in which case we say that it **stabilizes to output assignment** $y = O(C)$. Clearly an infinite computation stabilizes to at most one output assignment.

---

[2] With suitable stochastic assumptions on the rate at which interactions occur, it is possible to bound the expected number of interactions until the output stabilizes, a direction explored in [1].

The output of a finite computation is the output assignment of its last configuration. The output of an infinite computation that stabilizes to output assignment y is y; the output is undefined if the computation does not stabilize to any output assignment. Because of the nondeterminism inherent in the choice of encounters, the same initial configuration may lead to different computations that stabilize to different output assignments.

2.3   Graph Properties

We are interested in what properties of the network communication graph can be stably computed by protocols in this model. A **graph property** is a function P from graphs G to the set $\{0, 1\}$ where $P(G) = 1$ if and only if G has the corresponding property. We are interested in families of graphs, $\mathcal{G}_1, \mathcal{G}_2, \ldots,$ where $\mathcal{G}_n$ is a set of network graphs with n vertices. The **unrestricted** family of graphs contains all possible network communication graphs. The **all-pairs** family of graphs contains for each n just the all-pairs graph with n vertices. For every d, the family of d-**degree-bounded** graphs contains all the network communication graphs in which the in-degree and out-degree of each vertex is bounded by d. Similarly, the family of d-**colorable** graphs contains all the network communication graphs properly colorable with at most d colors.

We say that a protocol $\mathcal{A}$ stably computes the graph property P in the family of graphs $\mathcal{G}_1, \mathcal{G}_2, \ldots$ if for every graph G in the family, every infinite fair computation of $\mathcal{A}$ in G starting with the initial configuration stabilizes to the constant output assignment equal to $P(G)$. Thus the output of every device stabilizes to the correct value of $P(G)$.

3   Example: Is G a Directed Cycle?

In this section, we show that the property of G being a directed cycle is stably computable in the unrestricted family of graphs. Once a directed cycle is recognized, it can be organized (using leader-election techniques) to simulate a Turing machine tape of n cells for the processing of inputs, which vastly increases the computational power over the original finite-state devices, and is optimal with respect to the memory capacity of the network.

Theorem 1. *Whether G is a directed cycle is stably computable in the unrestricted family of graphs.*

Proof sketch: A weakly connected directed graph G is a directed cycle if and only if the in-degree and out-degree of each vertex is 1.

Lemma 1. *Whether G has a vertex of in-degree greater than 1 is stably computable in the unrestricted family of graphs.*

We give a protocol to determine whether some vertex in G has in-degree at least 2. The protocol is deterministic and has 4 states: $\{-, I, R, Y\}$, where $-$ is the initial state, I and R stand for "initiator" and "responder" and Y indicates that there is a vertex of in-degree at least 2 in the graph. The transitions are as follows, where x is any state and unspecified transitions do not change their inputs:

$$(1)\ (-,-) \mapsto (I,R)$$
$$(2)\ (I,R) \mapsto (-,-)$$
$$(3)\ (-,R) \mapsto (Y,Y)$$
$$(4)\ (Y,x) \mapsto (Y,Y)$$
$$(5)\ (x,Y) \mapsto (Y,Y)$$

The output map takes Y to 1 and the other states to 0. State Y is contagious, spreading to all states if it ever occurs. If every vertex has in-degree at most 1 then only transitions of types (1) and (2) can occur, and if some vertex has in-degree at least 2, then in any fair computation some transition of type (3) must eventually occur.

An analogous four-state protocol detects whether any vertex has out-degree at least 2. We must also guarantee that every vertex has positive in-degree and out-degree. The following deterministic two-state protocol stably labels each vertex with Z if it has in-degree 0 and P if it has in-degree greater than 0. The initial state is Z. The transitions are given by the following, where x and y are any states:

$$(1)\ (x,y) \mapsto (x,P)$$

To detect whether all states are labeled P in the limit, we would like to treat the outputs of this protocol as the inputs to a another protocol to detect any occurrences of Z in the limit. However, to do this, the protocol to detect any occurrences of Z must cope with inputs that may change before they stabilize to their final values. In the next section we show that all the Presburger predicates (of which the problem of Z detection is a simple instance) are stably computable with such "stabilizing inputs" in the unrestricted family of graphs, establishing the existence of the required protocol.

Similarly, there is a protocol that stably computes whether every vertex has out-degree at least 1. By running all four protocols in parallel, we may stably compute the property: does every vertex have in-degree and out-degree exactly 1? Thus, there is a protocol that stably computes the property of G being a directed cycle for the unrestricted family of graphs, proving Theorem 1. ∎

These techniques generalize easily to recognize other properties characterized by conditions on vertices having in-degrees or out-degrees of zero or one. A directed line has one vertex of in-degree zero, one vertex of out-degree zero, and all other vertices have in-degree and out-degree one. An out-directed star has one vertex of in-degree zero and all other vertices of in-degree one and out-degree zero, and similarly for an in-directed star. An out-directed arborescence has one vertex of in-degree zero and all other vertices have in-degree one, and similarly for an in-directed arborescence.

Theorem 2. *The graph properties of being a directed line, a directed star, or a directed arborescence are stably computable in the unrestricted family of graphs.*

In a later section, we generalize Theorem 1 to show that for any d there is a protocol to recognize whether the network communication graph has in-degree and out-degree bounded by d and to organize it as $O(n)$ cells of linearly ordered memory if so. We now turn to the issue of inputs.

4   Computing with Stabilizing Inputs

We define a model of stabilizing inputs to a network protocol, in which the input to each node may change finitely many times before it stabilizes to a final value. We are interested in what predicates of the final input assignment are stably computable. An important open question is whether any predicate stably computable in a family of graphs is stably computable with stabilizing inputs in the same family of graphs.

Though we do not fully answer this question, we show that a large class of protocols can be adapted to stabilizing inputs by generalizing two theorems from [1] to the case of stabilizing inputs: all Presburger predicates are stably computable with stabilizing inputs in the all-pairs family of graphs, and any predicate that can be stably computed with stabilizing inputs in the all-pairs family of graphs can be stably computed with stabilizing inputs in the unrestricted family of graphs.

We assume that there is a finite set of input symbols, and each device has a separate input port at which its current input symbol (representing a sensed value) is available at every computation step. Between any two computation steps, the inputs to any subset of the devices may change arbitrarily.

In the full paper, we define formally an extension of the basic stable computation model that permits protocols with stabilizing inputs. This extension is used to prove several results concerning such protocols, including:

Lemma 2. *Any Boolean combination of a finite set of predicates stably computable with stabilizing inputs in a family of graphs $\mathcal{G}_1, \mathcal{G}_2, \ldots$ is stably computable with stabilizing inputs in the same family of graphs.*

Theorem 3. *For every nondeterministic protocol $\mathcal{A}$ there exists a deterministic protocol $\mathcal{B}$ such that if $\mathcal{A}$ stably computes a predicate P with stabilizing inputs in a family of graphs, then $\mathcal{B}$ also stably computes P with stabilizing inputs in the same family of graphs.*

4.1   The Presburger Predicates

The Presburger graph predicates form a useful class of predicates on the multiset of input symbols, including such things as "all the inputs are a's", "at least 5 inputs are a's", "the number of a's is congruent to 3 modulo 5", and "twice the number of a's exceeds three times the number of b's" and Boolean combinations of such predicates. Every Presburger graph predicate is stably computable with unchanging inputs in the all-pairs family of graphs [1].

We define the Presburger graph predicates as those expressible in the following expression language.[3] For each input symbol $a \in X$, there is a variable $\#(a)$
that represents the number of occurrences of a in the input. A **term** is a linear
combination of variables with integer coefficients, possibly modulo an integer.
An **atom** is two terms joined by one of the comparison operators: $<, \leq, =, \geq, >$.
An **expression** is a Boolean combination of atoms.

Thus, if the input alphabet $X = \{a, b, c\}$, the predicate "all the inputs are
a's" can be expressed as $\#(b) + \#(c) = 0$, the predicate that the number of
a's is congruent to 3 modulo 5 can be expressed as $(\#(a) \bmod 5) = 3$, and the
predicate that twice the number of a's exceeds three times the number of b's by
$2\#(a) > 3\#(b)$.

In the full paper, we show:

Theorem 4. *Every Presburger graph predicate is stably computable with stabilizing inputs in the family of all-pairs graphs.*

Theorem 5. *For any protocol $\mathcal{A}$ there exists a protocol $\mathcal{B}$ such that for every
n, if $\mathcal{A}$ stably computes predicate P with stabilizing inputs in the all-pairs interaction graph with n vertices and G is any communication graph with n vertices,
protocol $\mathcal{B}$ stably computes predicate P with stabilizing inputs in G.*

The following corollary is an immediate consequence of Theorems 4 and 5.

Corollary 1. *The Presburger predicates are stably computable with stabilizing
inputs in the unrestricted family of graphs.*

5   Computing in Bounded-Degree Networks

In Section 3 we showed that there is a protocol that stably computes whether
G is a directed cycle. If G is a directed cycle, another protocol can organize a
Turing machine computation of space $O(n)$ in the graph to determine properties
of the inputs. Thus, certain graph structures can be recognized and exploited to
give very powerful computational capabilities.

In this section we generalize this result to the family of graphs with degree
at most d.

Lemma 3. *For every positive integer d, there is a protocol that stably computes
whether G has maximum degree less than or equal to d.*

Proof sketch: We describe a protocol to determine whether G has any
vertex of out-degree greater than d; in-degree is analogous, and the "nor" of
these properties is what is required. The protocol is nondeterministic and based
on leader election. Every vertex is initially a leader and has output 0. Each
leader repeatedly engages in the following searching behavior. It hops around

---

[3] These are closely related to the Presburger integer predicates defined in [1]. Details
will be given in the full paper.

the vertices of G and decides to check a vertex by marking it and attempting to place $d + 1$ markers of a second type at the ends of edges outgoing from the vertex being checked. It may decide to stop checking, collect a set of markers corresponding to the ones it set out, and resume its searching behavior. If it succeeds in placing all its markers, it changes its output to 1, hops around the graph to collect a set of markers corresponding to the ones it set out, and resumes its searching behavior.

When two leaders meet, one becomes a non-leader, and the remaining leader collects a set of markers corresponding to the ones that both had set out, changes its output back to 0, and resumes its searching behavior. Non-leaders copy the output of any leader they meet.

After the leader becomes unique and collects the set of markers that it and the last deposed leader had set out, then the graph is clear of markers and the unique leader resumes the searching behavior with its output set to 0. If there is no vertex of out-degree greater than d, the output will remain 0 (and will eventually be copied by all the non-leaders.) If there is some vertex of out-degree greater than d, the searching behavior eventually finds it and sets the output to 1, which will eventually be copied by all the non-leaders. ∎

Note that this technique can be generalized (using a finite collection of distinguishable markers) to determine the existence of any fixed subgraph in G. (For the lemma, the fixed subgraph is the out-directed star on $d + 2$ vertices.) Another variant of this idea (mark a vertex and try to reach the mark by following directed edges) gives protocols to determine whether G contains a directed cycle or a directed odd cycle.

Theorem 6. *There are protocols that stably compute whether G contains a fixed subgraph, or a directed cycle, or a directed cycle of odd length in the unrestricted family of graphs.*

For bounded-degree graphs, we can organize the nodes into a spanning tree rooted at a leader, which can then simulate a Turing machine tape of size $O(n)$ distributed across the nodes. This allows a population with a bounded-degree interaction graph to compute any function of the graph structure that is computable in linear space.

Theorem 7. *For every positive integer d, there is a protocol that for any d-bounded graph G stably constructs a spanning tree structure in G that can be used to simulate n cells of Turing machine tape.*

Proof sketch: The protocol is rather involved; details are deferred to the full paper. We give an outline of the protocol here.

The starting point is to label the vertices of G in such a way that no two neighbors of any vertex have the same label. A vertex can then send messages to a specific neighbor by waiting to encounter another vertex with the appropriate label.

To see that such a labeling exists, consider the graph G' obtained from G by ignoring the direction of edges and including an edge between any two nodes at distance 2 in G. A proper coloring of G' will give the required labeling of G, and the degree of G' is at most $4d^2$, so G' can be colored with $4d^2 + 1$ colors.

One part of the protocol eventually constructs a labeling of the desired kind by using leader election and a searching behavior that attempts to find two vertices at distance two that have the same label and nondeterministically relabel both. Eventually there will be no more relabeling.

The other part of the protocol attempts to use the constructed labeling to build a spanning tree structure in G. This is also based on leader election. Each leader begins building a spanning tree from the root, recording in each vertex the labels of the known neighbors of the vertex, and designating (by label) a parent for each non-root vertex. The leader repeatedly traverses its spanning tree attempting to recruit new vertices. When it meets another leader, one becomes a non-leader and the other begins building a new spanning tree from scratch.

The labeling portion of the protocol is running in parallel with the tree-construction, and it produces a "restart" marker whenever it relabels vertices. When a leader encounters a "restart" marker, it deletes the marker and again begins building a new spanning tree from scratch.

Eventually all the relabeling will be completed, and only one leader will remain, and the final spanning tree construction will not be restarted. However, it is important that the spanning tree construction be able to succeed given a correct labeling but otherwise arbitrary states left over from preceding spanning tree construction attempts. The leader repeatedly traverses the constructed spanning tree, setting a phase indicator (0 or 1) at each pass to detect vertices that are not yet part of the tree. An arbitrary ordering on the labels gives a fixed traversal order for the spanning tree, and a portion of each state can be devoted to simulating a Turing machine tape cell.

Because a leader cannot determine when it is unique, when the labeling is stable, or when the spanning tree is complete, it simply restarts computation in the tree each time it begins rebuilding a spanning tree. Eventually, however, the computation has access to n simulated cells of tape.   ∎

Itkis and Levin use a similar construction in their self-stabilizing protocols for identical nameless nodes in an asynchronous communication network (represented by an undirected, connected, reflexive graph G) with worst-case transient faults [21]. In their model, in addition to a finite number of bits of storage, each node x maintains a finite set of pointers to itself and its neighbors, and can detect whether a neighbor y has a pointer to x and/or y, and can set a pointer to point to the first (in a fixed ordering) neighbor with a given property. This additional information about the graph structure permits them to exploit storage $O(|V|)$ in all cases. By comparison, in our model there are no pointers and each device truly has only a constant amount of memory regardless of the size and topology of the network. For this reason, in an all-pairs graph of size n, only memory proportional to $\log n$ is achievable. So the bounded-degree restriction of Theorem 7, or some other limitation that excludes the all-pairs graph, appears to be necessary.

6   Discussion and Open Problems

It is open whether every predicate stably computable with unchanging inputs in a family of graphs is stably computable with stabilizing inputs in the same family of graphs. For the family of all-pairs graphs, both classes contain the Presburger predicates. It follows that if, in this family, some predicate is stably computable with unchanging inputs but not with stabilizing inputs, that predicate would not be Presburger. The existence of such a predicate would disprove a conjecture from [1].

A natural measure of the information-theoretic memory capacity of a graph G with a finite set of vertex labels L is the log of number of different isomorphism classes of G with vertex labels drawn from L. If L has at least two elements, the all-pairs graphs have memory capacity $\Theta(\log n)$, which can be exploited for computation in the setting of randomized interactions and computation with errors [1]. When the cardinality of L is large enough compared to d ($O(d^2)$ suffices), the memory capacity of d-degree bounded graphs is $\Theta(n)$, which we have shown above can be exploited by protocols for stable computation. Are there protocols to exploit the full-information theoretic memory capacity of arbitrary network communication graphs?

An important future direction is to study the running time of protocols in this model, perhaps under a stochastic model of pairwise interactions, as in the model of conjugating automata [1].

Acknowledgments

James Aspnes was supported in part by NSF grants CCR-9820888, CCR-0098078, CNS-0305258, and CNS-0435201. Michael J. Fischer and René Peralta were supported in part by NSF grant CSE-0081823. Hong Jiang was supported by NSF grant ITR-0331548.

The authors would like to thank David Eisenstat for helpful comments on the paper. The authors would also like to thank the anonymous referees for DCOSS 2005 for their many useful suggestions.

References

1. Angluin, D., Aspnes, J., Diamadi, Z., Fischer, M.J., Peralta, R.: Computation in networks of passively mobile finite-state sensors. In: PODC '04: Proceedings of the twenty-third annual ACM symposium on Principles of distributed computing, ACM Press (2004) 290–299
2. Angluin, D., Aspnes, J., Diamadi, Z., Fischer, M.J., Peralta, R.: Urn automata. Technical Report YALEU/DCS/TR–1280, Yale University Department of Computer Science (2003)
3. Diamadi, Z., Fischer, M.J.: A simple game for the study of trust in distributed systems. Wuhan University Journal of Natural Sciences **6** (2001) 72–82 Also appears as Yale Technical Report TR–1207, January 2001, available at URL ftp://ftp.cs.yale.edu/pub/TR/tr1207.ps.

4. Bernardinello, L., Cindio, F.D.: A survey of basic net models and modular net classes. In Rozenberg, G., ed.: Advances in Petri Nets: The DEMON Project. Volume 609 of Lecture Notes in Computer Science. Springer-Verlag (1992) 304–351

5. Esparza, J., Nielsen, M.: Decibility issues for Petri nets - a survey. Journal of Informatik Processing and Cybernetics **30** (1994) 143–160

6. Esparza, J.: Decidability and complexity of Petri net problems-an introduction. In Rozenberg, G., Reisig, W., eds.: Lectures on Petri Nets I: Basic models. Springer Verlag (1998) 374–428 Published as LNCS 1491.

7. Mayr, E.W.: An algorithm for the general Petri net reachability problem. SIAM J. Comput. **13** (1984) 441–460

8. Ginsburg, S., Spanier, E.H.: Semigroups, Presburger formulas, and languages. Pacific Journal of Mathematics **16** (1966) 285–296

9. Hopcroft, J., Pansiot, J.: On the reachability problem for 5-dimensional vector addition systems. Theoretical Computer Science **8** (1978) 135–159

10. Berry, G., Boudol, G.: The Chemical Abstract Machine. Theoretical Computer Science **96** (1992) 217–248

11. Ibarra, O.H., Dang, Z., Egecioglu, O.: Catalytic p systems, semilinear sets, and vector addition systems. Theor. Comput. Sci. **312** (2004) 379–399

12. Brand, D., Zafiropulo, P.: On communicating finite-state machines. Journal of the ACM **30** (1983) 323–342

13. Milner, R.: Bigraphical reactive systems: basic theory. Technical report, University of Cambridge (2001) UCAM-CL-TR-523.

14. Intanagonwiwat, C., Govindan, R., Estrin, D.: Directed diffusion: a scalable and robust communication paradigm for sensor networks. In: Proceedings of the 6th Annual International Conference on Mobile computing and networking, ACM Press (2000) 56–67

15. Madden, S.R., Franklin, M.J., Hellerstein, J.M., Hong, W.: TAG: a Tiny AGgregation service for ad-hoc sensor networks. In OSDI 2002: Fifth Symposium on Operating Systems Design and Implementation (December, 2002)

16. Fang, Q., Zhao, F., Guibas, L.: Lightweight sensing and communication protocols for target enumeration and aggregation. In: Proceedings of the 4th ACM International Symposium on Mobile ad hoc networking & computing, ACM Press (2003) 165–176

17. Zhao, F., Liu, J., Liu, J., Guibas, L., Reich, J.: Collaborative signal and information processing: An information directed approach. Proceedings of the IEEE **91** (2003) 1199–1209

18. Grossglauser, M., Tse, D.N.C.: Mobility increases the capacity of ad hoc wireless networks. IEEE/ACM Transactions on Networking **10** (2002) 477–486

19. Kahn, J.M., Katz, R.H., Pister, K.S.J.: Next century challenges: mobile networking for "Smart Dust". In: MobiCom '99: Proceedings of the 5th annual ACM/IEEE international conference on Mobile computing and networking, ACM Press (1999) 271–278

20. Chatzigiannakis, I., Nikoletseas, S., Spirakis, P.: Smart dust protocols for local detection and propagation. In: POMC '02: Proceedings of the second ACM international workshop on Principles of mobile computing, ACM Press (2002) 9–16

21. Itkis, G., Levin, L.A.: Fast and lean self-stabilizing asynchronous protocols. In: Proceeding of 35th Annual Symposium on Foundations of Computer Science, IEEE Press (1994) 226–239

Routing Explicit Side Information for Data Compression in Wireless Sensor Networks

Huiyu Luo and Gregory Pottie

Uiversity of California, Los Angeles, Ca, 90095, USA
{huiyu, pottie}@ee.ucla.edu

Abstract. Two difficulties in designing data-centric routes [2,3,4,5] in wireless sensor networks are the lack of reasonably practical data aggregation models and the high computational complexity resulting from the coupling of routing and in-network data fusion. In this paper, we study combined routing and source coding with explicit side information in wireless sensor networks. Our data aggregation model is built upon the observation that in many physical situations the side information that provides the most coding gain comes from a small number of nearby sensors. Based on this model, we propose a routing strategy that separately routes the explicit side information to achieve data compression and cost minimization. The overall optimization problem is NP hard since it has the minimum Steiner tree as a subproblem. We propose a suboptimal algorithm based on maximum weight branching and the shortest path heuristic for the Steiner tree problem. The worst case and average performances of the algorithm are studied through analysis and simulation.

1   Introduction

The need to lower the communication cost in wireless sensor networks has prompted many researchers to propose data-centric routing schemes that can utilize in-network data fusion to reduce the transmission rate. There are two major difficulties in designing such routes. First, the lack of reasonably practical data aggregation models has led researchers to use overly simplified ones [2,3,4,5]. For example, these models generally assume that sensors perform the same aggregation function regardless of the origin of the fused data. As a remedy, [5] suggests looking into models in which data aggregation is not only a function of the number of sources but also the identity of the sources. Second, the resulting optimization problem is often NP hard due to the coupling of routing and in-network data fusion [2,4]. Hence, algorithms that find exact solutions in polynomial time are unlikely to exist. In this paper, we try to build computationally useful models and devise heuristic algorithms for the combined routing and source coding problem.

Most previous work has considered using trees as the underlying routing structure [2,4,5] probably due to the fact that trees are the optimal solution to the shortest path problem and have been pervasive in network routing. However, in data-centric routing, trees are not necessarily optimal. In this paper, a

V. Prasanna et al. (Eds.): DCOSS 2005, LNCS 3560, pp. 75–88, 2005.

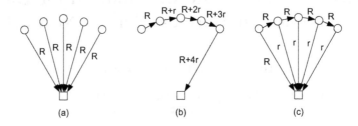

Fig. 1. Three routing strategies: (a) shortest path tree; (b) optimal tree; (c) designated side information transmission

simple strategy, which we call designated side information transmission (DSIT), is proposed. This method results in a non-tree routing structure and tends to distribute the traffic more evenly in the network. To motivate the idea and give a preview of the paper, consider the example depicted in Fig. 1. The edges between adjacent sensors (circles) have the weight $c_e = d$, and the edges connecting a sensor to the fusion center (square) have the weight $c_e = D$. The rate at which each sensor needs to transmit to the fusion center is R without any explicit side information and r if explicit side information from an adjacent sensor is available. We postulate that side information from other sensors can be used to help compress the data only when it is available at both the encoder (the sensor that generated the data) and decoder (fusion center). Assume $r \ll R$ and $d \ll D$. The objective is to minimize the cost $C = \sum_e c_e f_e$ of routing all the data to the fusion center, where f_e is the rate at which data is transmitted across edge e. Consider the following three strategies: (a) the shortest path tree is used; (b) we compress data using explicit side information and optimize over all spanning trees; (c) the data at each sensor is transmitted to an adjacent sensor to be used as explicit side information whenever the coding gain outweighs the transmission cost. This transmission to an adjacent sensor provides only explicit side information and needs not be relayed to the fusion center. The routes corresponding to the three strategies are shown in Fig. 1. Note that at least one sensor has to transmit at rate R to the fusion center so that all the data can be correctly recovered. The costs of the three strategies are:

$$C_a = 5RD$$
$$C_b = RD + 4rD + 4Rd + 6rd$$
$$C_c = RD + 4rD + 4Rd$$

The performance of (b) and (c) are about the same, and both are superior to that of (a). It is also evident that (c) results in more evenly distributed traffic than (b). This is because in DSIT, the communication to the fusion center is separated from the explicit side information transmission, and can be routed through any path.

There has been much recent research activity on data-centric routing. In [7], the interdependence of routing and data compression is addressed from the viewpoint of information theory. Clustering methods have been used by some researchers to aggregate data at the cluster head before transmitting them to the fusion center [8,9]. Since the cluster head is responsible for data aggregation and relaying, it consumes the most energy. Hence, dynamically electing nodes with more residual power to be cluster heads and evenly distributing energy consumption in network is a major issue in these schemes. In [3], a diffusion type routing paradigm that attaches attribute-value pairs to data packets is proposed to facilitate the in-network data fusion. The correlated data routing problems studied in [2, 4] are closely related to our work. In [2], the authors give a thorough comparison of data-centric and address-centric methods and a overview of recent effort in the field. [4] casts the data-centric routing problem as an optimization problem and seek solutions to it when different source coding schemes are applied. A similar optimization problem is also the subject of [5], where a grossly simplified data model is assumed.

The rest of the paper is organized as follows. In section 2, we present our network flow and data rate models. In section 3, an optimization problem is formulated out of the DSIT strategy, and a heuristic algorithm is proposed. The average performance of the heuristic algorithm is studied through simulations in section 4. Section 5 concludes the paper.

2   Network Models

2.1   Network Flows

The sensor network is modelled as a graph $\mathcal{G} = (\mathcal{N}, \mathcal{E})$. The node set $\mathcal{N}$ consists of a set $\mathcal{N}_s$ of n sensors and a special node t acting as the fusion center. We call a sensor active if it generates data. Denote by $\mathcal{N}_a$ the set of active sensors. Both active and non-active sensors can be relays. The edge set $\mathcal{E}$ represents m communication links. Here, we assume all the links are bi-directional and symmetric. If they are not, the network can be modelled as a directed graph, and the derivation in this paper will apply similarly. We also assume the network is connected so that data from any sensor can reach t. A weight c_e is associated with each edge $e \in \mathcal{E}$. It represents the cost (e.g. power) of transmitting data at unit rate across e. The flow f_e is defined as the rate at which data is transmitted across edge $e \in \mathcal{E}$. Data generated by node i and terminating at node j are denoted by f_e^{ij}. In particular, we define $f_e^i = f_e^{it}$. Clearly, $f_e = \sum_{i,j \in \mathcal{N}} f_e^{ij}$. Supposing $i, j \in \mathcal{N}_s$, denote by d_{ij} the minimum distance from i to j (i.e. the sum of edge weights along the shortest path from i to j), and d_i the minimum distance from i to t. The objective of our study is to minimize the total cost C while routing all the data from active sensors to the fusion center.

$$C = \sum_{e \in \mathcal{E}} c_e f_e \tag{1}$$

2.2   Source Coding with Explicit Side Information

Consider the problem of sampling a distributed field using wireless sensor networks. The measurements at sensors are coded and transmitted back to the fusion center, and used to reconstruct the field under some distortion constraint. There is likely to be a great deal of redundancy in the data collected by different sensors, since they are observing some common physical phenomenon. Denote by X_i the data stream produced by sensor i. (We assume X_i has been quantized and has a discrete alphabet.) Assume X_i satisfies the ergodic condition so that the results of statistical probability theory can be applied. In this paper, we consider source coding with explicit side information. In other words, only when the side information is available at both the encoder (the senor that generates the data) and decoder (fusion center) can it be used to compress the data. Supposing data stream X_k is entropy-coded using $X_1, X_2, \cdots, X_{k-1}$ as side information, we have the following:

$$f^k = H(X_k|X_1, X_2, \cdots, X_{k-1}) \tag{2}$$

Since the data rate f^k to the fusion center depends on the availability of data stream X_i, $i = 1, 2, \cdots, k-1$ at k, there are 2^{k-1} possibilities. In an attempt to simplify the data rate model and optimization, we assume that the the rate reduction provided by side information saturates as the number of helpers exceeds one:

$$f^k = \begin{cases} b_0^k & \text{no side information;} \\ \min_j b_1^{kj} & X_j \text{ is available at } k, \text{ and } j \in \mathcal{H}_k \end{cases} \tag{3}$$

in which b_0^k is the rate of coding X_k without any side information; b_1^{kj} ($b_1^{kj} \le b_0^k$) is the coding rate of X_k when only $X_j, j \in \mathcal{H}_k$ is available at k; $\mathcal{H}_k$ is the set of sensors whose data is correlated with sensor k's observations and can be used as its side information. When side information from more than one sensors is available, the one providing the most coding gain is used. In practice, the information on the set of helping sensors H_k and the rate reduction provided by their data can be obtained using specially designed coding schemes (e.g. [6]). In many physical situations, sensor measurements are highly correlated only in a small neighborhood. In others, although a large number of sensors have similar measurements, the reproduction fidelity constraints often permit thinning the number of active sensors so that again only a small number of sensors have high correlation. As a result, $\mathcal{H}_k$ generally comprises a small number of sensors that are close to sensor k. The quick saturation of coding gain provided by side information indicates when the data stream at a nearby sensor is available as side information, the additional coding gain provided by other sensors' observations is negligible. Note that b_1^{kj} can be about the same as or much less than b_0^k depending on source statistics. This greatly influences the route construction.

Since $H(X_k) - H(X_k|X_j) = I(X_k, X_j) \le H(X_j)$, we assume $b_0^k - b_1^{kj} \le b_0^j$. Therefore, $(b_0^k - b_1^{kj})d_k \le b_0^j d_k$, and there is no gain in feeding back explicit side information from t to sensors. The total cost of routing data to t can be decomposed as the sum of C_s representing the cost of routing side information and C_t the cost of transmitting data to t.

$$C = \sum_{e \in \mathcal{E}} c_e f_e = C_s + C_t \tag{4}$$

where

$$C_s = \sum_{i,j \in \mathcal{N}_s} \sum_{e \in \mathcal{E}} c_e f_e^{ij}, \quad C_t = \sum_{i \in \mathcal{N}_s} \sum_{e \in \mathcal{E}} c_e f_e^i \tag{5}$$

In applying source coding with explicit side information to sensor networks, we must avoid helping loops. In other words, if X_j's recovery relies on X_i, then X_j cannot be used as the side information for compressing X_i. To formalize this requirement, define a directed network $\mathcal{G}_h$ that consists of all the active sensor nodes. In addition, if X_i is used as side information for coding X_j, a directed edge (i, j) is formed from sensor i to sensor j. Then we have the following proposition:

Proposition 1: No helping loop is formed when using source coding with explicit side information if and only if the directed network $\mathcal{G}_h$ contains no directed cycles.

The proof is straightforward, and hence omitted. It is apparent that if the underlying routing structure is a directed acyclic network (DAG), the above proposition is automatically satisfied. For instance, spanning trees directed toward the fusion center are DAG's, so there will be no helping loops when using trees to route data. However, in this paper the rule of no directed cycles needs to be enforced explicitly.

3   Designated Side Information Transmission

3.1   Problem Formulation

In DSIT, we distinguish the data flow from sensor to the fusion center f_e^i and the transmission of explicit side information to other sensors f_e^{ij} $(i, j \in \mathcal{N}_s)$. The side information can only be provided by sensor to sensor transmissions f^{ij} not transmissions to the fusion center f^k. With the network model defined as in section 2, we formulate the following optimization problem.

Designated Side Information Transmission (DSIT)
GIVEN: A graph $\mathcal{G} = (\mathcal{N}, \mathcal{E})$ with weight c_e defined on each edge $e \in \mathcal{E}$, a special node $t \in \mathcal{N}$ acting as the fusion center, a set $\mathcal{H}_k$ of helping sensors and data rate function f^k as in Eq. (3) defined for each sensor $k \in \mathcal{N}_s = \mathcal{N} \setminus \{t\}$.

FIND: A set of routes transmitting the explicit side information among sensors such that the total cost of routing data to the fusion center $C = \sum_{e \in \mathcal{E}} c_e f_e$ is minimized.

We will see in ensuing discussion that the transmission of explicit side information has a Steiner tree problem embedded in it. Hence the overall problem is NP hard. As a result, we will focus on building a heuristic algorithm for the optimization.

3.2   Heuristic Algorithm

The total cost can be decomposed into the cost of routing explicit side information C_s and the cost of transmitting data to the fusion center C_t as in Eq. (4).
We first consider constructing routes from sensors to the fusion center. These
routes affect only C_t. In addition, as $f^k, k \in \mathcal{N}_a$ does not provide any side information, its routing is decoupled from the data aggregation process. Hence, the
shortest path should be used to achieve the minimum C_t:

$$C_t = \sum_{k \in \mathcal{N}_a} d_k f^k \qquad (6)$$

where d_k is the minimum distance from sensor k to t, and f^k is a function of
side information transmission. The design of routes for transmitting f^k must take
place before that for side information transmission because the latter depends
on the distance information from the former.

Designing routes for side information transmission is more complicated. First,
it has the minimum Steiner tree as a subproblem. This is illustrated by the
following problem instance. Given network $\mathcal{G} = (\mathcal{N}, \mathcal{E})$, we have a subset of the
active sensors $\mathcal{S} \subset \mathcal{N}_a$, and there is a sensor $u \in \mathcal{N}_a \setminus \mathcal{S}$. Assume $\mathcal{H}_k = u$ if
$k \in \mathcal{S}$, and $\emptyset$ otherwise. In addition, we assume that the rate function and edge
weights are defined such that the cost of transmitting side information from u to
any sensor in $\mathcal{S}$ using appropriately chosen routes is less than the cost reduction
resulting from the coding gain of side information. The optimization problem
becomes constructing a subtree that connects u and the sensors in $\mathcal{S}$, which is a
minimum Steiner tree. Therefore, the overall optimization problem is NP hard.
Second, we need to ensure that no helping loop can be formed while routing the
side information. This amounts to avoiding directed cycles in $\mathcal{G}_h$ according to
Proposition 1.

For a moment, we ignore the Steiner tree part, and use the shortest path
to route all the side information. This leads us to construct a network $\mathcal{G}_a$ as
follows. $\mathcal{G}_a$ includes the set of active sensors $\mathcal{N}_a$. In addition, for each ordered
pair of nodes $(i, j) \in \mathcal{N}_a$, create a directed edge (i, j) from sensor i to j and
assign the weight w_{ij} to represent the net coding gain resulting from routing
side information from i to j.

$$w_{ij} = \begin{cases} (b_0^j - b_1^{ji})d_j - d_{ij}b_0^i & i \in \mathcal{H}_j \\ -d_{ij}b_0^i & \text{otherwise} \end{cases} \qquad (7)$$

where b_0^j, b_1^{ji} and $\mathcal{H}_j$ are the data rates and set of helping sensors as in Eq. (3); d_j
and d_{ij} are the minimum distances defined in section 2. Denote by $\mathcal{A}_a$ the set of
directed edges with $w_{ij} > 0$. A branching on the directed graph $\mathcal{G}_a = (\mathcal{N}_a, \mathcal{A}_a)$
is a set of directed edges $\mathcal{B} \subseteq \mathcal{A}_a$ satisfying the conditions that no two edges in $\mathcal{B}$
enter the same node, and $\mathcal{B}$ has no directed cycle. It is evident that a branching
on $\mathcal{G}_a$ represents a feasible set of routes for side information transmission. No
two directed edges in $\mathcal{B}$ entering the same node ensures that a sensor uses side
information from at most one helper and no directed cycle avoids the helping

loop. The problem of minimizing the total cost is equivalent to maximizing the weight sum of the set of directed edges $\mathcal{B}$ that is a branching on $\mathcal{G}_a$, which is the so called maximum weight branching problem.

Maximum Weight Branching (MWB)
GIVEN: A directed graph $\mathcal{G}_a = (\mathcal{N}_a, \mathcal{A}_a)$ with weight w_e defined on each directed edge $e \in \mathcal{A}_a$.

FIND: A branching $\mathcal{B} \subseteq \mathcal{A}_a$ that maximizes $\sum_{e \in \mathcal{B}} w_e$.
    It has been shown that this problem can be solved efficiently [10]. Once the optimal branching $\mathcal{B}$ is determined, we revert to using Steiner trees. Define $\mathcal{S}_k$ as the set of sensors that receive side information from k based on the optimal branching $\mathcal{B}$. We use the shortest path heuristic proposed by [11] to construct the subtree that connects k and $\mathcal{S}_k$. This method has a worst case performance ratio of 2. Our heuristic algorithm is a combination of the maximum weight branching and the Steiner tree approximation. We state it as follows:

Designated Side Information Transmission Heuristic (DSIT Heuristic)
Given a network $\mathcal{G} = (\mathcal{N}, \mathcal{A})$ with edge weights and rate function properly defined, carry out the following steps.

1. Find the shortest path from each active sensor to the fusion center. These are the routes for transmitting data to the fusion center.
2. Construct a directed graph $\mathcal{G}_a = (\mathcal{N}_a, \mathcal{A}_a)$. $\mathcal{N}_a \subseteq \mathcal{N}$ consists of all active sensors, and $\mathcal{A}_a$ is the set of directed edges from i to j ($i, j \in \mathcal{N}_a$ and $i \neq j$) whose weight w_{ij} defined as in Eq. (7) is greater than zero.
3. Find the maximum weight branching on $\mathcal{G}_a$. Based on the optimal branching $\mathcal{B}$, determine the set of sensors $\mathcal{S}_k$ that each active sensor $k \in \mathcal{N}_a$ transmits side information to.
4. Run a shortest path heuristic for the Steiner tree problem to find the subtree that connects k and the sensors in $\mathcal{S}_k$.

3.3   Performance Analysis

Finding the maximum weight branching takes $O(m_a \log n_a)$ time, where $m_a = |\mathcal{A}_a|$ and $n_a = |\mathcal{N}_a|$. ($|\mathcal{S}|$ is the number of elements in finite set $\mathcal{S}$.) The shortest path heuristic for a Steiner tree requires $O(n_a n^2)$ time. The actual running time of the shortest path heuristic is in general much less because the number of nodes involved in constructing the shortest path is often a lot fewer than n. Regarding the performance of our heuristic algorithm compared to that of the optimal solution, we prove the following proposition.

Proposition 2: The ratio of the cost C_H resulting from our DSIT heuristic algorithm and the minimum cost C_{MIN} using the DSIT strategy is bounded by:

$$\frac{C_H}{C_{MIN}} \leq M \tag{8}$$

where $M = \max\{1, \max_{k \in \mathcal{N}_a} |\mathcal{S}_k^{opt}|\}$, the greater of one and the maximum number of sensors one sensor needs to transmit side information to in the optimal solution. The bound is tight in the sense that there is a network that attains the worst performance ratio.

Proof: First, we note that S_k^{opt} is in general not the same as the S_k in our heuristic algorithm. Consider the structure of an optimal solution. It can be given by the set of sensors S_k^{opt} that each $k \in \mathcal{N}_a$ sends explicit side information to. The side information circulated within the group of k sensors in S_k^{opt} is routed using the minimum Steiner tree. Denote by C_k^{ST} the sum of edge costs of these Steiner trees ($C_k^{ST} = 0$ if $S_k^{opt} = \emptyset$). The data is transmitted to the fusion center using the shortest path tree. Hence we can write the minimum cost as:

$$C_{MIN} = \sum_{k \in \mathcal{N}_a} f^k d_k + \sum_{k \in \mathcal{N}_a} b_0^k C_k^{ST} \tag{9}$$

Instead of the Steiner tree, consider relying on a shortest path tree to route the side information from k to the sensors in S_k^{opt}. Denote by C_k^{SPT} the sum of edge costs of such shortest path trees. The corresponding cost C' will be:

$$C' = \sum_{k \in \mathcal{N}_a} f^k d_k + \sum_{k \in \mathcal{N}_a} b_0^k C_k^{SPT} \tag{10}$$

Since $M_k C_k^{ST} \geq C_k^{SPT}$ [11], where $M_k = |S_k^{opt}|$, we have

$$\frac{C'}{C_{MIN}} \leq \max\{1, \max_{k \in \mathcal{N}_a} M_k\} = M$$

On the other hand, C_H is the optimal result of using the shortest path tree to route the side information. Therefore, $C_H \leq C'$. This gives rise to the bound in Eq. (8). To show the bound is tight, we look at the example in Fig. 2. The network setup is given in (a). The edge weights between sensors v_k and u_k ($k = 1, 2, 3$) is 1. Other edges have weight $\delta \ll 1$. All the sensors are active with data rate R without side information and 0 when side information is available. Denote $\mathcal{U} = \{u_1, u_2, u_3\}$, and $\mathcal{V} = \{v_1, v_2, v_3\}$. We assume $\mathcal{H}_k = \mathcal{U}$ when $k \in \mathcal{V}$

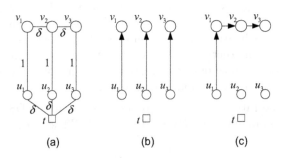

(a)   (b)   (c)

Fig. 2. A problem instance that approaches the worst performance ratio: (a) sensor network setup; (b) routes of side information transmission using DSIT heuristic; (c) routes of side information transmission in optimal solution

and $\emptyset$ when $k \in \mathcal{U}$. In Fig. 2, (b) and (c) illustrate how side information is transmitted in DSIT heuristic and optimal solutions. Therefore, $C_H = 3R + 3R\delta$ and $C_{MIN} = R + 5R\delta$. When $\delta \to 0$, the ratio C_H/C_{MIN} approaches $M = 3$ asymptotically. In a similar fashion, problem instances with arbitrary values of M can be devised.   Q.E.D.

The worst case scenario in the proof can be avoided by changing the heuristic algorithm to run multiple maximum weight branching and shortest path heuristic iterations. At each iteration, only one sensor is added to $S_k, k \in N_a$. However, this greatly increases the computational cost. Moreover, the pathological case in the proof rarely occurs in our assumed data rate model. The value of M is expected to be small as one's data helps mostly nearby sensors. Also since side information is often circulated within one's neighborhood, using shortest paths to approximate a Steiner tree introduces a moderate amount of error. What we are more interested in is the average behavior of the algorithm, which is examined through simulations in the next section.

4   Simulations

In our simulations, we place $(n + 1)$ nodes including the fusion center and n sensors in an $n_d \times n_d$ square, where $n_d = \lceil \sqrt{n+1} \rceil$. (Denote by $\lceil z \rceil$ the smallest integer such that $\lceil z \rceil \geq z$, and $\lfloor z \rfloor$ the largest integer such that $\lfloor z \rfloor \geq z$.) Supposing $\tilde{x}_i$ and $\tilde{y}_i, i = 1, \cdots, n + 1$, are random variables that are uniformly distributed in $[0, 1]$, the coordinates of node i is given by:

$$x_i = [(i \bmod n_d) - 1] + \tilde{x}_i$$
$$y_i = \lfloor (i - 1)/n_d \rfloor + \tilde{y}_i$$

We define a transmission radius r_c. If two nodes are no more than r_c away from each other, direct communication between the two nodes is allowed. Otherwise, a relay has to be used. Denote by d_e the length of edge e. When $d_e \leq r_c$, the edge weight c_e is proportional to d_e^α, where $\alpha = 2$ is the path loss factor. When the number of sensors increases, the network covers a larger area while maintaining the communication range and sensor to sensor spacing. A typical 100 node network constructed in this manner is depicted in Fig. 3. The node (in lower left corner) with a letter "t" next to it is the fusion center.

In our simulation, we assume that all the sensors are active. The helping set H_i of sensor i is defined as follows. Any pair of sensors that are no more than r_d away from one another has a probability of 0.5 to be in the helping sets of one another. Fig. 4 shows the resulting data correlation in the network. For simplicity, we assume the data rate function is the same for all the sensors:

$$f^k = \begin{cases} b_0 & \text{no side information} \\ \beta b_0 & \text{with side information} \end{cases} \tag{11}$$

where $i \in N_a$ and $0 \leq \beta \leq 1$. Fig. 5 shows the maximum weight branching on the network described in Fig. 3 and 4. The graph is a forest, and hence acyclic.

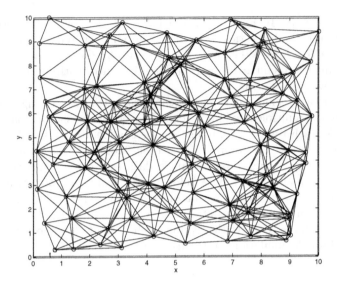

Fig. 3. A network of 100 nodes with $r_c = \sqrt{5}$. Two nodes are connected if direct transmission is allowed between the two

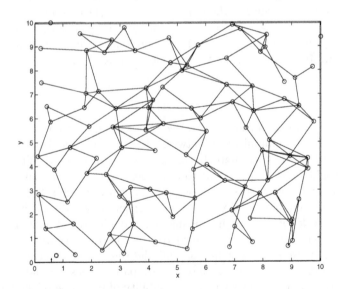

Fig. 4. This graph shows the correlation of data at different sensors when $r_d = 1.8$. Two nodes are connected by an edge if they are in the helping set H_i of one another

The root of each tree is indicated by a circle with a cross inside, from which there is a simple path to any other member of the tree. Based on this rule, the helping set S_k of each sensor k can be easily determined.

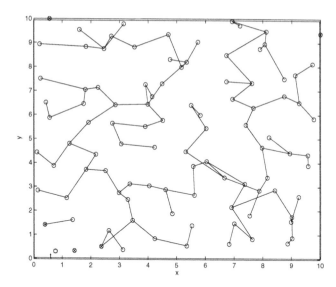

Fig. 5. A maximum weight branching on the network described in Fig. 3 and 4

We simulate for different network sizes and change the value of β. The performance of DSIT heuristic is compared to that of the shortest path tree, in which the same compression scheme based on explicit side information is used. Define the cost ratio $\mu = C_{DSIT}/C_{SPT}$. In Fig. 6, we plot μ against the number of nodes in the network. DSIT heuristic outperforms the shortest path tree in all cases. In addition, we observe that as the coding gain decreases (i.e. β increases), μ drops. This is expected considering that DSIT becomes the shortest path tree, which is also the optimum solution, when coding gain is zero. It is also noticed that μ increases as the number of nodes increases. This is explained by looking at a shortest path tree solution plotted in Fig. 7. The leaf nodes of a shortest path tree are generally far away from the fusion center while the source coding at these nodes receives no side information from other sensors. In contrast, in DSIT (Fig. 5), the nodes that receives zero side information (the roots of the subtrees) are mostly near the fusion center. As the network size increases, the leaf nodes become farther and farther away from the fusion center. Consequently, C_{SPT} rises faster than C_{DSIT}.

5   Discussion and Conclusion

The DSIT strategy relies heavily on our assumed network model, in particular, the assumptions that data streams are highly correlated only when they are from a small group of sensors close to one another, and thus the coding gain saturates when the number of helpers exceeds one. Therefore, this scheme may not be as effective in cases that deviate from these assumptions. Nonetheless, there

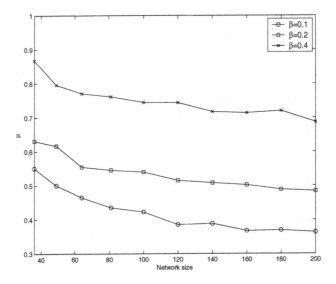

Fig. 6. Plot $\mu = C_{DSIT}/C_{SPT}$ against network size for $\beta = 0.1, 0.2$, and 0.4

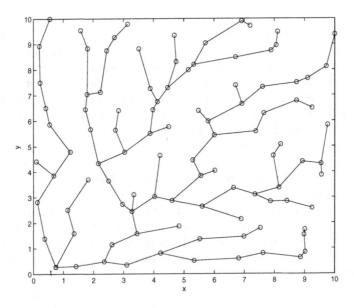

Fig. 7. The shortest path tree solution for the network described in Fig. 3 and 4

are practical reasons to consider this case. First, identifying the set of sensors that provides most coding gain and then processing side information incurs cost. The gain of an additional helper may not be enough to outweigh these costs. Second, using more than one helper increases the complexity of the model and the optimization. For example, it is possible to jointly code two helpers' data if they are to are be used as side information at the same sensor. If more than one helper has to be considered, we speculate that the problem can be approached in a multiple-step procedure. At each step, the number of helpers is restricted to at most one, and an algorithm similar to our heuristic scheme is used. This is an area that needs further research.

The decoupling of route design for f^k from side information transmission offers greater flexibilities than strategies based on trees. Unlike a tree structure that bundles the network flows, the transmission of f^k can virtually be routed to t through any path. As a result, traditional address-centric routing schemes that evenly distribute the traffic load and maximize node lifetime [12,13] can be applied. If the optimization objective is to maximize the network lifetime, we surmise the DSIT can take into account both the energy reserve of sensor nodes and source correlation. In contrast, address-centric routing focuses on node energy, and data-centric routing concentrates on source correlation only.

As we discussed in section 2, the number of sensors with highly correlated data can be brought down in a process of thinning the number of active sensors based on the reconstruction requirement. This pre-routing step makes in practice our procedure a two-phase operation. First determine the set of sensors that will participate in the fusion, then design the routes for transmitting the data to the fusion center. Currently, the first step is generally approached from a sampling point of view [14] trying to meet the distortion constraint, while route design attempts to minimize the energy consumption. It is of interest to ask whether a combined approach will yield better results. [15] is an interesting preliminary effort on that direction.

Acknowledgement

This material is based upon work supported by National Science Foundation (NSF) Grant #CCR-0121778.

References

1. G. Pottie and W. Kaiser, "Wireless sensor networks," • • • • • • • • • vol. 43, no. 5, pp. 51-58, May 2000.
2. B. Krishnamachari, D. Estrin, and S. Wicker, "Modelling data-centric routing in wireless sensor network," • • •   • • • • • •• •   • • • •• • ••• •   • • • • •• • • •• •   • • • • • •   • • • •
   •• • • • •
3. C. Intanagonwiwat, et al. "Directed diffusion for wireless sensor networking," •• • • • • •   • • •• • • • ••• ••• ••• vol. 11, no. 1, pp. 2-16, Feb 2003.
4. R. Cristescu, B. Beferull-Lozano, and M. Vetterli, "On network correlated data gathering," •• • •   •• • • • • • • Hongkong, March 7-11, 2004.
5. A. Goel and D. Estrin, "Simultaneous optimization for concave costs: single sink aggregation or single source buy-at-bulk," • • • • •• •   •   • • • •• • • •• •   • •   • • •• • ••
   • •• • •• •   • • 2003.

6. H. Luo, Y. Tong, and G. Pottie, "A two-stage DPCM scheme for wireless sensor networks," to appear in •• • ••• ••••• Philadelphia, USA.

7. A. Scaglione and S. Servetto, "On the interdependence of routing and data compression in multi-hop sensor networks," • • • ••• • • • ••••• • 2002.

8. S. Bandyopadhyay and E. J. Coyle, "An energy efficient hierarchical clustering algorithm for wireless sensor networks," ••• • ••••••• • 2003.

9. W. R. Heinzelman, A. Chandrakasan, and H. Balakrishnan, "Energy-efficient communication procotol for wireless microsensor networks," • ••• •• •••••• • •• •• • ••••• •••••••• Jan. 4-7, 2000.

10. R. E. Tarjan, "Finding optimum branchings," • ••• ••••• pp 25-35, vol. 7, 1977.

11. H. Takahashi and A. Matsuyama, "An approximate solution for the steiner problem in graphs," • ••••••••••••• 24, No. 6, pp 573-577, 1980.

12. J. H. Chang, L. Tassiulas, "Energy conserving routing in wireless ad-hoc networks," ••• • ••••••• • 2000.

13. S. Singh, M. Woo, C. S. Raghavendra, "Power-aware routing in mobile ad-hoc networks," • • • ••• • • • ••••• • ••••• Dallas, Texas.

14. R. Willett, A. Martin, and R. Nowak, "Backcasting: adaptive sampling for sensor networks," •• • • • 2004, Berkeley, Ca, USA.

15. D. Ganesan, R. Cristescu, and B. Beferull-Lozano, "Power-efficient sensor placement and transmission structure for data gathering under distortion constraints," •• • • • 2004, Berkeley, Ca, USA.

16. T. M. Cover and J. A. Thomas, • ••• •••• •• • •••••• ••••• • ••••••• John Wiley & Sons, 1991.

17. M. R. Garey and D. S. Johnson, • ••• •••••• ••• •• •••••••••••••• • • •••• •• ••• • •••••• ••• • •• •• • ••••••••• W. H. Freeman and Company, 1979.

18. F. K. Hwang, D. S. Richards, and P. Winter, • •• •••••• • • ••• • ••••• • North-Holland, 1992.

Database-Centric Programming for Wide-Area Sensor Systems

Shimin Chen[1], Phillip B. Gibbons[2], and Suman Nath[1,2]

[1] Carnegie Mellon University
{chensm, sknath}@cs.cmu.edu
[2] Intel Research Pittsburgh
phillip.b.gibbons@intel.com

Abstract. A wide-area sensor system is a complex, dynamic, resource-rich collection of Internet-connected sensing devices. In this paper, we propose *X-Tree Programming*, a novel database-centric programming model for wide-area sensor systems designed to achieve the seemingly conflicting goals of expressiveness, ease of programming, and efficient distributed execution. To demonstrate the effectiveness of X-Tree Programming in achieving these goals, we have incorporated the model into IrisNet, a shared infrastructure for wide-area sensing, and developed several widely different applications, including a distributed infrastructure monitor running on 473 machines worldwide.

1 Introduction

A wide-area sensor system [2, 12, 15, 16] is a complex, dynamic, resource-rich collection of Internet-connected *sensing devices*. These devices are capable of collecting high bit-rate data from powerful sensors such as cameras, microphones, infrared detectors, RFID readers, and vibration sensors, and performing collaborative computation on the data. A sensor system can be programmed to provide useful *sensing services* that combine traditional data sources with tens to millions of live sensor feeds. An example of such a service is a Person Finder, which uses cameras or smart badges to track people and supports queries for a person's current location. A desirable approach for developing such a service is to program the collection of sensors as a whole, rather than writing software to drive individual devices. This provides a high level abstraction over the underlying complex system, thus facilitating the development of new sensing services.

Recent studies [6, 11, 14, 18, 19, 32] have shown that declarative programming via a query language provides an effective abstraction for accessing, filtering, and processing sensor data. While their query interface is valuable, these models are tailored to resource-constrained, local-area wireless sensor networks [6, 14, 18, 19, 32] or provide only limited support, if any, for installing user-defined functions on the fly [6, 11, 18, 19, 32]. As a result, the programming models are overly restrictive, inefficient, or cumbersome for developing services on resource-rich, wide-area sensor systems. For example, consider a wide-area Person Finder service that for update scalability, stores each person's location in a database nearby that location, for retrieval only on demand. To

V. Prasanna et al. (Eds.): DCOSS 2005, LNCS 3560, pp. 89–108, 2005.

enable efficient search queries, the data can be organized into a location hierarchy with *filters* associated with each node of the hierarchy. These filters summarize the list of people currently within the node's subtree and are used to limit the scope of a search by checking the filters. Programming such filters, associating them with (all or parts of) a logical/semantic hierarchy, installing them on the fly, and using them efficiently within queries are not all supported by these previous models. Similarly, declarative programming models designed for wide-area, resource-rich distributed monitoring systems [12, 28, 31] do not support all these features.

In this paper, we present a novel database-centric approach to easily programming a large collection of sensing devices. The general idea is to augment the valuable declarative interface of traditional database-centric solutions with the ability to perform more general purpose computations on logical hierarchies. Specifically, application developers can write application-specific code, define on-demand (snapshot) and continuous (long-running) states derived from sensor data, associate the code and states with nodes in a logical hierarchy, and seamlessly combine the code and states with a standard database interface. Unlike all the above models (except for our earlier work [11]) that use a flat relational database model and SQL-like query languages, we use instead the XML hierarchical database model. Our experience in building wide-area sensing services shows that it is natural to organize the data hierarchically based on geographic/political boundaries (at least at higher levels of the hierarchy), because each sensing device takes readings from a particular physical location and queries tend to be scoped by such boundaries [13]. A hierarchy also provides a natural way to name the sensors and to efficiently aggregate sensor readings [11]. Moreover, we envision that sensing services will need a heterogeneous and evolving set of data types that are best captured using a more flexible data model, such as XML. This paper shows how to provide the above features within the XML data model.

We call our programming model *X-Tree Programming* (or *X-Tree* in short) because of its visual analogy to an Xmas tree: The tree represents the logical data hierarchy of a sensing service, and its ornaments and lights represent derived states and application-specific codes that are executed in different parts of the hierarchy. Sensor data of a sensing service is stored in a single XML document which is fragmented and distributed over a potentially large number of machines. Xpath (a standard query language for XML) is used to access the document as a single queriable unit. With X-Tree, user-provided code and derived states can be seamlessly incorporated into Xpath queries.

There are three main contributions of this paper. First, we propose X-Tree Programming, a novel database-centric programming model for wide-area sensor systems. Our X-Tree solution addresses the challenge of finding a sweet spot among three important, yet often conflicting, design goals: expressiveness, ease of programming, and efficient distributed execution. As we will show in Section 2, achieving these three goals in the same design is difficult. X-Tree's novelty comes from achieving a practical balance between these design goals, tailored to wide-area sensor systems. Second, we present important optimizations within the context of supporting X-Tree that reduce the computation and communication overheads of sensing services. Our caching technique, for example, provably achieves a total network cost no worse than twice the cost incurred by an optimal algorithm with perfect knowledge of the future. Third, we have implemented X-Tree within IrisNet [2, 11, 13], a shared infrastructure for wide-area sensing that we previously developed. We demonstrate the effectiveness of our solution through both controlled experiments and real-world applications on IrisNet, including a publicly available distributed infrastructure monitor application that runs on 473 machines worldwide. A rich collection of application tasks were implemented quickly and execute efficiently, highlighting the expressibility, ease of programming, and efficient distributed execution that X-Tree provides.

The rest of the paper is organized as follows. Section 2 examines two application examples, describes the challenges and overviews our solution. After Section 3 provides background information, Section 4 illustrates the programming interface, Section 5 describes our system support for distributed execution, and Section 6 discusses optimizations. Our experimental evaluation is in Section 7. Section 8 discusses related work. Finally, Section 9 concludes the paper.

2   Example Applications, Challenges, and Our Solution

This section describes two representative wide-area sensing services (we use *service* and *application* interchangeably) that we aim to enable. (Additional examples can be found in [8]). We highlight the desirable properties of an enabling programming model, challenges in achieving them, and our solution.

2.1   Applications

Person Finder. A person finder application keeps track of the people in a campus-like environment (e.g., a university campus) and supports queries for the current location of a person or the current occupants of a room. The application uses sensors like cameras, microphones, smart-badges, etc. along with sophisticated software (e.g., for face or voice recognition) to detect the current location of a person. For scalability, it is important that the sensor data is stored near their sources (i.e., stored by location) and is retrieved only on-demand. One way to implement this application would be to maintain a distributed database of all the people currently at each location. A query for some person would then perform a brute force search of the entire database; such a query would suffer from both a slow response time and high network overhead. A far more efficient implementation would organize the distributed database as a location hierarchy (e.g., the root of the hierarchy is the university, and the subsequent levels are campus, building, floor, and room) and then prune searches by using approximate knowledge of people's current locations. Such pruning can be implemented by maintaining a Bloom filter (a compressed bit vector representation of a set—similar to [23]) at every intermediate node of the hierarchy, representing the people currently within that part of the location hierarchy.

Infrastructure Monitor. A distributed infrastructure monitor [1] uses *software sensors* [24] to collect useful statistics (e.g., CPU load, available network bandwidth) on the infrastructure's host machines and communication network, and supports queries on that data. One way to scale such an application to a large number of hosts is to hierarchically organize the data. Figure 1 (right) shows part of the hierarchy used by IrisLog,

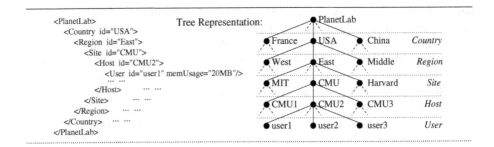

```
<PlanetLab>
  <Country id="USA">
    <Region id="East">
      <Site id="CMU">
        <Host id="CMU2">
          <User id="user1" memUsage="20MB"/>
          ... ...
        </Host>           ... ...
      </Site>             ... ...
    </Region>             ... ...
  </Country>              ... ...
</PlanetLab>
```

Fig. 1. An XML document representing IrisLog's logical hierarchy

an infrastructure monitoring service deployed on 473 hosts in PlanetLab [3]. Infrastructure administrators would like to use such an application to support advanced database operations like continuous queries and distributed triggers. Moreover, they would like to dynamically extend the application by incorporating new sensors, new sensor feed processing, and new aggregation functions, as needs arise.

2.2 Design Goals and Challenges

A programming model suitable for the above applications should have the following properties. First, it should have *sufficient expressive power* so that application code can use arbitrary sensor data and perform a rich set of combinations and computations on that data. For example, applications may perform complex tasks (e.g., face recognition) on complex data types (e.g., images), and/or combine application-specific states (e.g., Bloom filters) with standard database queries. Second, the model should support *efficient distributed execution* of application code, executing a piece of computation close to the source of the relevant data items. This exploits concurrency in the distributed environment, and saves network bandwidth because intermediate results (e.g., the location of a person) tend to be much smaller than raw inputs (e.g., images of the person). Finally, the model should be *easy to use*, minimizing the effort of application developers. Ideally, a developer needs to write code only for the core functions of the application. For example, suppose she wants to periodically collect a histogram of the resource usage of different system components, but the infrastructure monitor currently does not support computing histograms. Then it is desirable that she needs to write only the histogram computing function, and have it easily incorporated within the monitor.

While achieving any one of the above goals is easy, it is challenging to achieve all three in a single design. For example, one way to provide sufficient expressive power is to enable collecting all relevant data items in order to perform centralized processing, and using application code to maintain states (e.g., Bloom filters) outside of the database. However, this approach not only rules out distributed execution, but it requires developers to integrate *outside* states into query processing—a difficult task. To understand the difficulty, consider the Bloom filters in the person finder application. To employ pruning of unnecessary searches, an application developer would have to write code to break a search query into three steps: selecting the roots of subtrees within the hierarchy for which Bloom filters are stored using the database, checking the search key against the Bloom filters outside of the database, and then recursively searching any qualified subtrees again using the database. This is an onerous task.

Similarly, consider the goal of efficient distributed execution. Distributed execution of aggregation functions (mainly with an SQL-style interface) has been studied in the literature [4, 14, 18]. The approach is to implement an aggregation function as a set of accessor functions (possibly along with user-defined global states) and to distribute them. However, it is not clear how to distribute application code for a large variety of possible application tasks that may access and combine *arbitrary* data items and application-specific states. For example, under the existing approaches, it is difficult to associate user-defined states (e.g. filters) with *subsets* of sensor readings. One could argue that application developers should implement all aspects of the distributed execution of their code. However, this approach requires developers to track the physical locations of stored sensor data and manage network communications, thus violating the goal of ease of programming.

2.3   Our Solution

We observe that although there are a large variety of possible application tasks, many tasks perform similar kinds of computations. For example, a common computation paradigm is to combine a list of sensor inputs of the same type (e.g., numeric values) to generate a single result (e.g., a histogram). Other common computation paradigms include (1) computing multiple aggregates from the same set of data sources, and (2) performing a *group-by* query, i.e., grouping data into classes and computing the same aggregate for each class (e.g., computing the total CPU usage on all the machines in a shared infrastructure for every user). Therefore, our strategy is to *provide a higher level of automation for common computation paradigms*. In this regard, we are similar to previous approaches [4, 14, 18].

As mentioned in Section 1, X-Tree employs XML to organize data into a logical hierarchy, and the Xpath query language for querying the (distributed) XML database, and hence requires techniques suitable for a hierarchical data model, unlike previous approaches. To enable user-defined computations with XML and Xpath, it provides two components. First, for common computation paradigms, X-Tree provides a *stored function* component with a simple Java programming interface and extends the Xpath function call syntax for implementing and invoking application code. Our implementation of X-Tree (denoted the X-Tree system) automatically distributes the execution of this application code. Second, X-Tree provides a *stored query* component that allows application developers to define derived states and to associate Xpath queries and application codes with XML elements. In this way, developers can guide the distribution of their code in the logical hierarchy of an XML document for *arbitrary* application tasks, without worrying about the physical locations of sensor data and or any needed network communication. Note that the physical locations of the sensor data can change over time (e.g., for the underlying system's load balancing, caching, etc.). Our implementation of X-Tree works regardless of these dynamics and hides them from developers.

3   Background: XML Model and Distributed Query Processing

An XML document defines a tree: Each XML element (tag-pair, e.g., `<PlanetLab>`, `</PlanetLab>`) is a tree node, and its nested structure specifies parent-child relationships in the tree. Every XML element has zero or more attributes, which are name-value pairs. Figure 1 illustrates the XML document representing the logical hierarchy in Iris-Log. The root node is PlanetLab. It has multiple country elements as child nodes, which in turn are parents of multiple region elements, and so on. The leaf nodes represent user instances on every machine.

We can use Xpath path expressions to select XML elements (nodes) and attributes. In Xpath, "/" denotes a parent-child relationship, "//" an ancestor-descendant relationship, and "@" denotes an attribute name instead of an XML element name. For example, `/PlanetLab/Country[@id="USA"]` selects the USA subtree. An individual sensor reading, which is usually stored as a leaf attribute, can be selected with the whole path from root. `//User[@id="Bob"]/@memUsage` returns Bob's memory usage on every machine that he is using, as a list of string values. In order to compute the total memory usage of Bob, we can use the Xpath built-in function "sum": `sum(//User[@id="Bob"]/@memUsage)`. However, the handful of built-in functions hardly satisfy all application needs, and the original centralized execution mode suggested in the Xpath standard is not efficient for wide-area sensor systems.

IrisNet [2, 11, 13] supports distributed execution of Xpath queries, excluding functions. We highlight some of the features of IrisNet that are relevant to this paper; our description is simplified, omitting various IrisNet optimizations–see [11, 13] for further details. Sensor data of a service is conceptually organized into a single XML document, which is distributed across a number of host machines, with each host holding some fragment of the overall document. Sensing devices, which may also be hosts of XML fragments, process/filter sensor inputs to extract the desired data, and send update queries to hosts that own the data. Each fragment contains specially marked dummy elements, called *boundary elements*, which indicate that the true element (and typically its descendant elements) reside on a different host. IrisNet requires an XML element to have an `id` attribute unique among its siblings. Therefore, an element can be uniquely identified by the sequence of `id` attributes along the path from itself to the document root. This sequence is registered as a DNS domain entry, for routing queries.

To process an XML query, IrisNet extracts the longest query prefix with `id` attribute values specified and constructs an `id` sequence. Then, it performs a DNS lookup and sends the query to the host containing the element specified by the prefix; this host is called the *first-stop host*. The query is evaluated against the host's local fragment, taking into account boundary elements. In particular, if evaluating the query requires visiting an element x corresponding to a boundary element, then a subquery is formed and sent to a host storing x. Each host receiving a subquery performs the same local query evaluation process (including recursively issuing further subqueries), and then returns the results back to the first-stop host. When all the subquery results have been incorporated into the host's answer, this answer is returned.

In the following, we describe X-Tree Programming within the context of IrisNet. However, we point out that our solution is applicable in any XML-based database-centric approach that supports in-network query processing.

4   X-Tree Programming

This section describes the two components of X-Tree, stored functions and stored queries, for efficiently programming wide-area sensing services.

4.1   Stored Functions

The stored function component incorporates application-specific code. Its programming interface is shown in Figure 2. A stored function can be invoked the same way as a built-in function in a user query, as shown in Figure 2(a). The colon separated function name specifies the Java class and the method major name of the application code. The semantics is that the *Input_XPATH* expression selects a list of values from the XML document, the values (of type String or Node, but not both) are passed to the stored function as inputs, and the function output is the result of the invocation. Optional arguments to a stored function are typically constant parameters, but can be any Xpath expressions returning a single string value.

As shown in Figure 2(b), application developers implement three Java methods for a stored function: *init*, *compute*, and *final*, which enable the decomposition of the stored function computation into a series of calls to the three methods. For each output value of the *Input_XPATH* expression, the *init* method is called to generate an intermediate value. Intermediate values are merged by the *compute* method until a single intermediate value is left, which is then converted to the query result by the *final* method. The *args* array contains the values of the arguments in the query. As shown in Figure 2(c), our X-Tree system automatically performs this decomposition and distributes the execution of the methods to relevant hosts where the data are located.[1]

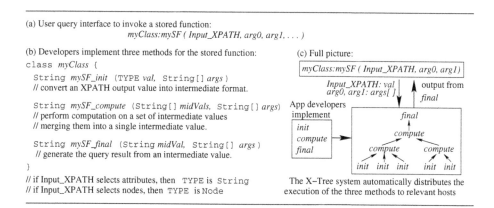

(a) User query interface to invoke a stored function:
        myClass:mySF (Input_XPATH, arg0, arg1, . . .)

(b) Developers implement three methods for the stored function:

```
class myClass {
   String mySF_init (TYPE val, String[] args )
   // convert an XPATH output value into intermediate format.

   String mySF_compute (String[] midVals, String[] args)
   // perform computation on a set of intermediate values
   // merging them into a single intermediate value.

   String mySF_final (String midVal, String[] args )
   // generate the query result from an intermediate value.
}
// if Input_XPATH selects attributes, then TYPE is String
// if Input_XPATH selects nodes, then TYPE is Node
```

(c) Full picture:

App developers implement
init
compute
final

The X-Tree system automatically distributes the execution of the three methods to relevant hosts

Fig. 2. Stored function programming interface

----

[1] For stored functions (e.g. median) that are difficult to decompose, application developers can instead implement a *local* method which performs centralized computation.

---

myClass:histogram (Input_XPATH, "bucket boundary 0", "bucket boundary 1", ...)
// A set of numeric values is selected to compute the histogram.

`class` *myClass* `{` // args[ ] specifies the histogram bucket boundaries.

  `String` *histogram_init* (`String` *val*,   `String[]` *args*)
  // determine bucket B for val, create an intermediate histogram with B's count set to 1 and all other counts set to 0.

  `String` *histogram_compute* (`String[]` *midVals*,   `String[]` *args*)
  // merge multiple intermediate histograms given by midVals[ ] by summing up the counts of corresponding buckets.

  `String` *histogram_final* (`String` *midVal*,   `String[]` *args*)
  // generate the query result from the final intermediate histogram.
`}`

Fig. 3. Implementation of a histogram aggregate

Stored functions support the ability to perform computation on a single list of values selected by an Xpath query. Examples are numeric aggregation functions, such as sum, histogram, and variance, and more complex functions, such as stitching a set of camera images into a panoramic image [8]. Figure 3 illustrates the implementation of a histogram aggregate. Here, the *Input_XPATH* query selects *attributes* and thus the *init* method uses String as the type of its first parameter. However, because arbitrary data structures can be encoded as Strings, the interface is able to handle complex inputs, such as images.

Compared to previous approaches for decomposing aggregation functions [4, 18], our approach supports more complex inputs. For example, it allows computing functions not just on values but also on XML *nodes* (Node as input type), which may contain multiple types of sensor readings collected at the same location (e.g. all kinds of resource usage statistics for a user instance on a machine). This improves the expressiveness of the query language; for example, several common computation paradigms (e.g., computing multiple aggregates from the same set of data sources, and performing group-by operations, as described in Section 6) can be specified within a given Node context of the logical hierarchy.

4.2   Stored Queries

Stored queries allow application developers to associate derived states with XML elements in a logical hierarchy. Naturally, states derived from a subset of sensor readings can be associated with the root of the smallest subtree containing all the readings.

These derived states can be either maintained automatically by our system or computed on demand when used in queries. As shown in Figure 4(a), application developers define a stored query by inserting a stored query sub-node into an XML element. The stored query has a name unique within the XML element. The query string can be any Xpath query. In particular, it can be a stored function invocation, and therefore developers can associate application codes with logical XML elements.

Figure 4(b) shows how to invoke an on-demand stored query. For each foo element, our system retrieves the stored query string. Then it executes the specified query within the context of the subtree rooted at the parent XML element (e.g., the foo element).

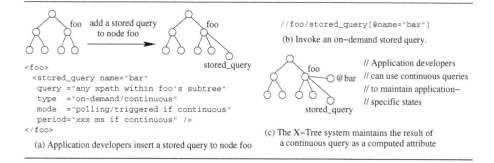

(a) Application developers insert a stored query to node foo

(b) Invoke an on-demand stored query.

(c) The X-Tree system maintains the result of a continuous query as a computed attribute

Fig. 4. Defining and using stored queries

For a continuous stored query, several additional attributes need to be specified, as shown in Figure 4(a). The query can be either in the polling mode or in the triggered mode. When the query is in the polling mode, the X-Tree system runs the query periodically regardless of whether or not there is a data update relevant to the query. When the query is in the triggered mode, the system recomputes the query result only when a relevant XML attribute is updated. As shown in Figure 4(c), the result of a continuous query is stored as a computed attribute in the database, whose name is the same as the stored query name. A computed attribute can be used in exactly the same way as a standard XML attribute. When adding a stored query, developers can also specify a duration argument. The X-Tree system automatically removes expired stored queries.

To support continuous stored queries, we implemented a continuous query scheme similar to those in Tapestry [27], NiagaraCQ [7], and Telegraph CACQ [20]. However, what is important for developers is that they can seamlessly use application-specific states in any queries, including stored function invocations and other stored query declarations, thus simplifying application programming.

4.3   Bottom-Up Composition of Application Tasks

Combining stored functions and stored queries, our solution supports bottom-up composition of application tasks that may combine arbitrary data items and application-specific states. This is because:

- Application-specific states can be implemented as computed attributes and used in exactly the same way as standard XML attributes.
- The *Input_XPATH* of a stored function may be an on-demand stored query invocation; in other words, an on-demand stored query higher in the XML hierarchy may process results of other on-demand stored queries defined lower in the XML hierarchy. This gives application developers the power to express arbitrary bottom-up computations using any data items in an XML document.

X-Tree Programming saves developers considerable effort. Developers need not worry about unnecessary details, such as the physical locations of XML fragments, network communications, and standard database operations. They can simply write code as stored functions and invoke stored functions in any user query. They can also associate derived states with any logical XML elements without worrying about the computation and/or maintenance of the states.

5   Automatically Distributed Execution of Application Code

Stored functions and stored queries can be dynamically added into applications. Developers upload their compiled Java code to a well-known location, such as a web directory. When a stored function invocation (or subquery) is received at a host machine, the X-Tree system will load the code from the well-known location to the host.

Given a stored function invocation, the X-Tree system automatically distributes the execution of the *init, compute,* and *final* methods to the relevant hosts where the data are located. The idea is to call the accessor methods along with the evaluation of the *Input_XPATH* query, which selects the input data.

As shown in Figure 5(a), the stored function invocation is sent to the first-stop host of the *Input_XPATH* query (hosting the leftmost fragment in the figure). The system employs the standard query processing facility (in our case, provided by IrisNet) to evaluate the *Input_XPATH* query against the local XML fragment. As shown in Figure 5(b), the results of querying the local fragment mainly consist of two parts: i) local input data items (squares in the figure), and ii) boundary elements (triangles in the figure) representing remote fragments that may contain additional input data.

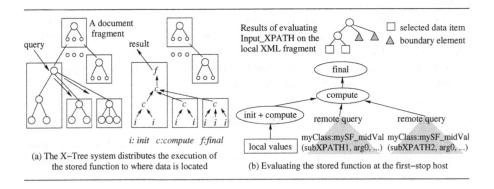

(a) The X-Tree system distributes the execution of the stored function to where data is located

(b) Evaluating the stored function at the first-stop host

Fig. 5. Automatically distributed execution for *myClass:mySF(Input_XPATH, arg0, ...)*

Next, the system composes a remote subquery for every boundary element, as shown in the shaded triangles in Figure 5(b). There are two differences between a remote subquery and the original stored function invocation. First, the function name is appended with a special suffix to indicate that an intermediate value should be returned. Second, a *subXPATH* query is used for the remote fragment. Note that the latter is obtained using the standard distributed query facility. The system then sends the subqueries to the remote fragments, which recursively perform the same operations. In the meantime, the system uses the *init* and *compute* methods to obtain a single intermediate value from the local data items.

Finally, when the intermediate results of all the remote subqueries are received, the system calls the *compute* method to merge the local result and all the remote results into a final intermediate value, and calls the *final* method to obtain the query result.

In summary, the X-Tree system automatically distributes the execution of stored functions to where the data are located for good performance. This scheme works regardless of the (dynamic) fragmentation of an XML document among host machines.

In addition to the above mechanism, application developers can use stored queries to guide the distributed execution of their code. They can define at arbitrary logical XML nodes stored queries that invoke stored functions. In this way, developers can specify the association of application-specific code to logical nodes in the XML hierarchy. Upon a reference of a stored query, the X-Tree system executes the stored function at the host where the associated logical XML node is located. Moreover, the stored function calls may in turn require results of other stored queries as input. In this way, developers can distribute the execution of their code in the logical XML hierarchy to support complex application tasks.

6   Optimizations

In this section, we first exploit the stored function Node interface to combine the computation of multiple aggregates efficiently and to support group-by. Then we describe a caching scheme that always achieves a total cost within twice the optimal cost.

6.1   Computing Multiple Aggregates Together and Supporting Group-by

In IrisLog, administrators often want to compute multiple aggregates of the same set of hosts at the same time. A naive approach would be to issue a separate stored function invocation for every aggregate. However, this approach uses the same set of XML nodes multiple times, performing many duplicate operations and network communications. Instead, like the usual mode of operation in any SQL-like language, the X-Tree system can compute all the aggregates together in a single query through a special stored function called *multi*. An example query, for the total CPU usage and maximum memory usage of all users across all hosts, is as follows:

```
myOpt:multi(//User, "sum", "cpuUsage", "max", "memUsage")
```

The *init*, *compute*, and *final* methods for *multi* are wrappers of the corresponding methods of the respective aggregate functions. A *multi*'s intermediate value contains the intermediate values for the respective aggregate functions. Note that *multi* can be used directly in any application to compute any set of aggregates.

Using similar techniques, we have also implemented an efficient group-by mechanism, which provides automatic decomposition and distribution for grouping as well as aggregate computations for each group. Please see [8] for details.

6.2   Caching for Stored Functions

In IrisNet, XML elements selected by an Xpath query are cached at the first-stop host in hopes that subsequent queries can be answered directly from the cached data. Because stored queries may invoke the same stored function repeatedly within a short interval, the potential benefits of caching are large. To exploit these potential benefits, we slightly modify our previous scheme for executing stored functions. At the first-stop (or any subsequent) host, the system now has three strategies. The first strategy is the distributed execution scheme as before. Because IrisNet cannot exploit cached *intermediate* values, this strategy does not cache data. The second strategy is to execute the *Input_XPATH* query, cache all the selected XML elements locally, and execute the stored function in a centralized manner by invoking all the methods locally. The third strategy is to utilize existing cached data if it is not stale, and perform centralized execution without sending subqueries, thus saving network and computation costs. Cached data becomes stale because of updates. In IrisNet, a user query may specify a tolerance time T to limit the staleness of cached data. Associated with a piece of cached data is its creation time and the piece is used to answer the query only if this time is within the last T time units.

Our system has to choose one of the strategies for an incoming query. For simplicity, we shall focus on improving network cost. Assume for a given stored function invocation, the centralized strategy costs K times as much as the distributed strategy, and the cost of a cache hit is 0. Moreover, we assume all queries have the same tolerance time T. This defines an optimization problem: find an algorithm for choosing the strategy to evaluate each incoming stored function so that the total cost is minimized.

To solve this optimization problem, we propose the algorithm in Figure 6(a). This algorithm does not require any future knowledge. It only requires the X-Tree system to keep per-query statistics so that the Y value can be determined. An example query pattern and the algorithm choices are shown in Figure 6(b). The algorithm performs distributed execution for the first K queries, then centralized execution for query $K + 1$, followed by a period of time T during which all queries are cache hits. Then the cached data is too stale, so the pattern repeats. An interesting, subtle variant on this pattern is when $> K$ consecutive queries use distributed execution (as shown in the rightmost part of the figure). This can arise because Y is calculated over a sliding time window. However, because the algorithm ensures that any $K+1$ consecutive distributed executions occur sparsely in a longer period of time than T, it is indeed better not to cache during these periods. We prove the following theorem in the full paper [8].

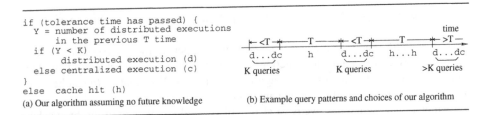

```
if (tolerance time has passed) {
    Y = number of distributed executions
        in the previous T time
    if (Y < K)
        distributed execution (d)
    else centralized execution (c)
}
else   cache hit (h)
```
(a) Our algorithm assuming no future knowledge

(b) Example query patterns and choices of our algorithm

Fig. 6. Optimization for caching

Theorem 1. *The algorithm in Figure 6(a) guarantees that the total cost is within twice the optimal cost.*

7   Evaluation

We have incorporated X-Tree Programming into IrisNet, and implemented the two applications discussed in Section 2. Although the Person Finder application is only a toy prototype, the Infrastructure Monitor application (IrisLog) has been deployed on 473 hosts in PlanetLab and has been publicly available (and in-use) since September 2003. The rich and diverse set of application-specific functions and states used by these applications (and others we studied [8]) supports the expressive power of X-Tree. The ease of programming using X-Tree is supported by the small amount of code for implementing these applications on our system: 439 lines of code for supporting Bloom filters in Person Finder and 84 lines of code for communicating with software sensors in IrisLog.

7.1   Controlled Experiments with Person Finder

We perform controlled experiments using the Person Finder application. For simplicity in understanding our results, we disabled IrisNet's caching features in all our experiments. We set up an XML hierarchy with 4 campuses in a university, 20 buildings per campus, 5 floors per building, 20 rooms per floor, and on average 2 people per room. Every room element contains a `name_list` attribute listing the names of the people in the room. We distribute the database across a homogeneous cluster of seven 2.66GHz Pentium 4 machines running Redhat Linux 8.0 connected by a 100Mbps local area network. The machines are organized into a three-level complete binary tree. The root machine owns the university element. Each of the two middle machines owns the campus and building elements for two campuses. Each of the four leaf machines owns the floor and room elements for a campus. In our experiments, we issue queries from a 550MHz Pentium III machine on our LAN and measure response times on this machine. Every result point reported is the average of 100 measurements.

Stored Functions.   In order to quantify the improvements in response times arising from our scheme for distributed execution of stored functions, we compute an aggregate function using two different approaches. The first approach uses the *init/compute/final* programming interface, so that the computation is automatically executed in a distributed fashion. The second approach extracts all the relevant input values from the database and performs a centralized execution[2].

In order to show performance under various network conditions and application scenarios, we vary a number of parameters, including network bandwidth, input value size

---

[2] We actually implemented this approach using the alternative *local* method in our Java programming interface, which is equivalent to an implementation outside of the XML database system that runs on the root machine.

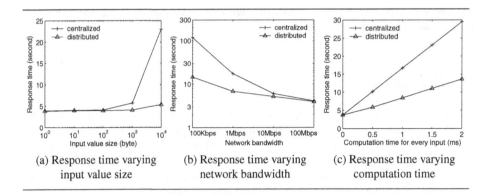

(a) Response time varying   (b) Response time varying   (c) Response time varying
        input value size   network bandwidth   computation time

Fig. 7. Distributed vs. centralized execution

to the stored function, and computation time of the function. The aggregation function we use models the common behavior of numeric aggregates (such as sum and avg), i.e. combining multiple input values into a single output value of similar size. For these experiments, every room element in the database contains a dummy attribute and the aggregations use this attribute. In order to make the size of input values a meaningful parameter to change, we choose to compute bit-by-bit binary OR on all the dummy attributes in the database and update every dummy attribute with a string of a given size before each experiment.

Figure 7(a) reports the response time of the two approaches while varying the length of every input value from 1 byte to 10,000 bytes. Centralized execution requires all input values to be transferred, while distributed execution only transfers intermediate results. As the input value size increases, the communication cost of the centralized approach increases dramatically, incurring large response time increases beyond 1000B. In contrast, distributed execution only suffers from minor performance degradations.

Figure 7(b) varies network bandwidth for the 100B points in Figure 7(a) in order to capture a large range of possible network bandwidth conditions in real use. The true (nominal) network bandwidth is 100Mbps. To emulate a 10Mbps network, we change the IrisNet network communication code to send a packet 10 times so that the effective bandwidth seen by the application is 1/10 of the true bandwidth. Similarly we send a packet 100 and 1000 times to emulate 1Mbps and 100Kbps networks. Admittedly, this emulation may not be 100% accurate since the TCP and IP layers still see 100Mbps bandwidth for protocol packets. Nevertheless, we expect the experimental results to reflect similar trends. As shown in Figure 7(b), when network bandwidth decreases, the performance gap between distributed and centralized execution increases dramatically. When network bandwidth is 1Mbps or lower, which is quite likely in a wide area network, distributed execution achieves over 2.5X speedups over the centralized approach.

Figure 7(c) studies the performance for computation-intensive aggregation functions. To model such a function, we insert a time-consuming loop into our aggregation

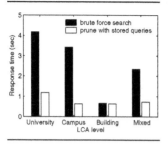

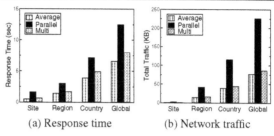

(a) Response time   (b) Network traffic

Fig. 8. Pruning vs. brute force   **Fig. 9.** Calling multiple aggregates

function so that this loop is executed once for every input value in both the distributed and the centralized approaches. Then we vary the total number of loop iterations so that the whole loop takes 0, 0.5ms, 1ms, 1.5ms and 2ms, respectively, which models increasingly computationally intensive aggregation functions. As shown in Figure 7(c), distributed execution achieves over 1.7X speedups when the computation time is at least 0.5ms per input. This is because distributed execution exploits the concurrency in the distributed database and uses all seven machines, while the centralized approach performs all the computation on a single machine.

Stored Queries. To study the benefit of stored queries, we compare the performance of the brute force search approach and a pruning approach enabled by stored queries. For the latter, we use continuous stored queries to maintain Bloom filters at the building elements, and user queries check the Bloom filters using Xpath predicates. At a building element, (sub)queries that do not pass the Bloom filter check are pruned.

We use a 550MHz Pentium III machine for generating background update requests that model the movements of people. In the model, a person stays in a room for a period of time, which is uniformly distributed between 1 second and 30 minutes, then moves to another room. When making a move, the person will go to a room on the same floor, in the same building, in different buildings of the same campus, and in different campuses, with probabilities 0.5, 0.3, 0.1, and 0.1, respectively.

Figure 8 shows the performance comparison. We measure response times for queries that look for a person in the entire university, in a particular campus, or in a particular building. The mixed workload is composed of 20% university level queries, 30% campus level queries, and 50% building level queries. Because the scope (university, campus, or building) of a query is presumed to be an end user's good guess of the person's location, we set up the queries so that a query would succeed in finding a person within the given scope 80% of the time.

As shown in Figure 8, the Bloom filter approach achieves dramatically better performance than the brute force approach for queries involving campus or university level elements, demonstrating the importance of stored queries. The building level results are quite close because pruning is less effective in a smaller scope and additional stored procedure overhead almost offsets the limited benefit of pruning.

7.2   Real World Experiments with IrisLog

Our workload consists of queries with four different scopes. The *global* queries ask for information about all the PlanetLab hosts (total 473 hosts).[3] The *country* queries ask about the hosts (total 290 hosts) within the USA. The *region* queries randomly pick one of the three USA regions, and refer to all the hosts (around 95 hosts per region) in that region. Finally, the *site* queries ask information about the hosts (around 4 hosts per site) within a randomly chosen USA site.

PlanetLab is a shared infrastructure; therefore all the experiments we report here were run together with other experiments sharing the infrastructure. We do not have any control over the types of machines, network connections, and loads of the hosts. Thus our experiments experienced all of the behaviors of the real Internet where the only thing predictable is unpredictability (latency, bandwidth, paths taken). To cope with this unpredictability, we ran each experiment once every hour of a day, and for each run we issued the same query 20 times. The response times reported here are the averages over all these measurements. We also report the aggregate network traffic which is the total traffic created by the queries, subqueries and the corresponding responses between all the hosts involved.

Calling Multiple Aggregates Using Multi. Figure 9 shows the performance of computing a simple aggregate (average), computing four different aggregates (average, sum, max, and min) using four parallel queries, and computing the same four aggregates using a single Multi query. For parallel queries, all the queries were issued at the same time, and we report the longest query response time.

From the figure, we see that the average query response time is small considering the number and the geographic distribution of the PlanetLab hosts. There exists a distributed tool based on Sophia [29] that can collect information about all PlanetLab hosts. Sophia takes minutes to query all the PlanetLab nodes [9]. In contrast, IrisLog executes the same query in less than 10 seconds.

Moreover, both the response time and the network overhead of the Multi operation are very close to those of a simple aggregation and are dramatically better than the parallel query approach. The Multi operation avoids the overhead of sending multiple (sub)queries, as well as the packet header and other common metadata in responses. It also avoids redundant selection of the same set of elements from the database.

We also studied the benefits on IrisLog of using our efficient group-by scheme. For a group-by query over all the nodes, our scheme achieves a 25% speedup in response time and an 81% savings in network bandwidth compared to the naive approach of extracting all the relevant data and computing group-by results in a centralized way [8].

8   Related Work

Sensor Network Programming. A number of programming models have been proposed for resource-constrained wireless sensor networks, including database-centric,

---

[3] Although IrisLog is deployed on 473 PlanetLab hosts, only 373 of them were up during our experiments. The query latency reported here includes the timeout period IrisLog experiences while contacting currently down hosts.

functional, and economic models. The database-centric programming models (e.g., Tiny-DB [18, 19], Cougar [6, 32]) provide an SQL-style declarative interface. Like X-Tree, they require decomposing the target function into init/compute/final operators for efficient distributed execution. However, the resource constraints of the target domain have forced these models to emphasize simplicity and energy-efficiency. In contrast, X-Tree is a more heavy-weight approach, targeted at resource-rich Internet-connected sensing devices, where nodes have IP addresses, reliable communication, plenty of memory, etc. The resource-rich target environment allows X-Tree, unlike the above systems, to sandbox query processing inside the Java virtual machine and to transparently propagate and dynamically load new aggregation operator code for query processing. Moreover, its XML data model allows posing queries in the context of a logical aggregation hierarchy. The functional programming models (e.g., programming with abstract regions [22, 30]) support useful primitives that arise in the context of wireless sensor network communication and deployment models. For example, the *abstract region* primitive captures the details of low-level radio communication and addressing. However, the requirements of wide-area sensing are different—generality is more important than providing efficient wireless communication primitives. Moreover, it is more natural to address wide-area sensors through logical hierarchy rather than physical regions. X-Tree aims to achieve these requirements. Proposals for programming sensor networks with economic models (e.g., market-based micro-programming's pricing [21]) are orthogonal to X-Tree. We believe that X-Tree can be used with such economic models, especially within a shared infrastructure (e.g., IrisNet [2, 13]) where multiple competing services can run concurrently.

Distributed Databases. Existing distributed XML query processing techniques [11, 26] support only standard XML queries. In contrast, X-Tree's query processing component supports user-defined operations. X-Tree leverages the accessor function approach for decomposing numeric aggregation functions [4, 18], and supports a novel scheme to automatically distribute the execution of stored functions. X-Tree's stored query construct has a similar spirit as the proposal for relational database fields to contain a collection of query commands [25]. The original proposal aims to support clean definitions of objects with unpredictable composition in a centralized environment. Because the logical XML hierarchy usually corresponds to real-life structures (such as geographical boundaries), X-Tree is able to support meaningful application-specific states computed from subsets of sensor readings. Moreover, stored queries can be seamlessly integrated into queries, and at the same time they can invoke application-specific code. This enables developers to compose arbitrary bottom-up computations, and to guide the distribution of application codes without knowing the physical layout of data.

Distributed Hierarchical Monitoring Systems. Astrolabe [28] allows users to use the SQL language to query dynamically changing attributes of a hierarchically-organized collection of machines. Moreover, user-defined aggregates can be installed on the fly. However, unlike X-Tree, it targets applications where the total aggregate information maintained by a single node is relatively small (≈ 1 KB), and the aggregates must be written as SQL programs. Hi-Fi [12] translates a large number of raw data streams into useful aggregate information through a number of processing stages, defined in terms of SQL queries running on different levels of an explicitly-defined machine hierarchy.

These processing stages can be installed on the fly. Both these systems target applications where aggregate data is continuously pushed toward the end users. Along with such *push-queries*, X-Tree targets *pull-queries* where relevant data is transferred over the network only when a query is posed. SDIMS [31] achieves a similar goal as Astrolabe by using a custom query language over aggregation trees built on top of a DHT. Moreover, it provides very flexible push vs. pull mechanisms. User-defined functions are more limited than with X-Tree, e.g., there does not appear to be an efficient means to perform bottom-up composition of distinct user-defined tasks. Finally, X-Tree differs from all three systems by using the XML data model and supporting a standard XML query language; thus it supports using a *logical* hierarchy that can be embedded on an arbitrary topology and a query language that incorporates the semantics of that hierarchy.

Parallel Programming. A number of programming models have been proposed to automatically parallelize computation within restricted target domains. For example, an associative function can be computed over all prefixes on an n element array in $O(\log(n))$ time on $n/\log(n)$ processors using parallel prefix computations [5, 17]. In the context of LANs, the MapReduce model [10], like X-Tree, requires programmers to decompose the high level task into smaller functions. The MapReduce implementation then efficiently and robustly parallelizes the execution of those functions into thousands of machines in a single cluster. X-Tree can be considered as a simplification and distillation of some of these models based on our requirements. In particular, X-Tree provides efficient in-network aggregation (through the *compute* function, which MapReduce lacks), supports a standard query processing language, provides location transparency, and is targeted toward wide-area networks.

9   Conclusion

In this paper, we present *X-Tree Programming*, a novel database-centric approach to easily programming a large collection of Internet-connected sensing devices. Our solution augments the valuable declarative interface of traditional database-centric approaches with the ability to seamlessly incorporate user-provided code for accessing, filtering, and processing sensor data, all within the context of the hierachical XML database model. We demonstrate the effectiveness of our solution through both controlled experiments and real-world applications, including an infrastructure monitor application on a 473 machine worldwide deployment. Using X-Tree Programming, a rich collection of application-specific tasks were implemented quickly and execute efficiently, simultaneously achieving the goals of expressibility, ease of programming, and efficient distributed execution. We believe that X-Tree Programming will enable and stimulate a large number of wide-area sensing services.

References

1. IrisLog: A Structured, Distributed Syslog. http://www.intel-iris.net/irislog.
2. IrisNet (Internet-scale Resource-Intensive Sensor Network Service). http://www.intel-iris.net/.

3. PlanetLab. http://www.planet-lab.org/.
4. F. Bancilhon, T. Briggs, S. Khoshafian, and P. Valduriez. FAD, a powerful and simple database language. In *Proc. VLDB 1987*.
5. G. E. Blelloch. Scans as primitive parallel operations. *ACM Transaction on Computers*, C-38(11), 1989.
6. P. Bonnet, J. E. Gehrke, and P. Seshadri. Towards sensor database systems. In *Proc. IEEE Mobile Data Management*, 2001.
7. J. Chen, D. J. DeWitt, F. Tian, and Y. Wang. NiagaraCQ: A scalable continuous query system for Internet databases. In *Proc. SIGMOD 2000*.
8. S. Chen, P. B. Gibbons, and S. Nath. Database-centric programming for wide-area sensor systems. Technical Report IRP-TR-05-02, Intel Research Pittsburgh, April 2005.
9. B. Chun. PlanetLab researcher and administrator, http://berkeley.intel-research.net/bnc/. Personal communication, November, 2003.
10. J. Dean and S. Ghemawat. MapReduce: Simplified data processing on large clusters. In *Proc. OSDI 2004*.
11. A. Deshpande, S. K. Nath, P. B. Gibbons, and S. Seshan. Cache-and-query for wide area sensor databases. In *Proc. SIGMOD 2003*.
12. M. J. Franklin, S. R. Jeffery, S. Krishnamurthy, F. Reiss, S. Rizvi, E. Wu, O. Cooper, A. Edakkunni, and W. Hong. Design considerations for high fan-in systems: The HiFi approach. In *Proc. CIDR'05*.
13. P. B. Gibbons, B. Karp, Y. Ke, S. Nath, and S. Seshan. Irisnet: An architecture for a worldwide sensor web. *IEEE Pervasive Computing*, 2(4), 2003.
14. J. Hellerstein, W. Hong, S. Madden, and K. Stanek. Beyond average: Toward sophisticated sensing with queries. In *Proc. IPSN 2003*.
15. P. R. Kumar. Information processing, architecture, and abstractions in sensor networks. Invited talk, *SenSys 2004*.
16. J. Kurose. Collaborative adaptive sensing of the atmosphere. Invited talk, *SenSys 2004*.
17. R. E. Ladner and M. J. Fischer. Parallel prefix computation. *J. of the ACM*, 27(4), 1980.
18. S. Madden, M. J. Franklin, J. M. Hellerstein, and W. Hong. TAG: A tiny aggregation service for ad-hoc sensor networks. In *Proc. OSDI 2002*.
19. S. Madden, M. J. Franklin, J. M. Hellerstein, and W. Hong. The design of an acquisitional query processor for sensor networks. In *Proc. SIGMOD 2003*.
20. S. Madden, M. Shah, J. M. Hellerstein, and V. Raman. Continuously adaptive continuous queries over streams. In *Proc. SIGMOD 2002*.
21. G. Mainland, L. Kang, S. Lahaie, D. C. Parkes, and M. Welsh. Using virtual markets to program global behavior in sensor networks. In *Proc. ACM SIGOPS European Workshop*, 2004.
22. R. Newton and M. Welsh. Region streams: Functional macroprogramming for sensor networks. In *Proc. ACM Workshop on Data Management for Sensor Networks*, 2004.
23. S. Rhea and J. Kubiatowicz. Probabilistic location and routing. In *Proc. INFOCOM 2002*.
24. T. Roscoe, L. Peterson, S. Karlin, and M. Wawrzoniak. A simple common sensor interface for PlanetLab. PlanetLab Design Notes PDN-03-010, 2003.
25. M. Stonebraker, J. Anton, and E. N. Hanson. Extending a database system with procedures. *ACM Transactions on Database Systems*, 12(3), 1987.
26. D. Suciu. Distributed query evaluation on semistructured data. *ACM Transactions on Database Systems*, 27(1), 2002.
27. D. B. Terry, D. Goldberg, D. Nichols, and B. M. Oki. Continuous queries over append-only databases. In *Proc. SIGMOD 1992*.
28. R. van Renesse, K. P. Birman, and W. Vogels. Astrolabe: A robust and scalable technology for distributed system monitoring, management, and data mining. *ACM Transactions on Computer Systems*, 21(2), 2003.

29. M. Wawrzoniak, L. Peterson, and T. Roscoe. Sophia: An information plane for networked systems. In *Proc. Hotnets-II*, 2003.
30. M. Welsh and G. Mainland. Programming sensor networks using abstract regions. In *Proc. NSDI 2004*.
31. P. Yalagandula and M. Dahlin. A scalable distributed information management system. In *Proc. Sigcomm'04*.
32. Y. Yao and J. Gehrke. Query processing in sensor networks. In *Proc. CIDR 2003*.

Using Clustering Information for Sensor Network Localization*

Haowen Chan, Mark Luk, and Adrian Perrig

Carnegie Mellon University
{haowenchan, mluk, perrig}@cmu.edu

Abstract. Sensor network localization continues to be an important research challenge. The goal of localization is to assign geographic coordinates to each node in the sensor network. Localization schemes for sensor network systems should work with inexpensive off-the-shelf hardware, scale to large networks, and also achieve good accuracy in the presence of irregularities and obstacles in the deployment area.

We present a novel approach for localization that can satisfy all of these desired properties. Recent developments in sensor network clustering algorithms have resulted in distributed algorithms that produce highly regular clusters. We propose to make use of this regularity to inform our localization algorithm. The main advantages of our approach are that our protocol requires only three randomly-placed nodes that know their geographic coordinates, and does not require any ranging or positioning equipment (i.e., no signal strength measurement, ultrasound ranging, or directional antennas are needed). So far, only the DV-Hop localization mechanism worked with the same assumptions [1]. We show that our proposed approach may outperform DV-Hop in certain scenarios, in particular when there exist large obstacles in the deployment field, or when the deployment area is free of obstacles but the number of anchors is limited.

1 Introduction

Many wireless sensor network applications require information about the geographic location of each sensor node. Besides the typical application of correlating sensor readings with physical locations, approximate geographical localization is also needed for applications such as location-aided routing [2], geographic routing [3], geographic routing with imprecise geographic coordinates [4, 5], geographic hash tables [6], and for many data aggregation applications.

Manually recording and entering the positions of each sensor node is impractical for very large sensor networks. To assign an approximate geographic coordinate to each sensor node, many automated localization algorithms have been developed. To obtain the information required for node locations, researchers proposed approaches that make

* This research was supported in part by CyLab at Carnegie Mellon under grant DAAD19-02-1-0389 from the Army Research Office, and grant CAREER CNS-0347807 from NSF, and by a gift from Bosch. The views and conclusions contained in this paper are those of the authors and should not be interpreted as representing the official policies, either expressed or implied, of Bosch, Carnegie Mellon University, NSF, the Army Research Office, the U.S. Government or any of its agencies.

V. Prasanna et al. (Eds.): DCOSS 2005, LNCS 3560, pp. 109–125, 2005.

different assumptions: (1) quantitative ranging/directionality measurements [7–11]; (2) long range beacons [12–16]; (3) centralized processing [17, 18]; and (4) a flat, unobstructed deployment area. We do not discuss protocols related to cases (1)–(3) because we are steering away from such assumptions.

Algorithms that assume a flat, unobstructed deployment area experience serious degradation in their position estimates in the presence of large obstacles and other irregularities in the deployment area. Most of the localization algorithms in current literature have been evaluated only for deployments clear of obstacles. However, such ideal deployments only represent the special case, while large obstacles are common in realistic settings. For a localization protocol to be practical, it is essential that it functions even in the presence of such irregularities. As can be seen in Figure 7 of Section 4, our algorithm has much better accuracy in recreating the topology of irregular deployments.

Generally, any algorithm that uses triangulation based on distance estimates to known *anchors* falls victim to errors caused by obstacles. An *anchor* is defined as a node that is aware of its own location, either through GPS or manual preprogramming during deployment. An example of a distance-triangulation protocol is the Ad-hoc Positioning System (APS) described by Niculescu and Nath [1]. They describe three methods of performing the distance estimate, the most widely cited of which is the DV-Hop method. DV-Hop uses a technique based on distance vector routing. Each node keeps the minimum number of hops to each anchor, and uses the hop count as an estimate of physical distance. Once a node has the estimated distance and location of 3 or more anchors, it performs least-squares error triangulation to estimate its own position. Nagpal et al. [19] describe a similar scheme but improve the accuracy of the distance estimation by using the average hop count of all the neighbors of a node as a distance estimate.

We present a localization scheme that requires no ranging or measuring equipment, no long range beacons, and no centralized processing, and is able to operate with arbitarily positioned anchor nodes. Furthermore, unlike DV-Hop, it makes no assumptions about the shape or internal topology of a deployment area: in particular, when the deployment area is occupied by large, well-spaced obstacles, our scheme significantly outperforms DV-Hop since it is able to re-create the physical topology of the network where DV-Hop cannot. Our scheme is based on the novel approach of first performing *sensor node clustering* on the network in order to create a regular structure of representative nodes (called *cluster-heads*). To the best of our knowledge, this is the first localization protocol that does not make any assumption on the sensor node's hardware, yet performs well in certain classes of irregular topologies.

2   Sensor Network Clustering

2.1   Clustering Goals

Performing clustering on a sensor network deployment prior to localization has two advantages. First, it creates a regular pattern from which location information can be extracted. Second, it helps reduce the amount of communication overhead since only the cluster-heads need to be involved in the initial phase of the localization.

When a sensor network is first deployed, we cannot assume any regularity in the spacing or the pattern of the sensor nodes. Figure 1(a) shows that after clustering, the

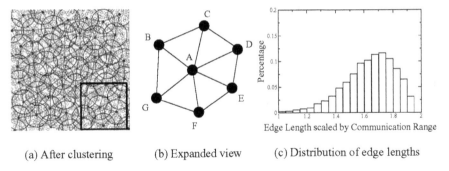

(a) After clustering   (b) Expanded view   (c) Distribution of edge lengths

Fig. 1. Effects of clustering.

cluster-heads are regularly spaced throughout the network. We call two cluster-heads *adjacent* if there exists some other sensor node that is within communication range of both cluster-heads. Figure 1(a) shows the graph created by setting each cluster-head as a vertex and connecting each pair of adjacent cluster-heads with an edge. We call this graph the *cluster-adjacency graph*. The edges in the graph are called *cluster-adjacency edges*. The cluster-adjacency graph is mesh-like and has few crossing edges. Further-more, the variance in the length of the edges between adjacent nodes is small. Fig-ure 1(c) is a histogram of edge length. Most edges fall within 1.3 to 1.7 r, with r being the maximum communication range between two nodes. The average edge length is 1.63 r and variance is 0.0309 r. Hence, the cluster-adjacency graph forms a regular structure from which location information can be extracted. This regularity is degraded slightly when the communication pattern is nonuniform (i.e., not a unit disk), or when the node density is low, but in general the regularity of cluster-head separation is always much greater than the distribution of physical distances between unclustered sensor nodes.

2.2   Modifying the ACE Algorithm

We chose to modify ACE [20] for use in our localization algorithm, because ACE al-ready produces clusters with highly regular separation. Our localization techniques are not confined to the clusters produced by ACE; we can also use any other clustering technique that produces clusters with highly regular separation, such as the algorithm proposed by Younis and Fahmy [21].

A brief description of ACE is as follows (for details, please refer to the original paper [20]). The algorithm proceeds in a fixed number of iterations. In each iteration, sensor nodes gradually *self-elect* to become cluster-heads if they detect that many nodes in their neighborhood do not belong to any cluster. To achieve regular separation, these clusters then *migrate* away from other clusters by re-selecting their respective cluster-heads.

We modified ACE to improve the regularity of the separation between cluster-heads. First, we increased the number of iterations for ACE from 3 to 5, trading off increased communication cost for increased regularity. We also modified the migratory mech-anism by approximating a *spring effect* between adjacent clusters. This effect causes clusters that are close together to migrate apart, and clusters that are too far apart to be attracted. During the migration phase, each cluster-head evaluates the potential *fitness score* of each candidate node C in its neighborhood. The score for each candidate C is calculated as the total number of sensor nodes belonging to the cluster if C was to become the next cluster-head, plus a modifier for each adjacent cluster of C. Let s be the estimated separation between C and the adjacent cluster-head in terms of maximum communication radius r; s can be estimated by counting the number of common nodes in both clusters – the more common nodes, the closer the two clusters are. The final modifier is calculated via the function $g(s)$:

$$g(s) = \begin{cases} d(3(0.125 - (\frac{1}{2}(s - 1.5)^2))) & \text{if } s \geq 1.5 \\ d(12(0.125 - (\frac{1}{2}(s - 1.5)^2))) - 1.125 & \text{if } s < 1.5 \end{cases}$$

Where d is the total number of nodes in the neighborhood of C. The constants used above are empirically derived – their exact values are not important to the correctness of the algorithm. Note that the function above reaches a maximum at $s = 1.5$, meaning that clusters that are estimated to be more than $1.5r$ apart will attract each other, while clusters that are less than $1.5r$ apart will repel each other.

3   Localization Procedure

In this section we describe how the localization algorithm proceeds after clustering is complete. We first describe a naive version of our general approach using several strong assumptions. In subsequent subsections, we will eliminate these assumptions and improve the accuracy of our basic algorithm using more complex approaches.

3.1   The Basic Cluster Localization Algorithm

In this section we describe a high level overview of our general approach. Each of the steps described in this section are naive approaches which we significantly improve in subsequent sections.

Locally-Aware Anchors. The algorithm starts from the anchor nodes, which are themselves cluster-heads and have knowledge of their geographical positions. We assume that these are *Locally-Aware Anchor Nodes*, able to determine the geographical positions of all the cluster-heads adjacent to themselves. This could be performed by installing ranging and direction-finding hardware on the anchor nodes, or more practically by pre-selecting the cluster-head nodes in their neighborhood and directly programming these coordinates into the anchors. This increases the hardware or installation overhead of the scheme, hence we eliminate this assumption in Section 3.4.

Expanding the Calibrated Set. The anchor nodes and their adjacent cluster-heads form an initial set of *calibrated nodes* which are aware of their positions. Given this base set of calibrated nodes, our algorithm will continually expand this set until all cluster-heads in the network have been calibrated. This is performed in a distributed manner where each cluster-head calibrates itself if two or more of its adjacent cluster-heads have successfully calibrated.

The self-calibration procedure uses the regularity of edge-lengths between cluster-heads to perform a position estimate. As an example, Figure 1(b) shows node A along with its adjacent neighbors, or its *cluster-head neighborhood*. If node A knows the topological configuration of its cluster-head neighborhood as well as the estimated physical positions of two neighbors, C and D, A can estimate its own position by assuming some pre-determined standard value l for the length of the edges AC and AD. After A is calibrated, node B can similarly estimate its position based on positions of C and A, further enabling node G to calibrate, and so on. In this manner, the set of calibrated nodes grows until all cluster-heads in the network are calibrated. We present an significantly improved method for position estimation in Section 3.3.

Refining the Position Estimate The initial position estimate is based on the early position estimates of two neighbors. As more information becomes available, more cluster-heads will be able to estimate their position, and some already-calibrated cluster-heads may further refine their position estimate. Each cluster-head reacts to this new information by recomputing its own position estimate. A simple way of improving the position estimate would be to repeat the initial position estimation once for each adjacent pair of calibrated cluster-heads in the calibrating node's neighborhood, and then taking the average position of these results. To prevent propagation of small changes, each node only rebroadcast its updated position if its difference from the previous position is larger than some threshold. When this occurs, we call it a *major* position update.

To improve the accuracy of this step, we have developed a more sophisticated algorithm for performing position refinement which we present in Section 3.2.

Termination. Each node continues refining its position until either of two conditions occurs:

- *The node has reached some maximum number of major position updates.* We count the number of major position updates, and when it reaches 10 in the case of repeated initial calibration (see Section 3.3) or 60 in the case of mesh relaxation (see Section 3.2) then the node terminates and accepts its current estimated position as its final position.
- *The node has not received any position updates during the past time period.* The amount of time to wait is chosen to be equal to the maximum time that position information needs to disseminate across the network, which is proportional to the diameter of the network. If the node has not received any new updates within this time frame after the last update, then there cannot be any further updates remaining in the system, indicating convergence.

Calibration of Follower Nodes. Thus far, locality calibration has only been performed on the cluster-heads. When the cluster-heads have been fully localized, there remains the final step of calibrating the non-cluster-head nodes (i.e., the *follower* nodes). Various methods exist for calibrating these nodes. In our algorithm, each node takes the average of the estimated positions of all the calibrated nodes within its communication range (including cluster-heads and other follower nodes). This produces a localization accuracy for the non-cluster-head nodes that is very close to the localization accuracy of the cluster-heads. When this step is complete, all the sensor nodes are localized.

3.2   Improved Position Refinement: Mesh Relaxation

We note that the goal of our algorithm is to solve for the geographical configuration of cluster-heads that is most likely, given the adjacency information of all cluster-heads and the position information of the anchors. We can approximate this solution by distributively solving for the global configuration in which the square of the difference of the length of each edge from the known average edge length l is minimized.

To solve this problem, we use *mesh relaxation*, an approximation algorithm for finding the least-squares solution to a set of pairwise constraints. Mesh relaxation has previously been studied for localization in robotics [22–24]. A general description of mesh relaxation is beyond the scope of this paper; we describe how the method is applied to our localization protocol.

Each cluster-head is modeled as a mass point, and the distance between each pair of adjacent cluster-heads is modeled as a spring of length equal to the average edge length l. The calculation thus becomes equivalent to a physical simulation. Consider a cluster-head A. It has some estimated coordinate $p_{A,t}$ at time t, and we wish to continue the simulation to update its position in time $t + \delta_t$. Let the set of A's adjacent cluster-heads be S. Each of the members of S will exert a force on A. According to Hooke's law, this force can be expressed as $F = k\Delta x$, with Δx defined as the displacement of a spring from its equilibrium length (set at the average edge length l), and k is the spring constant. Note that the value of k is irrelevant in this computation since we are looking for the point where all forces are equalized, which would be the same for any value of k. Hence, we let $k = 1$. The resultant force on A at time t is:

$$\boldsymbol{F}_A = \sum_{B \in S} (|\boldsymbol{d}_{B,A}| - l)\hat{d}_{B,A}$$

The variable $\boldsymbol{d}_{B,A}$ represents the 2-dimensional vector of the separation between the estimated positions of cluster-heads B from A at time t, i.e., $\boldsymbol{d}_{B,A} = p_{B,t} - p_{A,t}$. The variable $\hat{d}_{B,A}$ represents the unit vector in the direction of B from A, and l is the known average link length. We displace the position of A by $q\boldsymbol{F}_A$, a quantity proportional to the resultant force on A. Hence, the updated position of A is $p_{A,t+1} = p_{A,t} + q\boldsymbol{F}_A$. We iterate this process until the change in position is below a threshold c.

The above algorithm is naturally parallelizable onto the cluster-heads; each cluster-head A calculates the forces acting on itself based on the current estimated locations of the nodes in its cluster-head neighborhood S and updates its own estimated position, which is then sent as an update to all the members of S.

3.3   Improved Initial Calibration

While mesh relaxation produces accurate localization results, if it begins with a poor initial estimated position, it takes many iterations to converge. Hence, having an initial accurate estimate is essential to produce a workable algorithm. In this section we describe how to accurately make an initial estimate of a cluster-head's position based on the structure of the cluster-heads around it.

This algorithm consists of three steps. First, we acquire knowledge of a node's two-hop neighborhood to produce an ordered circular list of its adjacent cluster-heads. This list corresponds to either a counter-clockwise or a clockwise traversal of the set of adjacent cluster-heads. Then, we augment this list with some heuristic information about the relative separations of each cluster-head from its predecessor and successor in the circular list. Finally, the node calculates an estimate for its own position when two or more of its neighbors are calibrated.

Ordering the adjacent cluster-heads. To orient a given cluster-head A correctly within the topology, we need to extract an ordering in its set of adjacent cluster-heads S. This ordering will produce information that we will later use to derive a location estimate for A based on the location estimates of the members of S.

Figure 2(b) shows the neighborhood of cluster-head A. It has 5 adjacent cluster-heads, $\{B, C, D, E, F\}$. At the beginning of the protocol, A is aware of its neighborhood set (e.g., $\{D, C, F, B, E\}$) but not its order. The objective of this step of the algorithm is to derive an ordering on the set that corresponds to either a clockwise or counterclockwise traversal of the set, e.g., either (B, C, D, E, F) or (F, E, D, C, B). Note that the ordering is on a circle hence any cyclic shift of a correct sequence is still correct, e.g., (D, E, F, B, C).

We now introduce some terminology. Let the cluster-head for which we wish to derive the ordered circular list be the **calibrating node**. The **cluster-head neighborhood** of a cluster-head is the set of cluster-heads that are *adjacent* to it (recall that two cluster-heads are considered adjacent if there exists some node which is in communication range of both of them). If two members of the cluster-head neighborhood of the calibrating node are also adjacent, then we call them **directly linked** with respect to the calibrating node. Examples of directly linked neighbors of the calibrating node A in Figure 2(b) are B and C, or C and D. If two members of the cluster-head neighborhood are not adjacent, but they are adjacent to another cluster-head that is not the calibrating node, nor in the cluster-head neighborhood, then we call them **indirectly linked** with respect to the calibrating node. For example, in Figure 2(b), B and F are indirectly linked with respect to A since they are both adjacent to G. Finally, if two members of the cluster-head neighborhood have no adjacent cluster-heads in common besides the calibrating node, then they are **unlinked** with respect to the calibrating node. Unlinked pairs are also called **gaps** since they represent a discontinuity in the cluster-head neighborhood of a node.

At the beginning of the algorithm, each cluster-head communicates its neighborhood information to all the members of its cluster-head neighborhood. Thus every cluster-head is aware of its cluster-head topology up to two edges away. If the cluster-adjacency graph has no cross edges in its physical embedding, then it is straightforward to construct the ordered circular list of neighbors for a calibrating node. The calibrating node selects any neighbor as a starting point and traverses the set of neighbors by selecting the next neighbor that has a direct or indirect link to the current node, then appending it to the list. The selected neighbor then becomes the current node, and the process is iterated until the traversal returns to the starting point, or the current node has no direct or indirect links that have not already been traversed. In the latter case, traversal is restarted from the starting point in the opposite direction, and the nodes visited are pre-pended to the sequence. Ambiguities may arise if there are crossing edges, since in this case there may be more than one possible choice for the next node in the traversal. However this occurs sufficiently infrequently that this choice can be resolved easily. For example, we use the heuristic of choosing the node that has the most common neighbors with the calibrating node, and skipping over any alternative nodes (i.e., not including them in the traversal at all).

At the end of this step, we have constructed an ordered circular list of the cluster-head neighborhood of a calibrating node. This list represents an initial estimate of the local physical topology of cluster-heads around the calibrating node.

Assigning angles to adjacent cluster-heads. The next step in obtaining a physical mapping of this topology is to assign angular separations between subsequent members of the circular list. If all the cluster-heads in the list are all directly linked, we simply assign equal angular shares to each sector. For example, in Figure 2(a), each sector is given $60°$ for a total of $360°$.

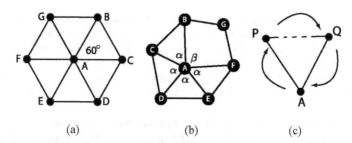

(a)   (b)   (c)

Fig. 2. a) Assignment of angular shares when there are 6 adjacent cluster-heads all directly linked. The cluster-heads are assumed to be equally distributed, hence each angle is $360°/6 = 60°$.
b) Assignment of 5 angular shares when there are 4 direct links and 1 indirect links. Angles opposite to a direct link is denotes as α, and β represents the angle opposite to an indirect link.
c) Orienting A's circular list direction. If A is clockwise from Q with respect to P, and P is clockwise from A with respect to Q, then Q must be clockwise from P with respect to A

If there are one or more indirect links, we wish to assign larger angular shares to sectors subtended by indirect links since indirect links are usually longer than direct links. Letting the length of a direct link be l, we estimate the length of an indirect link as $\sqrt{2}l$ based on the intuition that the vertices forming an indirect link form a quadrilateral with the typical shape of a square. Hence, if α is the angle assigned to a direct link, then we should assign approximately $\beta = \sqrt{2}\alpha$ to an indirect link. Figure 2(b) shows an example of a node having 5 neighbors where there are 4 direct links and 1 indirect link, resulting in $\alpha = 66.5°$ assigned to the direct sectors and $\beta = 94.0°$ assigned to the indirect links.

In the case where there is a gap in the circular list (e.g., at the edge of the deployment area or next to an obstacle), we use the heuristic of assigning $\alpha = 60°$ to angles subtended by direct links and $\beta = 90°$ to angles subtended by indirect links.

When this portion of the algorithm is completed, every node has an ordered circular list representing its cluster-head neighborhood, as well as an angle between each adjacent pair of members of the circular list, representing the estimated angular separation of the pair.

Performing the position estimate. At this point, the calibrating node is able to perform an initial position estimate if two or more of the nodes in its cluster-head neighborhood have already performed an initial position estimate. We call nodes which have successfully performed an initial position estimate, ***calibrated nodes***. A calibrated node not only has a position estimate but also has its circular list of its cluster-head neighborhood ordered in the canonical direction (i.e., physically clockwise or counterclockwise). For simplicity, we shall assume the canonical ordering is clockwise.

We first describe the algorithm using two calibrated nodes as reference nodes. Suppose the calibrating node A has two calibrated nodes P and Q in its cluster-head neighborhood. The first step in calibration is to orient A's circular list in the canonical clockwise direction. P and Q transmit to A their estimated positions (x_P, y_P) and (x_Q, y_Q) as well as their ordered cluster-head neighborhood list. These lists are ordered in the canonical clockwise direction. A observes its own position in these lists and deduces the ordering of its own list as follows. Figure 2(c) provides an illustration of the process. Suppose in the list of P, A occurs after Q. Furthermore, in the list of Q, P occurs after

A. Hence we know that A is clockwise from Q with respect to P, and P is clockwise from A with respect to Q, hence it must be that Q is clockwise from P with respect to A. Hence, in the ordered list of A, if the angular displacement of Q from P is greater than $180°$, then A needs to reverse its ordered list to put it in a clockwise order. The other case (where P is clockwise from Q with respect to A) follows an analogous argument.

Once A has determined the canonically correct ordering of its cluster-head neighborhood, it is now aware of which side of the line PQ it belongs. Hence, its initial position estimate can be calculated using basic trigonometry from the positions of P and Q and their estimated angular separation with respect to A. An estimate can be computed in several ways. We describe the method that we chose. The angle PAQ is known due to our angular assignment. Assuming that $AP = AQ$, we derive the angle QPA. Given this angle and the estimated position of P and Q, we can compute the angular bearing of A from P. We the compute A's estimated position with respect to P by assuming A's displacement from P is the known average edge length l. If there are multiple neighbors with known coordinates, we perform these operations once for each of them, i.e., for each P_i, compute the angular bearing of A from P_i and estimate A's position as a displacement of l along that bearing. After each estimate is computed, the final estimated position is calculated as the average (centroid) of all the estimates.

Repeated Initial Calibration. We have found that this empirical process of estimating position is highly accurate. In fact, we can use this algorithm for both the initial position calibration, and for position refinement instead of performing mesh relaxation. When new information arrives as neighboring nodes update their positions, we merely perform the same position estimation algorithm again to obtain the new estimate. This process achieves comparable performance with mesh relaxation while incurring less communication overhead.

3.4   Self-orienting Anchors

Thus far, we have assumed that the anchors are "locally aware" (i.e., know the physical locations of all cluster-heads in their neighborhood), and that all nodes are aware of the average edge length between any two adjacent cluster-heads. We now describe an optimisation to remove these assumptions.

In this scheme, each anchor picks an arbitrary orientation and sets the average edge length l to 1. It assigns estimated positions to all the cluster-heads in its neighborhood according to the angular share system described in Section 3.3. Calibration then proceeds with respect to each anchor as normal. When calibration is complete, each cluster-head has formed a location estimate with respect to each anchor's arbitrary coordinate system. Specifically, each anchor is now calibrated with respect to every other anchor's coordinate system. All the anchors exchange this information along with their known physical coordinates.

Now each anchor can proceed to orient and scale its coordinate system to best fit the estimated positions of every other anchor under its coordinate system with its known physical location. Specifically, consider some anchor A. Number the other anchors $1..m$. After all calibration is complete, each of the other anchors sends to A their respective estimated locations $e_1, e_2, \ldots, e_m$ under A's arbitrarily chosen coordinate system. Each of the other anchors also sends to A their respective actual physical locations, i.e., $p_1, p_2, \ldots, p_m$. Now, A finds a transform T characterised by a rotation θ, a scaling factor c, and a bit r indicating whether or not reflection is needed, such that T is the transform that yields the lowest sum of squared errors between Te_i and p_i for each of the other anchors:

$$T = \operatorname*{argmin}_{G} \sum_{i=1}^{m} (p_i - Ge_i)^2$$

At least 2 other anchors are needed to uniquely determine T. Once T is determined, it is then flooded to the rest of the network to allow the other cluster-heads to convert their estimated positions under A's coordinate system to actual physical locations.

This procedure will result in each cluster-head having several estimates of its position, one for each anchor. Based on the observation that position estimates increase in error with increasing hop distance from the anchor, each cluster-head uses the estimate associated with the closest anchor (in terms of cluster-head hop-count) and discards the others.

4   Results

Based on various combinations of the optimizations described in Section 3, we implemented three versions of our algorithm with various trade-offs:

1. Locally-Aware Anchors with Repeated Initial Calibration
2. Locally-Aware Anchors with Mesh Relaxation
3. Self-Orienting Anchors with Repeated Initial Calibration

With Locally-Aware Anchors, anchors are assumed to know the geographic positions of their immediate cluster-head neighborhood. This involves greater hardware or set-up cost. Self-Orienting Anchors do not make this assumption, and are only assumed to know their own geographic positions. The trade-off for removing this assumption is slightly lower accuracy and a higher communication cost.

In Repeated Initial Calibration, nodes are first calibrated using the method described in Section 3.3. When new information arrives and the nodes need to update their position estimates, they simply perform the initial calibration algorithm again to compute their new position. In Mesh Relaxation, the nodes are initialized similarly (i.e., using the technique of Section 3.3). However, as new information gets updated in the network, the nodes update their positions using mesh relaxation as described in Section 3.2. The trade-off is that mesh relaxation is more accurate than repeated initial calibration when using locally-aware anchors, but mesh relaxation requires more communication and takes a longer time. We used standard 32-bit floating point numbers during the simulated calculations, but we expect our results to hold also with lower levels of precision or with fixed-point computations.

We did not investigate the performance of self-orienting anchors with mesh relaxation, since these two methods did not interact well together and resulted in both higher communication overhead and less accuracy than self-orienting anchors with repeated initial calibration.

We provide a detailed quantitative analysis of each of the three versions of our algorithm. We evaluate our algorithms against the DV-Hop localization algorithm [1] with the smoothing optimization described by Nagpal et al. [19], because this is the only algorithm that also assumes no ranging/directional measurements, no long-range beacons, and no centralized processing. We investigated the performance of DV-Hop using normal anchors, as well as with Locally-Aware anchors for some scenarios. As we show from our results, although DV-Hop often has better accuracy in deployment settings with no obstacles and many well-placed anchors, our algorithms often outperform DV-Hop in more realistic settings in the presence of obstacles, irregularities and randomly placed anchors.

4.1 Base Simulation Assumptions and Parameters

Our base simulation setup is described for reference; how we vary the parameters of this base setup will be described later in each set of results. To evaluate the algorithms, we set up experiments using a deployment of 10,000 nodes over a square region of $20r \times 20r$ where r is the maximum communication radius. We do not assume that nodes are synchronized in time; nodes would periodically run an iteration of the algorithm regardless of the state of its neighbors. Anchor nodes are distributed randomly throughout the deployment. The base setup does not include obstacles.

To simulate irregular communication range, we used the DOI model (or Degree of Irregularity) described by He et al. [14]. The transmission range of a node is a random walk around the disc, bounded by the maximum range r_{max} and minimum range r_{min}. We chose to set $r_{min} = 0.5r_{max}$. Let the range of a node in the bearing θ (in degrees) be r_θ. We start with $r_0 = 0.5(r_{min} + r_{max})$ and compute each subsequent r_θ as a random walk, i.e., $r_\theta = r_{\theta-1} + X(r_{max} - r_{min})D$ where X is a random real value uniformly chosen in the range $[-1, 1]$ and D is the degree of irregularity (DOI). Note that r_{theta} is not allowed to exceed r_{max} or go below r_{min}. r_{theta} represents only the transmission range of a node; since our schemes require bidirectional communication, we require both nodes to be within each other's respective transmission ranges in order to be able to communicate.

The metric for localization is the accuracy of the estimated position, which is measured as the distance between a node's estimated position and its true location, divided by maximum communication range r. The accuracy of a particular trial is measured as the average error over all nodes in the deployment. We also measured how much the average error varies among different trials.

4.2 Varying Number of Anchor Nodes in Uniform deployment

We varied the number of anchors from 3 to 7 to observe how accuracy is improved with increasing number of anchors for each algorithm. We also studied how much the error varies over different trials.

Figure 3 shows the average error for all four of our algorithms as well as DV-Hop with normal anchors and with locally aware anchors. The average localization error for all algorithms improves as the number of anchors increases. However, while the performance of our algorithms remained relatively stable as the number of anchors were varied, DV-Hop showed a very high sensitivity to the number of anchors. With the minimum number of anchors (3), regardless of whether these anchors were locally aware, DV-Hop typically incurs higher error than any of our algorithms. With 7 anchors, however, DV-Hop's average error improves to roughly half of ours. This suggests that DV-Hop requires significantly more anchors than our protocols in order to be maximally effective. Furthermore, the rapid degradation of the performance of DV-Hop as the number of anchors decreases indicates that it is not robust in scenarios where an-

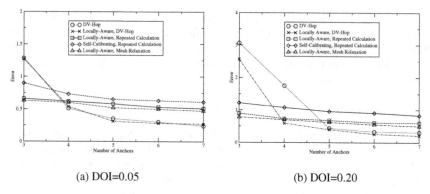

(a) DOI=0.05   (b) DOI=0.20

Fig. 3. Average Error with Varying Number of Anchors

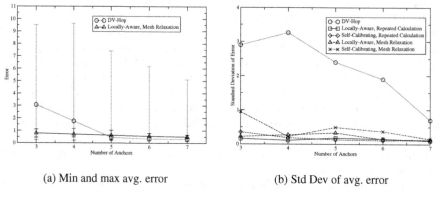

(a) Min and max avg. error   (b) Std Dev of avg. error

Fig. 4. Spread of Average Error for the different schemes, DOI=0.2

chor node failure could be a factor. On the other hand, our algorithms provide uniformly good performance even with a very small number of anchor nodes, indicating dependable performance even if anchor nodes are subject to failure.

Figure 3 also shows that our self-orienting algorithms exhibit slightly poorer but comparable performance to the algorithms with locally aware anchors. This is an indication that the algorithms for self-orientation are reasonably effective.

Figure 4a shows how the accuracy varies throughout different trials, with the error bars representing the minimum and maximum average error among all the trials. For clarity only one of our algorithms is shown; the other three exhibit similar behavior. Figure 4b graphs the standard deviation of the average error of each scheme over all our trials. We observe that although the expected error of DV-Hop becomes lower than our algorithms when five or more anchors are used, the variance is always much higher. In the best case, DV-Hop generates extremely accurate estimates. However, DV-Hop's error in the worst case scenario is significantly higher than the worst case error of our algorithm.

The intuition is that DV-Hop algorithm is more sensitive to the relative placement of anchors since it uses triangulation from anchors to estimate each node's position. Triangulation provides highly inaccurate results when anchors are placed in a co-linear

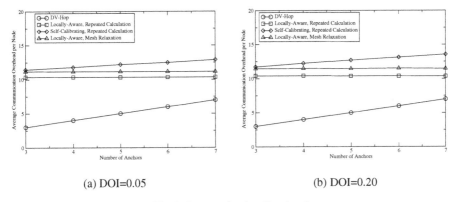

(a) DOI=0.05   (b) DOI=0.20

Fig. 5. Communication Overhead

fashion or are too close together. Clustering, on the other hand, has the advantage of not being significantly effected by positioning of anchor nodes. In certain cases, anchor nodes placed near each other actually improves performance. Random placement of anchor nodes thus proves to be a great advantage of cluster localization. As sensor networks become commodity technologies, random placement of anchors will be desirable because it allows for deployments by untrained personnel, instead of needing a specialized engineer to plan the process.

4.3 Communication Overhead

We measured the communication overhead of each of our schemes. We note that the overhead of performing clustering formed the bulk of our communications cost. Since the subsequent localization only involves cluster-heads, the communication cost for the network is low. An average of 9.11 communications per node were required for our modified ACE clustering protocol, while the localization schemes at most require about 3 more communications per node on top of that. Hence, all our schemes achieve communication costs comparable to DV-Hop.

4.4 Obstacles

In this section, we study deployments with obstacles. Previously, we showed that DV-Hop is often more accurate in the case where obstacles are absent from the deployment field and numerous anchors are present. However, in more realistic scenarios, obstacles of various size and shape can disrupt communication and consequently interfere with localization.

Our 2 types of obstacles are *walls* and *voids*. Walls are represented as a line segment with length of 250 units, or half the length of the deployment field. Walls can be oriented in any direction, and all communication through the wall is blocked. Voids are areas of various fixed shapes that are off-limits during deployment. Our experiments investigated the effect of irregularly placed walls and regularly-spaced voids on the various schemes.

Since DV-Hop counts the number of hops between nodes to estimate distance, it almost always overestimates distances when the 2 nodes are separated by some type of

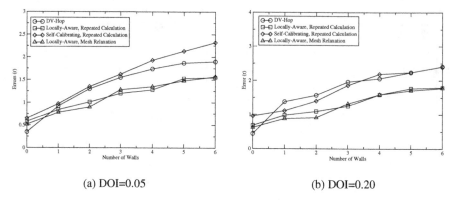

(a) DOI=0.05   (b) DOI=0.20

Fig. 6. Accuracy with obstacles, 5 anchor nodes

obstacle. This is because DV-Hop uses hop count as an estimator of physical distance. When an obstacle is between an anchor and a calibrating node, hop counts can be inflated leading to a large overestimate of the physical distance. This can negatively affect the accuracy of the position estimation.

Clustering, on the other hand, is not significantly affected by obstacles. As shown in Figure 7(c), the regular structure of cluster-heads are preserved around obstacles. Thus, localization based on clustering typically has much better performance than DV-Hop when faced with obstacles.

Figure 6 plots the accuracy of localization while increasing number of walls when using 5 anchor nodes. Locally-Aware schemes had the best performance, consistently outperforming DV-Hop whenever there are walls in the deployment area. The Self-Orienting schemes did not do as well in this scenario, yielding slightly worse performance compared with DV-Hop. In performing this series of tests, we observed the main flaw of clustering localization: our schemes perform by creating a map of the deployment area at the cluster-head level, which represents a relatively coarse granularity of resolution (about $1.5r$). Hence, our schemes perform best when here is sufficient space between obstacles to allow the regular but coarse structure of the clustering mesh to pass through the gap. If obstacles are placed close together (e.g., a gap of less than one communication radius), then the structure of cluster-heads through the gap may be too coarse to allow cluster-head localization to traverse the gap, leading to a failure in localization for certain segments of the deployment area. The self-calibrating scheme is particularly vulnerable to this effect since if an anchor is unable to find the estimated positions of at least two other anchors, it is unable to calibrate its own relative coordinate space and is thus almost useless for localization. Because the walls used in this experiment are extremely long and placed independently at random, their placement would occasionally create small gaps though which DV-Hop could perform localization but our cluster-based schemes could not. This explains the relatively small degree of the performance improvement of our schemes over DV-Hop in these scenarios.

Table 1 shows the average error for selected deployments of *voids*, which are more well-spaced obstacles which were designed with sufficient gaps between obstacle features to enable the coarse-grained clustering structure to pass through and map out the entire deployment area. The name of each deployment represents what type of void is

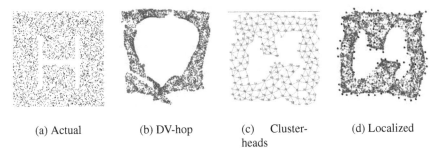

(a) Actual   (b) DV-hop   (c)   Cluster-   (d) Localized
                                           heads

Fig. 7. Large Obstacles

Table 1. Average error for selected irregular deployments (5 anchors, DOI=0.05)

	Cross	H	S	Thin S
DV-Hop	0.768	1.903	4.328	3.670
Locally Aware DV-Hop	0.686	1.650	5.145	3.575
Locally Aware, Mesh Relaxation	0.685	1.228	1.316	1.443
Locally Aware, R.I.C.	0.969	1.177	1.699	1.459
Self-Orienting, R.I.C.	1.235	1.469	2.951	2.919

in the deployment field. For example, cross is a large void in the middle of the deployment field in the shape of a cross. H is an obstacle in the shape of a large capital H as represented in Figure 7 and S is a similar large obstacle constructed in the shape of an S using straight line segments. Figure 7(a) shows the actual deployment of nodes for obstacle H. Figure 7(b) shows how DV-Hop is unable to reconstruct occluded areas. Figure 7(c) shows that our cluster-head localization phase yields a good reconstruction of the deployment area which leads to good localization accuracy for all nodes as shown in Figure 7(d).

The experimental results confirm our hypothesis that for well-spaced, deeply concave obstacles, clustering localization always performs significantly better than DV-Hop, much greater than the improvement shown in Figure 6 where the obstacles were not well-spaced. The Cross-shaped obstacle was sufficiently convex in shape that DV-Hop retained relatively good accuracy and performed roughly as well as cluster localization. However, both the H and S deployments possess deep cul-de-sacs which could not be accurately triangulated by DV-Hop (see Figure 7(b)). However, our clustering methods allowed our schemes to reconstruct the shape of the deployment area (see Figure 7(c)) which yielded significantly better accuracies. The self-orienting schemes suffered inaccuracies since occasionally some anchors were unable to deduce estimated positions of at least two other anchors and were thus unable to self-calibrate. However, this is not a fundamental weakness of clustering localization and can probably be addressed by more sophisticated self-calibration algorithms.

We hypothesize that the accuracy of cluster-localization is dependent on having sufficient clearance space between obstacles to allow the regular cluster-head mesh structure to pass through. To investigate the extent of this effect, we simulated our self-

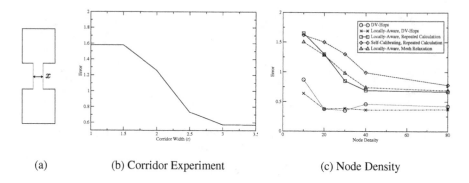

| (a) | (b) Corridor Experiment | (c) Node Density |

Fig. 8. Limitations of Cluster-Localization

calibrating scheme in a deliberately anomalous "dumbbell-shaped" deployment consisting of two large square deployment areas joined by a long narrow corridor (see Figure 8(a)). Only two anchors were placed in each each main deployment area; thus cluster-localization can only be successful if sufficient information can pass through the corridor to allow calibration of each anchor (unsuccessful nodes simply adopt the location of the nearest anchor). The results are shown in Figure 8(b) for a DOI of 0.2. As can be seen, accuracy degrades significantly when the corridor is too narrow to fit a sufficient number of cluster-heads for the reconstruction of the cluster-head topology; however once the corridor is sufficiently wide ($2.5r$ in this case), the cluster-head structure is able to reconstruct the shape of the corridor and yields accurate results.

We note that our schemes only yield high accuracy for sufficiently dense networks; Figure 8(c) shows that at low densities (less than 30 nodes per circle of radius r), the accuracy of cluster localization suffers while DV-Hop retains good performance. This is because at low densities, clustering is not as tight, and hence the number of adjacent cluster-heads is lower, leading to a more sparse cluster-head topology which yields less information for localization.

5   Conclusion

Localization continues to be an important challenge in today's sensor networks. In this paper, we propose to use clustering as a basis for determining the position information of sensor nodes. To the best of our knowledge, this is the first paper to consider this approach. Our clustering-based approach has many benefits: it is fully distributed, it provides good accuracy, it only requires that three randomly placed sensor nodes know their geographic position information, and it works with standard sensor node hardware without requiring any special hardware such as ultrasound or other ranging equipment. Moreover, our approach provides accurate position information even in topologies with walls and other concave structures, as long as the granularity of the obstacle features are on the same order as the separation between cluster-heads.

References

1. Niculescu, D., Nath, B.: Ad hoc positioning system (APS). In: Proceedings of IEEE GLOBECOM. (2001) 2926–2931
2. Ko, Y.B., Vaidya, N.: Location-aided routing (LAR) in mobile ad hoc networks. In: Proceedings of MobiCom, ACM (1998) 66–75
3. Karp, B., Kung, H.T.: GPSR: greedy perimeter stateless routing for wireless networks. In: Proceedings of MobiCom. (2000) 243–254
4. Newsome, J., Song, D.: GEM: Graph embedding for routing and data-centric storage in sensor networks without geographic information. In: Proceedings of SenSys. (2003) 76–88
5. Rao, A., Ratnasamy, S., Papadimitriou, C., Shenker, S., Stoica, I.: Geographic routing without location information. In: Proceedings of MobiCom. (2003) 96–108
6. Ratnasamy, S., Karp, B., Yin, L., Yu, F., Estrin, D., Govindan, R., Shenker, S.: GHT: a geographic hash table for data-centric storage. In: Proceedings of WSNA. (2002)
7. Capkun, S., Hamdi, M., Hubaux, J.P.: Gps-free positioning in mobile ad-hoc networks. Cluster Computing **5** (2002)
8. Savarese, C., Rabaey, J., Langendoen, K.: Robust positioning algorithms for distributed ad-hoc wireless sensor networks. In: Proceedings of the General Track: USENIX Annual Technical Conference. (2002) 317–327
9. Savvides, A., Han, C., Srivastava, M.B.: Dynamic fine grained localization in ad-hoc sensor networks. In: Proceedings of MobiCom. (2001) 166–179
10. Savvides, A., Park, H., Srivastava, M.B.: The n-hop multilateration primitive for node localization problems. Mobile Networks and Applications **8** (2003) 443–451
11. Ji, X., Zha, H.: Sensor positioning in wireless ad-hoc sensor networks with multidimensional scaling. In: Proceedings of IEEE Infocom. (2004)
12. Bahl, P., Padmanabhan, V.: Radar: an in-building RF-based user location and tracking system. In: Proceedings of IEEE Infocom. (2000)
13. Bulusu, N., Heidemann, J., Estrin, D.: GPS-less low cost outdoor localization for very small devices. IEEE Personal Communications Magazine **7** (2000) 28–34
14. He, T., Huang, C., Blum, B., Stankovic, J.A., Abdelzaher, T.: Range-free localization schemes for large scale sensor networks. In: Proceedings of MobiCom. (2003) 81–95
15. Priyantha, N.B., Chakraborty, A., Balakrishnan, H.: The Cricket location-support system. In: Proceedings of MobiCom. (2000)
16. Nasipuri, A., Li, K.: A directionality based location discovery scheme for wireless sensor networks. In: Proceedings of WSNA. (2002) 105–111
17. Doherty, L., Pister, K.S.J., Ghaoui, L.E.: Convex position estimation in wireless sensor networks. In: Proceedings of IEEE Infocom. (2001)
18. Shang, Y., Ruml, W., Zhang, Y., Fromherz, M.P.: Localization from mere connectivity. In: Proceedings of MobiHoc. (2003) 201–212
19. Nagpal, R., Shrobe, H., Bachrach, J.: Organizing a global coordinate system from local information on an ad hoc sensor network. In: Proceedings of IPSN. (2003)
20. Chan, H., Perrig, A.: ACE: An emergent algorithm for highly uniform cluster formation. In: Proceedings of EWSN. (2004)
21. Younis, O., Fahmy, S.: Distributed clustering in ad-hoc sensor networks: A hybrid, energy-efficient approach. In: Proceedings of IEEE Infocom. (2004)
22. Duckett, T., Marsland, S., Shapiro, J.: Learning globally consistent maps by relaxation. In: Proceedings of IEEE ICRA. (2000)
23. Golfarelli, M., Maio, D., Rizzi, S.: Elastic correction of dead-reckoning errors in map building. In: Proceedings of IEEE ICRA. (1998)
24. Howard, A., Matarić, M., Sukhatme, G.: Relaxation on a mesh: a formalism for generalized localization. In: Proceedings of IEEE IROS. (2001)

Macro-programming Wireless Sensor Networks Using *Kairos*

Ramakrishna Gummadi, Omprakash Gnawali, and Ramesh Govindan

University of Southern California,
Los Angeles, CA
`gummadi, gnawali, ramesh @usc.edu`
`http://enl.usc.edu`

Abstract. The literature on programming sensor networks has focused so far on providing higher-level abstractions for expressing *local* node behavior. *Kairos* is a natural next step in sensor network programming in that it allows the programmer to express, in a centralized fashion, the desired *global* behavior of a distributed computation on the entire sensor network. Kairos' compile-time and runtime sub-systems expose a small set of programming primitives, while hiding from the programmer the details of distributed-code generation and instantiation, remote data access and management, and inter-node program flow coordination. In this paper, we describe Kairos' programming model, and demonstrate its suitability, through actual implementation, for a variety of distributed programs—both infrastructure services and signal processing tasks—typically encountered in sensor network literature: routing tree construction, localization, and object tracking. Our experimental results suggest that Kairos does not adversely affect the performance or accuracy of distributed programs, while our implementation experiences suggest that it greatly raises the level of abstraction presented to the programmer.

1 Introduction and Motivation

Wireless sensor networks research has, to date, made impressive advances in platforms and software services [1, 2, 3]. The utility and practicality of dense sensing using wireless sensor networks has also been demonstrated recently [4, 5, 6]. It is now time to consider an essential aspect of sensor network infrastructure—support for *programming* wireless sensor network applications and systems components at a suitably high-level of abstraction. Many of the same reasons that have motivated the re-design of the networking stack for sensor networks (energy-efficiency, different network use models) also motivate a fresh look at programming paradigms for these networks.

Two broad classes of programming models are currently being investigated by the community. One class focuses on providing higher-level abstractions for specifying a node's *local behavior* in a distributed computation. Examples of this approach include the recent work on node-local or region-based abstractions [7, 8]. By contrast, a second class considers programming a sensor network *in the large* (this has sometimes been called *macroprogramming*). One line of research in this class enables a user to declaratively specify a distributed computation over a wireless sensor network, where

V. Prasanna et al. (Eds.): DCOSS 2005, LNCS 3560, pp. 126–140, 2005.

the details of the network are largely hidden from the programmer. Examples in this class include TinyDB [9, 10], and Cougar [11].

Kairos' programming model specifies the *global behavior* of a distributed sensornet computation using a *centralized* approach to sensornet programming. Kairos presents an abstraction of a sensor network as a collection of nodes (Section 3) that can all be tasked together simultaneously within a *single* program. The programmer is presented with three constructs: reading and writing variables at nodes, iterating through the one-hop neighbors of a node, and addressing arbitrary nodes. Using only these three simple language constructs, programmers *implicitly* express both distributed data flow and distributed control flow. We argue that these constructs are also natural for expressing computations in sensor networks: intuitively, sensor network algorithms process *named* data generated at individual nodes, often by moving such data to other nodes. Allowing the programmer to express the computation by manipulating *variables* at *nodes* allows us to almost directly use "textbook" algorithms, as we show later in detail in Section 3.2.

Given the single centralized program, Kairos' compile-time and runtime systems construct and help execute a node-specialized version of the compiled program for all nodes within a network. The code generation portion of Kairos is implemented as a language preprocessor add-on to the compiler toolchain of the native language. The compiled binary that is the single-node derivation of the distributed program includes runtime calls to translate remote reads and, sometimes, local writes into network messages. The Kairos runtime library that is present at every node implements these runtime calls, and communicates with remote Kairos instances to manage access to node state. Kairos is *language-independent* in that its constructs can be retrofitted into the toolchains of existing languages.

Kairos (and the ideas behind it) are related to shared-memory based parallel programming models implemented over message passing infrastructures. Kairos is different from these in one important respect. It leverages the observation that most distributed computations in sensor networks will rely on *eventual consistency* of shared node state both for robustness to node and link failure, and for energy efficiency. Kairos' runtime *loosely synchronizes* state across nodes, achieving higher efficiency and greater robustness over alternatives that provide tight distributed program synchronization semantics (such Sequential Consistency, and variants thereof [12]).

We have implemented Kairos as an extension to Python. Due to space constraints of this paper, we describe our implementation of the language extensions and the runtime system in detail in a technical report [13]. On Kairos, we have implemented three distributed computations that exemplify system services and signal processing tasks encountered in current sensor networks: constructing a shortest path routing tree, localizing a given set of nodes [2], and vehicle tracking [14]. We exhibit each of them in detail in Section 3 to illustrate Kairos' expressibility. We then demonstrate through extensive experimentation (Section 4) that Kairos' level of abstraction does not sacrifice *performance*, yet enables *compact* and *flexible* realizations of these fairly sophisticated algorithms. For example, in both the localization and vehicle tracking experiments, we found that the performance (convergence time, and network message traffic) and accuracy of Kairos are within 2x of the reported performance of explicitly distributed original versions, while the Kairos versions of the programs are more succinct and, we believe, are easier to write.

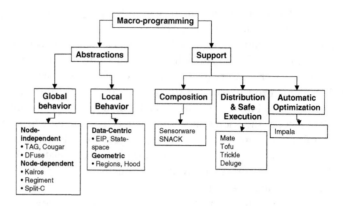

Fig. 1. Taxonomy of Programming Models for Sensor Networks

2   Related Work

In this section, we give a brief taxonomy (Figure 1) of sensornet programming and place our work in the context of other existing work in the area. The term "sensornet programming" seems to refer to two broad classes of work that we categorize as *programming abstractions* and *programming support*. The former class is focused on providing programmers with abstractions of sensors and sensor data. The latter is focused on providing additional runtime mechanisms that simplify program execution. Examples of such mechanisms include safe code execution, or reliable code distribution.

We now consider the research on sensor network programming abstractions. Broadly speaking, this research can be sub-divided into two sub-classes: one sub-class focuses on providing the programmer abstractions that simplify the task of specifying the node *local behavior* of a distributed computation, while the second enables programmers to express the *global behavior* of the distributed computation.

In the former sub-class, three different types of programming abstractions have been explored. For example, Liu *et al.* [15] and Cheong *et al.* [16] have considered node group abstractions that permit programmers to express communication within groups sharing some common group state. Data-centric mechanisms are used to efficiently implement these abstractions. By contrast, Mainland *et al.* [8] and Whitehouse *et al.* [7] show that topologically defined group abstractions ("neighborhoods" and "regions" respectively) are capable of expressing a number of local behaviors powerfully. Finally, the work on EIP [17] provides abstractions for physical objects in the environment, enabling programmers to express tracking applications.

Kairos falls into the sub-class focused on providing abstractions for expressing the global behavior of distributed computations. One line of research in this sub-class provides *node-independent* abstractions—these programming systems do not contain explicit abstractions for nodes, but rather express a distributed computation in a network-independent way. Thus, the work on SQL-like expressive but Turing-incomplete query systems (*e.g.*, TinyDB [10, 9] and Cougar [11]), falls into this class. Another body of work provides support for expressing computations over logical topologies [18, 19] or task graphs [20] which are then dynamically mapped to a network instance. This represents a plausible alternative to macroprogramming sensor networks. However, exporting the network topology as an abstraction can impose some rigidity in the programming model. It can also add complexity to maintaining the mapping between the logical and the physical topology when nodes fail.

Complementary to these approaches, *node-dependent* abstractions allow a programmer to express the global behavior of a distributed computation in terms of nodes and node state. Kairos, as we shall discuss later, falls into this class. As we show, these abstractions are natural for expressing a variety of distributed computations. The only other piece of work in this area is Regiment [21], a recent work. While Kairos focuses on a narrow set of flexible language-agnostic abstractions, Regiment focuses on exploring how *functional programming* paradigms might be applied to programming sensor networks in the large, while Split-C [22] provides "split" local-global address spaces to ease parallel programming that Kairos also provides through the remote variable access facility, but confines itself to the "C" language that lacks a rich object-oriented data model and a language-level concurrency model . Therefore, the fundamental concepts in these two works are language-specific.

Finally, quite complementary to the work on programming abstractions is the large body of literature devoted to systems in support of network programming. Such systems enable high-level composition of sensor network applications (Sensorware [23] and SNACK [24]), efficient distribution of code (Deluge [25]), support for sandboxed application execution (Maté [26]), and techniques for automatic performance adaptation (Impala [27]).

3   Kairos Programming Model

In this section, we describe the Kairos abstractions and discuss their expressibility and flexibility using three canonical sensor network distributed applications: routing tree construction, ad-hoc localization, and vehicle tracking.

3.1   Kairos Abstractions and Programming Primitives

Kairos is a simple set of extensions to a programming language that allows programmers to express the global behavior of a distributed computation. Kairos extends the programming language by providing three simple abstractions.

The first of these is the *node* abstraction. Programmers explicitly manipulate nodes and lists of nodes. Nodes are logically *named* using integer identifiers. The logical naming of nodes does *not* correspond to a topological structure. Thus, at the time of program composition, Kairos does not require programmers to specify a network topology. In Kairos, the node datatype exports operators like equality, ordering (based on node name), and type testing. In addition, Kairos provides a node_list iterator data type for manipulating node sets.

The second abstraction that Kairos provides is the list of *one-hop neighbors* of a node. Syntactically, the programmer calls a get_neighbors() function. The Kairos runtime returns the current list of the node's radio neighbors. Given the broadcast nature of wireless communication, this is a natural abstraction for sensor network programming (and is similar to *regions* [8], and *hoods* [7]). Programmers are exposed to the underlying network topology using this abstraction. A Kairos program typically is specified in terms of operations on the neighbor list; it may construct more complex topological structures by iterating on these neighbors.

The third abstraction that Kairos provides is *remote data access*, namely the ability to read from variables at named nodes. Syntactically, the programmer uses a `variable @node` notation to do this. Kairos itself does not impose any restrictions on which remote variables may be read where and when. However, Kairos' compiler extensions respect the scoping, lifetime, and access rules of variables imposed by the language it is extending. Of course, variables of types with node-local meaning (*e.g.,* file descriptors, and memory pointers) cannot be meaningfully accessed remotely.

Node Synchronization: Kairos' remote access facility effectively provides a shared-memory abstraction across nodes. The key challenge (and a potential source of inefficiency) in Kairos is the messaging cost of synchronizing node state. One might expect that nodes would need to synchronize their state with other nodes (update variable values at other nodes that have cached copies of those variables, or coordinate writes to a variable) often. In Kairos, only a node may write to its variable, thus mutually exclusive access to remote variables is not required; thereby, we also eliminate typically subtle distributed programming bugs arising from managing concurrent writes.

Kairos leverages another property of distributed algorithms for sensor networks in order to achieve low overhead. We argue that, for fairly fundamental reasons, distributed algorithms will rely on a property we call *eventual consistency*: individual intermediate node states are not guaranteed to be consistent, but, in the absence of failure, the computation eventually converges. This notion of eventual consistency is loosely molded on similar ideas previously proposed in well-known systems such as Bayou [28]. The reason for this, is, of course, that sensor network algorithms need to be highly robust to node and link failures, and many of the proposed algorithms for sensor networks use soft-state techniques that essentially permit only eventual consistency.

Thus, Kairos is designed under the assumption that *loose synchrony* of node state suffices for sensor network applications. Loose synchrony means that a read from a client to a remote object blocks *only* until the referenced object is initialized and available at the remote node and *not* on every read to the remote variable. This allows nodes to synchronize changed variables in a lazy manner, thereby reducing communication overhead. However, a reader might be reading a stale value of a variable, but because of the way distributed applications are designed for sensor networks, the nodes eventually converge to the right state. Where this form of consistency is inadequate, we provide a tighter consistency model, as described at the end of this section.

The Mechanics of Kairos Programming: Before we discuss examples of programming in Kairos, we discuss the mechanics of programming and program execution (Figure 2). As we have said before, the distinguishing feature of Kairos is that programmers write a single *centralized* version of the distributed computation in a programming language of their choice. This language, we shall assume, has been extended to incorporate

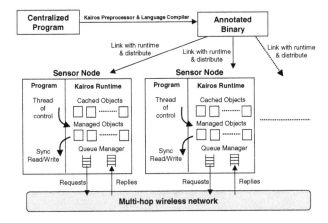

Fig. 2. Kairos Programming Architecture

the Kairos abstractions. For ease of exposition, assume that a programmer has written a centralized program **P** that expresses a distributed computation; in the rest of this section, we discuss the transformations on **P** performed by Kairos.

Kairos' abstractions are first processed using a *preprocessor* which resides as an extension to the language compiler. Thus, **P** is first pre-processed to generate annotated source code, which is then compiled into a binary **P_b** using the native language compiler. While **P** represents a global specification of the distributed computation, **P_b** is a node-specific version that contains code for what a single node does at any time, and what data, both remote and local, it manipulates.

In generating **P_b**, the Kairos preprocessor identifies and translates references to remote data into calls to the Kairos *runtime*. **P_b** is linked to the Kairos runtime and can be distributed to all nodes in the sensor network through some form of code distribution and node re-programming facility [29, 25]. When a copy is instantiated and run on each sensor node, the Kairos runtime exports and manages program variables that are owned by the current node but are referenced by remote nodes; these objects are called *managed objects* in Figure 2. In addition, it also caches copies of managed objects owned by remote nodes in its *cached objects* pool. Accesses to both sets of objects are managed through queues as asynchronous request/reply messages that are carried over a potentially multihop radio network.

The user program that runs on a sensor node calls *synchronously* into Kairos runtime for reading remote objects, as well as for accessing local managed objects. These synchronous calls are automatically generated by the preprocessor. The runtime accesses these cached and managed objects on behalf of the program after suspending the calling thread. The runtime uses additional background threads to manage object queues, but this aspect is transparent to the application, and the application is only aware of the usual language threading model.

```
1:  void buildtree(node root)
2:     node parent, self;
3:     unsigned short dist_from_root;
4:     node_list neighboring_nodes, full_node_set;
5:     unsigned int sleep_interval=1000;
       //Initialization
6:     full_node_set=get_available_nodes();
7:     for (node temp=get_first(full_node_set); temp!=NULL; temp=get_next(full_node_set))
8:        self=get_local_node_id();
9:        if (temp==root)
10:          dist_from_root=0; parent=self;
11:       else dist_from_root=INF;
12:       neighboring_nodes=create_node_list(get_neighbors(temp));
13:    full_node_set=get_available_nodes();
14:    for (node iter1=get_first(full_node_set); iter1!=NULL; iter1=get_next(full_node_set))
15:       for(;;)   //Event Loop
16:          sleep(sleep_interval);
17:          for (node iter2=get_first(neighboring_nodes); iter2!=NULL; iter2=get_next(neighboring_nodes))
18:             if (dist_from_root@iter2+1<dist_from_root)
19:                dist_from_root=dist_from_root@iter2+1;
20:                parent=iter2;
```

Fig. 3. Procedural Code for Building a Shortest-path Routing Tree

3.2   Examples of Programming with Kairos

We now illustrate Kairos' expressibility and flexibility by describing how Kairos may be used to program three different distributed computations that have been proposed for sensor networks: routing tree construction, localization, and vehicle tracking.

Routing Tree Construction: In Figure 3, we illustrate a *complete* Kairos program for building a routing tree with a given root node. We have implemented this algorithm, and evaluate its performance in Section 4. Note that our program implements shortest-path routing, rather than selecting paths based on link-quality metrics [30]: we have experimented with the latter as well, as we describe below.

The code shown in Figure 3 captures the essential functionality involved in constructing a routing tree while maintaining brevity and clarity. It shows how a centralized Kairos task looks, and illustrates how the Kairos primitives are used to express such a task. Program variable dist_from_root is the only variable that needs to be remotely accessed in lines 18-19, and is therefore a *managed object* at a source node and a *cached object* at the one-hop neighbors of the source node that programmatically read this variable. The program also shows how the node and node_list datatypes and their API's are used. get_available_nodes() in lines 6 and 13 instructs the Kairos preprocessor to include the enclosed code for each iterated node; it also provides an iterator handle that can be used for addressing nodes from the iterator's perspective, as shown in line 12. Finally, the program shows how the get_neighbors() function is used in line 12 to acquire the one-hop neighbor list at every node.

The event loop between lines 15-20 that runs at all nodes eventually picks a shortest path from a node to the root node. Our implementation results show that the path monotonically converges to the optimal path, thereby demonstrating progressive correctness. Furthermore, the path found is stable and does not change unless there are transient or permanent link failures that cause nodes to be intermittently unreachable.

This event loop illustrates how Kairos leverages eventual consistency. The access to the remote variable dist_from_root need not be synchronized at every step of the iteration; the reader can use the current cached copy, and use a lazy update mechanism to avoid overhead. As we shall see in Section 4, the convergence performance and the

```
1: void CooperativeMultilateration()

2:   boolean localized=false, not_localizable=false, is_beacon=GPS_available();
3:   node self=get_local_node_id();
4:   graph subgraph_to_localize=NULL;

5:   node_list full_node_set=get_available_nodes();
6:   for (node iter=get_first(full_node_set); iter!=NULL; iter=get_next(full_node_set)))
       //At each node, start building a localization graph
7:       participating_nodes=create_graph(iter);
8:       node_list neighboring_nodes=get_neighbors(iter);
9:       while ((!localized || !is_beacon) && !not_localizable)
10:         for (node temp=get_first(neighboring_nodes); temp!=NULL; temp=get_next(neighboring_nodes))
             //Extend the subgraph with neighboring nodes
11:           extend_graph(subgraph_to_localize, temp, localized@temp||is_beacon@temp?beacon:unknown);
             //See if we can localize the currently available subgraph
12:         if (graph newly_localized_g=subgraph_check(subgraph_to_localize))
13:           node_list newly_localized_l=get_vertices(newly_localized_g);
14:           for (node temp=get_first(newly_localized_l); temp!=NULL; temp=get_next(newly_localized_l))
15:             if (temp==iter) localized=true;
16:               continue;
             //If not, add nodes adjacent to the leaves of the accumulated subgraph and try again
17          node_list unlocalized_leaves;
18:         unlocalized_leaves=get_leaves(subgraph_to_localize);
19:         boolean is_extended=false;
20:         for (node temp=get_first(unlocalized_leaves); temp!=NULL; temp=get_next(unlocalized_leaves))
21:           node_list next_hop_l=get_neighbors(temp);
22:           for (node temp1=get_first(next_hop_l); temp1!=NULL; temp1=get_next(next_hop_l))
23:             extend_graph(subgraph_to_localize, temp1, localized@temp1||is_beacon@temp1?beacon:unknown);
24:             is_extended=true;
25:         if (!is_extended) not_localizable=true;
```

Fig. 4. Procedural Code for Localizing Sensor Nodes

message overhead of loose synchrony in real-world experiments is reasonable. We also tried metrics other than shortest hop count (such as fixing parents according to available bandwidth or loss rates, a common technique used in real-world routing systems [3]), and we found that the general principle of eventual consistency and loose synchrony can be applied to such scenarios as well.

Let us examine Figure 3 for the flexibility programming to the Kairos model affords. If we want to change the behavior of the program to have the tree construction algorithm commence at a pre-set time that is programmed into a base station node with id 0, we could add a single line before the start of the `for()` loop at line 7: `sleep(starting_time@0-get_current_time())`. The runtime would then automatically fetch the `starting_time` value from node 0.

Distributed Localization Using Multi-lateration: Figure 4 gives a complete distributed program for collaboratively fixing the locations of nodes with unknown coordinates. The basic algorithm was developed by Savvides *et al.* [2]. Our goal in implementing this algorithm in Kairos was to demonstrate that Kairos is flexible and powerful enough to program a relatively sophisticated distributed computation. We also wanted to explore how difficult it would be to program a "textbook" algorithm in Kairos, and compare the relative performance of Kairos with the reported original version (Section 4).

The goal of the "cooperative multi-lateration" algorithm is to compute the locations of all unknown nodes in a connected meshed wireless graph given ranging measurements between one-hop neighboring nodes and a small set of beacon nodes that already know their position. Sometimes, it may happen that there are not enough beacon nodes in the one-hop vicinity of an unknown node for it to mathematically laterize its location. The basic idea is to iteratively search for enough beacons and unknown nodes in the network graph so that, taken together, there are enough measurements and known co-ordinates to successfully deduce the locations of all unknown nodes in the sub-graph.

Figure 4 shows the complete code for the cooperative multi-lateration algorithm.[1] The code localizes non-beacon nodes by progressively expanding the subgraph, (subgraph_to_localize), considered at a given node with next-hop neighbors of un-localized leaf vertexes (unlocalized_leaves), and is an implementation of Savvides' algorithm [2]. The process continues until either all nodes in the graph are considered (lines 20-25) and the graph is deemed unlocalizable, or until the initiator localizes itself (using the auxiliary function subgraph_check()) after acquiring a sufficient number of beacon nodes. This program once again illustrates eventual consistency because the variable localized@node is a monotonic boolean, and eventually attains its correct asymptotic value when enclosed in an event loop. We also found an interesting evi-dence to the value of Kairos' centralized global program specification approach—we encountered a subtle logical (corner-case recursion) bug in the original algorithm de-scribed in [2] in a local (*i.e.,* bottom-up, node-specific) manner, that became apparent in Kairos.

Vehicle Tracking: For our final example, we consider a qualitatively different applica-tion: tracking moving vehicles in a sensor field. The program in Figure 5 is a straight-forward translation of the algorithm described in [14]. This algorithm uses probabilistic techniques to maintain belief states at nodes about the current location of a vehicle in a sensor field. Lines 14-16 correspond to step 1 of the algorithm given in [14–p. 7] where nodes diffuse their beliefs about the vehicle location. Lines 17-21 compute the probability of the observation z_{t+1} at every grid location given vehicle location x_{t+1} at time $t + 1$ (step 2 of the algorithm) using the latest sensing sample and vehicle dynam-ics. Lines 23-25 compute the overall posteriori probability of the vehicle position on the rectangular grid after incorporating the latest posteriori probability (step 3 of the algorithm). Finally, lines 26-40 compute the information utilities, I_k's, at all one-hop neighboring nodes k for every node, and pick that $k = \text{argmax}\_I_k$ that maximizes this measure (steps 4 and 5). This node becomes the new "master" node: *i.e.,* it executes the steps above for the next epoch, using data from all other nodes in the process.

This program illustrates an important direction of future work in Kairos. In this al-gorithm, the latest values of $p(z_{t+1}|\overline{x_{t+1}})[x][y]$@neighbors must be used in line 33 at the master because these $p(.)[x][y]$'s are computed at each sensor node using the latest vehicle observation sample. With our loose synchronization model, we cannot insure that the master uses these latest values computed at the remote sensor nodes because stale cached values may be returned instead by the master Kairos runtime, thereby ad-versely impacting the accuracy and convergence time of the tracking application. There are two possible solutions to this. One, which we have implemented currently in Kairos, is to provide a slightly tighter synchronization model that we call *loop-level synchrony*, where variables are synchronized at the beginning of an event loop (at line 11 of every iteration). A more general direction, which we have left for future work is to explore *temporal data abstractions*. These would allow programmers to express which samples

---

[1] Of course, we have not included the low-level code that actually computes the range estimates using ultrasound beacons. Our code snippet assumes the existence of node-local OS/library support for this purpose.

```
 1: void track_vehicle()
 2:    boolean master=true;
 3:    float z_{t+1}, normalizing_const;
 4:    float p(x_t|z̄_T)[MAX_X][MAX_Y], p(x_{t+1}|z̄_T)[MAX_X][MAX_Y],
         p(z_{t+1}|z̄_{T+1})[MAX_X][MAX_Y], p(x_{t+1} z^k_{t+1}|z̄_T)[MAX_X][MAX_Y], p(z^k_{t+1}|z̄_T), p(x_{t+1}|z̄_{T+1})[MAX_X][MAX_Y];
 5:    float max_I_k=I_k; node argmax_I_k, self=get_local_node_id();
 6:    node_list full_node_set=get_available_nodes();
 7:    for (node iter=get_first(full_node_set); iter!=NULL; iter=get_next(full_node_set))
 8:      for (int x=0; x<MAX_X; x++)
 9:        for (int y=0; y<MAX_Y; y++)
10:          p(x_t|z̄_T)[x][y]= \frac{1}{MAX\_X \times MAX\_Y};
11:    for(;;)
12:      sleep();
13:      if (master)
14:        for (int x=0; x<MAX_X; x++)
15:          for (int y=0; y<MAX_Y; y++)
16:            p(x_{t+1}|z̄_T)[x][y]= \sum_{0 \le x' < MAX\_X} \sum_{0 \le y' < MAX\_Y} \frac{\delta(\sqrt{x'^2+y'^2} - \sqrt{x^2+y^2} - v)p(x_t|z̄_T)}{\delta(\sqrt{x'^2+y'^2} - \sqrt{x^2+y^2} - v)};
17:      z_{t+1}=sense_z();
18:      normalizing_const=0;
19:      for (int x=0; x<MAX_X; x++)
20:        for (int y=0; y<MAX_Y; y++)
21:          p(z_{t+1}|x_{t+1})[x][y]= \frac{r}{\delta_a}\left[\Phi\left(\frac{a_{hi}-rz}{r\sigma}\right) - \Phi\left(\frac{a_{lo}-rz}{r\sigma}\right)\right];
22:          normalizing_const+=p(z_{t+1}|x_{t+1})[x][y]·p(x_{t+1}|z̄_T)[x][y];
23:      for (int x=0; x<MAX_X; x++)
24:        for (int y=0; y<MAX_Y; y++)
25:          p(x_{t+1}|z̄_{T+1})[x][y]= \frac{p(z_{t+1}|x_{t+1})[x][y]·p(x_{t+1}|z̄_T)[x][y]}{normalizing\_const};
26:      node_list neighboring_nodes=get_neighbors(iter);
27:      append_to_list(neighboring_nodes, self);
28:      max_I_k=-∞; argmax_I_k=self;
29:      for (node temp=get_first(neighboring_nodes); temp!=NULL; temp=get_next(neighboring_nodes))
30:        p(z^k_{t+1}|z̄_T)=0;
31:        for (int x=0; x<MAX_X; x++)
32:          for (int y=0; y<MAX_Y; y++)
33:            p(x_{t+1} z^k_{t+1}|z̄_T)[x][y]=p(z_{t+1}|z̄_{T+1})[x][y]@temp·p(x_{t+1}|z̄_T)[x][y];
34:            p(z^k_{t+1}|z̄_T)+=p(x_{t+1} z^k_{t+1}|z̄_T)[x][y];
35:        for (int x=0; x<MAX_X; x++)
36:          for (int y=0; y<MAX_Y; y++)
37:            I_k+= \log\left[\frac{p(x_{t+1} z^k_{t+1}|z̄_T)[x][y]}{p(x_{t+1}|z̄_T)[x][y]p(z^k_{t+1}|z̄_T)}\right]·p(x_{t+1} z^k_{t+1}|z̄_T)[x][y];
38:        if (max_I_k<I_k) argmax_I_k=temp;
39:      if (argmax_I_k!=self) master=false;
40:      master@argmax_I_k=true;
```

Fig. 5. Procedural Code for Vehicle Tracking

of the time series $p(.)[x][y]$ from remote nodes are of interest, while possibly allowing Kairos to preserve loose synchrony.

4   Kairos Evaluation

We have implemented the programming primitives discussed in the previous section in Python using its embedding and extendability API's [31], and have experimented with the three distributed algorithms described therein. More discussion about our implementation and evaluation can be found in [13]. Our testbed is a hybrid network of ground nodes and nodes mounted on a ceiling array. The 16 ground nodes are Stargates [32] that each run Kairos. In this setup, Kairos uses Emstar [33] to implement end-to-end reliable routing and topology management. Emstar, in turn, uses a Mica2 mote [34] mounted on the Stargate node (the leftmost picture in Figure 6 shows a single Stargate+Mica2 node) as the underlying network interface controller (NIC) to achieve realistic multihop wireless behavior. These Stargates were deployed in a small area (middle picture in Figure 6), making all the nodes reachable from any other node in

Fig. 6. Stargate with Mica2 as a NIC (left), Stargate Array (middle), and Ceiling Mica2dot Array (right)

a single physical hop (we created logical multihops over this set in the experiments below). The motes run TinyOS [35], but with S-MAC [36] as the MAC layer.

There is also an 8-node array of Mica2dots [37] mounted on a ceiling (rightmost picture in Figure 6), and connected through a multiport serial controller to a standard PC that runs 8 Emstar processes. Each Emstar process controls a single Mica2dot and is attached to a Kairos process that also runs on the host PC. This arrangement allows us to extend the size of the evaluated network while still maintaining some measure of realism in wireless communication. The ceiling Mica2dots and ground Mica2s require physical multihopping for inter-node communication. The Mica2dot portion of the network also uses physical multihopping for inter-node communication.

To conduct experiments with a variety of controlled topologies, we wrote a topology manager in Emstar that enables us to specify neighbors for a given node and blacklist/whitelist a given neighbor. Dynamic topologies were simulated by blacklisting/whitelisting neighbors while the experiment was in progress. The end-to-end reliable routing module keeps track of all the outgoing packets (on the source node) and periodically retransmits the packets until an acknowledgment is received from the destination. Hop-by-hop retransmission by S-MAC is complementary and used as a performance enhancement.

Routing Tree Performance: We implemented the routing tree described in Section 3.2 in Kairos, and measured its performance. For comparison purposes, we also implemented One Phase Pull (OPP) [38] routing directly in Emstar. OPP forms the baseline case because it is the latest proposed refinement for directed-diffusion that is designed to be traffic-efficient by eliminating exploratory data messages: the routing tree is formed purely based on interest–requests (interest messages in directed diffusion) that are flooded to the network and responses (data) are routed along the gradients setup by the interest. To enable a fair comparison of the Kairos routing tree with OPP, we also implemented reliable routing for OPP.

We varied the number of nodes in our network, and measured the time it takes for the routing tree in each case to stabilize (convergence time), and the overhead incurred in doing so. In the case of OPP, the resulting routing tree may not always be the shortest path routing tree (directed diffusion does not require that), while Kairos always builds a correct shortest path routing tree. So we additionally measure the "stretch" (the averaged node deviation from the shortest path tree) of the resulting OPP tree with respect to

Fig. 7. Convergence Time (left), Overhead (middle), and OPP Stretch (right) for the Routing Tree Program

the Kairos shortest path tree. Thus, this experiment serves as a benchmark for efficiency and correctness metrics for Kairos' eventual consistency model.

We evaluated two scenarios: first to build a routing tree from scratch on a quiescent network, and second to study the dynamic performance when some links are deleted after the tree is constructed. Figure 7 shows the convergence time ("K" is for Kairos, and "before" and "after" denote the two scenarios before and after link failures), overhead, and stretch plots for OPP and Kairos averaged across multiple runs; for stretch, we also plot the OPP standard deviation. It can be seen that Kairos always generates a better quality routing tree than OPP (OPP stretch is higher, especially as the network size increases) without incurring too much higher convergence time ($\sim$30%) and byte overhead costs ($\sim$2x) than OPP.

Localization: We have implemented the collaborative multilateration algorithm described in Section 3.2. Since we did not have the actual sensors (ultrasound and good radio signal strength measurement) for ToA (Time of Arrival) ranging, we hard-coded the pairwise distances obtained from a simulation as variables in the Kairos program instead of acquiring them physically. We believe this is an acceptable artifact that does not compromise the results below. We perturbed the pairwise distances with white Gaussian noise (standard deviation 20mm to match experiments in [2]) to reflect the realistic inaccuracies incurred with physical ranging devices.

We consider two scenarios in both of which we vary the total number of nodes. In the first case (left graph in Figure 8), we use topologies in which all nodes are localizable given a sufficient number and placement of initial beacon nodes, and calculate the

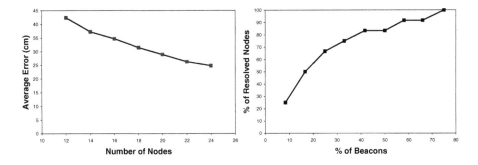

Fig. 8. Average Error in Localization (L) and Localization Success Rate (R)

Table 1. Performance of Vehicle Tracking in Kairos

K	Avg $\|\hat{x}_{MMSE} - x\|$	Avg $\|\hat{x} - \hat{x}_{MMSE}\|^2$	Avg Overhead (bytes)
12	42.39	1875.47	135
14	37.24	1297.39	104
16	34.73	1026.43	89
18	31.52	876.54	76
20	28.96	721.68	67
22	26.29	564.32	60
24	24.81	497.58	54

average localization error for a given number of nodes. The average localization error in Kairos is within the same order shown in [2–Figure 9], thereby confirming that Kairos is competitive here. Note that this error decreases with increasing network size as expected because the Gaussian noise introduced by ranging is decreased at each node by localizing with respect to multiple sets of ranging nodes and averaging the results. In the second scenario (right graph in Figure 8), we vary the percentage of initial beacon nodes for the full 24 node topology, and calculate how many nodes ultimately become localizable. This graph roughly follows the pattern exhibited in [2–Figure 12], thereby validating our results again.

Vehicle Tracking: For this purpose, we use the same vehicle tracking parameters as used in [14] (for grid size, vehicle speed, sound RMS, acoustic sensor measurement simulations, sensor placement and connectivity, and Direction-of-Arrival sensor measurements) for comparing how Kairos performs against [14]. A direct head-to-head comparison against the original algorithm is not possible because we have fewer nodes than they have in their simulations, so we present the results of the implementation in a tabular form similar to theirs. We simulate the movement of a vehicle along the Y-axis at a constant velocity. The motion therefore perpendicularly bisects the X-axis.

We do two sets of experiments. The first one is to measure the tracking accuracy as denoted by the location error ($\|\hat{x}_{MMSE} - x\|$) and its standard deviation ($\|\hat{x} - \hat{x}_{MMSE}\|^2$) as well as the tracking overhead as denoted by the belief state as we vary the number of sensors (K). The main goal here is to see whether we observe good performance improvement as we double the number of sensors from 12 to 24. As table 1 shows, this is indeed the case: the error, error deviation, and exchanged belief state all decrease, and in the expected relative order.

In the second experiment, we vary the percentage of sensor nodes that are equipped with sensors that can do Direction-of-arrival (DOA) based ranging, and not just direction-agnostic circular acoustic amplitude sensors. These results are described in [13].

5   Conclusion and Future Work

This paper should be viewed as an initial exploration into a particular model of macro-programming sensor networks. Our contribution in this paper is introducing, describing, and evaluating this model on its expressivity, flexibility, and real-world performance metrics. Kairos is not perfect in that, at least in its current incarnation, it does not fully shield programmers from having to understand the performance and robustness implications of structuring programs in a particular way; nor does it currently provide handles to let an application control the underlying runtime resources for predictability, resource management, or performance reasons. Finally, while Kairos includes a middleware communication layer in the runtime service that shuttles serialized program variables and objects across realistic multihop radio links, today, this layer lacks the ability to optimize communication patterns for a given sensornet topology. Therefore, we believe that Kairos opens up several avenues of research that will enable us to explore the continuum of tradeoffs between transparency, ease of programming, performance, and desirable systems features arising in macroprogramming a sensor network.

References

1. J. Elson, L. Girod, and D. Estrin, "Fine-grained network time synchronization using reference broadcasts," *OSDI, 2002.*
2. A. Savvides, C. Han, and S. Srivastava, "Dynamic Fine-Grained localization in Ad-Hoc networks of sensors," *MOBICOM, 2001.*
3. A. Woo, T. Tong, and D. Culler, "Taming the underlying challenges of reliable multihop routing in sensor networks," *SenSys, 2003.*
4. N. Xu, S. Rangwala, K. Chintalapudi, D. Ganesan, A. Broad, R. Govindan, and D. Estrin, "A wireless sensor network for structural monitoring," *SenSys, 2004.*
5. A. Mainwaring, J. Polastre, R. Szewczyk, D. Culler, and J. Anderson, "Wireless sensor networks for habitat monitoring," *WSNA, 2002.*
6. A. Cerpa, J. Elson, D. Estrin, L. Girod, M. Hamilton, and J. Zhao, "Habitat monitoring: application driver for wireless communications technology," *SIGCOMM Comput. Commun. Rev., 2001.*
7. K. Whitehouse, C. Sharp, E. Brewer, and D. Culler, "Hood: a neighborhood abstraction for sensor networks," *MobiSys, 2004.*
8. M. Welsh and G. Mainland, "Programming sensor networks using abstract regions," *NSDI, 2004.*
9. S. Madden, M. J. Franklin, J. Hellerstein, and W. Hong, "TAG: A tiny AGgregation service for ad-hoc sensor networks," *OSDI, 2002.*
10. S. Madden, M. J. Franklin, J. M. Hellerstein, and W. Hong, "The design of an acquisitional query processor for sensor networks," *SIGMOD, 2003.*
11. W. F. Fung, D. Sun, and J. Gehrke, "Cougar: the network is the database," *SIGMOD, 2002.*
12. S. V. Adve and K. Gharachorloo, "Shared memory consistency models: A tutorial," *IEEE Computer*, vol. 29, no. 12, pp. 66–76, 1996.
13. R. Gummadi, O Gnawali, and R Govindan, "Macro-programming Wirless Sensor Networks using Kairos," *USC CS Technical Report Number 05-848*, 2005, http://www.cs.usc.edu/Research/ReportsList.htm.
14. J. Reich J. Liu and F. Zhao, "Collaborative in-network processing for target tracking," *EURASIP, 2002.*
15. J. Liu, M. Chu, J. Liu, J. Reich, and F.Zhao, "State-centric programming for sensor and actuator network systems," *IEEE Perv. Computing, 2003.*
16. E. Cheong, J. Liebman, J. Liu, and F. Zhao, "Tinygals: a programming model for event-driven embedded systems," *SAC, 2003.*

17. T. Abdelzaher, B. Blum, Q. Cao, Y. Chen, D. Evans, J. George, S. George, L. Gu, T. He, S. Krishnamurthy, L. Luo, S. Son, J. Stankovic, R. Stoleru, and A. Wood, "Envirotrack: Towards an environmental computing paradigm for distributed sensor networks," *ICDCS, 2004.*

18. A. Bakshi, J. Ou, and V. K. Prasanna, "Towards automatic synthesis of a class of application-specific sensor networks," *CASES, 2002.*

19. A. Bakshi and V. K. Prasanna, "Algorithm design and synthesis for wireless sensor networks," *ICPP, 2004.*

20. R. Kumar, M. Wolenetz, B. Agarwalla, J. Shin, P. Hutto, A. Paul, and U. Ramachandran, "Dfuse: a framework for distributed data fusion," *SenSys, 2003.*

21. R. Newton and M. Welsh, "Region streams: Functional macroprogramming for sensor networks," *DMSN, 2004.*

22. http://www.cs.berkeley.edu/projects/parallel/castle/split-c/

23. A. Boulis, C. Han, and M. B. Srivastava, "Design and implementation of a framework for efficient and programmable sensor networks," *MobiSys, 2003.*

24. B. Greenstein, E. Kohler, and D. Estrin, "A sensor network application construction kit (SNACK)," *SenSys, 2004.*

25. J. W. Hui and D. Culler, "The dynamic behavior of a data dissemination protocol for network programming at scale," *SenSys, 2004.*

26. P. Levis and D. Culler, "Maté: a tiny virtual machine for sensor networks," *ASPLOS-X, 2002.*

27. T. Liu, C. M. Sadler, P. Zhang, and M. Martonosi, "Implementing software on resource-constrained mobile sensors: experiences with impala and zebranet," *MobiSys, 2004.*

28. D. B. Terry, M. M. Theimer, K. Petersen, Demers Demers, M. J. Spreitzer, and C. Hauser, "Managing update conflicts in Bayou, a weakly connected replicated storage system," pp. 172–183, Dec. 1995.

29. P. Levis, N. Patel, D. Culler, and S. Shenker, "Trickle: A self-regulating algorithm for code propagation and maintenance in wireless sensor networks," *NSDI, 2004.*

30. Douglas S. J. De Couto, Daniel Aguayo, John Bicket, and Robert Morris, "A high-throughput path metric for multi-hop wireless routing," in *Proceedings of the 9th ACM International Conference on Mobile Computing and Networking (MobiCom '03)*, San Diego, California, September 2003.

31. Guido van Rossum and Fred L. Drake Jr., editors, "Extending and embedding the python interpreter," http://docs.python.org/ext/ext.html.

32. http://www.xbow.com/Products/XScale.htm.

33. J. Elson, S. Bien, N. Busek, V. Bychkovskiy, A. Cerpa, D. Ganesan, L. Girod, B. Greenstein, T. Schoellhammer, T. Stathopoulos, and D. Estrin, "Emstar: An environment for developing wireless embedded systems software," *CENS-TR-9, 2003.*

34. http://www.xbow.com/Products/productsdetails.aspx?sid=72.

35. J. Hill, R. Szewczyk, A. Woo, S. Hollar, D. Culler, and K. Pister, "System architecture directions for networked sensors," *SIGOPS Oper. Syst. Rev., 2000.*

36. W. Ye and J. Heidemann, "Medium access control in wireless sensor networks," *ISI-TR-580, 2003.*

37. http://www.xbow.com/Products/productsdetails.aspx?sid=73.

38. J. Heidemann, F. Silva, and D. Estrin, "Matching data dissemination algorithms to application requirements," *SenSys, 2003.*

Sensor Network Calculus – A Framework for Worst Case Analysis

Jens B. Schmitt[1] and Utz Roedig[2]

[1] Distributed Computer Systems Lab,
University of Kaiserslautern, Germany
[2] Mobile & Internet Systems Laboratory,
University College Cork, Ireland

Abstract. To our knowledge, at the time of writing no methodology exists to dimension a sensor network so that a worst case traffic scenario can be definitely supported. In this paper, the well known network calculus is tailored so that it can be used as a tool for worst case traffic analysis in sensor networks. To illustrate the usage of the resulting sensor network calculus, typical example scenarios are analyzed by this new methodology. Sensor network calculus provides the ability to derive deterministic statements about supportable operation modes of sensor networks and the design of sensor nodes.

1 Introduction

1.1 Motivation

Decisions in daily life are based on the accuracy and availability of information. Sensor networks can significantly improve the quality of information as well as the ways of gathering it. For example sensor networks can help to get higher fidelity information, acquire information in real time, get hard-to-obtain information and reduce the cost of obtaining information. Therefore it is commonly assumed that sensor networks will be applied in many different areas in the future and they can be viewed as an important part in the vision of ubiquitous/pervasive computing [1].

Application areas for sensor networks might be production surveillance, traffic management, medical care or military applications. In these areas it is crucial to ensure that the sensor network is functioning even in a worst case scenario. It must be clear that the sensor network can support all possible communication patterns that might occur in the network without being overloaded. If a sensor network is used for example for production surveillance it must be ensured that messages indicating a dangerous condition are not dropped. If functionality in worst case scenarios cannot be proven, people might be in danger and the production system might not be certified by authorities.

As it may be difficult or even impossible to produce the worst case in a real world scenario or in a simulation in a controlled fashion an analytical framework

V. Prasanna et al. (Eds.): DCOSS 2005, LNCS 3560, pp. 141–154, 2005.

is desirable that allows a worst case analysis in sensor networks. Network calculus [4] is a relatively new tool that allows worst case analysis of packet-switched communication networks. Network calculus has successfully been applied to model wired IP-based networks built on QoS technologies like Integrated Services or Differentiated Services [2],[3].

1.2   Goals

Our long-term goal is to develop a *sensor network calculus* that allows an analytical investigation of performance-related characteristics of wireless sensor networks. This paper represents the first step towards such a framework. As sensor networks differ in many aspects from traditional wired IP-based networks, existing results from traditional network calculus cannot be transferred directly. In particular, sensor networks have different constraints, such as battery powered nodes and dynamic topologies. It is necessary to incorporate these constraints in the framework, so that they can be analyzed. For these reasons the following is presented in this paper:

> *An analytical framework based on network calculus to dimension sensor networks, in particular taking into account the various trade-offs and interdependencies between node power consumption, node buffer requirements and information transfer delay.*

1.3   Problem Scope

Different applications running on top of a sensor network might have different requirements regarding the information extracted from the field. One important requirement is a bound on the maximum *information transfer delay* for data delivery. If information is delayed too long on the transport path, the application cannot use the information as it is considered out-dated. At each hop of the transport path, a message can be delayed. If a sensor node that generates a message is some hops away from the sink, the message delay accumulates over the hops. If several messages are delayed in one node at the same time, buffer space for the messages must be available. Thus, information transfer delay is correlated with the *buffer requirements* of a sensor node. The delay in each node is caused by two interdependent aspects. First, the delay depends on the traffic that a node has to process (*arrival rate*). Second, the node needs a specific amount of time to receive, process and send a message (*service rate*). The first aspect depends on the *network topology*, as the traffic that enters a node might be generated by several other nodes. The second aspect is dominated by the reception delay caused by the common usage of *duty cycles* which defines the *power consumption* of a sensor node.

Most energy in a sensor node is used for communication and one very effective and generic way to solve the problem is to optimize power consumption on the data link layer by implementing duty cycles. Thereby, the receiver alters periodically between a power intensive idle mode (idle duration T_1) and a power saving sleep mode (sleep duration T_2) state. Thus, the duty cycle δ is defined as $\delta = T_1/(T_1 + T_2) \times 100[\%]$. To transmit, a sender has to catch the receiver in its idle phase. This is achieved by using a long preamble in front of each single message notifying the receiver of an incoming message. The need for the long preamble implies that the effective channel bandwidth is reduced and the information transfer delay is increased. The proportion between sleep and idle phase defines the power consumption of a node and the possible forwarding rate (service curve). The usage of duty cycles allows the lifetime of a node to be stretched from several days to over one year [6]. Therefore nearly all practical sensor networks in use today implement duty cycles as a method to extend the network lifetime [13],[8].

1.4   Outline

In the remaining paper it is shown how network calculus can be tailored and extended so that a worst case analysis of the relevant quantities in sensor networks is possible. Section 2 gives a brief summary of the network calculus and the basic sensor network calculus approach is presented. Section 3 shows in detail how the model is instantiated and applied. Section 4 presents some use cases that show how the model can be used to analyze common real-world scenarios. Section 5 presents related work and Section 6 concludes the paper.

2   Sensor Network Calculus

2.1   Background on Network Calculus

Network calculus is the tool to analyze flow control problems in networks with particular focus on determination of bounds on worst case performance. It has been successfully applied as a framework to derive deterministic guarantees on throughput, delay, and to ensure zero loss in packet-switched networks [4]. Network calculus can also be interpreted as a system theory for *deterministic* queueing systems, based on min-plus algebra. What makes it different from traditional queueing theory is that it is concerned with worst case rather than average case or equilibrium behaviour. It thus deals with bounding processes called arrival and service curves rather than arrival and departure processes themselves.

Next some basic definitions and notations are provided before some basic results from network calculus are summarized. In depth results can be found in [4].

Definition 1. *The input function $R(t)$ of an arrival process is the number of bits that arrive in the interval $[0, t]$. In particular $R(0) = 0$, and R is wide-sense increasing, i.e. $R(t_1) \leq R(t_2)$ for all $t_1 \leq t_2$.*

Definition 2. *The output function $R^*(t)$ of a system S is the number of bits that have left S in the interval $[0, t]$. In particular $R^*(0) = 0$, and R is wide-sense increasing, i.e. $R^*(t_1) \leq R^*(t_2)$ for all $t_1 \leq t_2$.*

Definition 3. *Min-Plus Convolution. Let f and g be wide-sense increasing and* $f(0) = g(0) = 0$. *Then their convolution under min-plus algebra is defined as*

$$(f \otimes g)(t) = \inf_{0 \leq s \leq t} \{f(t - s) + g(s)\}$$

Definition 4. *Min-Plus Deconvolution. Let f and g be wide-sense increasing and* $f(0) = g(0) = 0$. *Then their deconvolution under min-plus algebra is defined as*

$$(f \oslash g)(t) = \sup_{s \geq 0} \{f(t + s) - g(s)\}$$

Now, by means of the min-plus convolution, the arrival and service curve are defined.

Definition 5. *Arrival Curve. Let* α *be a wide-sense increasing function such that for* $t < 0$. α *is an arrival curve for an input function R iff* $R \leq R \otimes \alpha$. *It is also said that R is* α-*smooth or R is constrained by* α.

Definition 6. *Service Curve. Consider a system S and a flow through S with R and* R^*. *S offers a service curve* β *to the flow iff* β *is wide-sense increasing and* $R^* \geq R \otimes \beta$.

From these, it is now possible to capture the major worst-case properties for data flows: maximum delay and maximum backlog. These are stated in the following theorems.

Theorem 1. *Backlog Bound. Let a flow* $R(t)$, *constrained by an arrival curve* α, *traverse a system S that offers a service curve* β. *The backlog* $x(t)$ *for all t satisfies:*

$$x(t) \leq \sup_{s \geq 0} \{\alpha(s) - \beta(s)\} = v(\alpha, \beta) \tag{1}$$

$v(\alpha, \beta)$ is also often called the vertical deviation between α and β.

Theorem 2. *Delay Bound. Assume a flow* $R(t)$, *constrained by arrival curve* α, *traverses a system S that offers a service curve* β. *At any time t, the virtual delay* $d(t)$ *satisfies:*

$$d(t) \leq \sup_{s \geq 0} \{\inf\{\tau \geq 0 : \alpha(s) \leq \beta(s + \tau)\}\} = h(\alpha, \beta) \tag{2}$$

$v(\alpha, \beta)$ is also often called the horizontal deviation between α and β.

As a system theory network calculus offers further results on the concatenation of network nodes as well as the output when traversing a single node. Especially the latter for which now the min-plus deconvolution is used will be of high importance in the sensor network setting as it potentially involves a so-called *burstiness increase* when a node is traversed by a data flow.

Theorem 3. *Output Bound. Assume a flow $R(t)$ constrained by arrival curve α traverses a system S that offers a service curve β. Then the output function is constrained by the following arrival curve*

$$\alpha^* = \alpha \oslash \beta \geq \alpha \tag{3}$$

Theorem 4. *Concatenation of Nodes. Assume a flow $R(t)$ traverses systems S_1 and S_2 in sequence where S_1 offers service curve β_1 and S_2 offers β_2. Then the resulting system S, defined by the concatenation of the two systems offers the following service curve to the flow:*

$$\beta = \beta_1 \otimes \beta_2 \tag{4}$$

2.2   Sensor Network System Model

In this paper the common class of single base station oriented operation models is assumed. Within the traffic that is modeled only the sensor reports are taken into account. Traffic generated from the base station towards the nodes (e.g. interests [16] to set up the network structure and configure the nodes) is explicitly not taken into account. This is considered feasible based on the assumption that the traffic flowing towards the sensors is magnitudes lower than traffic caused by the sensing events. Furthermore, it is assumed that the routing protocol being used forms a tree in the sensor network.[1]Hence N sensor nodes arranged in a directed acyclic graph are given.

Each sensor node i senses its environment and thus is exposed to an input function R_i corresponding to its sensed input traffic. If sensor node i is not a leaf node of the tree then it also receives sensed data from all of its child nodes $child(i, 1), \ldots, child(i, n_i)$, where n_i is the number of child nodes of sensor node i. Sensor node i forwards/processes its input which results in an output function R_i^* from node i towards its parent node.

Now the basic network calculus components, arrival and service curve, have to be incorporated. First the arrival curve $\bar{\alpha}_i$ of each sensor node in the field has to be derived. The input of each sensor node in the field, taking into account its sensed input and its childrens input, is given by:

$$\bar{R}_i = R_i + \sum_{j=1}^{n_i} R_{child(i\ j)}^* \tag{5}$$

Thus, the arrival curve for the total input function for sensor node i is given by:

$$\bar{\alpha}_i = \alpha_i + \sum_{j=1}^{n_i} \alpha_{child(i\ j)}^* \tag{6}$$

---

[1] If multiple sinks are assumed one simple way to use the following methodology would be to analyze each tree resulting from a given sink in isolation and later on additively combine interesting quantities as for example buffer requirements at a certain sensor node.

Candidates for actual arrival curves on the sensed input are discussed in Section 3.1.

Second, the service curve has to be specified. The service curve depends on the way packets are scheduled in a sensor node which mainly depends on link layer characteristics (see Section 1.3). More specific, the service curve depends on how the duty cycle and therefore the energy-efficiency goals are set.[2] Again the discussion of actual candidates for sensor node service curves is deferred to Section 3.2 when the whole sensor network calculus framework has been presented. Finally, the output of sensor node i, i.e. the traffic which it forwards to its parent in the tree, is constrained by the following arrival curve:

$$\alpha_i^* = \bar{\alpha}_i \oslash \beta_i = \left(\alpha_i + \sum_{j=1}^{n_i} \alpha_{child(i\ j)}^* \right) \oslash \beta_i \qquad (7)$$

In order to calculate a network-wide characteristic like the maximum information transfer delay or local buffer requirements especially at the most challenged sensor node just below the sink (which is called node 1 from now on) an iterative procedure to calculate the network internal flows is required:

1. Let us assume that arrival curves for the sensed input α_i and service curves β_i for sensor node i, $i = 1, \ldots, N$, are given.
2. For all leaf nodes the output bound α_i^* can be calculated according to (3). Each leaf node is now marked as "calculated".
3. For all nodes only having children which are marked "calculated" the output bound α_i^* can be calculated according to (7) and they can again be marked "calculated".
4. If node 1 is marked "calculated" the algorithm terminates, otherwise go to step 3.

After the network internal flows are computed according to this procedure, the local worst case buffer requirements B_i and per node delay bounds D_i for each sensor node i can be calculated according to Theorem 1 and 2:

$$B_i = v(\bar{\alpha}_i, \beta_i) = \sup_{s \geq 0} \{ \bar{\alpha}_i(s) - \beta_i(s) \} \qquad (8)$$

$$D_i = h(\bar{\alpha}_i, \beta_i) = \sup_{s \geq 0} \{ \inf \{ \tau \geq 0 : \bar{\alpha}_i(s) \leq \beta_i(s + \tau) \} \} \qquad (9)$$

To compute the total information transfer delay $\bar{D}_i$ for a given sensor node i the per node delay bounds on the path $P(i)$ to the sink need to be added:

$$\bar{D}_i = \sum_{i \in P(i)} D_i \qquad (10)$$

---

[2] The service curve might further depend on whether more advanced sensor network characteristics like in-network processing, e.g. for aggregation or even prioritization of some traffic is provided.

The maximum information transfer delay in the sensor network can then obviously be calculated as

$$\bar{D} = \max_{i=1 \ldots N} \bar{D}_i \tag{11}$$

Discussion. Readers very knowledgeable in network calculus may wonder about the hop-by-hop calculation of the total delay as specified in (11) and whether it would not be possible to derive a network-wide service curve based on the concatenation result of Theorem 4. While due to the traffic aggregation inside the network the concatenation result cannot be applied directly, there is in fact a way to still derive a network-wide service curve based on modified service curves that take into account the effects of cross-traffic on a data flow [4]. However, the bounds achieved in this way are not necessarily lower than for (11). This depends on the actual parameters of arrival and service curves. Furthermore, we believe that the hop-by-hop calculation will lend itself better towards integrating in-network processing into future, more elaborate extensions of the model.

Often, sensor network applications may regard message transfer delay only as a constraint and primarily care about maximizing their lifetime. The length of the duty cycle, and thus the energy consumption properties of the sensor nodes, are incorporated into the service curve as will be discussed in Section 3.2. Hence, instead of calculating delay bounds and buffer requirements as described above, the calculations could also start with a given delay/buffer requirement and work out the length of the duty cycle and thus the power consumption level and therefore the network lifetime.

3   Instantiating the Model

Before progressing to some numerical examples the abstract sensor network calculus model needs to be instantiated with concrete arrival and service curves. In the following subsections these crucial aspects and their influence on the worst case behaviour of the system are discussed in a qualitative fashion before in Section 4 more quantitative results are presented.

3.1   Arrival Curve Candidates

Maximum Sensing Rate. The simplest option in bounding the sensing input at a given sensor node is based on its maximum sensing rate which is either due to the way the sensing unit is designed or limited to a certain value by the sensor network application's task in observing a certain phenomenon. For example, it might be known that in a temperature surveillance sensor system, the temperature does not have to be reported more than once per second at most. The arrival curve for a sensor node i corresponding to simply putting a bound on the maximum sensing rate is given by

$$\alpha_i(t) = p_i t = \gamma_{p_i\ 0}(t) \tag{12}$$

Note that the assumption is made that each sensor node has its individual arrival curve respectively maximum sensing rate.

This arrival curve can be used in situations where all sensor nodes are set up to periodically report the condition in a sensor field. Thereby each sensor has a maximum possible rate with which the sensing information can be reported.

Average Sensing Rate. Depending on the sensor network application the maximum sensing rate arrival curve might lead to very conservative bounds if the maximum sensing rate is only rarely the actual sensing rate. In this situation it would be much more useful if the arrival curve could be based on the average sensing rate. Additionally there should be permission of some short-term fluctuations if the sensing must be intensified for certain periods of high activity in the field. However, in order to avoid the use of the maximum sensing rate arrival curve it is crucial that the time during which the average sensing rate may be exceeded can be upper bounded. In many applications that should be possible since after some time the phenomenon will disappear again or has to be acted on such that it disappears again (e.g. in a sensor network that also comprises actuators). The arrival curve that captures the average sensing rate with short-term fluctuations for sensor node i is given by

$$\alpha_i(t) = s_i t + b_i = \gamma_{s_i \, b_i}(t) \qquad (13)$$

This affine arrival curve can be shown to be equivalent to the famous token/leaky bucket as it is known from traditional traffic control [4]. It allows sensing at a higher rate than s_i for short periods of time but in the long run only allows sensing at the average rate s_i.

This arrival curve can be used to describe situations in which sensors usually report with a low rate. If a phenomenon is detected in the vicinity of the sensor, the sensing rate is increased for a fixed amount of time.

Discussion. The set of sensible arrival curve candidates is certainly larger than the arrival curves described above. The more knowledge on the sensing operation and its characteristics is incorporated into the arrival curve for the sensing input the better the worst case bounds become. We consider it a strength of the sensor network calculus framework that it is open with respect to arbitrary arrival curves. On the other hand, the options presented above may be sufficient in a large number of sensor network scenarios.

3.2   Service Curve Candidates

The service curve captures the characteristics with which sensor data is forwarded by the sensor nodes towards the sink. It abstracts from the specifics and idiosyncracies of the link layer and makes a statement on the minimum service that can be assumed even in the worst case.

Rate-Latency Service Curve. A typical and well known example of a service curve from traditional traffic control in a packet-switched network is given by

$$\beta_{R \, T}(t) = R(t - T)^+ \qquad (14)$$

where the notation $(x)^+$ denotes x if $x \geq 0$ and 0 otherwise. This is often also called a rate-latency service curve and results from the use of many popular packet schedulers (for example Weighted Fair Queueing (WFQ) [11]) many of which can be generalized as guaranteed rate or latency rate schedulers [9], [10]. While for sensor networks there may often be neither a necessity nor the resources (e.g. energy, computational power, memory capacity) for a sophisticated scheduling algorithm like WFQ, the class of rate-latency service curves is still very interesting. This is due to the fact that the latency term nicely captures the characteristics induced by the application of a duty cycle concept. Whenever the duty cycle approach is applied there is the chance that sensed data or data to be forwarded just arrives after the last duty cycle (of the next hop!) is just over and thus a fixed latency occurs until the forwarding capacity is available again. In the simple duty cycle scheme presented in Section 1.3 this latency would need to be accounted for for all data transfers since the preamble length is fixed, in schemes where the data is repeated for a certain amount of time and a feedback from the receiver signals when the sender can stop this repetition, the latency would really represent a worst case scenario as just mentioned. For the forwarding capacity it is assumed that it can be lower bounded by a fixed rate which depends on transceiver speed, the chosen link layer protocol and the duty cycle. So, with some new parameters the following service curve at sensor node i is obtained:

$$\beta_i(t) = \beta_{f_i, l_i}(t) = f_i(t - l_i)^+ \tag{15}$$

Here f_i and l_i denote the forwarding rate respectively forwarding latency for sensor node i.

4   Sensor Network Calculus at Work

In this section some numerical examples for the previously presented sensor network calculus framework are described. These examples are chosen with the intention of describing realistic and common application scenarios, yet they are certainly simplifying matters to some degree for illustrative purposes. As mentioned in the previous sections, the sensor network calculus framework allows, from a worst case perspective, to relate the following local characteristics:

- *Sensing Activity*: this parameter is described in the framework by the *arrival curve* concept;
- *Buffer Requirements*: the buffer requirements of each node are described by the *backlog bound*;

to the following global characteristics:

- *Information Transfer Delay*: the delay in each node is described by the *delay bound*;
- *Network Lifetime*: the energy consumption is described by the *duty cycle* represented in the *service curve*.

The goal in using sensor network calculus is to determine specific values for these characteristics for a given application scenario. The scenario itself is characterized by further constraints such as *topology* and *routing*.

4.1   Basic Scenario

The intention of this example is to analytically explore the possible range of the characteristics discussed above in a realistic scenario. Thereafter it is analyzed in which operation range a state of the art sensor node could be used to form the sensor field.

Topology and Routing. The sensor field is assumed to be a grid, the distance between the sensors is d. Fig. 1 shows the lower half of a grid shaped sensor field with the base station (sink) located in its center. The size of the field is $8d \times 8d$, containing $N = 80$ sensors each with an idealized transmission range of $\sqrt{2}d$.

For the routing protocol, the Greedy Perimeter Stateless Routing (GPSR) protocol is used [12]. All nodes in GPSR must be aware of their position within a sensor field. Each node communicates its current position periodically to its neighbors through beacon packets. In the given static scenario, these beacons have to be transmitted only once. Upon receiving a data packet, a node analyzes its geographic destination. If possible, the node always forwards the packet to the neighbor geographically closest to the packet destination. If there is no neighbor geographically closer to the destination, the protocol tries to route around the hole in the sensor field. This routing around a hole is not used in the described topology. In Fig. 1 the resulting structure of the communication paths is shown.

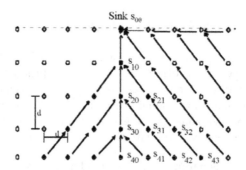

Fig. 1. Sensor Field with Grid Layout

Sensing Activity. It is assumed that the sensor field is used to collect data periodically from each of the sensors. Each sensor can report with a maximum report frequency of p. Thus, the maximum sensing rate arrival curve described by (12) is used to model the upper bound of the sensing activity of each node in the sensor field. A homogeneous field is assumed, hence

$$\alpha_i(t) = pt = \gamma_{p\,0}(t) \qquad (16)$$

Each node additionally receives traffic from its child nodes according to the traffic pattern implied by the topology and the routing protocol (see 1). Therefore, the arrival curve $\bar{\alpha}_i$ for the total input of a sensor node i is given by equation (6). Later it will be shown in detail how the relevant $\bar{\alpha}_i$ can be calculated.

Network Lifetime. To achieve a high network lifetime a duty cycle of $\delta = 1\%$ is set for the nodes in the network. As a sensor node, the Mica-2 [13] platform is assumed. Mica-2 supports a link speed of 19.2 kbit/s. The minimum idle time of the transceiver is $T_1 = 11[\text{ms}]$ (3ms to begin sampling, 8ms minimum preamble length), the corresponding sleep time is $T_2 = 1085[\text{ms}]$. Thus, a maximum packet forwarding rate of 0.89[packets/s] ($f = 258[\text{bit/s}]$) can be achieved.[3] The resulting latency for the packet forwarding is $l = T_1 + T_2$. This packet forwarding scheme can be described by the rate-latency service curve as described by equation (15) in Section 3.2:

$$\beta_i(t) = \beta_{f\,l}(t) = f(t - l)^+ = 258(t - 1.096)^+[\text{bit}] \tag{17}$$

Calculation. After defining the scenario, the sensor network calculus framework can now be used to evaluate the characteristics of interest and their interdependencies. Goal of the calculation is to determine these characteristics at the sensor node with the worst possible traffic conditions. In this example this is the node s_{10}. If the characteristics in this node are determined and the node is dimensioned to cope with them, all other nodes in the field (assuming homogeneity) are dimensioned properly as well.

To calculate the total traffic pattern, the algorithm described in Section 2.2 has to be used. First the output bound α_{40}^* of the leaf node s_{40} has to be calculated using (16), (17) and (3):

$$\alpha_{40} = \gamma_{p\,0}, \ \beta_{40} = \beta_{f\,l} = \beta, \ \alpha_{40}^* = \alpha_{40} \oslash \beta_{40} = \gamma_{p\,pl} \tag{18}$$

The output bound for node s_{40} is also the output bound for the other leaf nodes (e.g. $\alpha_{40}^* = \alpha_{41}^* = \alpha_{42}^* = \alpha_{43}^*$). Now the output bounds for the nodes one level higher in the tree can be calculated using equation (18), (16), (17) and (7):

$$\bar{\alpha}_{30} = \gamma_{p\,0} + 3\alpha_{40}^* = \gamma_{p\,0} + 3\gamma_{p\,pl} = \gamma_{4p\,3pl}, \ \alpha_{30}^* = \bar{\alpha}_{30} \oslash \beta = \gamma_{4p\,7pl} \tag{19}$$

The calculation can now be repeated until node s10 is reached: ...

$$\bar{\alpha}_{10} = \gamma_{p\,0} + 2\alpha_{21}^* + \alpha_{20}^* = \gamma_{16p\,34pl}, \ \alpha_{10}^* = \bar{\alpha}_{10} \oslash \beta = \gamma_{16p\,50pl} \tag{20}$$

After the arrival curve for node s_{10} is calculated, the worst case buffer requirements B_{10} and the information transfer delay D can be calculated according to equation (8) and (9):

---

[3] Values are taken from the TinyOS code (CC1000Const.h). The packet length is 36 bytes, the preamble length for 1% duty cycle is 2654 bytes.

$$B_{10} = v(\bar{\alpha}_{10}, \beta) = 50pl$$

$$D_{10} = h(\bar{\alpha}_{10}, \beta) = l + \frac{34pl}{f}, \quad D_{20} = h(\bar{\alpha}_{20}, \beta) = l + \frac{13pl}{f}$$

$$D_{30} = h(\bar{\alpha}_{30}, \beta) = l + \frac{3pl}{f}, \quad D_{40} = h(\bar{\alpha}_{40}, \beta) = l$$

$$D = D_{40} + D_{30} + D_{20} + D_{10} = 4l + \frac{50pl}{f}$$

Discussion. Now, after all nodes are calculated, it is possible to determine specific values for the characteristics of interest for the given application scenario. Furthermore it is possible to evaluate how these factors influence each other. As mentioned above, due to the channel speed and the selected duty cycle, the effective maximum forwarding speed is $f = 258[\text{bit/s}]$. The arrival rate of packets cannot be higher than the maximum forwarding speed. A higher arrival rate would result in an infinite queueing of packets. Therefore the sensing rate must be set such that $16p \leq f$. In the following, the highest possible integral sensing rate is assumed: $p = \lfloor f/16 \rfloor = 16[\text{bit/s}]$. This first result already shows the limits of this specific sensor field regarding its maximum sensing frequency. Translated in TinyOS packets with a standard size of 36 byte, the result shows that each sensor can only send a packet every 18 seconds.

The backlog bound at node s_{10} is now given by: $B_{10} = 50pl = 876.8[\text{bit}]$. This result can be translated into TinyOS packets with the standard size of 36 byte. In this case, $\lceil 3.04 \rceil = 4$ packets must be stored in the worst case in node s_{10}. As a Mica-2 node provides per default only a buffer space of one, a node modification would be necessary to support the described scenario in the worst case. The maximum information transfer delay is given by: $D = 4l + \frac{51pl}{f} = 7.85[\text{s}]$.

To improve the backlog bound and the information transfer delay, the duty cycle used in the nodes can be modified. Of course the improvements have to be paid in this case with a higher energy consumption in the nodes and thus a shorter network lifetime. If the duty cycle is set to 11.5%[4], a maximum packet forwarding rate of 0.54[packets/s] ($f = 2488[\text{bit/s}]$) can be achieved. The resulting delay for the packet forwarding is $l = T_1 + T_2 = 11 + 85 = 96[\text{ms}]$. Now the following is obtained: $B_{10} = 50pl = 76.8$. In this case now, only 1 TinyOS packets needs to be stored in node s_{10} even under worst case conditions. The information transfer delay is now given by: $D = 4l + \frac{51pl}{f} = 0.41[\text{s}]$.

---

[4] A duty cycle value offered by the TinyOS code for the Mica-2.

5   Related Work

To the knowledge of the authors, there are currently no tools available that allow a systematic and general analysis of the worst case traffic conditions in a sensor network. However, for the analytical investigation of very specific properties of a sensor network or sensor node prior work exists. Some research work proposes models that have the ability to predict the reliability of messages. In [14] for example it is shown how the reliability can be calculated and controlled by adjusting the message forwarding schemes. Another research field is the prediction of traffic patterns in sensor networks so that data aggregation can be performed. In [15] for example it is described how traffic patterns at nodes can be predicted and modified such that an efficient data aggregation is possible.

In this paper, network calculus is tailored for the purpose of establishing a very generic framework to analyze traffic patterns in a wireless sensor network. To our knowledge only one publication [17] exists in the context of wireless sensor networks that uses network calculus as analytical tool. The latter paper deals with the very different problem of developing a theoretically sound congestion control in distributed sensor networks. The authors make some basic observations on their flow controller using network calculus but do not consider to actually model the sensor network itself using network calculus which has been the goal of the paper at hand.

6   Conclusion and Outlook

An analytical framework based on network calculus to dimension sensor networks has been presented. Using real world examples it has been shown how to apply the framework for practical problems in sensor network dimensioning. More specifically, it has been demonstrated how the various trade-offs and interdependencies between node power consumption, node buffer requirements and information transfer delay can be described using the sensor network calculus framework.

In this paper, the focus has been set on network dimensioning. However, the presented framework can be used as well for e.g. on-line admission control. In such a case, the sensor network calculus framework can for example be used to dynamically decide if or if not an interest can be served by the sensor network. A further application might be in topology control, where the respective decisions may also be driven by how the system behaves in the worst case.

References

1. M. Weiser. The computer for the 21st century. Scientific American, 265(3):94-104, September 1991.
2. R. Braden, D. Clark, and S. Shenker. Integrated Services in the Internet Architecture: an Overview. RFC 1633, June 1994.

3. K. Nichols, S. Blake, F. Baker, and D. Black. Definition of the Differentiated Services Field (DS Field) in the IPv4 and IPv6 Headers. Proposed Standard RFC 2474, December 1998.
4. J.-Y. Le Boudec and P. Thiran. Network Calculus - A Theory of Deterministic Queueing Systems for the Internet. Springer, Lecture Notes in Computer Science, LNCS 2050, 2001.
5. Chipcon SmartRF CC1000 Data Sheet. Chipcon, www.chipcon.com. 2002.
6. Crossbow Technology INC. Smart Dust Training Seminar. Tutorial at the 1st IEEE European Workshop on Wireless Sensor Networks (EWSN2004), Berlin, Germany, January 2004.
7. J. Polastre. Design and Implementation of Wireless Sensor Networks for Habitat Monitoring. Masters Thesis. University of California, Berkeley. Berkeley, CA. May 23, 2003.
8. A. Barroso, J. Benson, T. Murphy, U. Roedig, C. Sreenan, S. Bellis, K. Delaney, B. OFlynn. The DSYS25 Sensor Platform. Demo at the SenSys2004. http://www.cs.ucc.ie/~url/pages/publications/files/sensys04barroso.pdf
9. P. Goyal, S. S. Lam, and H. Vin. Determining End-to-End Delay Bounds in Heterogeneous Networks. In Proceedings of Network and Operating System Support for Digital Audio and Video, 5th International Workshop, NOSSDAV95, Durham, New Hampshire, USA. Springer LNCS 1018, April 1995.
10. D. Stiliadis and A. Varma. Latency-Rate Servers: A General Model for Analysis of Traffic Scheduling Algorithms. IEEE/ACM Transactions on Networking, 6(5):611624, October 1998.
11. Alan Demers, Srinivasan Keshav, and Scott Shenker. Analysis and Simulation of a Fair Queueing Algorithm. In Proceedings of ACM SIGCOMM, pp. 3-12, August 1989.
12. B. Karp and H. T. Kung, "GPSR : greedy perimeter stateless routing for wireless networks," in Mobile Computing and Networking, pp. 243-254, 2000.
13. Crossbow Technology INC. Mica-2 Data Sheet. Crossbow,
14. http://www.xbow.com.
15. B. Deb, S. Bhatnagar and B. Nath. ReInForM: Reliable Information Forwarding using Multiple Paths in Sensor Networks. In Proceedings of the 28th Annual IEEE conference on Local Computer Networks (LCN 2003), Bonn, Germany, October 2003.
16. U. Roedig, A. Barroso, and C. J. Sreenan. Determination of Aggregation Points in Wireless Sensor Networks. In Proceedings of the 30th Euromicro Conference (EUROMICRO2004), Rennes, France, IEEE Computer Society Press, August 2004.
17. C. Intanagonwiwat, R. Govindan and D. Estrin. Directed Diffusion: A Scalable and Robust Communication Paradigm for Sensor Networks In Proceedings of the Sixth Annual International Conference on Mobile Computing and Networks (MobiCOM 2000), August 2000, Boston, Massachusetts.
18. J. Zhang, K. Premaratne and Peter H. Bauer. Resource Allocation and Congestion Control in Distributed Sensor Networks - a Network Calculus Approach. In Proceedings of Fifteenth International Symposium on Mathematical Theory of Networks and Systems, University of Notre Dame, August 12-16, 2002.

Design and Comparison of Lightweight Group Management Strategies in EnviroSuite*

Liqian Luo, Tarek Abdelzaher, Tian He, and John A. Stankovic

Department of Computer Science, University of Virginia,
Charlottesville, VA 22904
{ll4p, zaher, th7c, stankovic}@cs.virginia.edu

Abstract. Tracking is one of the major applications of wireless sensor networks. *EnviroSuite*, as a programming paradigm, provides a comprehensive solution for programming tracking applications, wherein moving environmental targets are uniquely and identically mapped to logical objects to raise the level of programming abstraction. Such mapping is done through distributed group management algorithms, which organize nodes in the vicinity of targets into groups, and maintain the uniqueness and identity of target representation such that each target is given a consistent name. Challenged by tracking fast-moving targets, this paper explores, in a systematic way, various group management optimizations including semi-dynamic leader election, piggy-backed heartbeats, and implicit leader election. The resulting tracking protocol, *Lightweight EnviroSuite*, is integrated into a surveillance system. Empirical performance evaluation on a network of 200 XSM motes shows that, due to these optimizations, Lightweight EnviroSuite is able to track targets more than 3 times faster than the fastest targets trackable by the original EnviroSuite even when 20% of nodes fail.

1   Introduction

The increasing popularity of sensor networks in large-scale applications such as environmental monitoring and military surveillance motivates new high-level abstractions for programming-in-the-large. As a result, several programming models that encode the overall network behavior (rather than per-node behaviors) have been proposed in recent years. Examples include virtual machines [1][2], and database-centric [3][4][5], space-centric [6][7], group-based [8][9], and environment-based [10] programming models, which offer virtual machine instruction sets, queries, sensor node groups and environmental events, respectively, as the underlying abstractions with which the programmer operates. These abstractions capture the unique properties of distributed wireless sensor networks and expedite software development.

---

* The work reported in this paper is funded in part by NSF under grants CCR-0208769 and ITR EIA-0205327.

V. Prasanna et al. (Eds.): DCOSS 2005, LNCS 3560, pp. 155–172, 2005.

An important category of sensor network applications involves tracking environmental targets. In these applications, some internal representation of the external tracked entity is maintained such as a state record, a logical agent, or a logical object representing the physical target. A given target in the environment should be represented uniquely and its identity should be preserved consistently over time. One way to ensure unique, consistent representation is to relay all sensor readings to a centralized base-station which runs spatial and temporal correlation algorithms to infer the presence of targets, assign them unique identities, and maintain such identities consistently. Such a centralized approach, however, is both inefficient and vulnerable. In addition to relying on a single point of failure, it results in excessive power consumption due to communication with a centralized bottleneck and may unduly increase latency, especially when targets move far away from the base-station.

To avoid these limitations, in EnviroSuite, we take the alternative approach of processing target data at or near the location where the target is sensed. Hence, appropriate distributed group management policies are needed to ensure the uniqueness and identity of target representation such that targets are given consistent names and sensors agree on which target they are sensing. This paper systematically investigates different system optimizations in the design of such group management algorithms in a sensor network. The resulting tracking system, *Lightweight EnviroSuite*, is used in a sensor network surveillance prototype that has since been transferred to the Defense Intelligence Agency (DIA). It is evaluated on a sensor network of 200 XSM motes [11]. Results from field tests of the overall system are provided, focusing on tracking performance. (Other performance aspects of the system such as efficacy of energy management algorithms will be reported elsewhere.) It is seen that realistic targets can indeed be tracked correctly despite environmental noise using low-range sensors. Our field test results show that even when 20% of nodes fail, Lightweight EnviroSuite is able to track targets more than 3 times faster than the fastest targets trackable by the original EnviroSuite, due to the optimizations described in this paper. The improved tracking coincides with reduced communication cost.

The remainder of the paper is organized as follows. Section 2 presents the background information and enumerates limitations of the original EnviroSuite. Section 3 explores group management strategies and their effects that constitute Lightweight EnviroSuite. Section 4 analyzes the performance results of a surveillance system that is constructed from Lightweight EnviroSuite. Finally, Section 5 supplies a summary and concludes the paper.

2   Background

Target tracking has received special attention in recent ad hoc and sensor networks literature. Many prior approaches (e.g., in the ubiquitous computing and communication domains) focused on tracking cooperative targets. Cooperative targets are those that allow themselves to be tracked typically by exporting a unique identifier to the infrastructure (such as a cell-phone number). Examples of cooperative targets include cell phones, RFID tags [12] and smart badges [13]. Since such devices are preconfigured with a unique identity, the tracking problem is generally reduced to that of locating the uniquely identified device and performing hand-offs if needed (e.g., in cellular phones). In contrast, our goal is to track non-cooperative targets such as enemy vehicles that do not broadcast self-identifying information. The presence and identity of such targets can only be inferred from sensory signatures, as opposed to direct communication with the target. Tracking non-cooperative targets is more challenging due to the difficulty in associating sensory signatures with the corresponding targets (e.g., all tanks look the same to our unsophisticated motes). The presence of target mobility further complicates the tracking problem.

One of the first research efforts on group management for non-cooperative target tracking has been conducted by researchers at PARC [14]. Their group management method dynamically organizes sensors into collaborative groups, each of which tracks a single target. Typical tracking problems such as multi-target tracking and tracking crossing targets are solved elegantly in a distributed way. Other approaches to target tracking include [15] which presents a particle filtering style algorithm for tracking using a network of binary sensors which only detect whether the object is moving towards or away from the sensor. A scalable distributed algorithm for computing and maintaining multi-target identity information in described in [16]. In [17] a tree-based approach is proposed to facilitate sensor node collaboration in tracking a mobile target.

The authors have investigated the tracking problem in several of their own prior publications. Similar to [8], in [18] we present a set of group management algorithms which form sensor groups at the locations of environmental events of interest and attach logical identities to the groups. Based on [18], EnviroTrack [19] proposes an environmental computing paradigm which facilitates tracking application development. EnviroSuite [20] further extends the paradigm to support a broader set of applications that are not limited to target tracking. Geared for tracking of fast-moving targets with low communication cost on available hardware platforms that have limited sensing and communication abilities, this paper proposes lightweight group management algorithms for EnviroSuite. Optimizations described in this paper may be applicable to other systems such as [8] as well.

The design of EnviroSuite assumes that each node can independently detect the potential presence of a target (subject to false alarms). For example, the presence of a magnetic signature, motion, and engine sound can be independently detected by each XSM mote to signify the potential presence of a nearby moving vehicle in a desert surveillance scenario. It is further assumed that sensor readings do not interfere with each other. Hence, tracking reduces to the problem of correct mapping of nodes that detect target signatures to the actual physical target identities responsible for these signatures. EnviroSuite therefore organizes nodes that detect targets of interest into groups, each representing one target.

Different from traditional centralized tracking schemes, the data association between targets and groups in EnviroSuite is done in a distributed way. Namely, all nodes that sense a target and can communicate directly assume that they sense the same target and consequently join the same group. The resulting aggregate behavior is that connected regions of sensors that sense the same signature are fused into the same group. Observe that when the target moves, group membership changes reflecting the changing set of sensors that can sense it at the time. A leader is elected for the group among the current members. A leadership hand-off algorithm ensures that group state is passed to each new leader.

In this paper, we focus on point targets (i.e., those approximated by a point in space, such as a vehicle), as opposed to diffuse region targets (given by an area, such as a chemical spill). It is assumed that the communication range of a node is sufficiently larger than its sensing range. Hence, all group members sensing a point target are within each other's radio range. Consequently, the data dissemination scheme within each group can simply use local broadcast to share sensory information. The leader performs data fusion in an application-specific manner to collect higher-level target information. Geographic forwarding [21] is used for communication with destinations external to the group. For example, in the evaluation section, each group leader estimates target position by averaging locations of group members and sends the result to remote base-stations periodically. Extensions of this scheme to diffuse region targets are described elsewhere [20].

This section briefly reviews the EnviroSuite programming paradigm and its core component, *MGMP* (*multi-target group management protocol*). It also analyzes and evaluates the limitations of MGMP when facing the practical requirements of a typical surveillance system.

2.1   EnviroSuite Abstractions and Challenges

EnviroSuite [20] is an object-based framework that supports *environmentally immersive programming* for sensor networks. Environmentally immersive programming refers to an object-based paradigm in which logical objects and objects representing physical environmental entities are seamlessly combined. Hence, EnviroSuite differs from other object-based systems in that its objects may be representations of elements in the external environment. At the implementation level, such objects are maintained by the corresponding group leaders. Upon detection of external elements of interest, nodes detecting the element self-organize into a group. The group leader in EnviroSuite dynamically creates an object instance to represent the tracked target. The group management protocol maintains a unique and identical mapping between object instances (or group leaders) and the corresponding environmental elements they track, such that object instances float across the network geographically following the elements they represent. This co-location is ideal for the execution of location sensitive object code that carries out sensing and actuation tasks. Objects encapsulate the aggregate state of the elements they represent (collected and stored by group leaders), making such state available to their methods. They are therefore the units that encapsulate program data, computation, communication, sensing and actuation. Object instances are destroyed when their corresponding environmental elements leave the network. This occurs naturally when the membership of the corresponding sensor group is reduced to zero.

Objects can be point objects (created for mobile targets that dynamically change their geographical locations), region objects (mapped to static or slowly moving regions), or function objects (not mapped to an environmental element). EnviroSuite is able to support both point objects and region objects in the same framework due to their similarities. Namely, (i) the corresponding external elements are detected by a group of geographically continuous nodes, and (ii) the aggregate state of the elements are collected by group leaders. However, the focus of this paper is on target tracking applications. Hence, we discuss mainly maintenance of point objects.

The biggest problem faced in EnviroSuite is the challenge of maintaining a unique and identical mapping between each object and the corresponding environmental element despite of distribution and possible mobility in the environment. In the rest of this paper, a *target* refers to a geographically continuous activity in the physical environment that persists over some interval of time. *Object uniqueness* dictates that each target be represented and that it be represented by exactly one logical object instance. *Object identity* dictates that the mapping between targets and objects be immutable. In other words, a target is always mapped to the same object instance identified by its *object ID*.

The problem of object uniqueness and identity is complicated by several factors. One is the need for seamless object migration across nodes as the target moves. Another is that sensor nodes that become aware of an external target should be able to tell whether it is a target previously seen by other neighboring sensors or not. Otherwise, an incorrect target list will be collectively maintained or an incorrect mapping will result between targets and objects. In the following we describe a solution to these problems.

2.2   MGMP and Its Limitations

The core component of EnviroSuite previously proposed by us is a set of multi-target group management protocols, named MGMP, which resolves the object uniqueness and identity problem. When predefined target signatures are detected by a set of nearby nodes, MGMP reacts by creating a *group* attached with a unique object ID. The set of nodes become group *members*, whose task is to periodically sense, calculate and report predefined object *attributes* (such as temperature, location, etc.) to the group *leader*. These reports are called *member reports*. A single leader is elected among these members to uniquely represent the group as one object to the external world. To avoid electing a node that is imminently going out of sensing range which results in yet another election, it is preferred to elect nodes near the target. Though current leader election doesn't enforce such preference, it can be easily adapted to do so if distances to the target can be inferred from detection results.

The leader is responsible for the maintenance of object attributes. It records member reports keeping only the most recent one from each member. It periodically creates a digest of the reports, and either keeps it as the internal state or

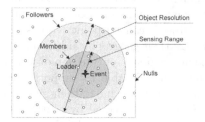

Fig. 1. Node state transitions in MGMP   **Fig. 2.** States of nodes around a target

sends it to the external world (e.g., base stations) based on user specification. These reports are called *leader reports*. The leader is also responsible for object uniqueness and identity maintenance. It periodically sends leader heartbeats to nodes within half an *object resolution* (in meters, defined by users) to advertise the object ID as well as its internal states. By design, half the object resolution must be larger than twice the sensing range such that every member can receive leader heartbeats. Nodes within half the object resolution, which cannot sense, but are aware of the target through heartbeats, are called group *followers* as distinguished from members. Follower nodes are prevented from spawning new groups. Follower nodes are centered around the leader since leader locations provide a good approximation of target positions. Though centering around the target would be a better solution, given the fact that both communication range and sensing range are irregular, the extra complexity required to do so is not worth it.

Nodes dynamically join or leave the group whenever they detect or lose the target. If a leader loses the target, it sends out a `RESIGN` message to request a leadership handoff. Upon the reception of such messages, the most current members reelect a new leader to take over leadership. Figure 1 illustrates the complete node state transitions in MGMP and Figure 2 depicts a typical node state distribution around a target. MGMP employs two important strategies to enhance robustness in the face of failures:

1. **Dynamic Leader Election:** Leaders are always dynamically elected among all current members. There are no pre-designated leader candidates. When leader election starts, each member sets a timer at random from 0 up to *maximum back-off time* and, when the timer expires, claims its leadership by messages, the reception of which terminates other members' timers as well as their unsent messages. This strategy ensures robustness to node failures.
2. **Periodic Leader Heartbeats:** Leaders periodically send out heartbeats so that leader failure can be detected by neighboring nodes.

The main goal of our deployed system is to alert a military command and control unit of the occurrence of targets of interest in hostile regions. Targets of interest may include civilian persons (unarmed), armed persons or vehicles. The system is required to obtain and report current positions of such targets to a remote base station to create tracks in real time. Several application requirements must be satisfied to make this system useful in practice. First, the application must have the ability to track typical military vehicles with velocities varying from 5 mph to 35 mph. Object uniqueness and identity must be ensured. Second, in our application, real-time updates on target trajectory must be sent to a base-station to be used by other devices such as cameras, which requires high accuracy and low reporting latency. Given the severely constrained bandwidth of current mote platforms, the communication cost should be minimized to reduce communication latency and maximize information throughput.

Does MGMP satisfy these requirements? We first try to answer the question through experimental results on TOSSIM [22]; a simulator for TinyOS [23] that emulates the execution of application code on the motes. Our experiments consist of 120 nodes deployed in a 30×4 grid 10 meters apart. Sensing range is set to be 1 grid length and communication range is 3 grid lengths. These settings reflect our real system where sensor devices are deployed in a grid 10 meters apart, sensing range is around 10 meters, and communication range is approximately 30 meters. We simulate a target moving across the field in a straight line to test tracking performance. The target is tracked with a sensor polling period of 0.02 s. Consistent with the real system requirements, members report to leaders their own locations twice a second. Leaders triangulate received locations to estimate target position and report estimations to the base station (located in a corner node) twice a second. The same testbed is used in later sections to evaluate new schemes.

Figure 3 shows the number of objects formed for the single target during its presence in the field. The uniqueness of target representation requires that only one object be formed. As is seen in figure, this is not always the case. The number of objects formed is one at lower target velocities, but it increases as target velocity increases. This violation is due to the difficulty in reaching agreement on target identity quickly enough which leads some sensors to believe that they are seeing different targets. The effects of maximum back-off time are more subtle. As is seen from Figure 3, if maximum back-off time is too small such as 0.2 s, the number of objects generated can be large since multiple nodes may become leaders and create new groups at the same time to represent the same target. If it is too large, fast targets may move out of the sensing range of a node before its back-off timer expires, so that the current object is lost and spurious ones are created. In theory, it is possible to derive the appropriate back-off time analytically for a particular target velocity. The main idea is that leader migration (via election and hand-off) should be faster than target speed for the target to never escape its tracking group. Nevertheless, such a derivation would have to be experimentally validated since it is difficult to account for various imperfections such as the irregularity of the sensing and communication ranges and the non-uniform distributions of nodes in practice.

Observe that no back-off timer value in Figure 3 can maintain object uniqueness at target speeds more than 1 grid length per second (grid/s) or 22 mph (since grid length is 10 meters). These results are far from the desired perfor-

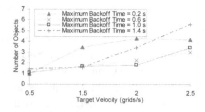
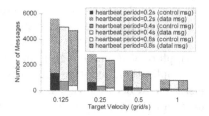

Fig. 3. Number of objects for varied maximum back-off time and target velocities

Fig. 4. Number of messages for varied heartbeat periods and target velocities

mance (35 mph). The overhead of dynamic leader election is the main reason why better results cannot be achieved, since long leader-election delays slow down the migration of groups, thus making fast-moving targets untrackable.

Figure 4 depicts the number of *control messages* (leader heartbeats and other group maintenance messages) and *data messages* (member reports and leader reports) sent during the presence of a target. The number of data messages decreases with increasing target velocity, because it is proportional to the duration of target presence due to periodicity of member reports and leader reports. The number of control messages exhibits similar trends since heartbeats that dominate control messages are also periodic. Obviously, control messages also decrease with longer heartbeat periods. However, minimizing communication cost by indefinitely increasing heartbeat periods is not feasible since longer heartbeat periods increase the vulnerability to message loss.

The above observations give insights into improvements to EnviroSuite that enhance tracking performance and reduce communication cost while maintaining robustness to failures. These improvements are described next.

3   Group Management Strategies in Lightweight EnviroSuite

This section explores in more detail the performance problems of current strategies, proposes a series of new strategies, and applies them one by one to Enviro-Suite to verify their individual effects on the current system. These new group management strategies, as a whole, constitute a very practical and efficient version of EnviroSuite, called *Lightweight EnviroSuite*.

3.1   Semi-dynamic Leader Election

Dynamic leader election, as the main factor that limits tracking performance of MGMP, affects the maintenance of object uniqueness and identity in two ways. First, it causes long leader handoff delays. In dynamic leader election, all members are competitors for leadership. Hence, consensus has to be achieved among all on a single leader. Obviously, the more members participate, the slower

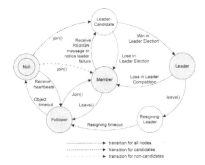

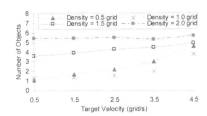

Fig. 5. Node state transitions for semi-dynamic leader election

Fig. 6. Number of objects for varied target velocities and candidate densities

the consensus. Second, it increases the possibility of message collisions since all members are exchanging messages to compete for leadership.

A better solution is to allow only a portion of all members to compete for leadership, which we call *semi-dynamic leader election*. Semi-dynamic leader election includes an initialization phase which pre-elects a portion of the nodes to be *candidates* (the potential competitors for leadership); others become *non-candidates*. The pre-election of candidates is similar to dynamic leader election in EnviroSuite. Each node sets a random timer and, when the timer expires, claims itself as a candidate. Nodes within distance x that receive this claim message become non-candidates. Ideally, the algorithm elects at least one candidate within any circular area of radius x. We call this x the *candidate density*. The node state transition changes accordingly as shown in Figure 5. Transitions to `Leader-Candidate` occur only when the corresponding nodes are candidates. `Null` nodes become `Members` instead of `Leader-Candidate` when they are non-candidates. Since only `Leader-Candidate` nodes attend leader election, these changes make candidates the only ones competing for leaderships.

Figure 6 illustrates the number of objects created for targets with different velocities when different candidate densities are set. Semi-dynamic leader election allows for a smaller maximum back-off time (set to 0.2 s in the following experiment) in leader election due to a reduced number of competitors. The $Density = 0.5\,grid$ curve performs the same as dynamic leader election since all nodes are candidates (maximum back-off time is set to 0.6 s for better performance in this case). As seen from Figure 6, a proper candidate density, say 1.0 grid, makes the semi-dynamic scheme outperform the dynamic one.

Observe that, a very low candidate density results in worse performance than dynamic leader election since candidates are so scarce that, in most groups, no leader is elected to maintain objects. We call the phenomenon a *leader desert*. The dark grey circle in Fig. 7 shows a leader desert where no candidate exists. If the target moves further to the right and gets detected by the nearest candidate outside the follower set, a spurious object is created by the candidate since it is not aware of the existing object. Even when candidate density is 1.0 grid, a leader desert still appears occasionally, which hurts object uniqueness slightly.

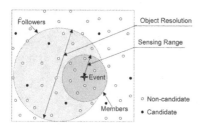
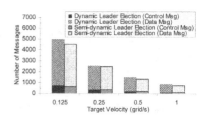

Fig. 7. Leader desert

Fig. 8. Comparison of number of messages for different target velocities

This explains why semi-dynamic leader election performs a little worse when target velocity is 0.5 grid/s. The leader desert problem is solved in later sections. As a side-effect, semi-dynamic leader election also results in lower communication cost since fewer control messages are sent to compete for leadership, which is shown in Fig. 8.

On the disadvantage side, robustness to node failures is expected to degrade when using semi-dynamic leader election due to a higher vulnerability to failures of leader candidates. However, this can be partially compensated by executing candidate pre-election more frequently. We discuss the overall failure robustness of the new scheme in later sections.

3.2   Piggy-Backed Heartbeat

Periodic leader heartbeat entails big overheads that are not affordable in applications with severe bandwidth constraints. Yet, it plays the most critical role in MGMP. First, it recruits followers to prevent these boundary nodes from creating spurious groups. Second, its periodicity makes leader failures perceivable, and thus recoverable. Third, the periodicity also improves robustness to message loss. Therefore, the challenge is how to reduce overhead while retaining the advantages of frequent heartbeats.

Fortunately, another component in MGMP exhibits the behavior of sending periodic messages; namely, *object attribute collection*. Members periodically send sensed attribute data to leaders and leaders periodically aggregate received data, process it, and send results to the external world if required. If heartbeats can be piggy-backed into these member reports and leader reports, periodic heartbeat becomes almost free. We call this new scheme *piggy-backed heartbeat*, where heartbeats are transformed into *leader heartbeats* (heartbeats piggy-backed into leader reports) and *member heartbeats* (heartbeats piggy-backed into member reports). Leader desert is no longer an obstacle to object state dissemination, since members take over this task during leader absences. Since object uniqueness is ensured through leader uniqueness, members are only allowed to repeat heartbeats originated from leaders and leaders are still the only authority that may update object information.

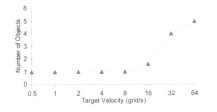

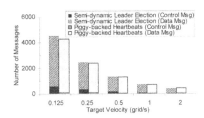

Fig. 9. Object maintenance in semi-dynamic leader election

Fig. 10. Comparison of communication overhead

In the piggy-backed heartbeat scheme, member heartbeats and leader heartbeats are treated differently: only the reception of leader heartbeats transits pre-elected candidates from state `Null` to `Follower`, while member heartbeats transits them into an intermediate state, called `Null-Follower`. If a node detects a target while in this state, it transits to Leader-Candidate to compete for leadership. This is unlike a regular `Follower`, which becomes a `Member` upon target detection. Without these changes, member heartbeats may transit all potential leaders to `Follower` state and then to `Member` upon the detection of the target, making the group follow a target without any leaders.

The aforementioned efforts make the piggy-backed heartbeat scheme a big improvement in object maintenance. Compared with the maximum trackable velocity seen in MGMP (1 grid/s), this improved version maintains object uniqueness and identity for targets with velocities up to 8 grid/s as shown in Figure 9. Figure 10 suggests another big improvement in reducing control messages.

3.3   Implicit Leader Election

It is possible to further reduce the protocol costs by employing an *implicit leader election* scheme. An assumption is made in this scheme that monitoring tasks are periodic and that after each period monitoring results are communicated by each tracking group to the external world. This assumption is reasonable since periodicity is a typical property of sensor network applications. The scheme allows candidates to start the execution of leader tasks such as data aggregation, whenever they detect the target. Note that, their task periods are unsynchronized since nodes usually do not begin to sense the target at exactly the same time. As a result, multiple but limited potential leaders are executing tasks in a group. At any point in time, if the node that first reaches the end of a task period sends out a result report, other neighboring nodes including other potential leaders simply accept the results and become *inactive* in their current task periods, which prevents them from reporting the same redundant results when finishing their periods. Hence, the external world sees the illusion of a single group leader.

Figure 11 illustrates an example. Candidate A senses the target from time 0 to 2.5 and B from 0.25 to 3.5. The length of task period is 1. Both A and B are initially active. At time 1, A reaches the end of its period and sends a result report. Receiving this report, B admits A as the current leader, accepts

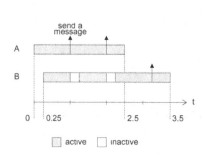

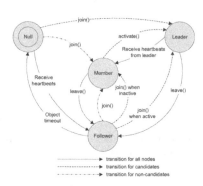

Fig. 11. An example of implicit leader election

Fig. 12. Node state transitions for implicit leader election

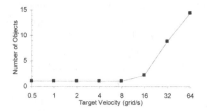

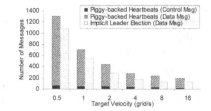

Fig. 13. Object maintenance in implicit leader election

Fig. 14. Comparison of communication overhead

A's results and makes itself inactive, which makes B silent at time 1.25 when it finishes its task period. Similarly B is still silent in the next period since A sends a message first. Finally, A quits the leader competition at time 2.5 when it loses the target. B continues A's work and reports the results at the end of its third period (time 3.25). This way, tasks are executed continuously between different leaders, the results of which are exposed to the external world at a rate $(3/3.5 \approx 0.9\,\text{report/s})$ that is very near to the defined rate $(1\,\text{report/s})$.

Figure 12 depicts the new node state transition graph. `activate()` is called by each node when beginning a new task period. Different from all previous versions, intermediate states (`LeaderCandidate`, `ResigningLeader`) no longer exist since implicit leader election eliminates the need for nodes to stay at the `LeaderCandidate` state sending `CANDIDATE` messages to compete for leadership and to stay at `ResigningLeader` sending `RESIGN` messages to start new leader election. The elimination of control messages further improves communication performance as shown in Figure 14. Meanwhile, a comparable performance in object maintenance (maximum trackable velocity 8 grid/s) is achieved as Figure 13 shows.

Note that implicit leader election can not be applied to applications where duplicate execution of tasks is not allowed or where tasks are not periodic. However, since the EnviroSuite compiler [20] has the ability to dynamically select modules during compilation based on programmer's application definition, a version without implicit leader election can be composed in such cases.

Overall, due to the optimizations mentioned above, Lightweight EnviroSuite achieves roughly an order of magnitude improvement in maximum trackable velocity (8 grid/s compared with 1 grid/s in EnviroSuite). The number of messages per second for targets at 0.5, 1 and 2 grid/s is reduced 25%, 12% and 37%, respectively.

3.4   Failure Tolerance

This section discusses the overall robustness of Lightweight EnviroSuite to typical failures in the sensor network: message loss and node failures. Message loss harms object maintenance by making the existence of the current leader unnoticed by other candidates, which may result in multiple nodes sending duplicate leader reports in the same period to the external world. However, the external world can always recognize duplicate reports through version numbers attached in the reports. Message loss can also prevent nodes on a group's outer boundaries from getting heartbeats. Consequently, these nodes may not become aware of the object and create spurious objects. However, Lightweight EnviroSuite allows members to disseminate heartbeats, which maintains a comparatively high heartbeat frequency and makes the possibility that a node fails to get any heartbeats very low. At the same time, objects attach their ages, which increase with the increase of finished task periods, to heartbeats. Object information from younger objects is discarded in favor of older ones. Therefore, spurious objects are eventually terminated due to their young ages.

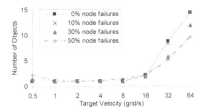

Fig. 15. Performance of object mainte-   **Fig. 16.** Performance of object maintenance for varied message loss   nance for varied node failures

Lightweight EnviroSuite is also robust to node failures. As stated earlier, it can go through leader deserts without losing object information or terminating task execution. In a similar way, it is able to overcome node deserts smaller than half an object resolution. Although we may temporarily lose track of the target inside a node desert, the target and its associated object can be picked up again in most cases after passing the node desert.

Experimental results confirm our conclusions as shown in Figure 15 and Figure 16. Lightweight EnviroSuite shows consistently good performance in object maintenance for targets with velocities up to 4 grid/s. However, after velocities exceed 8 grid/s, surprisingly bigger message loss results in fewer objects. This is because the number of objects is counted based on leader reports at the base station and a higher message loss results in fewer received leader reports, and thus fewer observed objects. For node failures up to 50%, Lightweight EnviroSuite exhibits comparable performance at velocities between 1 and 4 grid/s. However, when target velocities are as low as 0.5 grid/s, higher node failures do hurt performance. That is because higher node failures result in larger node or leader deserts. Slower targets may fail to cross the deserts before followers forget about the object.

4   System Evaluation

We integrated Lightweight EnviroSuite into an energy-efficient surveillance system, called *Vigilnet* [24], subsequently transitioned to the DIA. In December 2004, in the process of technology transition, we deployed 200 XSM motes running the Vigilnet system on sandy and grassy roads with a 3-way intersection and collected performance data in field tests. Figure 17 depicts the deployment of the system. Nodes are approximately deployed in a grid 10 meters apart, covering one 300-meter road and one 200-meter road. Each rectangular dot represents one XSM mote in the field. Several base-stations were deployed. Some nodes are missing in the GUI because they are turned off to emulate failures.

The XSM mote extends the MICA2 platform [25] by improved peripheral circuitry, new types of sensors and better enclosures. It communicates approximately 30 meters when deployed on grassy ground. The primary goal of the field test is to evaluate system ability to detect, classify and track one or multiple moving targets, which can be either SUVs, persons or persons carrying a ferrous object (suggestive of a weapon).

 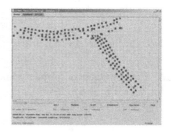

Fig. 17. System deployment

4.1   Overview of Vigilnet

Vigilnet is implemented on top of TinyOS. Figure 18 shows the layered architecture of Vigilnet. Components colored in dark grey are those implemented by Lightweight EnviroSuite.

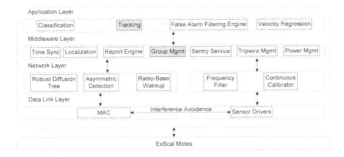

Fig. 18. System architecture of Vigilnet

Time synchronization (Time Sync), localization (Localization), and communication (MAC, Robust Diffusion Tree, Asymmetric Detection, and Report Engine) services constitute the lower-level components that are the basis for implementing higher-level services. Power management (Radio-Base Wakeup, Sentry Service, Power Mgmt, and Tripwire Mgmt), target classification (Sensor Drivers, Frequency-filter, Continuous Calibrator, Classification, and False Alarm Filtering Engine) and tracking (Group Mgmt, Tracking, and Velocity Regression) comprise main higher-level services. Target classification detects and classifies three types of targets with the help of collaborative group management provided by Lightweight EnviroSuite. Tracking components are responsible for estimating target positions and calculating target velocities.

Overall the system consists of 21,457 lines of source code, among which 2,884 are contributed by Lightweight EnviroSuite. The executable binary of Vigilnet occupies 85,926 bytes of code memory and 3,154 bytes of data memory, which can easily fit into XSMs equipped with 4KB data memory and 128KB code memory.

4.2   Tracking Performance Evaluation

Consistent with simulation, tracking modules in Vigilnet report to the base station estimations of target positions twice every second to provide sufficient data for false alarm processing and classification. A spanning-tree based routing [24] is used to disseminate such reports. The communication latency of these reports plays a critical role in achieving good tracking performance. Therefore, we suggest that when the system scales up to cover bigger fields, multiple base stations should be deployed. The false alarm filtering engine component executed in the base mote filters these reports and slows down the report rate to upper layers to once every 3 seconds due to bandwidth limitations. Target velocity calculation takes such reports as inputs.

Table 1 lists the comparison of tracking performance between Vigilnet equipped with the original EnviroSuite (measured at a previous field test conducted in August 2003) and Vigilnet equipped with Lightweight EnviroSuite. As is seen, without Lightweight EnviroSuite the maximum trackable velocity is about 5 to 10 mph, while the new Vigilnet system tracks targets up to 35 mph.

Table 1. Tracking performance comparison

Target vel.	Vigilnet with EnviroSuite	Vigilnet with Lightweight EnviroSuite
5 mph	successful	successful
10 mph	partially successful	successful
20 mph	failed	successful
30 mph	untested	successful
35 mph	untested	partially successful

Table 2. Tracking performance of Vigilnet with Lightweight EnviroSuite

Target type		Avg. tracking error	Std. dev. of tracking errors	Actual vel.	Calculated vel.
walking person		6.19 meter	3.28 meter	3±1 mph	2.9 mph
running person		6.67 meter	3.89 meter	7±1 mph	6.9 mph
vehicle		7.06 meter	3.98 meter	10±1 mph	10.5 mph
vehicle		5.91 meter	3.02 meter	20±1 mph	23.5 mph
two	1	5.58 meter	4.76 meter	10±1 mph	9.2 mph
vehicles	2	6.33 meter	3.52 meter	10±1 mph	9.9 mph

Note that, the success of tracking is an end-user metric measured by the accuracy of position and velocity calculations, which depend on several factors besides EnviroSuite group management protocols. A track is said to be successful only when the final calculated velocity within a 20% error. Due to the limited length of the field and the fixed report rate (once every 3 seconds), velocity calculation does not perform well when the velocity reaches 35 mph. However, the tracking performance of Lightweight EnviroSuite itself is actually better than the reported results for the integrated system.

Table 2 shows in more details the tracking performance of the new Vigilnet system. As seen, tracking errors are between 5.5 meters and 7.5 meters. These results were collected with 20% of the nodes randomly turned off to emulate failures. In all listed targets whose velocities vary from 3 mph to 20 mph, the maximum error of velocity calculation is less than 10%, which reflects the good tracking performance supplied by Lightweight EnviroSuite.

To give a more concrete view of the tracking performance of Lightweight EnviroSuite, Figure 19 shows the tracking trajectories for the following scenarios: (i) one vehicle drives across the field from left to right; (ii) two vehicles keep a distance of about 50 meters before they separate (the first one goes from left to right and turns right at the intersection and the second one goes from left to right). In the one-vehicle-tracking case, the rugged trajectory in the center of the horizontal road shows explicitly that existing node failures do affect tracking accuracy. The two-vehicle-tracking case proves the ability of Lightweight EnviroSuite to track multiple targets with the same sensory signatures as long as they keep a distance (50 meters) that is more than half an object resolution (set to 30 meters in the system). This was deemed sufficient by the client for operational use.

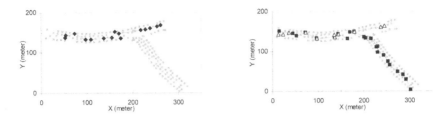

Fig. 19. Tracking trajectories for one vehicle and two vehicles

Field test results show that Lightweight EnviroSuite successfully improves the maximum trackable speed from near 10 mph to near 35 mph. Given our 10-meter-apart grid deployment and 20% node failures, tracking errors are still as small as about 6 meters and the maximum error in velocity calculation doesn't exceed 10%. These results from physically deployed systems validate that Lightweight EnviroSuite is practical, effective, and efficient on current hardware with limited communication and sensing capabilities. In this paper, we did not supply an experimental comparison between EnviroSuite and other high-level sensor network programming systems. This is, in part, due to the difficulty in porting application code across the different systems. If applications are re-implemented (as opposed to ported), it is hard to separate the effects of application-level implementation decisions from the inherent strengths and weaknesses of the underlying programming frameworks when interpretting performance comparison results.

5   Summary

Research on programming paradigms and frameworks in sensor networks has been very active in recent years. This paper presents our effort in improving an existing programming paradigm (EnviroSuite) into a more practical and efficient version (Lightweight EnviroSuite) that can be directly utilized to build practical large-scale systems with realistic requirements. The resulting version is shown to be efficient via experimental testing on a physically deployed large-scale surveillance system consisting of 200 XSM motes. This is one of the first attempts to use high-level sensor network programming languages in building and deploying real sensor network applications.

References

1. Boulis, A., Srivastava, M.B.: A framework for efficient and programmable sensor networks. In: OPENARCH '02. (2002)
2. Levis, P., Culler, D.: Mate: a virtual machine for tiny networked sensors. In: ASPLOS. (2002)
3. Li, S., Son, S.H., , Stankovic, J.: Event detection services using data service middleware in distributed sensor networks. In: IPSN '03. (2003)

4. Madden, S.R., Franklin, M.J., Hellerstein, J.M., , Hong, W.: The design of an acquisitional query processor for sensor networks. In: SIGMOD. (2003)
5. Yao, Y., Gehrke, J.E.: The cougar approach to in-network query processing in sensor networks. In: Sigmod Record, Volume 31, Number 3. (2002)
6. Welsh, M., Mainland, G.: Programming sensor networks using abstract regions. In: NSDI '04. (2004)
7. Newton, R., Welsh, M.: Region streams: functional macroprogramming for sensor networks. In: DMSN '04: Proceeedings of the 1st international workshop on Data management for sensor networks, New York, NY, USA, ACM Press (2004) 78–87
8. Liu, J., Liu, J., Reich, J., Cheung, P., , Zhao, F.: Distributed group management for track initiaition and maintenance in target localization applications (2004)
9. Whitehouse, K., Sharp, C., Brewer, E., Culler, D.: Hood: A neighborhood abstraction for sensor networks. In: MobiSYS '04. (2004)
10. Liu, J., Chu, M., Liu, J., Reich, J., Zhao, F.: State-centric programming for sensor-actuator network systems. In: IEEE Pervasive Computing. (2003)
11. XSM motes: (http://www.cast.cse.ohio-state.edu/exscal/index.php?page=main)
12. Stockman, H.: The active badge location system. In: Proceedings of the IRE. (1948) 1196–1204
13. Want, R., Hopper, A., Falcao, V., Gibbons, J.: The active badge location system (1992)
14. Liu, J., Chu, M., Liu, J., Reich, J., Zhao, F.: Distributed state representation for tracking problems in sensor networks. In: IPSN '04. (2004)
15. Aslam, J., Butler, Z., Constantin, F., Crespi, V., Cybenko, G., Rus, D.: Tracking a moving object with a binary sensor network. In: SenSys '03, New York, NY, USA, ACM Press (2003) 150–161
16. Shin, J., Guibas, L., Zhao, F.: A distributed algorithm for managing multi-target identities in wireless ad-hoc sensor networks. In: IPSN '03. (2003)
17. Zhang, W., Cao, G.: Optimizing tree reconfiguration for mobile target tracking in sensor networks. In: INFOCOM '04. (2004)
18. Blum, B., Nagaraddi, P., Wood, A., Abdelzaher, T., Son, S., Stankovic, J.: An entity maintenance and connection service for sensor networks. In: MobiSys '03. (2003)
19. Abdelzaher, T., Blum, B., Cao, Q., Evans, D., George, J., George, S., He, T., Luo, L., Son, S., Stoleru, R., Stankovic, J., Wood, A.: Envirotrack: Towards an environmental computing paradigm for distributed sensor networks. In: ICDCS '04. (2004)
20. Luo, L., Abdelzaher, T., He, T., Stankovic, J.A.: Envirosuite: An environmentally immersive programming framework for sensor networks. (In: Submitted to ACM Transactions on Embedded Computing Systems)
21. Karp, B.: Geographic Routing for Wireless Networks. PhD thesis, Harvard University (2000)
22. Levis, P., Lee, N., Welsh, M., , Culler, D.: Tossim: Accurate and scalable simulation of entire tinyos applications. In: SenSys '03. (2003)
23. Hill, J., Szewczyk, R., Woo, A., Hollar, S., Culler, D., Pister, K.: System architecture directions for networked sensors. In: ASPLOS-IX, New York, NY, USA, ACM Press (2000) 93–104
24. He, T., Krishnamurthy, S., Stankovic, J.A., Abdelzaher, T., Luo, L., Stoleru, R., Yan, T., Gu, L., Hui, J., Krogh, B.: Energy-efficient surveillance system using wireless sensor networks. In: MobiSYS '04, New York, NY, USA, ACM Press (2004) 270–283
25. UC Berkeley. MICA motes: (http://www.tinyos.net/scoop/special/hardware/)

Design of Adaptive Overlays for Multi-scale Communication in Sensor Networks

Santashil PalChaudhuri*, Rajnish Kumar**, Richard G. Baraniuk***,
and David B. Johnson*

*Rice University, Department of Computer Science, Houston, TX 77005-1892
**Georgia Institute of Technology, College of Computing, Atlanta, GA 30332-0280
***Rice University, Department of Electrical and Computer Engineering,
Houston, TX 77005-1892

Abstract. In wireless sensor networks, energy and communication bandwidth
are precious resources. Traditionally, layering has been used as a design principle
for network stacks; hence routing protocols assume no knowledge of the application behavior in the sensor node. In resource-constrained sensor-nodes, there
is simultaneously a need and an opportunity to optimize the protocol to match
the application. In this paper, we design a network architecture that efficiently
supports multi-scale communication and collaboration among sensors. The architecture complements the previously proposed Abstract Regions architecture for
local communication and collaboration. We design a self-organizing hierarchical overlay that scales to a large number of sensors and enables multi-resolution
collaboration. We design effective Network Programming Interfaces to simplify
the development of applications on top of the architecture; these interfaces are
efficiently implemented in the network layer. The overlay hierarchy can adapt to
match the collaboration requirements of the application and data both temporally
and spatially. We present an initial evaluation of our design under simulation to
show that it leads to reduced communication overhead, thereby saving energy. We
are currently building our architecture in the TinyOS environment to demonstrate
its effectiveness.

1 Introduction

Sensor networks [1] consist of a large number of small, low-powered wireless nodes
with limited computation, communication, and sensing abilities. Their ubiquitous, on-
demand sensing capabilities have enabled numerous new applications, from vibration
monitoring throughout buildings in active earthquake zones to air pollution tracking
to microclimate investigations in tropical rain forests. In a battery-powered sensor network, energy and communication bandwidth are a scarce resources. Thus there is a need
and opportunity to adapt the networking to match the application in order to minimize
the resources consumed and extend the life of the network.

Sensor network applications have several characteristics that distinguish them from
other networks (such as LANs or ad hoc wireless communication networks) and make
matching the networking to the application challenging. For example, different applications demand a wide range of different communication patterns among the nodes,

V. Prasanna et al. (Eds.): DCOSS 2005, LNCS 3560, pp. 173–190, 2005.

including data aggregation, dissemination, attribute-based routing, local collaboration, and multiple resolution. Several networking protocols have been developed to handle these kinds of communication efficiently. However, since each involves widely different communication abstractions and trade-offs, no single protocol can be optimized for all applications. Moreover, many distributed signal processing applications demand multiple communication patterns. Finally, designing or matching protocols to applications is a very difficult task for applications developers.

In this paper, we propose an *adaptive network architecture* that matches the communication characteristics of many different applications by optimizing based on application feedback. We design a hierarchical overlay to handle aggregation, dissemination, and multiple resolution, and we leverage the Abstract Region [2] architecture to handle local collaboration between nodes. These overlays coexist and serve different purposes; together they efficiently support a wide range of different application data communication patterns. To simplify the application design, we provide a set of Network Programming Interfaces to abstract the details of low-level communication. Applications specify their communication needs through these interfaces; the architecture then uses this information to optimize the communication data flow.

Many applications, such as large-scale collaborative sensing, distributed signal processing, and data assimilation, require the sensor data to be available at multiple resolutions, or allow fidelity to be traded-off for energy efficiency. We form a self-organizing network hierarchy that can scale to very large numbers of nodes using multi-scale data communication. Our multi-scale hierarchical overlay adapts to form clusters such that data communication becomes efficient. While a self-organized hierarchy has been known to scale well, ours is the first proposal to align the network hierarchy with the application data flow.

After overviewing related work in Section 2, we describe our hierarchical overlay in Section 3. Section 4 details our initial design evaluation. We conclude and suggest directions for future work in Section 5.

2   Related Work

Various protocols and architectures have been proposed over the years for sensor networks. Earliest were the diffusion class [3] of algorithms, which are effective in aggregation and dissemination communication abstractions but cannot support multi-resolution communication required by many applications. There are various ways of implementing these diffusion algorithms depending on the application behavior, such as two-phase pull, one-phase pull, and push. The specific diffusion behavior can be chosen by the application to match its requirements [4]. GARUDA [5] is an architecture that handles reliable delivery of downstream data under various notions of reliability.

Fractional Cascading [6] and DIMENSIONS [7] have recently been proposed to handle multi-scale data communication. We extend their design, which is essentially for storage and retrieval of sensor data, to have more general applicability. Their architecture is based on a regular grid structure and assumes regular data sampling. However, practical multiscale transforms need to accommodate networks with arbitrary irregular placement of sensors; we achieve this using our self-organizing hierarchy. Our architecture adapts to the node collaboration requirements to make the networking more efficient, which is not addressed in these two previous approaches.

SDIMS [8] is another hierarchical approach that is more targeted toward wired networks but can be adapted for wireless sensor networks. This approach provides a flexible API for configuration but does not address the adaptibility to application requirements. It does have a tunable interface for a tradeoff between latency and overhead for added flexibility.

The goal of Abstract Regions [2] is to simplify the application design by providing abstract interfaces to hide the details of low-level communication. It proposes the concept of neighborhood as a programming unit, and shows how various applications can be efficiently written using it. Many applications need multi-resolution data though, and our architecture addresses the abstraction requirement for such applications. The Abstract Region concept of neighborhood is thus complementary with our approach, and both together cover a much wider range of application requirements. Hood [9] is another approach to providing a programming abstraction, but it is also targeted toward neighborhood-based programming models only.

3   Hierarchical Overlay

We design the hierarchical overlay for efficient aggregation, dissemination, and multiple resolution of application data. In this overlay, a self-organizing hierarchical clustering is formed, inspired by the self-organization component of protocols like Safari [10] and L+ [11].

The hierarchy is a recursive organization of nodes into cells, cells into supercells, and so on, based on an autonomous self-election of a subset of the nodes into *drums*, and iteratively drums self-electing to become higher level drums, and so on. The drum is also called the *parent* for all nodes within it's cell. Figure 1 shows an example cell hierarchy. In the figure, nodes 1, 2, 3, 4, 5, and A group together to form a cell (called *fundamental cell*) with node A being the drum for the cell. Each of the drums in the network, namely nodes A, B, C, and D form a higher level cell with node C being the drum for that higher-level cell (called a *super-cell*). This hierarchy formation goes on iteratively until all the nodes come under one highest level cell. The self-selected drums aid in this hierarchy formation by sending periodic beacon packets.

Each node can be thought of as a level 0 cell and a level 0 drum (level 0 drums do not send beacon packets). Every level k drum is at the same time also a level i drum, for all $i < k$. This hierarchy formation algorithm is distributed, with no central coordination. Drums of the same level are roughly uniformly spaced, with higher level drums more sparse than lower level drums. A *coordinate* of any drum at level i is the concatenation of the coordinate of the level $i + 1$ drum with which it associates, along with a unique identifier. This unique identifier can be any random string large enough to avoid collisions. We use an address assignment technique proposed in TreeCast [12], whereby the nodes are given compact addresses minimizing the length of address strings. Various sensor applications require the identifiers of nodes along with their values, and this technique leads to efficient encoding of the identity (and location) of the nodes.

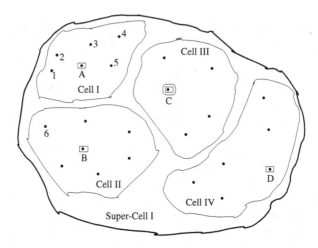

Fig. 1. Cluster hierarchy

In the initial startup period, each drum sends beacon packets (*beacons*) periodically, containing the beacon sequence number, the drum level, and a hop count, which aids in the hierarchy formation. These beacons are forwarded by all nodes within the hop count limit or those within the cell defined by the parent of the drum sending the beacon. In Figure 1, the beacons from node B reach all the nodes in super-cell I. The beacons sent during this initial phase also give the shortest path from any drum of level i to its associated higher level drums and the drums in the same level within it's super-cell.

Algorithm 1 shows the *Cluster Formation* algorithm. A drum of level i is denoted by $drum_i$, and $drum_0$ denotes the nodes. Beacons sent by $drum_i$ are denoted by $beacon_i$. $\mathcal{D}_1$ is the fundamental cell diameter and is dependent of the cluster size required by the application.

The drums wait a random time between 0 and T_{max}, before deciding whether to become a higher level drum or associate with another drum. The association scope of the drums, determined by the hop limit of the beacons, increase geometrically with the level of drum. Decreasing α increases the number of levels in the hierarchy while reducing the number of nodes in each level, whereas increasing the α has the opposite effect. Lines 8-10 of the algorithm ensure that the beacons of any drum reach all nodes within its super-cell. Lines 11-13 ensure that no drums of same level form too close to each other, as that reduces the efficiency of the clustering. Lines 14-16 make the hierarchy evolve such that the drums are always associated with closest higher level drum. The number of levels formed in the hierarchy is of $O(\log(N))$, where N is the number of nodes in the network. As a simple analysis of the cost of the algorithm, if instantaneous propagation is assumed, then all $level_i$ drums are separated by $\mathcal{D}_i$ hops, with very high probability, and so the total startup phase is of $O(T_{max} \log(N))$. Therefore the latency can be made smaller by reducing T_{max} to an optimal value.

For null *Selectors*, the hierarchy formed can be analyzed using the Random Sequential Adsorption (RSA) model, since under the instantaneous propagation approximation, the hierarchy formation conforms to the RSA model [10]. The average number of level 1 drums formed is

---

Algorithm 1. Cluster Formation

1: **repeat**
2:      $Drum_i$ *wait for a random time up to* T_{max}
3:      **if** $Drum_i$ does not hear a higher level drum beacon, or is not the highest level drum **then**
4:          *Steps up to* $level_{i+1}$ *and starts sending periodic beacons with hop limit of* $\mathcal{D}_{i+1} = \alpha \times \mathcal{D}_i = \alpha^i \times \mathcal{D}_1$
5:      **else**
6:          *Associates with the nearest* $drum_{i+1}$.
7:      **end if**
8:      **if** $Drum_i$ hears any non-duplicate beacon by a drum in it's super-cell **then**
9:          $Drum_i$ *forwards the beacon.*
10:     **end if**
11:     **if** $Drum_i$ hears any $beacon_i < \mathcal{D}_i$ hops away **then**
12:         *The drum with the lower id steps down to* $level_{i-1}$
13:     **end if**
14:     **if** $Drum_i$ hears any $beacon_{i+1}$, which is closer than current $beacon_{i+1}$ **then**
15:         $Drum_i$ *associates itself with the closer drum*
16:     **end if**
17: **until** All nodes are assigned stable coordinates

$$n = 0.547 \left(\frac{N}{\pi \frac{\mathcal{D}_1^2}{4} \rho} \right) \tag{1}$$

where N is the number of nodes in the network, ρ is the node density in hop-metric sense, its value depending on the transmission range and spatial density of nodes. When applied to multiple levels of the hierarchy, this gives us

$$\frac{n_i}{n_{i+1}} = \left(\frac{\mathcal{D}_{i+1}}{\mathcal{D}_i} \right)^2 \tag{2}$$

where n_i is the number of level i drums.

3.1   Adaptive Hierarchy

During startup, each drum sends beacon packets periodically, which are forwarded by all nodes based on a policy of geographical (hop count) proximity. These beacons then induce a cell hierarchy that is proximity based. In various sensor network applications, proximity is a good measure of correlation and hence of data interchange. So, the above approach of cell structure formation matches the collaboration between and nodes.

In many other applications though, the collaboration and communication take place between nodes based on various other constraints. For example, nodes with similar magnetic field or temperature readings within a neighboring scope might need to communicate more often. So, if the clustering hierarchy matches the collaborative set of nodes, communication can be efficiently abstracted and implemented. We use some application specified filters, called *Selectors*, to align this hierarchy to match the collaboration and communication sets of nodes.

Algorithm 2. Recluster (*Selectors*)

1: *Drum sends out beacons with Selectors*
2: *Nodes matching the Selectors (re)select the Drum as their parent*
3: **if** Any node become orphan or cell is sub-optimal **then**
4:     *It sends Solicit Beacons with specified Selectors*
5:     *Nodes matching the Selectors respond with Beacons*
6:     *Choose the drum most suitable with respect to the Selectors, and become a child of the drum*
7: **end if**

We define a *Selector* as a tuple of ⟨*attribute, value, operator*⟩, where *attribute* is any application specified variable, and *value* is a valid element from the range of the attribute. The *operator* is a binary operation (such as $>$, $<$, or $=$) with *value* being one of the operands. We extend the definition of a Selector to form:

Selectors = Selectors $\wedge$ Selectors | Selectors $\vee$ Selectors | Selectors | Selector | null

The values for the attributes are assumed to be shared between the application, sensor hardware, and the networking layer at a node, which enables the Selectors to be evaluated at the networking layer. An operator needing a time-series of previous attribute values might entail the sharing of whole data structures of application computed values. Currently only scaler operators are supported; more complex operators could have application-defined call-back functions to enable them to be evaluated by the application.

The beacons of drums are forwarded by a node if the hop count in the beacon packet is less than a specific value and the specified *Selectors* evaluates to true for the attribute values in that node. For an empty *Selectors*, the effect is to forward the beacons based only on hop count.

Reclustering of the network hierarchy to adapt it closely to the communication flow is initiated by the application locally in cells where adaptation is needed. This is best judged by the application, as the networking layer does not have any knowledge of the application logic or how the sensed values influence the communication. Algorithm 2 shows the *Recluster Algorithm*. The parameters for cluster formation are changed locally to reflect current communication patterns. The *Selectors* encode the criterion for the new cluster formation in the *Split Phase* of the algorithm. After reclustering, some nodes might become *orphans* (nodes without parent) or some cells might be smaller than optimal. These nodes or cells then merge with neighboring cells meeting the criterion in the *Merge Phase* of the algorithm. This reclustering is triggered by the application locally, and only occurs in the cells needing it for efficiency. The rest of the clusters in different areas are not changed. Hence, this reclustering takes place locally only where necessary and invokes no long range messaging.

3.2   Network Programming Interfaces

We design a set of address-free Network Programming Interfaces (NPIs) to adhere to the paradigm that communication for the typical sensor network applications should be expressed without referencing specific nodes [13]. The interest is in data over space and time, rather than individual node values. The subject of direct one-to-one communication has been extensively studied in the literature, and we will not propose any new protocol for this but will leverage the existing body of work. We provide primitives to ease the programmability in a sensor network, by capturing the interfaces that are needed by sensor applications in general. Abstract Regions [2] proposed a flexible means of node addressing, by supporting data sharing using a tuple-space-like programming model. Their approach is similar to the MPI approach for parallel machines, by hiding the details of the sharing primitives. We support similar primitives and extend them for our multi-scale architecture, although sharing is explicit using put and get primitives, to provide a more efficient implementation of the programming model. The two groups of interfaces we support are *discovery* and *communication*.

Each of the interfaces can be implemented in either blocking or non-blocking fashion. In non-blocking mode, the operation is invoked through a command, and when the operation is complete, a callback is invoked on the original requesting component. In blocking mode, the operation may block and then resume on an interrupt either by a timer or message arrival. The blocking mode is significantly easier to program, as in the non-blocking mode the programmer has to handle the synchronization and callback explicitly. TinyOS [14] supports the non-blocking concurrency model, but a lightweight thread-like abstraction called Fiber [2] has been implemented recently as a blocking model.

Figure 2 shows the *virtual* hierarchy schematic of Figure 1, which will help explain the interfaces. All the levels above level 0 are virtual, and the nodes only exist at the lowest level.

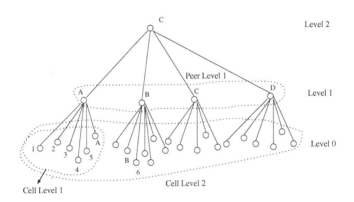

Fig. 2. Virtual hierarchy

Discovery Interfaces. Discovery interfaces can be invoked by any node to give itself information about related nodes. This information is gathered periodically, with a period as specified by the application, and the application is informed of any changes in the information. This procedure is continuous, either triggered by node failures or additions. The information might contain the identifiers of the set of nodes, their locations, link quality and number of hops to each of them, and resource (e.g., remaining battery or available sensors) present in each of them. The information learned from these discovery interfaces can be used to configure the Selectors with any specified criterion. For example, the Selectors might be specified to filter nodes within a specified geographical distance or with higher than a specific link quality. There are three possible interfaces supported. As a node can be simultaneously in different levels, a level is specified for each interface to specify the level for which the information is required. The format for representing the gathered information is dependent on the implementation.

- *Parent Information (Level)*: Returns information about the parent of the node at the specified level. For example in Figure 2, this interface when invoked on node 6 with level 1 returns information about node B, and with level 2 returns information about node C.
- *Peer Information (Level)*: Returns information about the peers of the node at the specified level. For example in Figure 2, this interface when invoked on node B at level 1, returns information on the set of nodes A, B, C and D.
- *Cell Information (Level)*: Returns information about the cell of the node at the specified level. For example in Figure 2, this interface when invoked on node A, returns information on the set of nodes 1, 2, 3, 4, 5 and A.

Communication Interfaces. In our multi-scale architecture, we support both kinds of communication models: *Put* and *Get*. In Put, a node sends data to its cell, parent, or peers, whereas in Get, a node solicits data from its cell, parent, or peers. The Put interfaces correspond to the push paradigm, and the Get interfaces correspond to the pull paradigm that has been proposed by the diffusion type of algorithms [3]. We support both types, as different applications might be optimized using different paradigms, as pointed out by by Heidemann et. al [4]. In some situations where the data generation rate is infrequent and unknown, polling using Get will be inefficient; using Put by the source of the data when the data is generated will be optimal. In another scenario, where the data generation rate is high and consumption rate is low, pushing data using Put will entail redundant data communication; using Get by the consumer of the data when the data is required is optimal in this case. The Put Interface can be implemented in multiple ways: stored locally, sent immediately to the designated scope, or cached at different intermediate locations. Similarly, the Get Interface implementation might involve either fetching remote data or local retrieval. The specific implementation depends on the application characteristics.

We also support *Reduction* interfaces that use an associative operator (such as *sum*, *max*, or *min*) to reduce an attribute across all the nodes in a specified region. This Reduction interface can be implemented using Get and Put, but efficient implementations can take advantage of local reductions while propagating the values. This abstraction also provides ease of programmability.

There are three groups of primitives a node might address: its parent, its peers, or its cell. This leads to six different interfaces for Put and Get. Reduction interfaces are done on either cells or peers. All examples below refer to Figure 2. For node A, the parent is node C; the peers at level 1 are B, C and D; and the peers at level 0 as well as the cell at level 1 are nodes 1, 2, 3, 4, and 5.

- *PutParent (Attribute, Value)*: The value of the attribute is sent to the parent node. When called on Node A, the data is sent to node C.

- *PutCell (Level, Selectors, Attribute, Value)*: Level can be at most one level higher than the node using this interface. So, a node of $level_0$ can send message to the fundamental cell. In general, a $drum_i$ can send message to all nodes in the same $level_{i+1}$ cell. For Node A, PutCell called with level 1 delivers data to nodes 1, 2, 3, 4 and 5, while called with level 2 delivers data to all the nodes marked by Cell level 2. The targeted nodes can filter the receipt of the Data using the *Selectors*.
- *PutPeer (Level, Selectors, Attribute, Value)*: The level can be at most same as the level of the node using the interface. This interface provides the same functionality as PutCell for $level_0$ nodes. For node A, level 1 delivers the data to nodes B, C and D (Peer Level 1).
- *GetParent (Attribute)*: The value of the attribute is solicated from the parent node.
- *GetCell (Level, Selectors, Attribute)*: Level can be at most one level higher than the node using this interface. In this interfaces, Data is received from the cell nodes matching the *Selectors*.
- *GetPeer (Level, Selectors, Attribute)*: The level can be at most same as the level of the node using the interface. This interface provides same functionality as GetCell for $level_0$ nodes.
- *ReduceCell (Level, Selectors, Attribute, Operator)*: This interface is applied on the attribute for all nodes in the cell specified by level and *Selectors*. And the reduced attribute value is stored locally. For example for operator *max*, the maximum attribute value within the cell is returned by this interface.
- *ReducePeer (Level, Selectors, Attribute, Operator)*: Similar interface where the scope is all the peer nodes at the specified level.

3.3   Efficient Communication Operations

The Network Programming Interface in the previous section is used by the applications to form a clustered hierarchy and to adapt it for efficient communication. In this section, we describe mechanisms for efficiently supporting the different communication interfaces. We assume the existence of bidirectional wireless links, which is true for most commonly used wireless MAC protocols.

- *With the parent node*: The drum beacons that are used to form the cluster hierarchy is utilized to route from and to parent node, by following the path or reverse path of the beacons respectively. If the path breaks, due to nodes in the path moving away or dying, then local route repair is done to find a new route. The drum whose path breaks, sends out beacons for a short interval to repair the broken path.
- *With the peer nodes:* At any level, the peer nodes need to be able to communicate with each other efficiently. This is achieved by expanding the scope of the beacons for the drums. The beacon packet of a level n drum is also forwarded by all nodes in the level $n + 1$ cell of the originating drum. The reverse path is followed to reach each peer. This is only done in the startup period or when a path breakage is detected. Multicasting at network or MAC layer (if possible) is done to prevent duplicate packets along common part of the paths. For example in Figure 1, beacon packets from node A in cell I is flooded to the whole super-cell I. And hence B, C and D know of the shortest path to A.

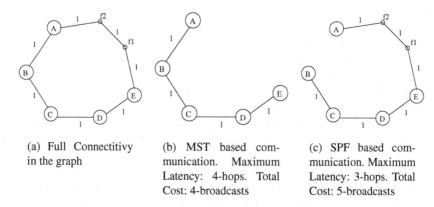

(a) Full Connectitivy in the graph

(b) MST based communication. Maximum Latency: 4-hops. Total Cost: 4-broadcasts

(c) SPF based communication. Maximum Latency: 3-hops. Total Cost: 5-broadcasts

Fig. 3. Example topology connecting peer nodes (A, B, C, D and E) and forwarding nodes (f1 and f2)

The above technique leads to minimum latency communication from any node to the rest of it's peers, using shortest path. But, it also leads to higher cost in terms of number of forwards. Alternatively a Minimum Spanning Tree can be formed between the peer nodes. This leads to lesser number of forwards, but also leads to higher maximum latency. Figure 3 shows an example topology to illustrate this. This tradeoff can be exposed for the application to choose from.

– *Within the cell*: Communication from any node to the whole cell can be achieved by simple flooding within the scope of the cell, whereby each node forwards a packet exactly once. However, this is very sub-optimal and leads to many redundant broadcasts specially in a dense network. We describe next our strategy for optimal cell flooding.

Optimal Cell Flooding. Typical broadcasting using simple flooding, where each node forwards a packet exactly once, leads to broadcast storm problems [15] and is very energy inefficient. To form a more efficient flooding algorithm, there has been substantial work with regard to carefully choosing the forwarding nodes to reduce the number of forwards without reducing it's effectiveness. Williams and Camp [16] categorized the techniques recently. In probability-based methods, nodes forward with some variable probability parameter chosen randomly or depending on the number of broadcasts heard. In area-based methods, distance or location of the nodes are taken into account by the node before deciding on forwarding. Both of these methods are completely localized without the need for any coordination. In neighbor knowledge methods, a distributed algorithm forms a Connected Dominating Set (CDS) to choose a subset of nodes to be forwarders. This has more overhead than the previous methods, but leads to more optimal flooding due to better knowledge about the neighborhood. Ideally, a Minimal Connected Dominating Set (MCDS) will give the most efficient set of nodes to forward packets such that all nodes are reached. Building a MCDS is an *NP-Complete* problem, however, and the problem gets harder in sensor networks in the absence of global knowledge. There has been a significant body of work on approximating the MCDS using heuristics [17].

There are two types of neighbor knowledge methods proposed: *Relaying* in which a node determines the forwarding status of it's neighbors, and *Pruning* in which a node makes a local decision on it's forwarding decision. Multipoint relaying is an efficient approach for relaying, and has been used in the OLSR ad hoc network routing protocol [18]. The re-broadcasting nodes are explicitly chosen by the upstream nodes, either via "hello" packets, or within the header of each broadcast packet. In relaying, there is either additional overhead in each packet to designate the forwarding nodes, or relevant state that has to be maintained by each node. The statelessness of protocols has prime importance in an unreliable network of sensor nodes, as a stateless protocol never operates on out of date state. In the pruning methods, nodes decide on their own locally whether to forward or not, leading to better reliability in the face of failure. There is automatic correction for small changes and robustness to big changes. Nodes can go to sleep independently in pruning methods, but there needs to be additional coordination in relaying methods so that no delegated forwarder is sleeping. In the absence of MAC layer multicast for pruning, nodes have to broadcast every packet for which all the neighbors have to process the packet before knowing that they are not designated forwarders. And finally, a comparison paper [16] showed very similar performance for both of the methods. We chose to use pruning method for the above reasons.

Various approaches for pruning based broadcasting use knowledge of k-hop neighborhood information, m-hop last visited nodes information for each packet, and priority between nodes. Larger value of k leads to more optimal forwarding set, but also entails higher cost for maintaining this neighborhood information. Larger value of m is also useful, but entails packet overhead. Wu and Dai [19] have proposed a generic scheme to cover all pruning based approaches. A node determines it's forward status by finding existing replacement paths between all pairs of its k-neighbors. If all the replacement path nodes have higher priority values than itselt or is already visited, then the node chooses to be a non-forwarder. Else, it forwards.

In designing an efficient flooding algorithm, the factors we chose were the following. The computational complexity is $O(k^2)$ for such algorithms, thereby dictating a small value of k. Also, recent proposals for sensor network MAC protocols [20, 21] maintain a 2-hop neighborhood to deal with efficient assignment of conflict-free slots. We choose value of k to be equal to 2. The information for m equal to 1 is available at no cost when the packet is received, by looking at the source. A greedy approach is taken to prioritize the nodes based on their degree of connectivity. Our solution is similar to the approach taken in the Scalable Broadcast Algorithm [22]. In a more or less static network, the neighborhood information is invariant. On detection of neighborhood change, this two-hop neighborhood is recalculated by all nodes broadcasting their neighborhood nodes. Our *Cell Flood Algorithm* is shown in Algorithm 3.

$maximum(|\mathcal{N}_X - \mathcal{N}_A|\, for\, X \in \mathcal{N}_A \bigcap \mathcal{N}_B)$ is the maximum number of additional nodes that can be covered by any node which has received a broadcast from $node_A$ and is in the neighborhood of $node_B$. Greedy approach is in choosing the broadcasting node in line 5 of the algorithm, by favoring the node with maximum additional coverage. In Line 6, S is updated every time the node receives a broadcast of the packet from any

---

Algorithm 3. Cell Flood Algorithm

Require: $Node_B$ receives a broadcast packet from a neighbor $node_A$
Require: $\mathcal{N}_B$ be the set of 1-hop neighbors of $node_B$.
Require: $\mathcal{S}_B$ is a set of nodes in the 1-hop neighborhood which have not yet received the packet

1: **if** $Node_B$ has handled the same broadcast packet before **then**
2:     $Node_B$ *silently drops the packet*
3: **else**
4:     $Node_B$ calculates a timer proportional to the ratio
$$\left(\frac{maximum(|\mathcal{N}_X - \mathcal{N}_A| \, for \, X \in \mathcal{N}_A \cap \mathcal{N}_B)}{|\mathcal{N}_B - \mathcal{N}_A|} \right)$$
5:     *Till timer expires, node$_B$ updates $\mathcal{S}_B$ when it hears any other neighbor broadcast, using the neighborhood information*
6:     **if** $\mathcal{S}_B = \emptyset$ **then**
7:         $Node_B$ *silently drops the packet*
8:     **else**
9:         $Node_B$ *rebroadcasts the packet with some jitter*
10:     **end if**
11: **end if**

---

neighboring node. Elements of $\mathcal{S}$ which are present in the $\mathcal{N}$ of the neighboring node are removed. If $\mathcal{S}$ becomes empty, the packet it dropped, else the packet is forwarded.

Proof of full coverage: Every node in the network checks it's neighborhood to determine whether all neighbors have received a packet, and forwards the packet if there is any uncovered node in the neighborhood. Hence, all the neighbors of any node in the network receives the packet. As the full network is an union of the neighborhood of all the nodes, hence all the nodes in the network are covered.

Optimality of coverage: The proposed greedy heuristic leads to a approximate MCDS for the graph. Analysis of the approximation bound gives $log(n)$ times the cardinality of the optimal MCDS solution, where n is the number of nodes in the graph. The proof is similar to the optimality proof for Multipoint forwarding [23] .

Optimal Selectors Implementation. *Selectors* filter can be specified in some of the communication interfaces. If the filter is *null*, then the communication operations are efficiently implemented as elaborated previously. If this filter is not *null* but has a *Selectors*, then this scope (termed *Selectors* scope) is a subset of the scope with *null* selector (termed *null* scope). The communication operation can be done assuming *null* selector, and filtered at each node. This is not most efficient, specially is the *Selector* scope is significantly less than *null* scope. We implement the scoping of the *Get* and *Put* operations using the following technique.

    If operations with any *Selectors* filter is invoked the first time, the communication has to be delivered in the *null* scope as there is no knowledge of the location of the nodes matching the *Selectors*. But, for frequently invoked *Selectors* filters, an optimized implementation is done to cover the matching nodes only. If any particular *Selectors* if invoked frequently at any particular node, a trigger is set. The first packet being delivered after this trigger is set includes a *DoReinforment* flag. All the nodes matching this *Selectors* filter, sends a *Reinforced* message back to the source, specifying the *Selectors*. All the nodes through which this *Reinforced* message passes back are part of the forward set of nodes for the specified *Selectors*. All the subsequent packets with this *Selectors*, has a *OnlyReinforced* flag and is forwarded by the forward set of nodes only. This is remembered for a specific interval, greater than the period specified in the communication interface, if present. Periodically the *DoReinforcement* flag is set again and forwarded by all nodes to account for any change in topology or interest.

3.4   Multi-scale Application

To demonstrate the effectiveness of the network programming interfaces, we describe a distributed wavelet compression algorithm which can effectively use the interfaces to optimize the communication. Multi-resolution data analysis, processing and compression is useful for various sensor network applications. A lot of previous work on wavelet-based processing in sensor networks have assumed regularly-spaced data.

Recently, Wagner et. al. [24] have proposed a haar wavelet based multi-scale data analysis which enables irregular wavelet transform. In the bottom-up approach of that algorithm, all the sensors in the fundamental cell sends their sensor readings to the drum for that cell using *PutParent*. The locations and identifiers of the nodes are also available to the drum through the Discovery interface. The drum calculates the *scaling coefficient* describing the average reading of the cluster and the *wavelet coefficient* encoding the deviations from the average readings. These scaling coefficients are then passed up the hierarchy using the Parent interfaces, and similarly computed. In the top-down approach of the algorithm, querying is from the top-level and drilling down until the requisite resolution of the data is obtained. This is achieved using the GetPeer interfaces. Finest resolution is obtained if query goes down until the fundamental cell level.

For any compression, maximum efficiency with minimal loss is achieved when many data points are similar enough to be represented with one data point. When compressing a field of sensor data, various regions in space might have similar readings. So, if there is one cluster for each region, the region might be efficiently represented by a single value. But, if one cluster encompasses two different regions with divergent values, compression is not so efficient. The reclustering technique is used here such that each cluster has similar values and hence efficient representation. Locally near the region boundaries clusters are aligned with the regions.

4   Initial Design Evaluation

We have performed an initial evaluation of the design of our architecture by simulating the adaptive overlay formation in the *ns-2* simulator. We are currently implementing the protocol and interfaces in TinyOS. A full evaluation of our architecture will then be possible by modifying existing and new sensor network applications to use this architecture.

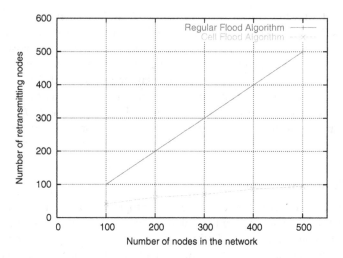

Fig. 4. Number of retransmitting nodes with increasing network size

4.1   Cluster Operations

In this section, we evaluate communication operations. In particular, we show the effectiveness of the Cell Flood Algorithm. To flood a particular cell with or without Selectors, this algorithm builds an approximate Minimum Connected Dominating Set. Thereby, the number of forwarding nodes is reduced without compromising the quality of the flood. Figure 4 shows the number of retransmitting nodes with increasing network size. The area of the network is kept constant while increasing the number of nodes, thereby increasing the node density. For a regular flooding algorithm, where each node forwards a packet exactly once, the number of forwarders is exactly equal to the number of nodes. In our Cell Flood Algorithm, the number of retransmitting nodes grows very slowly with increasing network size. The percentage of retransmitting nodes decreases with increasing network size, thereby showing good scalability.

4.2   Hierarchy Formation

In this section, we show the effectiveness of our hierarchy formation. Figure 5 shows the hierarchy formed for an example topology. 500 nodes are evenly distributed in an area of 3000 meters by 3000 meters. The radio range is taken to be 250 meters. Each part of Figure 5 shows the first and second level clusters formed, along with rays connecting each node to it's parent. Figure 5(a) shows the clustering for $\mathcal{D}_1$ equal to 1 hop, Figure 5(b) for $\mathcal{D}_1$ equal to 2 hops, and Figure 5(c) for $\mathcal{D}_1$ equal to 3 hops. As the number of hops allowed in the fundamental cell increases, the size of it increases along with decrease in the number of levels in the hierarchy.

Figure 6 illustrates the number of drums at each level of the hierarchy as a percentage of the total number of nodes in the network. This is shown for increasing networks sizes, with the node density kept constant. As shown in the analysis in Equation 2, the number of drums in increasing levels decreases quadratically. Larger the network, lesser

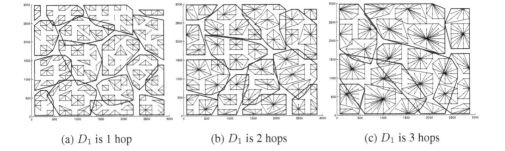

(a) D_1 is 1 hop   (b) D_1 is 2 hops   (c) D_1 is 3 hops

Fig. 5. Example topology with different values of D_1

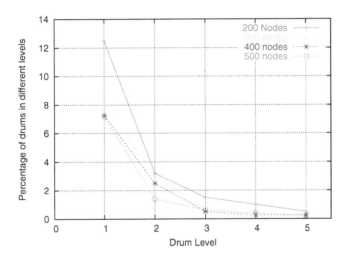

Fig. 6. Percentage of nodes which are drums are various levels

is the percentage of drums at each level. In this scenario, the value of α, which is the ratio between the beacon hop limits for consecutive drum levels, is taken as 2.

Figure 7 shows the latency for cluster formation with increasing network size. Here again, in all the networks sizes, the density is kept constant. The higher the level of drum, the longer it takes to stabilize. This is because the higher level drums are further spread apart and have larger beaconing intervals, which makes any change in higher levels propagate more slowly. The startup latency increases slowly with increasing network size. For the scenario sizes experimented with, the drum level 3 for all network sizes get stable at the same time. The cost of cluster formation arises from the beaconing during this phase, and hence is directly proportional to the length of time it takes to stabilize.

This also illustrates the effect of local change for adapting the hierarchy to communication requirements. The local clustering changes take place at a lower latency as shown in the figure. Therefore adaptivity of the clustering can be achieved with low latency, and hence also with low cost.

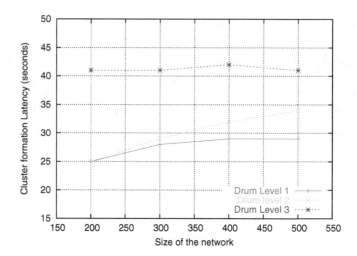

Fig. 7. Cluster formation latency with increasing size of network

5   Conclusion and Future Work

In this paper, we have motivated the need for a multi-scale architecture for sensor networks. Apart from enabling multi-resolution collaboration, a clustering hierarchy allows the network to scale to a very large number of sensors. Our architecture design adapts to the communication and collaboration requirements of the application, reducing communication energy and bandwidth usage. Our architecture also provides an abstraction to its low-level networking aspects, thereby simplifying application design.

We are currently implementing the architecture in TinyOS. We are using the fibers as blocking threads to implement Abstract Regions. This will enable us to deploy real applications and quantify the utility of our abstractions and adaptation interface. Currently, our evaluation is limited by the constraints of the *ns-2* simulator.

In this paper we have not addressed sensor network reliability or QoS requirements. These are important aspects that deserve attention. We plan to develop abstractions with tunable parameters through which the application can control the trade-off between resource usage and accuracy/reliability. Abstract Regions currently provides a tuning interface, but it entails the application to specify low-level parameters such as number of retransmissions. We intend to build tunable parameters at a higher level, which will enable the application to set goals, which will be automatically translated by the networking layer into the low-level parameters.

The Selectors implementation is currently fairly straightforward. The need to share attributes between hardware, application, and networking layer remains. This has been tackled previously in various approaches towards cross-layer design. However, there is ample scope for formalizing these cross-layer interactions and implementing them in an efficient way. Very complicated non-scalar Selectors can be supported by a fallback function provided by the application or by a loadable kernel module written by the application writer.

References

1. Akyildiz, I., Su, W., Sankarasubramaniam, Y., Cayirci, E.: Wireless sensor networks: A survey. Computer Networks **38** (2002) 393–422
2. Welsh, M., Mainland, G.: Programming sensor networks using abstract regions. In: NSDI. (2004) 29–42
3. Intanagonwiwat, C., Govindan, R., Estrin, D.: Directed diffusion: A scalable and robust communication paradigm for sensor networks. In: Proceedings, Sixth Annual Int. Conf. on Mobile Computing and Networking (MobiCOM '00), Boston, Massachussetts, USA (2000) 56–67
4. Heidemann, J., Silva, F., Estrin, D.: Matching data dissemination algorithms to application requirements. In: SenSys '03: Proceedings of the 1st international conference on Embedded networked sensor systems, ACM Press (2003) 218–229
5. Park, S.J., Vedantham, R., Sivakumar, R., Akyildiz, I.F.: A scalable approach for reliable downstream data delivery in wireless sensor networks. In: MobiHoc '04: Proceedings of the 5th ACM international symposium on Mobile ad hoc networking and computing, ACM Press (2004) 78–89
6. Gao, J., Guibas, L.J., Hershberger, J., Zhang, L.: Fractionally cascaded information in a sensor network. In: IPSN'04: Proceedings of the third international symposium on Information processing in sensor networks, ACM Press (2004) 311–319
7. Ganesan, D., Greenstein, B., Perelyubskiy, D., Estrin, D., Heidemann, J.: An evaluation of multi-resolution storage for sensor networks. In: SenSys '03: Proceedings of the 1st international conference on Embedded networked sensor systems, ACM Press (2003) 89–102
8. Yalagandula, P., Dahlin, M.: A scalable distributed information management system. In: SIGCOMM '04: Proceedings of the 2004 conference on Applications, technologies, architectures, and protocols for computer communications, ACM Press (2004) 379–390
9. Whitehouse, K., Sharp, C., Brewer, E., Culler, D.: Hood: a neighborhood abstraction for sensor networks. In: MobiSYS '04: Proceedings of the 2nd international conference on Mobile systems, applications, and services, ACM Press (2004) 99–110
10. Du, S., Khan, M., PalChaudhuri, S., Post, A., Saha, A., Druschel, P., Johnson, D.B., Riedi, R.: Self-organizing hierarchical routing for scalable ad hoc networking. Technical report, Rice (2004)
11. Chen, B., Morris, R.: L+:scalable landmark routing and address lookup for multi-hop wireless network. Technical Report MIT-LCS-TR-837, Laboratory for Computer Science Massachusetts Institute for Technology (2002)
12. PalChaudhuri, S., Du, S., Saha, A., Johnson, D.: Treecast: A stateless addressing and routing architecture for sensor networks. In: Proceedings of the 4th International Workshop on Algorithms for Wireless, Mobile, Ad Hoc and Sensor Networks (WMAN). (2004)
13. Culler, D., Shenker, S., Stoica, I.: Creating an architecture for wireless sensor networks. In: http://today.cs.berkeley.edu/SNA/. (2004)
14. Hill, J., Szewczyk, R., Woo, A., Hollar, S., Culler, D., Pister, K.: System architecture directions for networked sensors. In: ASPLOS-IX: Proceedings of the ninth international conference on Architectural support for programming languages and operating systems, ACM Press (2000) 93–104
15. Ni, S.Y., Tseng, Y.C., Chen, Y.S., Sheu, J.P.: The broadcast storm problem in a mobile ad hoc network. In: MobiCom '99: Proceedings of the 5th annual ACM/IEEE international conference on Mobile computing and networking, ACM Press (1999) 151–162
16. Williams, B., Camp, T.: Comparison of broadcasting techniques for mobile ad hoc networks. In: MobiHoc '02: Proceedings of the 3rd ACM international symposium on Mobile ad hoc networking & computing, ACM Press (2002) 194–205

17. Blum, J., Ding, M., Thaeler, A., Cheng, X.: Connected Dominating Set in Sensor Networs and MANETs. In: Handbook of Combinatorial Optimization (Editors D.-Z. Du and P. Pardalos), Kluwer Academic Publisher (2004) 329–369
18. Clausen, T., (editors), P.J., Adjih, C., Laouiti, A., Minet, P., Muhlethaler, P., Qayyum, A., L.Viennot: Optimized link state routing protocol (olsr). RFC 3626 (2003) Network Working Group.
19. Wu, J., Dai, F.: Broadcasting in ad hoc networks based on self-pruning. In: Proceedings of Infocom '03. (2003)
20. Bao, L., Garcia-Luna-Aceves, J.J.: A new approach to channel access scheduling for ad hoc networks. In: MobiCom '01: Proceedings of the 7th annual international conference on Mobile computing and networking, ACM Press (2001) 210–221
21. Rajendran, V., Obraczka, K., Garcia-Luna-Aceves, J.J.: Energy-efficient collision-free medium access control for wireless sensor networks. In: Sensys '03: Proceedings of the first international conference on Embedded networked sensor systems, ACM Press (2003) 181–192
22. Peng, W., Lu, X.C.: On the reduction of broadcast redundancy in mobile ad hoc networks. In: Poster at MobiHoc '00: Proceedings of the 1st ACM international symposium on Mobile ad hoc networking & computing, IEEE Press (2000) 129–130
23. Qayyum, A., Viennot, L., Laouiti, A.: Multipoint relaying for flooding broadcast messages in mobile wireless networks. In: HICSS '02: Proceedings of the 35th Annual Hawaii International Conference on System Sciences (HICSS'02)-Volume 9, IEEE Computer Society (2002) 298
24. Wagner, R., Sarvotham, S., Baraniuk, R.: A multiscale data representation for distributed sensor networks. In: Proceedings of the IEEE International Conference on Acoustics, Speech and Signal Processing (ICASSP), Philadelphia (2005)

Fault-Tolerant Self-organization in Sensor Networks*

Yi Zou and Krishnendu Chakrabarty

Department of Electrical and Computer Engineering,
Duke University, Durham, NC 27708, USA
{yz1, krish}@ee.duke.edu

Abstract. Sensor nodes in a distributed sensor network can fail due to a variety of reasons, e.g., harsh weather conditions, sabotage, battery failure, and component wear-out. Since many wireless sensor networks are intended to operate in an unattended manner after deployment, failing nodes cannot be replaced or repaired during field operation. Therefore, by designing the network to be fault-tolerant, we can ensure that a wireless sensor network can perform its surveillance and tracking tasks even when some nodes in the network fail. In this paper, we describe a fault-tolerant self-organization scheme that designates a set of backup nodes to replace failed nodes and maintain a backbone for coverage and communication. This scheme has been implemented on top of an energy-efficient self-organization technique for sensor networks. The proposed fault-tolerance-node selection procedure can tolerate a large number of node failures, without losing either sensing coverage or communication connectivity.

1   Introduction

Wireless sensor networks can be deployed to provide continuous surveillance and monitoring over a designated area of interest [1]. Many wireless sensor nodes have low cost and small form factors [1]; therefore, they can be deployed in large numbers with high redundancy. Since nodes are deployed in a redundant fashion, not every node in the network needs to be active for sensing and communication all the time. The operational lifetime of the network can be increased by selecting only a subset of nodes active, while keeping the other nodes in a sleep state. Fewer active nodes also places less demand on the limited network bandwidth.

Since a wireless sensor network should ideally perform surveillance tasks in an unattended manner, it needs to operate as long as possible, even when many

---

* This research was supported by DARPA, and administered by the Army Research Office under Emergent Surveillance Plexus MURI Award No. DAAD19-01-1-0504. Any opinions, findings, and conclusions or recommendations expressed in this publication are those of the authors and do not necessarily reflect the views of the sponsoring agencies.

V. Prasanna et al. (Eds.): DCOSS 2005, LNCS 3560, pp. 191–205, 2005.

sensor nodes fail. This motivates our work on fault-tolerant self-organization. Most recent work aims to provide fault tolerance in the deterministic deployment of sensor nodes [6, 13]. Much less attention has been devoted to distributed protocols that can replace failing nodes in the network with spare nodes. Failing sensor nodes result in coverage loss and breakage in communication connectivity, hence there is a need for a node replacement protocol and self-organization scheme that designates nodes as fault tolerance (spare) nodes.

This paper presents a fault-tolerant self-organization scheme that ensures communication connectivity and sensing coverage when nodes fail, either sequentially or simultaneously. We first present analytical results to characterize the extent of redundancy needed for fault tolerance. We then describe a distributed scheme that achieves fault tolerance by selecting fault tolerance nodes that can replace failing nodes. We show that the proposed approach provides communication connectivity and sensing coverage even when up to Ω nodes fail, where Ω is a user-defined parameter.

The paper is organized as follows. In Section 2, we briefly describe related prior work. In Section 3, we present the background and assumptions used in this paper work. Section 4 describes fault tolerance for communication connectivity. Section 5 addresses the fault tolerance for sensing coverage. We present simulation results for the proposed distributed self-organization technique in Section 6. Section 7 concludes the paper and outlines directions for the future work.

2   Related Work

Energy-efficient self-organization in wireless sensor networks has received considerable attention in the literature [1, 9, 12, 14, 15]. Energy considerations have been used to find a set of (active) nodes that can form a backbone for the network [2, 3, 15, 16, 19]. The selection of active nodes to guarantee both sensing coverage and communication connectivity has been studied in [10, 17, 18, 21]. A recent approach distinguishes connectivity from sensing, and determines the configuration of the nodes with considerations in both communication connectivity and sensing coverage [18].

Fault-tolerance in distributed sensor networks has received relatively less attention [6, 13]. Problems studied include the characterization of sensor fault modalities, fault-tolerance in multiple-sensor fusion [13], and reliable information dissemination [6]. Recent work for fault-tolerance in wireless sensor networks can be categorized as being focused on fault detection [4, 7, 8] or fault-tolerant operations [11, 20]. In [7], the authors present various fault tolerance techniques at different levels, including the physical layer for communication, hardware components of a sensor node, system software such as the embedded operating system, middleware, and applications. In [8], the authors consider faults in node sensor measurements and develop a distributed Bayesion algorithm to detect and correct such faults. [4] also addresses a similar fault detection problem, and presents a crash identification mechanism. In [20], the authors show that a sensor network with n nodes is asymptotically connected if each node is directly connected to at least $5.1774 \log n$ neighboring nodes. [11] shows that for a wireless sensor network with n nodes, the connectivity probability with up to k failing nodes is at least $e^{e^{\alpha}}$ when the transmission radius r satisfies $n\pi r^2 \geq \ln n + (2k-1) \ln \ln n - 2 \ln k! + 2\alpha$. However, prior work has not characterized the redundancy necessary for fault tolerance, and no distributed self-organization protocol has directly considered this issue.

3   Preliminaries

The discussions in this paper are based on the following assumptions. The ad hoc sensor network is deployed with sufficient nodes such that the network is connected. All sensor nodes have the same maximum communication range r_c and maximum sensing range r_s. We represent the surveillance field by a 2D grid, whose dimension is given as $X \times Y$. Let $\mathcal{G} = \{g_1, g_2, \cdots, g_m\}$ be the set of all grid points, and $m = |\mathcal{G}| = XY$. We use S to denote the set of n sensor nodes that have been placed in the sensor field, i.e., $|S| = n$. A node with id k is referred to as s_k ($s_k \in S, 1 \leq k \leq n$). Let d_i^k be the distance between the grid point g_i and the sensor node s_k. In the graph representation $G(V, E)$ for a set S of nodes, we use the vertex $v \in V$ in the graph model interchangeably with its corresponding node $s \in S$. We model sensing coverage using the probability p_i^k that a target at grid point g_i is detected by a node s_k. p_i^k is defined as: $p_i^k = e^{-\alpha d_i^k}$ if $d_i^k \leq r_s$ and $p_i^k = 0$ otherwise, where α is a parameter representing the physical characteristics of the sensor [21, 22]. Assume that S_i is the set of nodes that can detect grid point g_i; thus the detection probability $p_i(S_i)$ for grid point g_i is evaluated as: $p_i(S_i) = 1 - \prod_{s_k \in S_i}(1 - p_i^k)$.

Note that all sensors are assumed to be statistically independent in their detection processes. We assume that there exists a subset of active sensor nodes that provides a backbone for sensing coverage and communication connectivity. This can be achieved using techniques described in [10, 17, 18, 21]. The failure of these active nodes can result in loss of connectivity and/or loss of sensing coverage. We use S_a (S_s) to denote the set of active (sleeping) nodes determined by such active nodes selection algorithms, and the following discussion assumes that S_a and S_s have already been determined. We consider the threshold p_{th} to be a parameter underlying a successful sensing coverage over the sensor field. The following conditions are implicitly satisfied: 1) $\forall g_i \in \mathcal{G}$ and $S_i \subseteq S_a$, $p_i(S_i) \geq p_{th}$; 2) $\forall s_k \in S_a$ is connected.

We focus here on fault-tolerant self-organization, where both sensing coverage and connectivity are preserved with the support from the designated fault tolerance (FT) nodes when active nodes fail. We refer to this as the fault-tolerance-nodes-selection (FTNS) problem. The proposed distributed FTNS algorithm is executed after S_a and S_s are determined, where a set S_t of nodes is designated to be FT nodes (backup nodes for active nodes). These FT nodes provide fault tolerance for the existing active nodes. They need not be active unless the active nodes that they are supporting fail. They can run in a power-saving mode and periodically query whether the active nodes are still alive using very limited bandwidth.

Note that simultaneous failures of nodes in S_a and S_t may result in loss in sensing coverage or breakage in communication connectivity since FT nodes are not backed up by nodes in S_t. However, if only FT nodes fail or FT nodes and their non-neighboring active nodes fail, the sensing coverage and communication connectivity are still guaranteed. Furthermore, the proposed distributed algorithm can be applied in a repeated manner to select more FT nodes for the previously selected FT nodes.

We assume that the number of nodes initially deployed in the sensor field is sufficient to achieve fault-tolerant operations, i.e., we have enough sleeping nodes available to select as FT nodes. Some observations and additional definitions are listed below:

- It is trivial to see that if all failing nodes are sleeping nodes, the existing active nodes can tolerate the failure of up to $|S_s|$ nodes.
- We define the maximum number of active nodes that can fail simultaneously without losing sensing coverage or communication connectivity as the *degree of fault tolerance (DOFT)*, denoted by Ω ($\Omega \geq 1$).
- The nodes that are selected from the set of sleeping nodes to obtain a Ω-DOFT wireless sensor network are referred to as Ω-fault-tolerant (Ω-FT) nodes. We denote the set of Ω-FT nodes as S_t^Ω.
- Let $S_t^0 = \phi$ and $S_a^\Omega = S_t^\Omega \cup S_a$. It follows that S_a^Ω provides a solution to the Ω-DOFT FTNS problem. In other words, a Ω-DOFT FTNS-derived sensor network is still connected and provides undiminished coverage of the surveillance area if any Ω active nodes fail.

4   Connectivity-Oriented Fault Tolerance

In this section, we focus on the analysis of fault tolerance for communication connectivity. The discussion on fault tolerance for sensing coverage is presented in next section.

4.1   An Upper Bound on the Number of Fault Tolerance Nodes

We first consider the case of 1-DOFT, i.e., $\Omega = 1$. Let N_k be the set of neighbors for s_k, N_k^a be the set of active neighbors, and N_k^s be the set of sleeping neighbors. Let Δ_k be the number of neighboring nodes for s_k, Δ_k^a be the number of active neighboring nodes for s_k, and Δ_k^s be the number of sleeping neighboring nodes for s_k. In other words, $\Delta_k = |N_k|, \Delta_k^a = |N_k^a|$, and $\Delta_k^s = |N_k^s|$. It is trivial to see that $\forall s_k \in S, \Delta_k \geq 1$ otherwise S is not connected. Thus communication connectivity is not affected if any node in S_s fails. This is also true if multiple nodes in S_s fail. Therefore any number of sleeping nodes in S_s can fail either sequentially or simultaneously. This implies that only active nodes need to be considered as failing nodes for the analysis of connectivity fault tolerance.

It can be seen that, if $\exists s_k \in S$ such that $\Delta_k = 1$, then Ω-DOFT ($\Omega \geq 1$) cannot be achieved for the network since when this neighbor node of s_k fails, s_k is disconnected from the rest of the network [22]. For any wireless sensor network with S_a ($S_a \neq \phi$), $\forall s_k \in S$, s_k is connected to at least one node in S_a, i.e., $\Delta_k^a \geq 1$. Therefore, $\Delta_k \geq \Delta_k^a \geq 1$. In the sensor network with S_a as backbone for both sensing and communication, if $s_k \notin S_a$, i.e., s_k is a sleeping node, we can expect $\Delta_k > 1$ due to the need for sensing coverage; otherwise an active node must be located exactly at the same location as s_k. This observation leads to a lower bound on the node density required in the sensor field for fault tolerance. This lower bound can be used as a necessary condition for the fault-tolerant sensor node deployment.

Consider a total of n nodes with communication radius as r_c each in a sensor field with area A. In order to achieve Ω-DOFT ($\Omega \geq 1$), a lower bound on the total number of nodes n in the sensor field is given by $n \geq \frac{3A}{\pi r_c^2}$. The proof, which can be found in [22], is straightforward and is therefore omitted. For example, consider the extreme case of $A = \pi r_c^2$. For this case, we must have $n \geq 3$. This is obviously true since if there are only two nodes, neither of them can fail. In the following discussion, we assume that the initial sensor deployment has provided a sufficient number of nodes for fault tolerance. Our goal is to designate extra sleeping nodes as back-up nodes, i.e., FT nodes, to provide fault tolerance when currently-selected active nodes fail. We also need to minimize the number of FT nodes. Before we present bounds on the number of FT nodes needed to achieve Ω-DOFT, we state the following theorem. The proof of the theorem can be found in [22].

Theorem 1. *Let $s_k \in S$ be a node in the sensor network. Let the region that lies within the communication range r_c of s_k be A_k^* and let S^* be the set of nodes within A_k^*. Assume that all nodes in S^* are connected to each other, i.e., $\forall s_i, s_j \in S$, there exists a routing path from s_i to s_j. In order to ensure communication connectivity between the nodes in S^* if s_k fails, it is sufficient to have 10 nodes (not counting s_k) in A_k^*.*

Based on Theorem 1, we can derive the upper bound on the number of FT nodes needed within the communication region of an arbitrarily-chosen node. Assume that N_k is the set of neighbor for $s_k \in S$. Consider the special case where $S = N_k \cup \{s_k\}$, i.e., all nodes in $S \setminus \{s_k\}$ are neighbors of s_k. Suppose the nodes in N_k are not connected. When s_k fails, $\exists s_i, s_j \in N_k$ such that no routing can be formed between s_i and s_j. Thus fault tolerance can only be achieved if there is sufficient node density in the network. Let Γ_k be the number of FT nodes required for an arbitrarily-chosen node s_k in a 1-DOFT sensor network. Next we present a sufficient condition on Γ_k in the following theorem.

Theorem 2. *The network is 1-DOFT with respect to the failure of any node $s_k \in S_a$ if $\forall s_k \in S_a$, the nodes in N_k are connected and $\Gamma_k \geq 10$.*

The proof of Theorem 2 can be found in [22]. Generally, we have $\forall s_k \in S_a, |\Delta_k^a| \geq 1$ since $|S_a| > 1$, which implies the following corollary [22].

Corollary 1. *When the number of active nodes is greater than one, i.e., $|S_a| > 1$, the sensor network is 1-DOFT with respect to the failure of any node $s_k \in S_a$ if $\forall s_k \in S_a$, N_k is connected and there are 9 or more FT neighbor nodes for s_k.*

Corollary 1 shows that Δ_k^a is a measure of the communication connectivity support provided by the active neighbors of s_k when s_k fails. In fact, $\Delta_k^a > 0$ implies that there exists built-in fault tolerance for s_k. The fault tolerance provided by the active neighbors in N_k^a decreases the maximum number of FT nodes needed when s_k fails. Note that the above is true only for $\Omega = 1$ since when $\Omega > 1$, nodes in N_k^a may also fail at the same time when s_k fails. Both Theorem 2 and Corollary 1 assume that when $s_k \in S_a$ fails, the selected FT nodes for s_k do not fail. Since FT nodes are selected to provide fault tolerance for active nodes in S_a, their own failures are not considered in the analysis. However, the same procedure of selecting FT nodes for active nodes in S_a can be applied repeatedly to select more FT nodes in a sequential manner for more failing nodes.

Our goal in this paper is to develop a distributed self-organization algorithm, where nodes rely only on single-hop knowledge within their neighborhood. Therefore, we allow each active node $s_k \in S_a$ to select FT nodes only from its sleeping neighbors. Recall that we denote the set of FT nodes in a Ω-DOFT sensor network FT nodes as S_t^Ω. Let N_k^Ω be the set of FT neighbors for an arbitrarily-chosen $s_k \in S_a$ in a Ω-DOFT network. Obviously, $N_k^\Omega \subseteq N_k^s$ and $\Gamma_k = |N_k^\Omega|$. When each active node finds its corresponding N_k^Ω, the set S_t^Ω is determined, i.e., $S_t^\Omega = \cup_{\forall s_k \in S_a} N_k^\Omega$, where the total number of FT nodes in this Ω-DOFT sensor network is $|S_t^\Omega|$. Next, we derive an upper bound on the total number of FT nodes needed for the entire sensor network. For 1-DOFT case, this bound is obtained directly from Theorem 3, as shown in below.

Theorem 3. *Consider a wireless sensor network consisting of n nodes each with communication radius r_c. Let the set of nodes be denoted by S. Assume that all nodes in S are connected, i.e., $\forall s_i, s_j \in S$, there exists a routing path from s_i to s_j. Let $G(V, E)$ be the connected graph corresponding to S, i.e., $|V| = |S|$ and v_k is the vertex representing $s_k \in S$, where $\forall u, v \in V, (u, v) \in E$ if $d(u, v) \leq r_c$. Assume that S_a is the set of (active) backbone nodes. The subgraph corresponding to S_a is denoted by $G_a(V_a, E_a)$, where G_a is a CDS of G. Let S_t^1 be the set of nodes selected as FT nodes to achieve 1-DOFT. An upper bound on the total number of FT nodes needed to achieve 1-DOFT is given as*

$$|S_t^1| \leq \begin{cases} 10, & \text{if } |V_a| = 1; \\ 9|V_a| - |E_a|, & \text{if } |V_a| > 1. \end{cases} \tag{1}$$

The proof the theorem can be found in [22]. Next, we consider a more general fault tolerance scenario where $\Omega > 1$. Note that we assume $|S_a| \geq \Omega$ for the analysis of Ω-DOFT; otherwise Ω-DOFT is not meaningful. In the following, we determine the number of nodes Γ_k needed for an arbitrarily-chosen active node to achieve Ω-DOFT in its communication region. Note that for a Ω-DOFT sensor network, if $\exists s_k \in S_a$ such that $\Omega > \Delta_k^a$, the DOFT in the communication region of s_k is at most $\Delta_k^a + 1$. However, when Ω-DOFT is achieved for the entire sensor network, fault tolerance with the maximum number of failing nodes in the communication region of s_k is automatically achieved. In the following, we assume that $\Delta_k^a \geq \Omega - 1$ to simplify the discussion. Note also that since $\Omega > 1$, we have $|S_a| > 1$. Therefore we can ignore the special case where only one node is active and all other nodes are placed within its communication range.

Theorem 4. *The network is Ω-DOFT ($\Omega > 1$) with respect to failures of any Ω nodes inside the communication region of an arbitrarily-chosen $s_k \in S_a$ ($1 < \Omega \leq \Delta_k^a + 1$), if the nodes in N_k are connected and $\Gamma_k \geq \Omega$. Moreover, Γ_k is lower-bounded by the following:*

$$\Gamma_k \geq \begin{cases} \Omega + 9, & \text{if } s_k \text{ fails and } \Omega = \Delta_k^a + 1; \\ \Omega + 8, & \text{if } s_k \text{ fails and } \Omega < \Delta_k^a + 1; \\ \Omega, & \text{if } s_k \text{ does not fail.} \end{cases} \quad (2)$$

The proof of Theorem 4 is given in [22]. Bounds on the total number of FT nodes needed to achieve Ω-DOFT ($\Omega > 1$ and $|S_a| \geq \Omega > 1$) can be derived in a similar way, where details of this part can be found in [22]. Again, note that it is possible that $\exists s_k \in S_a$ such that $\Delta_k^a + 1 < \Omega$. In this case, since the maximum number of failing nodes within the communication region of s_k is at most $\Delta_k^a + 1$, Ω-DOFT for s_k refers to the failure of up to $\Delta_k^a + 1$ nodes inside the communication region of s_k, and the failure of $\Omega - (\Delta_k^a + 1)$ nodes outside the communication region of s_k.

4.2   Lower Bound on the Number of Fault Tolerance Nodes

To reduce energy consumption, it is desirable to minimize the number of FT nodes needed, i.e., to minimize the size of S_t^Ω. In this section, we present a lower bound on the number of FT nodes needed to achieve the required Ω-DOFT ($\Omega \geq 1$) in wireless sensor networks. Let $N_k^f \subseteq N_k^a$ be the set of failing active neighbors of s_k, i.e., $S_f = \cup_{s_k \in S_a} N_k^f$. Let $N_k^\Omega \subseteq N_k^s$ be the set of FT nodes for s_k, i.e., $S_t^\Omega = \cup_{s_k \in S_a} N_k^\Omega$.

We know from previous subsections that $\forall s_k \in S_a$, FT nodes of s_k keep all neighbors nodes of s_k in N_k connected. This implies that the subgraph representing N_k^Ω is a CDS of the subgraph representing N_k. When $\Omega = 1$, the minimization of $|S_t^\Omega|$ is equivalent to finding the MCDS for the subgraph representing N_k for each active node $s_k \in S_a$. However, since no failing active node has any failing active neighbors for $\Omega = 1$, such an MCDS for s_k also contains existing active neighbors in N_k^a as existing dominating nodes. Let S_t^1 be the set of nodes selected as FT nodes to achieve 1-DOFT in the sensor network. It then easy to see that a lower bound on the total number of FT nodes needed to achieve 1-DOFT, i.e., $|S_t^1|$, is given by:

$$|S_t^1| \geq \begin{cases} 1, & \text{if } |S_a| = 1; \\ 0, & \text{if } |S_a| > 1. \end{cases} \quad (3)$$

Note that the best case of $|S_t^1| = 0$ when $|S_a| > 1$ rarely happens in practice, because it requires that neighbors of any active node are also neighbors of at least another active node. This implies that all nodes are within a circle of radius r_c. Since $|S_a| > 1$, this makes the other $|S_a| - 1$ nodes unnecessary. It is possible to have several such nodes but if $|S_a|$ is very large, there will be a significant energy overhead for these nodes. When $\Omega > 1$, the analysis is more complicated because when an active node s_k fails, some active neighbors in N_k^a may also fail at the same time. To simplify the discussion, we define function $\mathcal{M}$ as $\bar{S}_a = \mathcal{M}(S, S_a)$, where 1) $S_a \subseteq \bar{S}_a$; 2) The subgraph representing $\bar{S}_a$ is a connected dominating set (CDS) of the graph representing S; 3) For all possible sets that satisfies 1) and 2), $\bar{S}_a$ has the smallest size.

We refer to determining $\bar{S}_a$ as a constrained minimum connected dominating set (constrained MCDS) problem. Note that if $S_a = \phi$, then $\bar{S}_a$ is the MCDS of S. To achieve Ω-DOFT ($\Omega \geq 1$) in the wireless sensor network, we need to find the set of FT nodes S_t^Ω such that $S_t^\Omega = \bigcup_{\forall S_f \subseteq S_a, |S_f| \leq \Omega} \mathcal{M}(S \setminus S_f, S_a \setminus S_f)$. Let $N_k^f \subseteq N_k^a$ be the set of failing active neighbors of s_k. We can obtain a lower bound on the number of FT nodes needed to achieve Ω-DOFT ($\Omega > 1$) as follows [22]:

$$|S_t^\Omega| \geq \Big| \bigcup_{\forall |S_f| \leq \Omega, S_f \subseteq S_a} \mathcal{M}(S \setminus S_f, S_a \setminus S_f) \Big| \qquad (4)$$

Note that if $\Omega = |S_a|$, $S_a \setminus S_f = \phi$, then $|S_t^\Omega| \geq |\mathcal{M}(S \setminus S_a, \phi)|$.

4.3   Connectivity-Oriented Selection of Fault Tolerance Nodes

Since the CDS and MCDS problems for a given graph are $\mathcal{NP}$-complete [5,3,19], finding the constrained MCDS described earlier to achieve Ω-DOFT as shown in Equation (4) is also $\mathcal{NP}$-complete. When only single-hop knowledge is available, for any $s_k \in S_a$, there are a total of $\sum_{i=1}^\Omega \binom{|N_k^a|}{i}$ possible combinations of failing nodes for s_k; as a result, the total number of possible combination of failing nodes for all the active nodes is $\sum_{\forall s_k \in S_a} \left(\sum_{i=1}^\Omega \binom{|N_k^a|}{i} \right)$. Each evaluation requires the finding of the MCDS for neighbors of the failing node. Even though failing active nodes may share many neighbors, a thorough evaluation in this way is still computationally very expensive.

For a wireless sensor network with a set S_a of active nodes serving as a backbone, the maximum number of nodes that can fail is $|S_a|$. We propose the following distributed procedure to achieve fault tolerance for the simultaneous failure of up to $|S_a|$. The proposed distributed procedure is based on the algorithm from [19]. The procedure contains three steps as shown in Fig. 1.

In Step 1 of Fig. 1, each active node selects a FT node for any of its disconnected active neighbors. We refer to this type of FT nodes as gateway FT nodes since they provide alternative routing paths for active neighbors of the failing node. When that potential failing node actually fails, the network traffic from the failing node to its active neighbors can still be delivered. Though the first type of FT nodes are able to take care of the routing data originating from failing active nodes, they are not necessarily connected among each other and are not necessarily connected to sleeping neighbors of the failing active node.

Step 2 in Fig. 1 deals with this problem by using a modified version of the algorithm proposed in [19], which proposed a distributed approach for constructing the CDS for a connected but not completely connected graph. In the worst case, when all nodes in S_a fail at the same time, the subgraph representing FT nodes should be a CDS of the subgraph representing S_s. We can therefore utilize the algorithm proposed in [19] with the target graph representing S_s. Note that in Step 2, we have already found gateway FT nodes, therefore Step 2 needs only check for connectivity of disconnected FT nodes. To ensure that the proposed distributed procedure is also applicable to more general scenarios, Step 3 is added to handle the case that the subgraph representing S_s is a completely connected graph. Also note that the proposed connectivity-oriented fault tolerance nodes selection procedure in Fig. 1 is designed in a distributed manner since it requires only local knowledge for each node. The local knowledge can be obtained upon the initialization stage of sensor networks by methods such as integrating the neighbor information into the HELLO message. We next prove that the proposed distributed procedure achieves $|S_a|$-DOFT for a wireless sensor network with the set of active nodes given by S_a.

Theorem 5. *Consider a wireless sensor network with n nodes, each with communication radius r_c, denoted by the set S. Assume that all nodes in S are connected, i.e., $\forall s_i, s_j \in S$, there exists a routing path from s_i to s_j. Assume that S_a is the set of active nodes as a backbone that keeps all nodes connected. Assume that S_t is the set of FT nodes obtained from the distributed FT nodes selection procedure given by Fig. 1. The set S_t achieves Ω-DOFT in this wireless sensor network, where $\Omega = |S_a|$.*

Proof: Since the maximum number of nodes that can fail is $|S_a|$, we only need to consider the case that the selected FT nodes in S_t are able to keep the network fully connected when all nodes in S_a fail. Let $G_s(V_s, E_s)$ be the subgraph representing $S_s = S \setminus S_a$ and $G_t(V_t, E_t)$ be the subgraph representing S_t. To prove that G_t is a CDS of G_s, we first show that G_t is connected, then we show that for any $v \in V_s$, v is either in V_t or adjacent to a vertex in V_t.

Consider any $u, v \in V_t$. Since G_s is connected, $\exists P(u, v)$ as the shortest path from u to v in G_s, where $P(u, v) \subseteq V_s$ is the set of the vertices in the path. If $|P(u, v)| = 2$, the theorem is trivially proved. Assume $|P(u, v)| \geq 3$, and let

---

Distributed FT nodes selection procedure

Step 1. $\forall s_k \in S_a$, for each pair of disconnected active neighbors, s_k selects $s_i \in N_k^s$ as a FT node if s_i connects both.
Step 2. $\forall s_k \in S_s$,
    Step 2.1. if has two disconnected FT neighbors, then s_k assigns itself as FT node;
    Step 2.2. if has two disconnected FT neighbor node and sleeping node, then s_k assigns itself as FT node;
Step 3. $\forall s_k \in S_s$, if s_k has no FT neighbors, then s_k assigns itself as FT node.

Fig. 1. Distributed connectivity-oriented fault tolerance nodes selection procedure

$P(u, v) = \{u, u_1, u_2, \cdots, v\}$. Consider predecessor vertices of u in $P(u, v)$, i.e., u_1. Since $u \in V_t$, from Step 2 in Fig. 1, u_1 has to be in V_t, irrespective of whether u_2 is in V_t. The same argument holds for u_2. Doing this repeatedly, we have $\forall w \in P(u, v), w \in V_t$, i.e., $P(u, v) \in V_t$. Next, $\forall v \in V_s$, from Step 3 in Fig. 1, v has at least one FT neighbor. Therefore, G_t is a CDS of G_s.   ∎

5   Coverage-Centric Fault Tolerance

In Section 4, we have discussed the Ω-DOFT problem for fault-tolerant communication connectivity of up to Ω active nodes failing simultaneously ($\Omega > 1$). However, we should also take fault tolerance for sensing coverage into account to achieve the surveillance goal over the field of interest. This implies that the nodes selected as FT nodes must be able to provide enough sensing coverage over the areas that were originally under the surveillance of the Ω failing active nodes.

5.1   Loss of Sensing Coverage

Recall the collective coverage probability for a grid point g_i defined in Section 3. Since only active nodes in S_a perform communication and sensing tasks, the collective coverage probability for g_i is actually from nodes in S_i^a, where $S_i^a \subseteq S_i$ is the set of active nodes that can detect g_i. When nodes fail in the network, the set of active nodes that can detect g_i, i.e., S_i^a, changes with time, which subsequently changes the sensing coverage over that grid point. Let $q_i(S)$ be a mapping from a set S of nodes to the coverage probability for grid point g_i, $p_i(t)$ be a mapping from a time instant t to the coverage probability for grid point g_i, and $\mathcal{S}(t)$ be a mapping from a time instant t to a set of nodes. Then $\mathcal{S}_i(t)$ is the set of nodes that can detect grid point g_i at time instant t. For example, if at time instant t, only nodes in the subset S_i^a, i.e., active nodes, detect grid point g_i, therefore $\mathcal{S}_i(t) = S_i^a$ and $p_i(t) = q_i(\mathcal{S}_i(t)) = q_i(S_i^a)$. Therefore, as shown in Section 3, coverage probability of g_i under the fault tolerance constraint is a function of time given by $p_i(t) = q_i(\mathcal{S}_i^a(t)) = 1 - \prod_{s_k \in \mathcal{S}_i^a(t)}(1 - p_i^k)$, where $\mathcal{S}_i^a(t)$ is the set of active nodes that can still detect g_i at time instant t. Therefore, the goal is to ensure that the selected FT nodes and existing active nodes, i.e., $S_a^\Omega = S_a \cup S_t^\Omega$, are able to keep the sensor field adequately covered whenever up to Ω active nodes fail. Thus, successful sensing coverage over the sensor field for FTNS in wireless sensor networks is indicated by $\forall g_i \in \mathcal{G}, \; p_i(t) \geq p_{th}$, where p_{th} is the coverage probability threshold defined in Section 3. Theorem 6 shows the relationship between the loss of sensing coverage and the fault-tolerant operation in wireless sensor networks.

Theorem 6. *Consider a wireless sensor network with n nodes, each with communication radius r_c, denoted by the set S. Assume that all nodes in S are connected, i.e., $\forall s_i, s_j \in S$, there exists a routing path from s_i to s_j. Let $\mathcal{G}$ be the set of all the grid points in the sensor field. Let S_i be the set of nodes that* *can detect the grid point $g_i \in \mathcal{G}$ initially after the deployment. Let $\mathcal{S}_i(t)$ be the set of nodes that can detect g_i at time t, and $\mathcal{S}_i^f(t)$ be the set of failing active nodes for g_i at time t. Throughout the operational life time of a sensor network, $\forall g_i \in \mathcal{G}$, the following must be satisfied for any time instant t:*

$$p_f(t+1) \leq \frac{p_i(t) - p_{th}}{1 - p_{th}}. \tag{5}$$

where $p_i(t) = 1 - \prod_{s_k \in \mathcal{S}_i(t)}(1 - p_i^k)$ and $p_f(t+1) = 1 - \prod_{s_k \in \mathcal{S}_i^f(t+1)}(1 - p_i^k)$.

Proof: Consider time instants t and $t+1$. Obviously we have $\mathcal{S}_i(t) \subseteq S_i$ and $\mathcal{S}_i(t) = \mathcal{S}_i(t+1) \cup \mathcal{S}_i^f(t+1)$. Then we have

$$p_i(t) = 1 - \prod_{s_k \in \mathcal{S}_i(t)}(1 - p_i^k) = 1 - \prod_{s_k \in \mathcal{S}_i(t+1)}(1 - p_i^k) \prod_{s_k \in \mathcal{S}_i^f(t+1)}(1 - p_i^k).$$

Let $p_f(t) = 1 - \prod_{s_k \in \mathcal{S}_i^f(t)}(1 - p_i^k)$ and $p_f(t+1) = 1 - \prod_{s_k \in \mathcal{S}_i^f(t+1)}(1 - p_i^k)$. So

$$p_i(t) = 1 - (1 - p_i(t+1))(1 - p_f(t+1)) = p_f(t+1) + p_i(t+1)(1 - p_f(t+1)).$$

Therefore, $p_i(t+1) = \frac{p_i(t) - p_f(t+1)}{1 - p_f(t+1)}$, thus

$$p_i(t+1) \geq p_{th} \quad \Rightarrow \quad \frac{p_i(t) - p_f(t+1)}{1 - p_f(t+1)} \geq p_{th},$$

which implies that

$$p_f(t+1) \quad \leq \quad \frac{p_i(t) - p_{th}}{1 - p_{th}}. \qquad \blacksquare$$

From the proof of Theorem 6, we see that $p_f(t+1)$ represents the sensing coverage loss at time $t+1$ at grid point g_i caused by the failing nodes in $\mathcal{S}_i^f(t+1)$. To satisfy the coverage probability threshold requirement, $p_f(t+1)$ must not exceed $\frac{p_i(t) - p_{th}}{1 - p_{th}}$. In other words, if we can bound the coverage loss $p_f(t)$ below $\frac{p_i(t) - p_{th}}{1 - p_{th}}$ during the operational lifetime of the sensor network for all grid points on the field, the sensor network is able to tolerate up to Ω nodes failing simultaneously. When $p_i(t)$ drops, the bound on the coverage loss from failing nodes at the next time instant, i.e., $p_f(t+1)$, becomes tighter since $\frac{p_i(t) - p_{th}}{1 - p_{th}}$ decreases when $p_i(t)$ decreases. This can also be used as a warning criteria to inform the base station whether a current node may lose sensing coverage over its sensing area.

Note that the fault tolerance problem for sensing coverage differs from the fault tolerance problem for communication connectivity discussed in Section 4 since there is no direct relationship between the number of failing nodes and the coverage loss $p_f(t)$. For example, for g_i with $|\mathcal{S}_i(t)| = 1$, $p_i(t)$ might be the same as $p_j(t)$ for g_j where $|\mathcal{S}_j(t)| = 1, 2, 3$ or even higher. This is due to the fact that for any grid point g_i, $p_i(t)$ is not directly related to the number of nodes that can detect g_i but rather to the distances from these nodes to g_i defined in Section 3.

5.2   Distributed Approach

We next propose a coverage-centric fault tolerance algorithm that can be executed in a distributed manner, and requires much less computation than the centralized case. Without loss of generality, assume $r_c \geq 2r_s$, i.e., $S_i \subseteq N_k$. For grid point $g_i \in A_k$ corresponding to node $s_k \in S_i^a \subseteq S_a$, the maximum coverage loss happens when all nodes in S_i^a fail. In this case, the coverage loss for g_i, denoted as $q_i(S_i^a)$, is given as $q_i(S_i^a) = 1 - \prod_{s_k \in \cup S_i^a}(1 - p_i^k)$. Let $S_i^\Omega \subseteq S_i^s$ be the set of FT nodes for grid point g_i. The coverage compensation from S_i^Ω, denoted as $q_i(S_i^\Omega)$, is given as $q_i(S_i^\Omega) = 1 - \prod_{s_k \in S_i^\Omega}(1 - p_i^k)$. Let $q_i(S_i^a \cup S_i^\Omega)$ be the coverage from both active nodes and the FT nodes for g_i. Similarly, $q_i(S_i^a \cup S_i^\Omega) = 1 - \prod_{s_k \in S_i^a \cup S_i^\Omega}(1 - p_i^k)$. Assuming that the maximum coverage loss happens at time instant $t+1$, i.e., $\mathcal{S}_i(t) = S_i^a \cup S_i^\Omega, \mathcal{S}_i(t+1) = S_i^\Omega$, and $\mathcal{S}_f(t+1) = S_i^a$, then accordingly, we have corresponding expressions as $p_i(t) = q_i(S_i^a \cup S_i^\Omega), p_i(t+1) = q_i(S_i^\Omega)$, and $p_f(t+1) = q_i(S_i^a)$. From Equation (5), if the following is satisfied for all grid points in the sensing area of s_k, i.e., A_k, then the node s_k is able to tolerate the maximum number of failing active nodes within its own sensing area without losing sensing coverage. This is shown as: $q_i(S_i^a) \leq \frac{q_i(S_i^a \cup S_i^\Omega) - p_{th}}{1 - p_{th}}$ ($\forall g_i \in A_k$). This evaluation procedure is per grid point, which can be executed on either a sleeping node or an active node. For any $g_i \in \mathcal{G}$, only one node needs to perform the selection of FT nodes for g_i. This implies that total number of nodes required for executing such evaluation procedure is $\lceil \frac{|\mathcal{G}|}{|A_k|} \rceil$ or $\lceil \frac{A}{\pi r_s^2} \rceil$ where A is the area of the surveillance field (assuming that either $r_c \geq 2r_s$ or $\lceil \frac{2r_s}{r_c} \rceil$-hop knowledge is available). Fig. 2 shows the pseudocode for the coverage-centric fault tolerance nodes selection algorithm.

Note that in Fig. 2, there is no need to calculate $q_i(S_i^a)$ every time since it is available from the previous stage when S_a is determined [22]. Further computation can be reduced by temporarily storing the $q_i(S_k^\Omega)$ for the current grid point for evaluation at the next grid point, where $q_i(S_i^\Omega \cup S_k^\Omega)$ can be obtained by $q_i(S_i^\Omega \cup S_k^\Omega) = q_i(S_i^\Omega) + q_i(S_k^\Omega) - q_i(S_i^\Omega)q_i(S_k^\Omega)$.

Procedure • •••• ••• •• ••••• • • • •••••• (s$_k$)

---

01 Set $S_k^\Omega = \phi$;
02 **For** $\forall g_i \in A_k$
03   **If** $q_i(S_i^a) \leq \frac{q_i(S_i^\Omega) - p_{th}}{1 - p_{th}}$ **Continue;**   **End**
04   **For** $\forall S_i^\Omega \subseteq S_i^s \setminus S_k^\Omega$
05   **If** $q_i(S_i^a) > \frac{q_i(S_i^\Omega \cup S_k^\Omega) - p_{th}}{1 - p_{th}}$ **Continue;**   **End**
06   Set $S_k^\Omega = S_k^\Omega \cup S_i^\Omega$;   **Break;**
07   **End**
08 **End**

Fig. 2. Pseudocode for the distributed coverage-centric fault tolerance nodes selection

Note that similar to the connectivity-oriented fault tolerance nodes procedure in Fig. 1, the coverage-centric fault tolerance nodes selection procedure in Fig. 2 is also designed in a distributed manner. However, it must be noted that the one-hop communication neighbors of a node might be different from its sensing neighbors due to the relationship of the node's sensing radius and communication radius [21, 22]. If $r_c < 2r_s$, it is necessary for s_k to know locations of nodes up to $\lceil \frac{2r_s}{r_c} \rceil$ hops from it to evaluate its coverage redundancy [21, 22].

6   Simulation Results

We next evaluate the proposed distributed FTNS procedure using MatLab and the data collected in ns2 for distributed coverage-centric active nodes selection (CCANS) procedure described in [21]. The data from CCANS contains locations of sensor nodes after deployment and their final state decisions. There are 150, 200, 250, 300, 350, and 400 nodes in each random deployment, respectively, on a 50×50 grid representing a 50m $\times$ 50m sensor field. All nodes have the same maximum communication radius $r_c = 20$m and maximum sensing range $r_s = 10$m. The value of Ω is set to the number of active nodes, i.e., Ω corresponds to the the maximum number of active nodes that can fail simultaneously. Figures 3(a)– 3(c) show the simulation results for distributed fault-tolerance self-organization procedure.

Fig. 3(a) shows the results obtained for connectivity-oriented selection of FT nodes. Note that the percentage of FT nodes decreases nearly at the same rate as the percentage of active nodes. This is because the connectivity-oriented FT nodes selection algorithm is executed in a distributed manner and each node uses only one-hop knowledge. Note also that the percentage of FT nodes is lower than the percentage of active nodes determined by CCANS. This is because CCANS considers both communication connectivity and sensing coverage in selecting

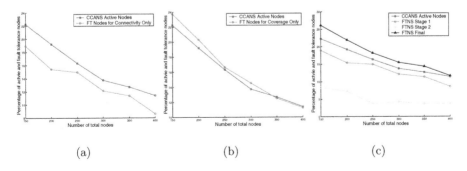

(a)   (b)   (c)

Fig. 3. Simulation results: (a) Percentage of fault tolerance nodes for communication connectivity only (b) Percentage of fault tolerance nodes for sensing coverage only (c) Percentage of fault tolerance nodes for the FTNS algorithm (with both coverage and connectivity concerns)

active nodes. Fig. 3(b) shows the results for coverage-centric selection of FT nodes. The percentage of FT nodes is nearly the same as the percentage of active nodes obtained from the CCANS algorithm. This is due to the fact that we have $r_c = 2r_s$ in this scenario. As shown in [22], when $r_c = 2r_s$, the connectivity is automatically guaranteed by the subset of nodes needed to maintain sensing coverage.

Fig. 3(c) presents the result for the distributed FTNS algorithm. It consists of two stages, which are referred to as "FTNS Stage 1" and "FTNS Stage 2", respectively, in Fig. 3(c). The FTNS Stage 1 corresponds to the connectivity-oriented selection of FT nodes and FTNS Stage 2 corresponds to the coverage-centric selection of FT nodes. The FT nodes that have already been selected in Stage 1 are checked first in Stage 2 to see if they already provide enough sensing coverage for fault tolerance. This decreases the number of FT nodes needed for Stage 2 of coverage-centric FT nodes selection. From Fig. 3(c), the total number of FT nodes selected for both connectivity and coverage is only slightly higher than the number of original active nodes.

7   Conclusions

In this paper, we have investigated fault tolerance for coverage and connectivity in wireless sensor networks. Fault tolerance is necessary to ensure robust operation for surveillance and monitoring applications. Since wireless sensor networks are made up of inexpensive nodes and they operate in harsh environments, the likely possibility of node failures must be considered. We have characterized the amount of redundancy required in the network for fault tolerance. Based on an analysis of the redundancy necessary to maintain communication connectivity and sensing coverage, we have proposed the FTNS algorithm for fault-tolerant self-organization. FTNS is able to provide a high degree of fault tolerance such that even when all of these active nodes fail simultaneously, coverage and connectivity in the network are not affected. We have implemented FTNS in MatLab and presented representative simulation results.

References

1. I. F. Akyildiz, W. Su, Y. Sankarasubramaniam, and E. Cayirci, "A survey on sensor networks," •• • • ••• • •• •••••••, pp. 102–114, 2002.
2. K. M. Alzoubi, P. J. Wan, and O. Frieder, "Distributed heuristics for connected dominating sets in wireless ad hoc networks," ••• •• • •• •••, vol. 4, pp. 1–8, 2002.
3. V. Bharghavan and B. Das, "Routing in ad hoc networks using minimum connected dominating sets," • •••• ••• •• ••• , pp. 376–380, 1997.
4. S. Chessa and P. Santi, "Crash faults identification in wireless sensor networks," ••• •••••• ••• • •••••••••, vol. 25, no. 14, pp. 1273–1282, 2002.
5. M. R. Garey and D. S. Johnson, ••• •••••• ••• ••••••••••••• • ••••• •• ••• •••••• •• $\mathcal{NP}$•••• •••••••••, W. H. Freeman and Co., 1979.

6. S. S. Iyengar, M. B. Sharma, and R. L. Kashyap, "Information routing and reliability issues in distributed sensor networks," •• • • •••••• •••••• •••••••, vol.
   40, no. 2, pp. 3012–3021, 1992.
7. F. Koushanfar, M. Potkonjak, and A. Sangiovanni-Vincentelli, "Fault tolerance in
   wireless ad-hoc sensor networks," • •••• •• • • ••••••, 2002,
8. B. Krishnamachari and S. S. Iyengar, "Distributed Bayesian algorithms for fault-
   tolerant event region detection in wireless sensor networks," •• • • ••••••• ••• •
   ••••••, vol. 53, pp. 241–250, March 2004.
9. C. Gui and P. Mohapatra, "Power conservation and quality of surveillance in target
   tracking sensor networks," • •••• • ••• •• , pp. 129–143, 2004.
10. H. Gupta, S. R. Das, and Quinyi Gu, "Connected sensor cover: self-organziation of
    sensor networks for efficient query execution," • •••• • ••• ••, pp. 189–200, 2003.
11. X. Y. Li, P. J. Wan, Y. Wang, and C. W. Yi, "Fault tolerant deployment and
    topology control in wireless networks," • •••• •• • • • •••• ••, pp. 117–128, 2003.
12. J. Polastre, J. Hill and D. Culler, "Versatile low power media access for wireless
    sensor networks," • •••• •••••, pp. 95–107, 2004.
13. L. Prasad, S. S. Iyengar, R. L. Rao, R. L. Kashyap, "Fault- tolerant sensor inte-
    gration using multiresolution decomposition," • •••••• • •••• , vol. 49, 1994.
14. K. Seada, M. Zuniga, A. Helmy, and B. Krishnamachari. "Energy-efficient for-
    warding strategies for geographic routing in lossy wireless sensor networks," • ••••
    ••••••, pp. 108–121, 2004.
15. S. Slijepcevic and M. Potkonjak, "Power efficient organization of wireless sensor
    networks," • •••• •• • , pp. 472–476, 2001.
16. I. Stojmenovic, M. Seddigh, and J. Zunic, "Dominating sets and neighbor elimina-
    tion based broadcasting algorithms in wireless networks," • •••• •• • • •• ••• • ••••
    •••••• •••••••, 13(1), pp.14–15, 2002.
17. D. Tian and N. D. Georganas, "A node scheduling scheme for energy conservation
    in large wireless sensor networks," • ••••• ••• • • • ••• • • ••, vol. 3, pp.271-
    290, 2003.
18. X. R. Wang, G. L. Xing, Y. F. Zhang, C. Y. Lu, R. Pless, and C. Gill, "Integrated
    coverage and connectivity configuration in wireless sensor networks," Proc. • ••••
    •• • • ••••••, pp.28–39, 2003.
19. J. Wu, "Extended dominating-set-based routing in ad hoc wireless networks with
    unidirectional links," •• • • •••••••••• •••• ••• • •• ••, vol. 22:1-4, pp.327-340, 2002.
20. F. Xue and P. R. Kumar, "The number of neighbors needed for connectivity of
    wireless networks," • ••••••• • ••• •••• vol. 10, no. 2, pp. 169–181, 2004.
21. Y. Zou and K. Chakrabarty, "A distributed coverage- and connectivity-centric
    technique for selecting active nodes in wireless sensor networks," accepted for pub-
    lication in •• • • ••••••• •• ••••••, 2005.
22. Y. Zou, "Coverage-driven sensor deployment and energy-efficient information pro-
    cessing in wireless sensor networks," Ph.D Thesis, Duke University, Dec. 2004.

TARA: Thermal-Aware Routing Algorithm for Implanted Sensor Networks[*]

Qinghui Tang, Naveen Tummala, Sandeep K.S. Gupta[1],
and Loren Schwiebert[2]

[1] Arizona State University, Tempe AZ 85287, USA
{qinghui.tang, naveen.tummala, sandeep.gupta}@asu.edu
[2] Wayne State University, Detroit, MI 48202
loren@cs.wayne.edu

Abstract. Implanted biological sensors are a special class of wireless sensor networks that are used in-vivo for various medical applications. One of the major challenges of continuous in-vivo sensing is the heat generated by the implanted sensors due to communication radiation and circuitry power consumption. This paper addresses the issues of routing in implanted sensor networks. We propose a thermal-aware routing protocol that routes the data away from high temperature areas (hot spots). With this protocol each node estimates temperature change of its neighbors and routes packets around the hot spot area by a withdraw strategy. The proposed protocol can achieve a better balance of temperature rise and only experience a modest increased delay compared with shortest hop, but thermal-awareness also indicates the capability of load balance, which leads to less packet loss in high load situations.

1 Introduction

An implanted biomedical sensor is a device that detects, records and transmits information regarding a physiological change in the biological environment. It also finds its usages in various medical applications like retinal prosthesis [1], and cancer detection. Sensors work cooperatively by exchanging information and monitoring environmental changes, so wireless communication is necessary for reliable data communication amongst sensors.

The routing problem in wireless sensor networks has been well studied. Most of the routing protocols in wireless sensor networks are designed to satisfy power efficiency or delay constraints. None of them consider the possibility of hazardous effects resulting from communication radiation and power dissipation of implanted sensors. In our previous work [2], we show the communication scheduling among sensors should consider the cluster leadership history and sensor locations to minimize thermal effect on the surrounding tissues.

Radio signals used in wireless communication produce electrical and magnetic fields. The human tissue will absorb radiation and experience temperature rise

---

[*] This work is supported in part by NSF grants ANI-0086020.

V. Prasanna et al. (Eds.): DCOSS 2005, LNCS 3560, pp. 206–217, 2005.

when exposed to electromagnetic fields. Even with modest heating, some organs which are very sensitive to temperature rise due to lack of blood flow to them, are prone to thermal damage (e.g., lens cataract [3]). Continuous operation of sensor circuitry also contributes to the temperature rise of tissues. Specific Absorption Rate (SAR, unit is W/kg) is a measure of the rate at which radiation energy is absorbed by the tissue per unit weight. The relationship between radiation and SAR is given by

$$SAR = \frac{\sigma |E|^2}{\rho} \ (W/kg) \ , \tag{1}$$

where E is the induced electric field by radiation, ρ is the density of tissue and σ is the electrical conductivity of tissue. Many countries and organizations set strict standards for peak values of SAR. Experiments show exposure to an SAR of 8W/kg in any gram of tissue in the head or torso for 15 minutes may have a significant risk of tissue damage [4].

In this work, our purpose is to reduce the possibility of overheating and we are the first to consider thermal influence in a sensitive application environment. We demonstrate the thermal effects caused by sensors implanted in a biological body. We first obtain a radiation model of a communication antenna. With the Finite-Difference Time-Domain method (FDTD) we can calculate the temperature rise and SAR of each implanted sensor node by using Pennes bioheat equation [5]. Then we propose Thermal-Aware Routing Algorithm (TARA) to handle packet transmission in the presence of temperature hot spots. Extensive simulations were conducted to verify and compare its performance with shortest hop algorithm. The smaller maximum temperature rise and smaller average temperature rise of TARA indicate that TARA is a safer routing solution for implanted applications. Although our protocol has higher transmission delay, it also introduces less traffic congestion into the network because thermal-awareness of TARA technically equals to load balancing capability (the more traffic a node handles, the more energy is consumed and the more heat is generated. Balancing temperature rise equals to balancing traffic load).

In Section 2, we give a brief overview of related work and classification of the hot spot routing problem. We describe our system model and proposed thermal-aware routing protocol in Section 3. We also discuss how to calculate SAR and temperature rise in Section 3. Simulation results and comparison with shortest hop algorithm is presented in Section 4. We conclude in Section 5.

2   Related Work

Ad-hoc routing has been studied extensively for ad-hoc networks and sensor networks. In most situations, energy-efficiency, delay constraints, and transmission hops are design constraints. Those algorithms are not appropriated for in vivo biomedical sensors since they did not consider that even though some routes are eligible choices for light traffic and shorter delay, the temperature is already so high that further operation of those forwarding nodes is unbearable.

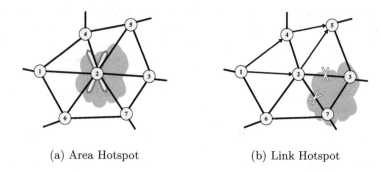

(a) Area Hotspot   (b) Link Hotspot

Fig. 1. Difference between area hot spot and link hot spot (a) all links are broken, node 2 is not accessible (b) some links are broken, node 2 is accessible

Thus we realized there are existing two hot spot routing problems in the networking field, namely, **area hot spot** and **link hot spot**. We believe current research on ad-hoc routing or sensor network routing focus only on link hot spot problem where only some network links of a node suffer congestion and packet loss, whereas other links are still available. Figure 1 shows the difference between area hot spot and link hot spot. In the former, the area surrounding node 2 is the hot spot region, and all links connecting with node 2 are disconnected thus node 2 is isolated from network. Whereas in link hot spot as shown in Figure 1(b), only some links connecting node 2 with the outside network are disconnected; node 2 is accessible and qualified as a forwarding node (e.g., route 6->2->5 is still available).

It may seem that since a load balancing routing algorithms distributes packets evenly, implying even distribution of power consumption due to communication and data processing, it can also minimize temperature rise. However, this is not always true. For instance, as shown in Figure 1(b), route $1 \rightarrow 4 \rightarrow 5$ and $1 \rightarrow 2 \rightarrow 5$ can be used to balance traffic. If we assume node 2 already suffers from temperature rise due to previous packet forwarding from node 4 to node 7, then node 2 is no longer available for load balancing and forwarding packets between node 1 and node 5.

Some heuristic protocol, e.g., the clustering-based protocol LEACH [6] can evenly distribute the energy consumption among the sensors in the network on a long term basis (large time scale). But on short term (small time scale), since the rotation of cluster head is made randomly, it is highly possible that some high temperature nodes are selected as the cluster lead. Similarly, the "Energy Equivalence Routing" proposed by Ding *et al.* [7] adjusts routing path periodically to balance energy consumption among sensors, but may not satisfy short time temperature rise criteria. Woo *et al.* [8] suggested link connectivity status (based on wireless signal quality) should be maintained in a neighborhood table regardless of cell density. In our study, link connectivity also depends on environment temperature and node distribution density plays a critical role in temperature change. Estrin *et al.* [9] mentioned that the deployment of sensors becomes a form of environmental contamination, and the emission of active sensors could be harmful or disruptive to organisms. Thus environmental impact of sensor networks needs to be examined in more details. Our protocol considers such new constraint of implanted applications, and tries to reach load balancing and temperature rise balance in a smaller time scale to reduce bio-safety risk. We expect our work will motivate further research attention in this field. Also we believe there are some similar scenarios where communication or operation of network will cause environmental disturbance which is undesired:

- In **drug development** or **food processing**, some enzymes are used as catalysts or ingredients. Enzymes are extremely sensitive to the change of environment, such as pH value and temperature. Embedded sensors are used for reagent delivery, and quality monitoring or quality control.
- In **protein crystal growth**, Micro-Electro Mechanical Systems (MEMS) and sensors are embedded to control and monitor the growth process [10]. Temperature is one of the critical factors leading to successful crystal growth. Obviously we hope to minimize the heating influence resulting from the operation of embedded MEMS and sensors.

3   Thermal Aware Routing Protocol

3.1   System Model

The temperature surrounding the sensor nodes can be measured by using a temperature sensor inside the sensor node's circuitry, but this will enlarge the sensor size and increase complexity of the sensor circuit. So in our model we assume that we cannot measure the temperature surrounding the sensor node. Temperature is estimated by observing sensor activities. In our model, we assume that the dielectric and perfusion properties of the tissues are known. The locations of the biosensors are predefined as they are physically implanted rather than randomly dropped as in some other sensor applications.

As shown in Figure 2, distributed sensors are implanted inside biological tissues, and continuously generate packets that need to be sent to the edge of the network, where a **gateway node** aggregates data and transmits to a **base station**, which is located outside the body. Each sensor has an omnidirectional antenna and no sensor node is disconnected from the network. We assume no collision happens due to simultaneously transmission.

3.2   Temperature Estimation

We identify two major sources that cause the heating effects: the radiation from the sensor node antenna and the power dissipation of sensor node circuitry.

Radiation from the Antenna. We assume that the sensor node has a short dipole antenna consisting of a short conducting wire of length dl with a sinusoidal

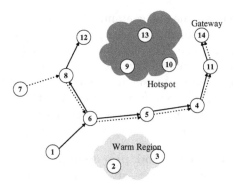

Fig. 2. System model: all data packets are forwarded to the gateway. Nodes inside the hotspot region are disconnected from other nodes

drive current I. To analyze the effect of radiation on the tissue, we assume the tissue to be homogeneous with no sharp edges or rough surfaces. The space around the antenna is divided into near field and far field. The extent of the near field is given by $d_0 = \frac{\lambda}{2\pi}$, and λ is the wavelength of the RF used by wireless communication. SAR in the near field and far field is given by [11]:

$$SAR_{NF} = \frac{\sigma\mu\omega}{\rho\sqrt{\sigma^2 + \epsilon^2\omega^2}} \left(\frac{Idl\sin\theta e^{-\alpha R}}{4\pi} \left(\frac{1}{R^2} + \frac{|\gamma|}{R} \right) \right)^2 \qquad (2)$$

and [12]:

$$SAR_{FF} = \frac{\sigma}{\rho} \left(\frac{\alpha^2 + \beta^2}{\sqrt{\sigma^2 + \omega^2\epsilon^2}} \frac{Idl}{4\pi} \right)^2 \frac{\sin^2\theta e^{-2\alpha R}}{R^2}. \qquad (3)$$

where R is the distance from the source to the observation point, θ is the angle between the observation point and the x-y plane, and γ is the propagation constant[1].

We assume our two-dimensional control volume located at the x-y plane and the control volume is perpendicular to the small dipole, thus we can safely assume the radiation pattern is omnidirectional on the 2D plane and $\sin\theta = 1$. The near field or the far field radiation of the sensor's transmitter causes the heating of the tissue because of the absorption of the radiation. The Specific Absorption Rate is used to estimate the potential heat effect on the human tissue.

Power Dissipation by Sensor Node Circuitry. The power dissipation of sensor circuitry will raise the temperature of sensor nodes. The power consumed by the sensor circuitry divided by the volume of sensor, is the power dissipation

---

[1] Given by $\gamma = \alpha + j\beta$, phase constant β is given as [13] $\beta = \omega\sqrt{\frac{\mu\epsilon}{2}} \left[\sqrt{1 + \left(\frac{\sigma}{\omega\epsilon}\right)^2} + 1 \right]^{1/2}$ (rad/m).

density, denoted as P_c, which depends on its implementation technology and architecture. In our analysis, we have considered the typical power consumption for a regular sensor circuitry operation.

3.3   Calculating Temperature Rise

The above mentioned sources of heating the tissue can cause a rise of temperature of the sensor and its surrounding area. The rate of rise in temperature is calculated by using the Pennes bioheat equation [5] as follows:

$$\rho C_p \frac{dT}{dt} = K \, \nabla^2 \, T + \rho SAR - b(T - T_b) + P_c + Q_m \; (W/m^2) \, . \qquad (4)$$

In this equation, ρ is the mass density, C_p is the specific heat of the tissue, K is the thermal conductivity of the tissue, b is the blood perfusion constant, which indicates how fast the heat can be carried away by blood flow inside the tissue, and T_b is the temperature of the blood and the tissue.

On the left side of Eq. (4), $\frac{dT}{dt}$ is the rate of temperature rise in the control volume. Terms on the right side indicate the heat accumulated inside the tissue. $K \, \nabla^2 \, T$ and $b(T - T_b)$ are the heat transfer due to the conduction and the blood perfusion, respectively. ρSAR, P_c and Q_m are the heat generated due to the radiation, the power dissipation of circuitry, and the metabolic heating, respectively

Finite-Difference Time-Domain (FDTD) [14] is an electromagnetic modeling technique that discretizes the differential form of time and space, which can also be used for heating application. The entire problem space is discretized into small grids. Each grid is marked with a pair of coordinates (i, j). Due to space limitation, we show only the result of the new bioheat equation after some manipulations:

$$T^{m+1}(i, j) = \left[1 - \frac{\delta_t b}{\rho C_p} - \frac{4\delta_t K}{\rho C_p \delta^2} \right] T^m(i, j) \qquad (5)$$

$$+ \frac{\delta_t}{C_p} SAR + \frac{\delta_t b}{\rho C_p} T_b + \frac{\delta}{\rho C_p} P_c$$

$$+ \frac{\delta_t K}{\rho C_p \delta^2} \left[\begin{array}{c} T^m(i+1, j) + T^m(i, j+1) \\ +T^m(i-1, j) + T^m(i, j-1) \end{array} \right] \, ,$$

where $T^{m+1}(i, j)$ is the temperature of the grid (i, j) at time $m + 1$, and δ_t is discretized time step, δ is the discretized space step (i.e., the size of the grid).

From (5), we can find the temperature of the grid point (i, j) at time $m + 1$, which is a function of the temperature at grid point (i, j) at time m, as well as a function of the temperature of surrounding grid points ($(i + 1, j)$, $(i, j + 1)$, $(i - 1, j)$, and $(i, j - 1)$) at time m. Once we know the properties of the tissue, the properties of blood flow, and the power or heat absorbed by the tissue, we can estimate the temperature at a given time and whether the heat effects would cause any damage to the surrounding tissues.

3.4   Protocol Description

In our protocol the forwarding is based on localized information of the temperature and hop counts to the destination. High temperature node identified as hot spot by its neighbors, and this information will be spread to other nodes by withdrawn packets. Once the temperature drops down, the neighboring node will inform other nodes about the new availability of the path.

Setup Phase. In the setup phase, by exchange neighborhood information, each sensor collects information to form its own neighbor nodes list, and also knows the number of hops to other nodes and who is the next hop for a specified destination address.

Data Forwarding. Each node sends a reading to the gateway node in either of these two circumstances: when there is a change in the sensed reading or when a base station sends a signal to the node asking for the data. It is assumed that each node knows the location of the gateway. When a node receives a packet that is destined to the gateway, it selects the next node based on the minimal temperature criteria. Any packets destined to the hot spot node will be buffered until the estimated temperature drops. If the buffered packet exceeds its time constraint, it is dropped. Any packet whose destination is not a hot spot node but the next hop is hot spot will be withdrawn and returned to the previous hop node.

Hot spot Detection. Each node listens to its neighbors' activity, counting the packets they transmitted and received. Based on this activity each node can evaluate the communication radiation and power dissipation of its neighbors, and estimate their temperature changes by using FDTD. Once the temperature of one neighbor exceeds a predefined threshold, the node would mark that neighbor as a hot spot. Also, if no activity happens the temperature of the neighbor will gradually drop, and the neighbor will be removed from the hot spot list once its temperature is beneath some threshold.

   To reduce network overhead and save energy the node will not actively broadcast the hot spot information to other nodes. But if the temperature drops beneath the threshold, the node will notify its neighbor of this new availability information.

Withdrawal. In certain circumstances, a node might receive a packet where it does not have any next hop to forward to or all the nodes in the forwarding set are in the hot spot region. During such a situation, the node returns the packet to the previous node. The previous node tries to forward the packet to another available next hop node, or returns the packet to its previous node if no next hop available. Hot spot information will be carried with packets to inform precedent nodes about the hot spot. If a node (namely, **A**) send a withdraw packet to another node (namely, **B**), when temperature drops down later, node A is responsible for notifying B that the clear of the hot spot.

Example. As shown in Figure 2, when node 1 has a packet for node 14, it chooses node 6 since node 2 is inside the warm region, which means that the

Table 1. Parameters and their values used in Simulation

Parameter	Property	value
ε_r	Relative permittivity at 2MHz or 2.45GHz	826 or 52.73
P_{NL}	Power consumption of non-leader sensor node	$1mW$
C_p	Specific heat	3600 $\left[\frac{J}{kg^\circ C}\right]$
K	Thermal conductivity	0.498 $\left[\frac{J}{ms^\circ C}\right]$
b	Blood perfusion constant	2700 $\left[\frac{J}{m^3 s^\circ C}\right]$
T_b	Fixed blood temperature	$37\ ^\circ C$
ρ	Mass density	1040 $\frac{kg}{m^3}$
σ	Conductivity at 2MHz or 2.45GHz	0.5476 or 1.7388 $\left[\frac{S}{m}\right]$
P_L	Power consumption of Leader node	$5\ mW$
T_1	Leader Time	$600\ sec$
δ	Control Volume cell size	$0.005\ m$
δ_t	Time step of FDTD	$10\ sec$
I_0	Current provided to sensor node antenna	$0.1\ A$

temperature is above the normal temperature but not as critical as hot spots. When the packet reaches node 12 through the path 1→6→8→12, computation of temperature at node 12 reveals that the temperature of the next hop is too high to perform any forwarding operation. Then node 12 returns the packet along its origin route, until a node has a next hop available to forward data (node 6 in this case). During this withdrawal phase the packet direction is set to be negative. Any node receiving a negative direction packet knows that it is a back routed packet and updates within itself the hot spot information obtained from the packet (the high temperature at node 9 and node 13). Now if the node 7 has a packet for node 14, it sends the packet to node 8, which should forward the packet to node 12. But the hot spot information obtained from the previous withdrawal packet makes node 8 select node 6 as a next hop.

4   Simulation

We modeled a two-dimensional area with 12 nodes implanted. The simulation program was developed on Matlab. We use the tissue properties and wave propagation characteristics shown in Table 1. We used various scenarios with different node locations and took the average to plot the graphs. We compared shortest hop protocol with our approach as it demonstrates the trade-off. Maximum temperature rise and average temperature rise are used as metrics for measuring the performance of the proposed protocol.

Maximum Temperature Rise is the highest temperature rise inside the whole network area. It captures the effectiveness of a routing protocol to direct data away from the hot spots. As shown in Figure 3, the maximum temperature rise of TARA is much lower than the shortest hop, which can be attributed

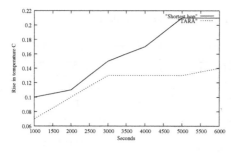

Fig. 3. Maximum temperature rise of TARA (• •••• ••• •) is lower than that of Shortest hop (•• ••• ••• •) because TARA directs packets away from hotspot area

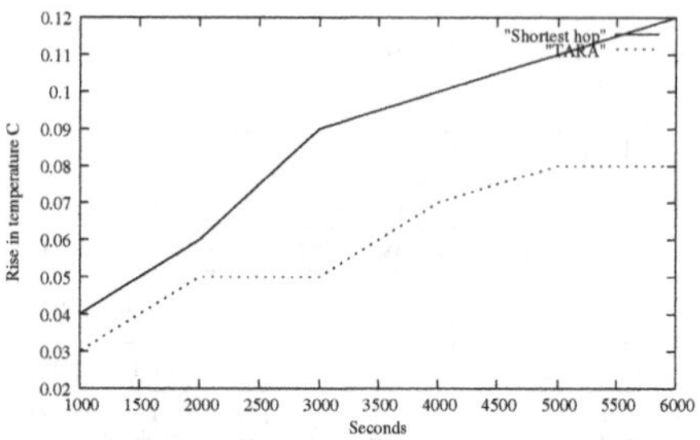

Fig. 4. Average temperature rise of TARA (• •••• ••• •) is lower than that of Shortest path (•• ••• ••• •) because the latter excessively uses the same path

to TARA's selecting the next node that has the least temperature residue and avoiding further deterioration of the overheated area.

Average Temperature Rise is used to estimate the average amount of heat generated at each point inside the network. As shown in Figure 4, TARA has a lower average temperature than the shortest hop whose continuing usage of the same shortest path creates an overheated area.

Notice that in Figure 3, TARA experiences a flat temperature history during time period from 3000 seconds to 5000 seconds. This can be explained as: because part of the network reaches the temperature threshold, TARA is trying to route packets through other low temperature area, thus the peak temperature of network is keeps unchanged; at the same time period, average temperature is still rising as shown in Figure 4 since network operation is continuously introducing heat into the network.

Figure 5 and Figure 6 demonstrate the temperature rise distribution inside the network. Figure 5 demonstrates that temperature rise is relatively evenly distributed in the whole area since the thermal-awareness of TARA avoids introducing the overheated area. Several temperature peaks of Figure 6 indicate the unawareness of the overheated area of the Shortest Hop algorithm, which may lead to thermal damage to human tissues. We also notice that the gateway node at coordinates (6, 6) is always experiencing the highest temperature rise since it is responsible for gathering and forwarding all data to basestation. In practice, the gateway node is located near the surface of the body, where better heat conduction and ventilation will counteract the overheating problem.

Delay Performance. In this section, we explain the simulation results of our protocol implemented on **Crossbow MICA2** sensor motes. The aim is to demonstrate the trade-off between delay/throughput and temperature rise. We use the average delay and percentage of packets meeting the deadline as the metrics to demonstrate the performance of protocols. First, we introduce only one constant flow of traffic, travelling across the network. As shown in Figure 7, for various deadline constraints, more packets of Shortest Hoop meet the dead-

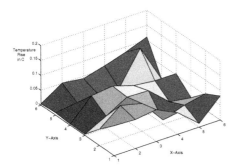

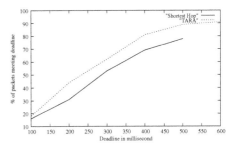

Fig. 5. Temperature distribution of TARA: temperature rise is relatively even in the whole area since the thermal-awareness of TARA avoids introducing the overheated area

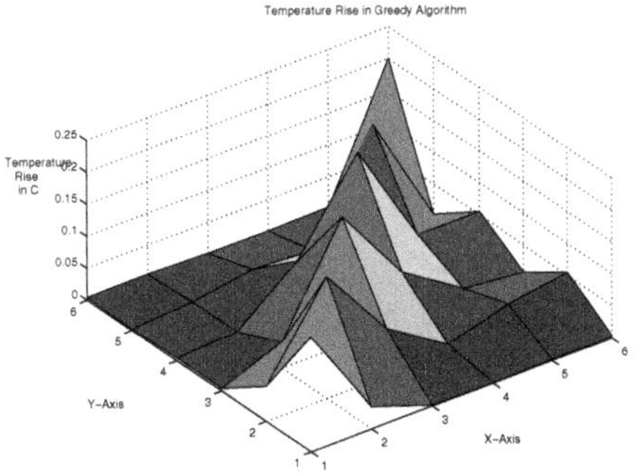

Fig. 6. Temperature distribution of Shortest Hop: several temperature peaks indicate the unawareness of the overheated area

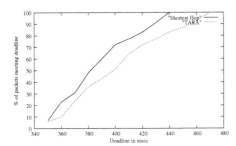

Fig. 7. Performance of TARA and Shortest Hop with only one constant traffic flow traveling across the network, TARA experiences a higher loss due to larger delay

Fig. 8. Delay performance with low traffic: unlike Shortest Hop, thermal-awareness of TARA equals to load balancing, thus fewer packets are lost due to congestion

line than that of TARA because its packets travel along the shortest path, and TARA experiences a higher loss due to larger delay.

Secondly, we introduce randomly generated traffic of about 100 bytes/second from different nodes of the network. From Figure 8 we can see more packets of TARA meet the delay deadline than that of Shortest Hop. This is because the greedy approach of Shortest Hop results in congestion in some shortest paths, whereas TARA's thermal-awareness also indicates its load balancing capability which results in less congestion and less packet loss.

When the deadline is shorter than 400 millisecond, under the same percentage of packets meeting deadline, e.g., 70%, our proposed protocol only experiences a modest increase in packet delivery times (50 millisecond) while evenly distributing load among the sensors to satisfy the temperature rise constraint.

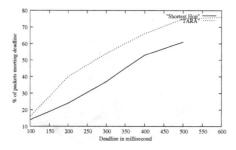 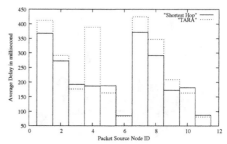

Fig. 9. Delay performance with high traffic: unlike Shortest Hop, thermal-awareness of TARA equals to load balancing, thus fewer packets are lost due to congestion

Fig. 10. Although Shortest Hop experiences higher packet loss, but those survived packets have smaller tranmission delay compared with TARA

Figure 9 shows the percentage of packets meeting the deadlines with a higher traffic of about 250 bytes/second. TARA still has a higher percentage of packets meeting the deadline than shortest-hop routing. Figure 8 and Figure 9 also conclude that TARA can achieve load balancing and temperature rise balance simultaneously.

Although the shortest hop algorithm has higher packet losses, from another perspective, those packets who survive the congestion and the buffer overflow have smaller delay. To demonstrate this, we also calculated the average delay for packets generated from different nodes. As shown in Figure 10, average packet delay of Shortest Hop is smaller that TARA. But it also shows that TARA from different source nodes only suffers a modest delay than that of Shortest Hop routing. This means TARA is promising at reaching temperature rise balancing without significant degradation on delay performance.

5   Conclusion

This paper presents a new thermal-aware routing protocol TARA for implanted biosensor networks. TARA takes into consideration the heating factors resulting from the operation of sensor nodes and routes data to minimize thermal effects in a heat-sensitive environment. It also introduces less traffic congestion in the network because thermal-awareness of TARA technically equals to a load balancing capability. TARA is compared with a shortest-hop routing protocol. The smaller maximum temperature rise and average temperature rise of TARA indicate that TARA is a safer routing solution for implanted applications without significant degradation on delay performance. Next, we will improve the TARA protocol to consider more design constraints, such as multipath routing [15], energy consumption, link distance among sensors and the channel estimation mechanism. We will extend our investigation by comparing TARA with other routing protocols in terms of energy-efficiency, overhead, recovery time, etc.

References

1. Schwiebert, L., Gupta, S.K.S., Auner, P.S.G., Abrams, G., Lezzi, R., McAllister, P.: A biomedical smart sensor for visually impaired. In: IEEE Sensors 2002, Orlando, FL., USA (2002)
2. Tang, Q., Tummala, N., Gupta, S.K.S., Schwiebert, L.: Communication scheduling to minimize thermal effects of implanted biosensor networks in homogeneous tissue. IEEE Tran. Biomedical Eng. (accepted for publication)
3. Hirata, A., Ushio, G., Shiozawa, T.: Calculation of temperature rises in the human eye for exposure to EM waves in the ISM frequency bands. In: IEICE Trans. Comm. Volume E83-B. (2000) 541–548
4. International Electrotechnical Commission IEC: Medical Electrical Equipment, Part 2-33: Particular Requirement for the Safety of Magnetic Resonance Systems for Medical Diagnosis IEC 60601-2-33. 2nd edn. (1995)
5. Pennes, H.H.: Analysis of tissue and arterial blood temperature in the resting human forearm. Journal of Applied Physiology **1.1** (1948) 93–122
6. Heinzelman, W.R., Chandrakasan, A., Balakrishnan, H.: Energy-efficient communication protocol for wireless microsensor networks. In: Proc. 33rd Hawaii Int'l Conf. System Sciences. Volume 8., IEEE Computer Society (2000) 8020
7. Ding, W., Iyengar, S.S., R. Kannan and, W.R.: Energy equivalence routing in wireless sensor networks. Journal of Microcomputers and Applications: Special issue on Wireless Sensor Networks (2004)
8. Woo, A., Tong, T., Culler, D.: Taming the underlying challenges of reliable multihop routing in sensor networks. In: SenSys '03, ACM Press (2003) 14–27
9. Estrin, D., Michener, W., Bonito, G.: Environmental cyberinfrastructure needs for distributed sensor networks. A report from a national science foundation sponsored workshop (2003)
10. Tseng, F.G., Ho, C.E., Chen, M.H., Hung, K.Y., Su, C.J., Chen, Y.F., Huang, H.M., Chieng, C.C.: A chip-based-instant protein micro array formation and detection system. In: Proc. NSTI Naotech'04. Volume 1., Boston (2004) 39–42
11. Prakash, Y., Lalwani, S., Gupta, S.K.S., Elsharawy, E., Schwiebert, L.: Towards a propagation model for wireless biomedical applications. In: IEEE ICC 2003. Volume 3. (2003) 1993–1997
12. National Council on Radiation and Measurements: A Practical Guide to the Determination of Human Exposure to Radio Frequency Fields. Report no. 119. NCRP, Bethseda, MD (1993)
13. Ulaby, F.T.: Fundamentals of Applied Electromagnetics. Prentice-Hall (1999)
14. Sullivan, D.M.: Electromagnetic Simulation Using the FDTD Method. IEEE Press (2000)
15. Ganesan, D., Govindan, R., Shenker, S., Estrin, D.: Highly-resilient, energy-efficient multipath routing in wireless sensor networks. SIGMOBILE Mob. Comput. Commun. Rev. **5** (2001) 11–25

Multiresolutional Filtering of a Class of Dynamic Multiscale System Subject to Colored State Equation Noise

Peiling Cui[1,2], Quan Pan[2], Guizeng Wang[1], and Jianfeng Cui[2]

[1] Department of Automation, Tsinghua University, Beijing 100084, China
[2] Department of Automatic Control, Northwestern Polytechnical University,
Xi'an, Shaanxi 710072, China

Abstract. In this paper, modeling and estimation of a class of dynamic multiscale system subject to colored state equation noise is proposed. The colored state noise vector is augmented in the system state variables, the state space projection equation is used to link the scales, and then a new system model is built. The new model is in a form suitable for the application of the Kalman filter equations. Haar-wavelet-based model and estimation algorithm are given. Monte Carlo simulation results demonstrate that the proposed algorithm is effective and powerful in this kind of multiscale estimation problem.

1 Introduction

The multiscale autoregressive (MAR) framework [1-6] was developed to model a variety of random processes compactly and estimate them efficiently. MAR estimates the stationary processes from numerous measurements by multiscale techniques, which aim at saving much computation compared with the traditional linear minimum mean-square-error estimation. Many advanced systems are mostly observed by several sensors independently at different scales. The resolution and sampling frequencies of the sensors are supposed to decrease from sensor 1 to sensor J, the real state at each scale is $x_j, j = 1, 2, \ldots, J$, respectively. The state equation is described by a partial differential equation at the finest scale. An important practical problem in the above systems is to find a state estimator given the observations. This problem has been studied in recent years due to the numerous applications associated with it [7]-[16].

In [15], an algorithm for optimal and dynamic multiresolutional distributed filtering is derived. The wavelet transform is utilized as a bridge linking the signals at different resolution levels. In [16], an optimal estimation of a class of dynamic multiscale systems is discussed. The sampling frequencies of the sensors are supposed to decrease by a factor of two. That paper introduces the state space projection equation, and fuses the information at all scales by the measurement equation augmentation.

In aforementioned papers, they all assumed that state equation noise is white. However, this is a real limitation in many practical situations. When the noise is

V. Prasanna et al. (Eds.): DCOSS 2005, LNCS 3560, pp. 218–227, 2005.

colored, the state equation needs to be adjusted before Kalman filtering techniques can be employed. Very little work has been proposed for solving the estimation problem of dynamic multiscale system subject to colored state equation noise. It is the main focus of this paper.

The colored state noise vector is augmented in the system state variables. The state space projection equation is used to link the scales, and then a new system model is built. This model is in a form suitable for the application of the standard Kalman filter equations [17]-[18]. Haar-wavelet-based model and estimation algorithm are given. The good estimation accuracy of the proposed method is further evidenced by the Monte Carlo simulation results.

The paper is organized as follows: In Section II, we describe our wavelet realization approach by Haar wavelet transform, and a theorem is given for the filtering outputs at each scale. We show the performance of our algorithm by the Monte Carlo simulation in Section III. Finally we provide some concluding remarks in Section IV.

2 Haar-Wavelet-Based Modeling and Estimation Algorithm

For convenience, let the sampling rate decrease from sensor 1 to sensor J by a factor of two. Obviously, sensor 1 corresponds to the finest scale. The state at all scales in time interval ΔT is called a state block, and the measurement a data block. In every ΔT, the state estimation must be updated when a new data block is available. We hope the approximation of any node at any scale is accomplished in time interval ΔT, not using the state nodes outside of it. We choose the Haar wavelet [19]. This choice is motivated by the particularly simple realization of the Haar wavelet transform in our multiscale framework by using a bintree structure. Haar wavelet is the simplest and most widely used one with low-pass filter $\left[\sqrt{2}/2, \sqrt{2}/2\right]$.

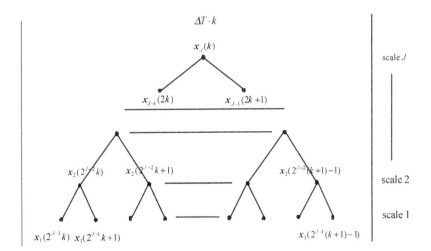

Fig. 1. Tree structure of the dynamic multiscale system state nodes at time interval $k\Delta T$

For clarity, we unify the notations of state nodes firstly[16]. Fig.1 shows the system bintree structure at time interval $k\Delta T$. $x_J(k)$ is denoted as the state of scale J, $x_{J-1}(2k)$ and $x_{J-1}(2k+1)$ the states at scale $J-1$. Analogically, at scale j there are 2^{J-j} state nodes, which are denoted as $x_j(2^{J-j}k)$, $x_j(2^{J-j}k+1)...x_j(2^{J-j}(k+1)-1)$. Assuming the multiscale system state structure satisfies the dyadic structure of Haar wavelet, the node $x_j(2^{J-j}k+m_j)$ can be expressed with the finest scale nodes as follows

$$x_j(2^{J-j}k+m_j) = \left(\frac{\sqrt{2}}{2}\right)^{J-1} \sum_{i=0}^{2^{J-1}-1} x_1(2^{J-1}k+2^{J-1}m_j+i)$$

where $m_j = 0,1,...,2^{J-j}-1$.

2.1 State Equation

For simplicity, we assume the system to be time-invariant. The discrete state transition equation of the finest scale at time interval $(k+1)\Delta T$ is

$$x_1(2^{J-1}(k+1)) = Ax_1(2^{J-1}(k+1)-1)+B\xi(2^{J-1}(k+1)-1)$$

where $x_1(\bullet)\in R^{N_x}$, state transition matrix $A\in R^{N_x \times N_x}$, noise stimulus matrix $B\in R^{N_x \times u}$, the dimension of colored noise $\xi(\bullet)$ is u, and

$$\xi(2^{J-1}(k+1)-1) = D\xi(2^{J-1}(k+1)-2)+w(2^{J-1}(k+1)-2)$$

where $w(\bullet)\in R^u$ is a Gaussian white process with covariance q.

Define a N_x+u state vector such that it includes the original system state variables and the colored state noise vector, then

$$\begin{bmatrix} x_1(2^{J-1}(k+1)) \\ \hline \xi(2^{J-1}(k+1)) \end{bmatrix} = \begin{bmatrix} A & B \\ \hline O & D \end{bmatrix}\begin{bmatrix} x_1(2^{J-1}(k+1)-1) \\ \hline \xi(2^{J-1}(k+1)-1) \end{bmatrix} + \begin{bmatrix} O \\ \hline w(2^{J-1}(k+1)-1) \end{bmatrix}$$

Denoting

$$x_1^a(2^{J-1}(k+1)-1) = \begin{bmatrix} x_1(2^{J-1}(k+1)-1) \\ \hline \xi(2^{J-1}(k+1)-1) \end{bmatrix}, \quad A^a = \begin{bmatrix} A & B \\ \hline O & D \end{bmatrix},$$

$$w^a(2^{J-1}(k+1)-1) = \begin{bmatrix} O \\ \hline w(2^{J-1}(k+1)-1) \end{bmatrix}$$

where $x_1^a(\bullet)\in R^{N_x+u}$, $A^a \in R^{(N_x+u)\times(N_x+u)}$, $w^a(\bullet)\in R^{N_x+u}$. The system state equation by augmenting the state is

$$x_1^a(2^{J-1}(k+1)) = A^a x_1^a(2^{J-1}(k+1)-1)+w^a(2^{J-1}(k+1)-1) \qquad (1)$$

where

$$Cov\left(w^a(2^{J-1}(k+1)-1)\right) = \begin{bmatrix} O & O \\ \hline O & q \end{bmatrix}$$

Then

$$x_1^a(2^{J-1}(k+1)+m_1) = (A^a)^{m_1+1}\bullet x_1^a(2^{J-1}(k+1)-1)+\sum_{i=-1}^{m_1-1}(A^a)^{m_1-i-1}w^a(2^{J-1}(k+1)+i)$$

where $m_1 = 0,1,...,2^{J-1}-1$. Letting

$$\bar{x}(k) = col(x_1^a(2^{J-1}k), x_1^a(2^{J-1}k+1), \cdots, x_1^a(2^{J-1}(k+1)-1))$$

$$\bar{w}(k) = col(w^a(2^{J-1}(k+1)-1), w^a(2^{J-1}(k+1)), ..., w^a(2^{J-1}(k+2)-2))$$

$$\bar{A}(m_1) = (A^a)^{m_1+1}, \bar{B}(m_1) = [(A^a)^{m_1},(A^a)^{m_1-1}, ..., I, O, ..., O]$$

where $\bar{x}(k)$ is $2^{J-1}(N_x+u)\times 1$ matrix, col denotes arranging the data in the bracket into column vector, $\bar{w}(k) \in R^{2^{J-1}(N_x+u)\times 1}$, $\bar{A}(m_1) \in R^{(N_x+u)\times N_x+u)}$, $\bar{B}(m_1) \in R^{(N_x+u)\times 2^{J-1}(N_x+u)}$ is with zero elements on the last $(2^{J-1}-m_1-1)\times(N_x+u)$ columns.

Letting

$$\bar{A} = \begin{bmatrix} O & \cdots & O & \bar{A}(0) \\ O & \cdots & O & \bar{A}(1) \\ \vdots & \ddots & \vdots & \vdots \\ O & \cdots & O & \bar{A}(2^{J-1}-1) \end{bmatrix} \quad \bar{B} = \begin{bmatrix} \bar{B}(0) \\ \bar{B}(1) \\ \vdots \\ \bar{B}(2^{J-1}-1) \end{bmatrix},$$

where $\bar{A} \in R^{2^{J-1}(N_x+u)\times 2^{J-1}(N_x+u)}$, $\bar{B} \in R^{2^{J-1}(N_x+u)\times 2^{J-1}(N_x+u)}$. Then we have

$$\bar{x}(k+1) = \bar{A}\bar{x}(k) + \bar{B}\bar{w}(k) \tag{2}$$

2.2 Measurement Equation

From the state space projection equation [16], we have

$$x_j = P_j x_1$$

where P_j is the state space projection operator, suppose that discrete measurement equation at scale j is

$$z_j(2^{J-j}k+m_j) = C_j x_j(2^{J-j}k+m_j) + v_j(2^{J-j}k+m_j) \tag{3}$$

where the dimension of $z_j(\cdot)$ is N_z, C_j is measurement matrix, $v_j(\cdot)$ is a Gaussian white process with covariance $R_j(\cdot)$, $v_j(\cdot)$ and $w(\cdot)$ are mutually independent.

Then

$$z_j(2^{J-j}k+m_j) = C_j P_j x_1(2^{J-j}k+m_j) + v_j(2^{J-j}k+m_j)$$

$$= [C_j P_j \mid O]\begin{bmatrix} x_1(2^{J-j}k+m_j) \\ \rule{2cm}{0.4pt} \\ \xi(2^{J-j}k+m_j) \end{bmatrix} + v_j(2^{J-j}k+m_j)$$

where $O \in R^{N_z \times u}$ is a matrix with zero elements.

$$\Upsilon_j(m_j) = \begin{bmatrix} \underbrace{0\cdot\Lambda,...,0\cdot\Lambda}_{m_j 2^{J-1}} & \underbrace{(\frac{F}{2})^{J-1}\cdot\Lambda...,(\frac{F}{2})^{J-1}\cdot\Lambda}_{2^{J-1}} & \underbrace{0\cdot\Lambda,...,0\cdot\Lambda}_{(2^{J-j}-m_j-1)2^{J-1}} \end{bmatrix}$$

where $\Lambda = [I\ O]$, I is $N_x \times N_x$ identity matrix, $O \in R^{N_x \times u}$ is zero matrix. $\Upsilon_j(m_j)$ is $N_x \times 2^{J-1}(N_x+u)$ matrix.

At scale j, the state node $x_j(2^{J-j}k+m_j)$ can be written as

$$x_j(2^{J-j}k+m_j) = \Upsilon_j(m_j)\bar{x}(k)$$

then Eq.(3) becomes

$$z_j(2^{J-j}k+m_j) = C_j\Upsilon_j(m_j)\bar{x}(k)+v_j(2^{J-j}k+m_j)$$

Denoting

$$\bar{z}_j(k) = col\left(z_j(2^{J-j}k),z_j(2^{J-j}k+1),...,z_j(2^{J-j}(k+1)-1)\right)$$

$$\bar{C}_j = \begin{bmatrix} C_j\Upsilon_j(0) \\ C_j\Upsilon_j(1) \\ \vdots \\ C_j\Upsilon_j(2^{J-j}-1) \end{bmatrix}$$

$$\bar{v}_j(k) = col\left(v_j(2^{J-j}k),v_j(2^{J-j}k+1),...,v_j(2^{J-j}(k+1)-1)\right)$$

and the covariance matrix of $\bar{v}_j(k)$

$$\bar{R}_j(k) = diag[R_j(\bullet),R_j(\bullet),...,R_j(\bullet)]$$

Then

$$\bar{z}_j(k) = \bar{C}_j\bar{x}(k)+\bar{v}_j(k)$$

Denoting

$$\bar{z}(k) = col\left(\bar{z}_j(k),\bar{z}_{j-1}(k),...,\bar{z}_1(k)\right)$$

$$\bar{C} = col\left(\bar{C}_J,\bar{C}_{J-1},...,\bar{C}_1\right)$$

$$\bar{v}(k) = col\left(\bar{v}_j(k),\bar{v}_{j-1}(k),...,\bar{v}_1(k)\right)$$

Then the augmented measurement equation is obtained

$$\bar{z}(k) = \bar{C}\bar{x}(k)+\bar{v}(k) \tag{4}$$

the covariance matrix of $\bar{v}(k)$ is

$$\bar{R}(k) = diag\left[\bar{R}_J(k),\bar{R}_{J-1}(k),...,\bar{R}_1(k)\right].$$

2.3 Kalman Filter

Eqs. (2) and (4) can be written together

$$\begin{cases} \bar{x}(k+1) = \bar{A}\bar{x}(k)+\bar{B}\bar{w}(k) \\ \bar{z}(k) = \bar{C}\bar{x}(k)+\bar{v}(k) \end{cases}$$

where $\bar{w}(k)$ and $\bar{v}(k)$ are white noise. Assuming that the filter is stable, the Linear Minimum Mean-square Error (LMMSE) $\hat{\bar{x}}(k)$ of state $\bar{x}(k)$ can be obtained by performing Kalman filtering. In the following, the optimal estimation of the state node at the finest scale is given.

Denoting

$$T_j(m_j) = \left[\underbrace{0\cdot\Lambda,...,0\cdot\Lambda}_{m_j 2^{J-1}}\quad\underbrace{\Lambda,...,\Lambda}_{2^{J-j}}\quad\underbrace{0\cdot\Lambda,...,0\cdot\Lambda}_{(2^{J-j}-m_j-1)2^{J-1}}\right]$$

where $\Lambda = [I\ O]$, I is $N_x \times N_x$ identity matrix, O is $N_x \times u$ zero matrix. $T_j(m_j)$ is $N_x \times 2^{J-1}(N_x+u)$ matrix, then the LMMSE estimation of state node $x_j(2^{J-j}k+m_j)$ is $T_j(m_j)\cdot\hat{\bar{x}}(k)$.

2.4 Filtering Outputs at Each Scale

Filtering outputs at each scale can be obtained according to the theorem 1 shown below. The detailed proof can be referred to reference [16].

Theorem 1. Suppose $\hat{\overline{x}}(k)$ is the LMMSE of $\overline{x}(k)$, then the LMMSE of node $x_j(2^{J-j}k+m_j)$ is $\Upsilon_j(m_j) \cdot \hat{\overline{x}}(k)$.

3 Simulation Results

For verifying the validity of our algorithm, consider the following constant-velocity dynamic system with position-only measurements at two scales.

$$\begin{cases} x_1(2k+2) = Ax_1(2k+1) + B\xi(2k+1) \\ z_j(2^{2-j}k+2^{2-j}-1) = C_jx_j(2^{2-j}k+2^{2-j}-1) + v_j(2^{2-j}k+2^{2-j}-1) \end{cases}$$

where $j = 1,2, \xi(\bullet)$ is colored noise, and

$$\xi(2k+1) = \psi\xi(2k) + w(2k)$$

$w(k)$ and $v_j(\bullet)$ are Gaussian white noises with zero mean, and

$$\begin{cases} E(w(k)w(l)^T) = q\delta_{kl}, E(v_j(k)v_j(l)^T) = r_j\delta_{kl} \\ E(v_j(k)w(l)^T) = 0, E(w(k)v(l)^T) = 0 \end{cases}$$

Letting

$$A = \begin{bmatrix} 1 & T \\ 0 & 1 \end{bmatrix}, B = \begin{bmatrix} \frac{1}{2}T^2 & T \end{bmatrix}^T, C_1 = C_2 = \begin{bmatrix} 1 & 0 \end{bmatrix},$$

$$T = 1, q = 1, r_1 = 46, r_2 = 40, \psi = 0.5.$$

T is the sampling rate, and $x_1 = [\text{position, velocity}]'$.

Fig.2 shows a sequence of the true state and the estimated state at scale 1. Fig.3 shows a sequence of the true state and the estimated state at scale 2. Fig.4, Fig.5, Fig.6 and Table 1 give the results of Monte Carlo Simulation (100 runs). Fig.4 compares the measurement noise RMS with the estimation error RMS at two scales. The noise compression ratio at scale 1 and scale 2 are 5.1865dB and 2.0254dB, respectively.

Fig.5 and Fig.6 show the influence of q and ψ on the noise compression ratio that is denoted by y-axis. We can see that, The noise compression ratios at two scales decrease with the increasing of q and ψ.

Table 1 shows the influence of r_1 and r_2 on the noise compression ratio. It can be seen that, the noise compression ratio increases at scale 1 and decreases at scale 2 with the decreasing of r_2 while r_1 is unchanged. When r_2 remains unchanged and r_1 decreases, the ratio increases at scale 2 and decreases at scale 1. The ratio at two scales increase with the decreasing of r_1 and r_2.

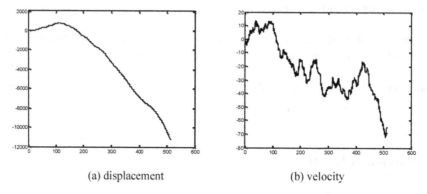

(a) displacement   (b) velocity

Fig. 2. True state (dotted) and the estimated state (solid) at scale 1

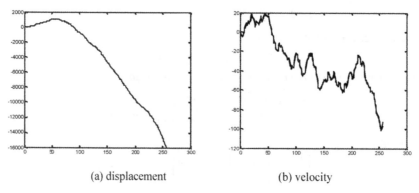

(a) displacement   (b) velocity

Fig. 3. True state (dotted) and the estimated state (solid) at scale 2

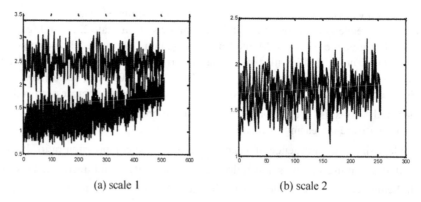

(a) scale 1   (b) scale 2

Fig. 4. Measurement noise RMS (dotted) and the estimation error RMS (solid)

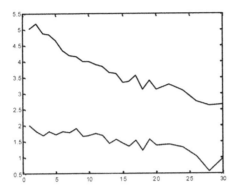

Fig. 5. The influence of q on the noise compression ratio of scale 1 (solid) and scale 2 (dotted)

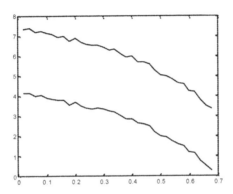

Fig. 6. The influence of ψ on the noise compression ratio of scale 1 (solid) and scale 2 (dotted)

Table 1. The influence of r_1 and r_2 on the noise compression ratio

Parameters					Noise compression ratio	
ψ	q	T	r_1	r_2	Scale 1(dB)	Scale 2(dB)
0.5	1	1	56	50	5.0387	2.0041
0.5	1	1	56	45	5.3545	1.9045
0.5	1	1	56	40	5.9295	1.9194
0.5	1	1	50	50	4.9169	2.3066
0.5	1	1	45	50	4.8509	2.7186
0.5	1	1	40	50	4.3298	2.7169
0.5	1	1	35	50	4.3136	3.2719
0.5	1	1	46	40	5.1865	2.0254

4 Conclusion

In this paper, modeling and estimation of a class of dynamic multiscale system is proposed. The system state is described with a partial differential equation, and is observed by multiple sensors in a closed subspace sequence of the state space. The noise in the state equation is colored noise. The colored state noise vector is augmented in the system state variables. The state space projection equation is used to link the scales. The model of new system satisfies the Kalman filter condition. Haar-wavelet-based model and estimation algorithm are given. Monte Carlo simulation results verify the validation of our algorithm.

Acknowledgement

This research has been supported by National Natural Science Foundation of China under grant 60172037.

References

1. Basseville, M., Benveniste, A., Chou, K., Golden, S., Nikoukhah, R., Willsky, A. S.: Modeling and Estimation of Multiresolution Stochastic Processes. IEEE Trans. Information Theory (1992) 38: 766-784
2. Chou, K., Willsky, A. S., Benveniste, A.: Multiscale Recursive Estimation, Data Fusion, and Regularization. IEEE Trans. Automatic Control (1994) 39: 464-478
3. Chou, K., Willsky, A. S., Nikoukhah, R.: Multiscale Systems, Kalman Filters, and Riccati Equations. IEEE Trans. on Automatic Control (1994)39: 479-492
4. Daoudi, K., Frakt, A. Willsky, A. S.: Multiscale Autoregressive Models and Wavelets. IEEE Trans. Information Theory (1999) 45: 828-845
5. Luettgen, M., Karl, W., Willsky, A. S. Tenney, R.: Multiscale Representations of Markov Random Fields. IEEE Trans. on Signal Processing (1993) 41: 3377-3396
6. Frakt, A.: Internal Multiscale Autoregressive Processes, Stochastic Realization, and Covariance Extension. PhD thesis, Massachusetts Institute of Technology (1999)
7. Hong, L.: Centralized and Distributed Kalman Filtering in Multi-coordinate Systems with Uncertainties. Proceedings of the IEEE Aerospace and Electronics Conference (1990) 389 -394
8. Hong, L., Chen, G., Chui, C. K.: A Filter-Bank Based Kalman Filtering Technique for Wavelet Estimation and Decomposition of Random Signals. IEEE Trans. on Circuits and Systems II (1998) 45(2):237–241
9. Hong, L.: Multiresolutional Filtering using Wavelet Transform. IEEE Trans. on Aerospace and Electronic Systems (1993) 29(4): 1244-1251
10. Hong, L.: Distributed Filtering using Set Models. IEEE Trans on Aerospace and Electronic Systems (1992) 28(4): 1144-1153

11. Hong, L.: Adaptive Distributed Filtering in Multicoordinated Systems. IEEE Trans. on Aerospace and Electronic systems (1991) 27(4): 715-724
12. Hong, L.: An Optimal Reduced-Order Stochastic Observer-Estimator. IEEE transactions on Aerospace and Electronic systems (1992) 28(2): 453-461
13. Hong, L.: Optimal Multiresolutional Distributed Filtering. Proceedings of the 31th Conference on Decision and Control (1992): 3105-3110
14. Hong, L., Werthmann, Werthmann, J. R., Bierman, G. S., Wood, R. A: Real-Time Multiresolution Target Tracking, Signal and Data Processing of Small Target, Orlando, FL (1993) : 233-244
15. Hong, L.: Multiresolutional Distributed Filtering. IEEE Trans. on Automatic Control. (1994) 39(4): 853-856
16. Zhang, L.: The Optimal Estimation of a Class of Dynamic Multiscale Systems. PhD Thesis, Northwestern Polytechnic University, Xi'an, PRC (2001)
17. Anderson, B. D. O., Moore, J. B.: Optimal Filtering. Englewood Cliffs, N. J., Prentice-Hall, Inc. (1979)
18. Chen, C. T.: Linear System Theory and Design. New York: Holt, Rinehart and Winston (1970)
19. Daubechies, I.: Ten Lectures on Wavelets. CBMS-NSF Series in Appl. Math., SIAM (1992)

Design and Analysis of Wave Sensing Scheduling Protocols for Object-Tracking Applications

Shansi Ren, Qun Li, Haining Wang, and Xiaodong Zhang

College of William and Mary,
Williamsburg, VA 23187 USA
{sren, liqun, hnw, zhang}@cs.wm.edu

Abstract. Many sensor network applications demand tightly-bounded object detection quality. To meet such stringent requirements, we develop three sensing scheduling protocols to guarantee worst-case detection quality in a sensor network while reducing sensing power consumption. Our protocols emulate a line sweeping through all points in the sensing field periodically. Nodes wake up when the sweeping line comes close, and then go to sleep when the line moves forward. In this way, any object can be detected within a certain period. We prove the correctness of the protocols and evaluate their performances by theoretical analyses and simulation.

1 Introduction

Detecting and tracking moving objects is a major class of applications in sensor networks. Many of these applications demand stringent object detection quality. A simple solution to meeting required object detection quality is full sensing coverage, in which intruding objects can be timely detected . However, full sensing coverage requires that a large portion of the sensors remain awake continuously, resulting in energy inefficiency and short system lifetime. As an alternative approach, probabilistic sensing coverage has been recently proposed in [5, 12]. Instead of staying awake all the time, nodes can periodically rotate between active state and sleeping state to conserve energy while meeting the object detection quality requirement. Compared with full sensing coverage, probabilistic sensing coverage leads to significant energy conservation and much longer system lifetime.

The previous work on probabilistic coverage focused on providing average-case object detection quality for surveillance applications. Thus far, little attention in the literature has been paid to designing protocols that guarantee bounded worst-case object detection quality. Many military surveillance applications, however, often demand such stringent object detection quality requirements, e.g., an enemy vehicle must be detected in one minute. To address this problem, we propose wave sensing, a new sensing scheduling scheme under probabilistic coverage to provide bounded worst-case object detection quality. In this scheme, at any moment, active nodes on the sensing field form connected curves. These curves move back and forth both horizontally and vertically across the field, so that every geographical point is scanned at least once within a limited amount of time. In this way, the wave sensing scheme can guarantee that an object

V. Prasanna et al. (Eds.): DCOSS 2005, LNCS 3560, pp. 228–243, 2005.

is detected with certainty (i.e., 100% probability) in a given observation duration, and the distance it traversed before detection is bounded.

Specifically, we develop three wave sensing schedules to provide guaranteed worst-case object detection quality in terms of sufficient phase and worst-case stealth distance[1]. We prove several properties of these schedules, and mathematically analyze their average-case object detection quality. We also investigate energy consumption of the wave protocols. Our protocol design and analyses provide insights into the interactions between object detection quality, system parameters, and network energy consumption. Given a worst-case object detection quality requirement, we are able to optimize average-case object detection quality as well as network energy consumption by choosing appropriate network parameters. We evaluate the performance of the wave protocols through extensive simulation experiments.

The rest of the paper is organized as follows. We formulate the object detection and tracking problem in Section 2. Three wave sensing scheduling protocols are presented and analyzed in Section 3. In Section 4, we investigate average-case object detection quality of the protocols under a model for wave sensing protocols. We evaluate the performance of the protocols in Section 5. Section 6 sketches related work. Finally, we conclude our work in Section

2   Object Detection and Tracking Problem Formulation

We assume that sensors are randomly and independently deployed on a square sensing field. The field can be completely covered when all nodes are active. Considering the vastness of the sensing field, the size of a moving object is negligible. In a sensing schedule, a node periodically wakes up and goes to sleep to conserve energy.

We define two metrics to characterize average object detection quality:
- *Detection Probability* (DP). The detection probability is defined as the probability that an object is detected in a given observation duration.
- *Average Stealth Distance* (ASD). The average stealth distance is defined as the average distance an object travels before it is detected for the first time.

For worst-case object detection quality of the network, we have the following two metrics:
- *Sufficient Phase* (SP). The sufficient phase is defined as the smallest time duration in which an object is detected with certainty no matter where the object initially appears on the field.
- *Worst-case Stealth Distance* (WSD). The worst-case stealth distance is defined as the longest possible distance that an object travels before it is detected for the first time.

Taking energy constraints into account, we further define the following metric:
- *Lifetime*. The system lifetime is the elapsed working time from system startup to the time when the object detection quality requirement cannot be met for the first time, with the condition that live nodes continue their current sensing periods.

---

[1] Sufficient phase and worst-case stealth distance are formally defined in Section 2.

In our previous protocol design of the random and the synchronized schedules in [12], we focused on how to meet the requirements of detection probability, average stealth distance and energy consumption. However, worst-case object detection quality metrics, such as sufficient phase and worst-case stealth distance, are not bounded in these protocols. Given an observation duration, an object can escape the detection; it can also travel an infinite distance even its average stealth distance is small. In practice, many applications demand stringent requirements on worst-case object detection quality. For example, an object must be detected in 10 seconds with certainty. Therefore, we aim to design sensing scheduling protocols that achieve a bounded sufficient phase and worst-case stealth distance, while minimizing energy consumption of the system.

3   Design of Wave Sensing Protocols

In this section, we present three wave sensing protocols. The main idea behind these protocols is as follows. When the distance between any two nodes is less than their sensing diameter $2R$, their sensing ranges intersect and form a connected region. If currently-active nodes make up a connected stripe with two ends on opposite borders of the field, the stripe divides the field into two halves. Under such a circumstance, in a sufficiently long time duration, an object can be detected when it crosses this stripe. For any specified continuous curve with two ends on opposite borders of the field, it is always possible to find a set of nodes whose sensing ranges completely cover this curve under the assumption that the field are completely covered when all nodes wake up. To further reduce the detecting time, we allow the curve to move so that every geographical point on the field can be scanned at least once, without leaving any sensing hole in a limited amount of time. We define the curve (line) to be covered as the *active curve (line)*, and define the union of the sensing ranges of all active sensors that cover the active curve (line) as the *hot region*.

We aim to design protocols that ensure: 1) the hot region should have no sensing hole in it; 2) the hot region should be as thin as possible in order to save the network energy consumption; 3) the active curve should move circularly like sea waves, so that the object can be detected rapidly and energy consumption variance among nodes is small.

3.1   Line Wave Protocol Design

In this protocol, we make two assumptions. First, we assume that every node on the field has a timer that is well synchronized with others. The global timer synchronization techniques of [8] can be used in this protocol. Second, we assume that every node is aware of its own geographical location on the field through some method, either by GPS or some other localization techniques.

Line Wave Protocol Description. In the line wave protocol, the active curve is a straight line, as shown in Fig. 1. This protocol is specified as follows.

1. At the system startup time, all nodes synchronize their timers, and obtain their geographical coordinates. There are two active lines on the two opposite borders of the field moving towards the center. All nodes are informed of the initial positions,

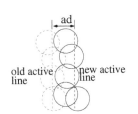

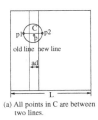

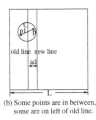

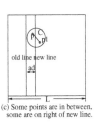

(a) All points in C are between two lines.

(b) Some points are in between, some are on left of old line.

(c) Some points are in between, some are on right of new line.

old active line   new active line

Fig. 1. Line wave protocol illustration

Fig. 2. There is no sensing hole in hot regions of the line wave protocol

the settling time, and the advancing distance (ad) of the active lines. Note that $ad < 2R$.

2. Every node computes current positions of the active lines based on its timer and the information of the active lines it obtained. Then, it calculates if its sensing range intersects the active lines. If there is intersection, this node wakes up.

3. After the active lines have stayed at their current positions for their settling time, they move forward with a distance ad towards the field center. When they reach the center, they go back to the field borders. Step 2 repeats.

Note that a sleeping node periodically wakes up to receive new messages addressed to it. Sensing tasks are distributed to all nodes, thus energy consumption variance among nodes is kept small.

Bounded Sufficient Phase and Worst-Case x Stealth Distance. We define the x *stealth distance* as the distance a moving object travels on x-axis before it is detected. A *handoff* is defined as the process when active lines advance to their new positions, all nodes covering the new lines wake up, and those nodes covering old lines only go to sleep.

Theorem 1. *In the line wave protocol, the sufficient phase of any moving object is bounded by $2P$, where P is the wave sensing period. In other words, the moving object can always be detected in $2P$.*

Proof: Consider the handoff process of an active line in the line wave protocol. Suppose this line moves from left to right. We denote the old active line as ol, and denote the new active line as nl. Note that the distance between ol and nl is less than $2R$. We first prove that there is no sensing hole in the union of ol's hot region and nl's hot region.

We use contradiction in the proof. Suppose there is a sensing hole H in the sensing range union. Consider a point $p \in H$. When all nodes on the field wake up, the field can be completely covered. Thus, there must exist a sensor s that can cover p when it wakes up. Denote the circle of s's sensing range as C. We know that C either intersects ol or nl or both. If not, there could be several cases.

– All points in C are between ol and nl, as shown in Fig. 2(a). Consider the diameter of C on x-axis. Because the diameter has a length of $2R$, while the distance between

ol and nl is less than $2R$, the diameter must intersect at least one of these two lines. Then s must be active. This contradicts the assumption that p is not covered.

- Part of points in C are between ol and nl, and part of points in C are on the left of ol, or on the right of nl. Then we choose a point p between ol and nl, and a point on the left of ol, as shown in Fig. 2(b), or a point on the right of nl, as shown in Fig. 2(c). Denote this point as $p1$. We draw a line to connect p and $p1$, then this line must intersect either ol or nl. Thus, s should be active. This contradicts the assumption that p is not covered.
- All points in C are totally on the left of ol, or on the right of nl. This is impossible, because it contradicts the assumption that p is covered by s.

Therefore, there is no sensing hole in the hot regions of ol and nl.

We next prove an object can be detected in $2P$. As shown in Fig. 3, in a handoff process, the field is divided by the four active lines into five regions, denoted as A, B, C, D, and E, respectively. Consider the initial position $O1$ of the moving object.

- If $O1 \in A \cup E$, after one scanning period P, $O1 \in C$. After another period P, $C = \phi$. This means that the trajectory of this object must have intersected one of the active lines, thus has been detected.
- If $O1 \in B \cup D$, it is covered by the sensing ranges of active sensors, thus has already been detected.
- If $O1 \in C$, then after one scanning period P, $C = \phi$. This means that the trajectory of this object has encountered one of the active lines at least once, thus has been detected.

In summary, the object can always be detected in $2P$. □

Lemma 1. *The worst-case x stealth distance is less than $2vP$. If the object moves along a straight line with an even speed v, the worst-case x stealth distance is less than L.*

Proof: According to Theorem 1, the object is detected in $2P$ time. The distance that the object travels in $2P$ is $2vP$.

Suppose that the object travels along a straight line, and it takes $2P$ to detect this object. In the first P, when the object is behind one of the active lines and is chasing that line, it can travel at most $\frac{L}{2}$ on x-axis without being detected. In the second P, the object is between the two active lines, the maximum x distance it can travel is $\frac{L}{2}$. Therefore, the object can travel at most L on x-axis before being detected. □

3.2   Stripe Wave Protocol Design

One restriction of the line wave protocol is the precision requirement on node coordinates. To relax this constraint, we design a stripe wave protocol. In this protocol, stripes, instead of lines, are covered by active sensors, as shown in Fig. 4. When the stripe width is larger than the required coordinate precision, object detection quality can be achieved.

In the stripe wave protocol, nodes wake up if their sensing ranges intersect active stripes. The width of active stripes is twice of their advancing distance. In this way, there is an overlap between the old stripe and the new stripe. All the other procedures remain the same as those of the line wave protocol.

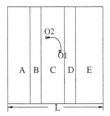

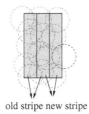

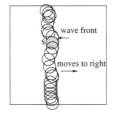

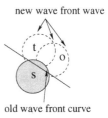

Fig. 3. The object is detected in the line wave protocol

Fig. 4. The sensing stripe handoff in the stripe wave protocol

Fig. 5. The wave front moves in the distributed wave protocol

Fig. 6. An active sensor activates a set of nodes to cover its wave front curve

Sufficient Phase and Worst-Case x Stealth Distance of Stripe Wave Protocols. If one active stripe stays in a place for the same amount of time, and advances the same distance in the same direction as the active lines of the line wave protocol. Then for any point p on the field, if p is covered in the line protocol, it is also covered in the stripe protocol. From Theorem 1 and Lemma 1, we can have the following two corollaries.

Corollary 1. *In the stripe wave protocol, the sufficient phase of a moving object is at most $2P$, where P is the wave sensing period. In other words, the moving object can always be detected in a duration of $2P$.*

Corollary 2. *The worst-case x stealth distance is less than $2vP$. If the object moves along a straight line with an even speed, then the worst-case x stealth distance is less than L.*

3.3   Distributed Wave Protocol

As described above, the line wave and stripe wave protocols require global clock synchronization. To relax such requirement, we design a distributed wave protocol that can be implemented locally on each individual node, and does not need timer synchronization among nodes.

Hot Regions and Wave Fronts. In this distributed wave protocol, there are two continuous active curves with two ends on two opposite borders of the field. These two curves scan the sensing field periodically, so that every point can be covered at least once during one wave sensing period. A set of sensors wake up to cover these curves. We define the *hot region* of an active curve as the union of sensing ranges of active sensors covering this curve, and define the *wave front* of the curve as the boundary of its hot region in its moving direction. Fig. 5 illustrates the wave front of a hot region moving to the right. Since the active curve is continuous, the wave front is continuous as well. For an active curve scanning the field from left to right, its wave front also moves from left to right.

Active Curves Move Forward. Here we describe how an active curve moves forward in our distributed wave protocol. Consider an active sensor s that has part of its sensing circle on the wave front of the active curve. As shown in Fig. 6, we define the wave front curve of s as the part of its sensing circle on the wave front. Before s goes to sleep, it finds all nodes whose sensing ranges intersect its wave front curve to wake up. After those sensors become active, part of their sensing circles become part of the new wave front. For example, in Fig. 6, before s goes to sleep, it finds node t and node o to wake up because the sensing ranges of t and o intersect the wave front curve of s. In this way, the wave front of an active curve always moves forward, eventually it reaches the center vertical line of the field. The same process repeats afterwards.

Note that a sensing hole is a set of continuous points on the field that have not been covered in one scanning period. We have the following theorem.

Theorem 2. *In the distributed wave protocol, the wave front of an active curve can scan the whole sensing field in an finite time without leaving any sensing hole.*

Proof: Consider an active sensor s that has part of its sensing circle on the wave front. We claim that s can always find a set of sensors that have not waked up in current scanning period to cover s's wave front curve.

Let $P(t)$ be the set of points on the field that have been sensed at time t in current scanning period, then we have $P(t) \subset P(t + \Delta t)$, where Δt is a time increment. In other words, the wave front always moves forward and does not go back. For any point p on the field, $p \in P(t) \Rightarrow p \in P(t + \Delta t)$. This implies that if a point $p \in P(t)$, then p is behind the wave front at time $t + \Delta t$. Therefore, the sensors that had already waked up and gone back to sleep in the past cannot cover points on the current wave front. Since any point on the field is within the sensing range of some sensor, there must exist a set of sensors that can cover s's wave front. Thus, we can find sensors that had not waked up to cover s's front wave curve at time $t + \Delta t$. On the other hand, according to the design of this distributed wave protocol, the wave front is continuous with two ends on the opposite borders of the field. Therefore, no sensing hole will be created in this distributed protocol. □

Lemma 2. *In one scanning period, every node on the field wakes up exactly once, and consumes the same amount of energy given that they stay awake for the same amount of time.*

Proof: We assume that no two sensor nodes are located at the same geographical coordinates. We only consider one of the active curves, since the proof can be applied to the other curve due to the symmetry. When a node goes to sleep, it always activates those nodes ahead of the wave front to cover its wave front curve. We use induction to prove that the nodes behind the wave have already waked up once.

- Base. At system start-up time, the active curve is on a side border of the square. Only a set of nodes wake up to cover this curve, and all other nodes have not waked up yet.
- Induction step. Suppose at time t, all nodes behind the wave front have waked up once and only once. Consider the next earliest moment that one active node on the wave front goes to sleep. It activates all nodes that can partly cover its wave front curve. Because these newly-activated sensors are ahead of the wave front, they were in sleeping mode before time t, and have just waked up at t.

On the other hand, if a node has not waked up yet, its sensing range must intersect the wave front curve of some node m at some moment t', where t' is less than the scanning period. Therefore, it will be activated by node m at some moment.   □

We directly obtain the following conclusion from Lemma 2.

Corollary 3. *The scanning period of this distributed wave protocol is less than $wt \cdot n$, where wt is the active duration of nodes in one scanning period and n is the total number of nodes on the sensing field.*

3.4   Partially-Covered Sensing Field

In the partially-covered sensing field, only if the object moves in the covered region, can it be detected in a bounded time with a bounded worst-case stealth distance before detection by using three sensing protocols proposed above.

4   Modeling and Analysis of Wave Sensing Scheduling Scheme

From the earlier protocol design, we know that the scanning period P of the line wave and stripe wave protocols is independent of sensor locations on the field. However, in the distributed wave protocol, the scanning period P is decided by the sensor locations, not controlled by the protocol itself. In this section, we first establish a model for the line wave and stripe wave protocols, then we analyze average-case object detection quality and energy consumption properties of the wave schedules under this model.

In the model, nodes are assumed to be densely deployed on a square field so that the field can be completely covered when all nodes wake up. The field is divided into many identical stripes (in the one-dimensional (*1-d*) wave schedule) or squares (in the two-dimensional (*2-d*) wave schedule), in each of which there are active lines periodically moving through with a constant speed, as shown in Fig. 7 and Fig. 8. The parameters of the model are listed in Fig. 11.

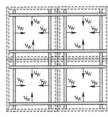
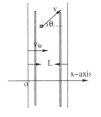
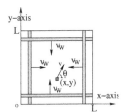

Fig. 7. *1-d* wave schedule illustration

Fig. 8. *2-d* wave schedule illustration

Fig. 9. DP and ASD analysis of the *1-d* wave schedule

Fig. 10. DP and ASD analysis of the *2-d* wave schedule

In the *1-d* wave schedule, the sensing field is divided into multiple parallel vertical stripes with widths of L, as shown in Fig. 7. Two active lines start from the two borders

parameter	meaning
d	density of sensors
R	sensing radius of a sensor
v	speed of a moving object
P	scanning period of the scheme
f	active ratio of sensors in P
v_w	wave line scanning speed
t_a	observation duration
wt	line (curve) settling time
L	side length of a stripe or a square

	DP	ASD
$d \uparrow$	→	→
$R \uparrow$	→	→
$v \uparrow$	↑	↑
$t_a \uparrow$	↑	→
$wt \uparrow$	↓	↑

Fig. 11. Model parameters of the wave protocols

Fig. 12. DP and ASD change when parameters increase in the model

of each stripe moving towards the stripe center with a constant speed v_w. Once they reach the center, the process starts over again. In the *2-d* wave schedule, the sensing field is divided into multiple equally-sized squares with side lengths of L, as shown in Fig. 8. Four active lines start from the four borders of the square moving towards the square center. Once they reach the center, they return to the borders, and the process repeats. Note that the width of hot regions is negligible considering the vastness of the stripes and the squares.

From the model, we know that $P = \frac{L}{2v_w}$, and $f = \frac{4R}{L}$.

4.1   *1-d* Wave Schedule Analysis

Now we mathematically analyze the DP and ASD of the *1-d* wave schedule under our model.

Detection Probability. Consider the sensing stripe where the object is located. We view this stripe as a Euclidean coordinate space with an origin on its left border, as shown in Fig. 9. Note that L is negligible compared to the height of the stripe. Two active lines move from the two borders of the stripe towards the center with a speed of v_w. We define $mod(z_1, z_2)$ as the remainder of z_1 divided by z_2 for any two variables z_1 and z_2.

Suppose at system startup time 0, the object is located at (x, y), and the two active lines are on the two borders of the stripe. Let t be the time when we start observation, where $0 \leq t \leq 2P$. At time t, the x-coordinates of the two active lines are $mod(v_w t, \frac{L}{2})$ and $L - mod(v_w t, \frac{L}{2})$. Since the object can move in any direction, we denote the angle between its moving direction and the x-axis as θ. Then the x-coordinate of the object is $x + vt \cos \theta$ at time t. If the object and one of the active lines meet at time t', then we know $x + vt' \cos \theta = mod(v_w t', \frac{L}{2})$, or $x + vt' \cos \theta = L - mod(v_w t', \frac{L}{2})$.

Define the meeting function $t_i(x, \theta)$ as the time it takes for the ith wave line to meet the object from the system startup time, where $i = 1, 2$. We know whether the object is detected depends on the observation duration t_a. If $t_a < t_1(x, \theta)$ and $t_a < t_2(x, \theta)$, the object cannot be detected. Otherwise, it can be detected.

Suppose after t_m time, the moving object is detected. Denote t_{m1} and t_{m2} as the time for the object to meet the left active line and the right active line, respectively. Then we have $x + vt_{m1}\cos\theta = mod(v_w t_{m1}, \frac{L}{2})$ and $x + vt_{m2}\cos\theta = L - mod(v_w t_{m2}, \frac{L}{2})$. Thus, $t_m(x,\theta,t) = min(t_{m1}(x,\theta), t_{m2}(x,\theta)) - t$.

We define a new boolean detecting function $D(\theta, x, t)$ as follows:

$$D(\theta, x, t) = \begin{cases} 1, & \text{when } t_a \geq t_m(x,\theta,t). \\ 0, & \text{when } t_a < t_m(x,\theta,t). \end{cases}$$

To get the detection probability DP, we integrate $D(\theta, x, t)$ over t, x, and θ, respectively. Therefore, we can get the following theorem.

Theorem 3. *In the* 1-d *wave schedule,* $DP = \frac{\int_0^{2P} dt \int_0^L dx \int_0^{2\pi} D(\theta, x, t)d\theta}{4\pi PL}.$

Average Stealth Distance. Consider the distance dis this object travels in the observation duration t_a, we have $dis = vt_m(x,\theta,t)$. To derive the average stealth distance, we integrate this dis over the ranges of the three variables θ, x, and t, respectively. Then we have the following lemma.

Lemma 3. *In the* 1-d *wave schedule,* $ASD = \frac{\int_0^{2P} dt \int_0^L dx \int_0^{2\pi} vt_m(x,\theta,t)d\theta}{4\pi PL}.$

Sufficient Phase and Worst-case x Stealth Distance. From the analyses of the line wave and stripe wave protocols, we have the following corollaries.

Corollary 4. *The sufficient phase in* 1-d *wave schedule is less than* $2P$.

Corollary 5. *In the* 1-d *wave schedule, the worst-case x stealth distance is upper bounded by* $min(2vP, L)$.

4.2   2-d Wave Schedule Analysis

In the *2-d* wave schedule, the sensing field is divided into repetitive grid squares so that an object can be detected more quickly.

Detection Probability. Suppose we start our observation at time t, where $0 \leq t \leq 2P$. We consider the square where the moving object is located, and view it as a Euclidean coordinate space with an origin on its left bottom corner. Assume the object is located at (x, y) with a moving speed of v, where $0 \leq x \leq L, 0 \leq y \leq L$. Denote the angle between its moving direction and the x-axis as θ.

At time t', the location functions of the four active lines are: $x(t') = mod(v_w t', \frac{L}{2})$; $x(t') = L - mod(v_w t', \frac{L}{2})$; $y(t') = mod(v_w t', \frac{L}{2})$; and $y(t') = L - mod(v_m t', \frac{L}{2})$. The coordinates of the object are: $x + vt'\cos\theta, y + vt'\sin\theta$. If the object is not detected, it must be inside the square whose borders are the four active lines, as shown in Fig. 10. Thus, we have $mod(v_w t', \frac{L}{2}) < x + vt'\cos\theta < L - mod(v_w t', \frac{L}{2})$, and $mod(v_w t', \frac{L}{2}) < y + vt'\sin\theta < L - mod(v_w t', \frac{L}{2})$. Otherwise, if these conditions are not satisfied, the object is detected.

Denote t_{m1}, t_{m2}, t_{m3}, and t_{m4} as the time when the object meets the four active lines respectively, then we have: $x + vt_{m1}\cos\theta = mod(v_w t_{m1}, \frac{L}{2})$; $x + vt_{m2}\cos\theta = L - mod(v_w t_{m2}, \frac{L}{2})$; $y + vt_{m3}\sin\theta = mod(v_w t_{m3}, \frac{L}{2})$; and $y + vt_{m4}\sin\theta = L - mod(v_w t_{m4}, \frac{L}{2})$. The time it takes to detect the object when starting observation from t can be calculated as: $t_m(x, y, \theta, t) = min(t_{m1}, t_{m2}, t_{m3}, t_{m4}) - t$.

We define a detection boolean function

$$D(x, y, \theta, t) = \begin{cases} 1, & \text{if } t_a \geq t_m(x, y, \theta, t). \\ 0, & \text{if } t_a < t_m(x, y, \theta, t). \end{cases}$$

Similar to the *1-d* wave analysis, we can get the following theorem immediately.

Theorem 4. *In the 2-d wave schedule,* $DP = \dfrac{\int_0^{2P} dt \int_0^{2\pi} d\theta \int_0^L dx \int_0^L D(x,y,\theta,t) dy}{4\pi PL^2}$.

Average Stealth Distance. The average detecting time is $\overline{DT}(x, y, \theta, t) = \dfrac{\int_0^{2P} dt \int_0^{2\pi} d\theta \int_0^L dx \int_0^L t_m(x,y,\theta,t) dy}{4\pi PL^2}$. Therefore, we have the following lemma.

Lemma 4. *In the 2-d wave schedule,* $ASD = \dfrac{\int_0^{2P} dt \int_0^{2\pi} d\theta \int_0^L dx \int_0^L vt_m(x,y,\theta,t) dy}{4\pi PL^2}$.

Sufficient Phase and Worst-Case Stealth Distance. Similar to the analyses in the line wave and stripe wave protocols, we have the following bounds for sufficient phase and worst-case stealth distance.

Lemma 5. *In the 2-d wave schedule, the sufficient phase of an object is not greater than $2P$. In other words, when $t_a > 2P$, the object is detected with certainty.*

Proof: Consider the initial location $O1$ of the object when we start our observation. We define interior squares as the shrinking squares with the active lines as their borders.

- If $O1$ is inside an interior square, the object will meet one of the active lines after a period P, thus will be detected.
- Suppose $O1$ is outside all interior squares. After one period P, if it meets the active lines, it is detected. If it is not detected in one P, then it enters one interior square. After another P, this object will be detected.

Therefore, its sufficient phase is not greater than $2P$.   □

Corollary 6. *In the 2-d wave schedule, the worst-case stealth distance is less than $2vP$. If the object moves along a straight line with a constant speed, then its worst-case stealth distance is less than $\sqrt{2}L$.*

Proof: The first part is a direct conclusion from Lemma 5. Suppose the object moves a time of $2P$ before being detected. In the first P, the object can move a distance of at most $\frac{\sqrt{2}L}{2}$, which is the half of the diagonal length of the square. In the second P, the object can move at most $\frac{\sqrt{2}L}{2}$ as well. Therefore, the object can move at most $\sqrt{2}L$.   □

4.3 Node Remaining Energy and System Lifetime

Let T be the continuous working time of a node, and all nodes have the same T. Then in the wave schedules, in the ith period, nodes have remaining energy in the range $[T - ifP, T - (i-1)fP]$ for $1 \leq i \leq \frac{T}{fP}$. A node will last for $\frac{T}{fP}$ periods, thus its working time is $\frac{T}{fP} \cdot P = \frac{T}{f}$. This implies that the system lifetime is $LT = \frac{T}{f}$.

5 Evaluation of Wave Protocols

We conduct extensive simulation experiments to verify our analyses and to evaluate the wave protocols. We assess the average-case object detection quality based on the simulation results of DP and ASD.

In our experiments, we generate a 200×200 grid field, and randomly place $d \times 40,000$ sensors on it. One constraint on these sensors is that when all of them are active, their sensing ranges should be able to cover the whole sensing field. A small object moves along a straight line with a constant speed v. We generate two active sensing lines or stripes at the two borders of the field moving towards the center periodically. We run each simulation for hundreds of times. We use the ratio of times of detection over the number of experiments to estimate DP, and use the average non-detecting distance to estimate ASD. Since we have given upper bounds on the SP and the WSD in the protocol design part, we do not evaluate them in our experiments. Effects of system parameters on DP and ASD of the line wave and stripe wave protocols are listed in Fig. 12.

5.1 Comparison of the Three Wave Protocols

Different from the line wave and stripe wave protocols, the wave sensing period of the distributed wave protocol depends on geographical locations of the nodes. We compare the wave sensing period of these three protocols under the following parameter setting: $d = 0.3, R = 1.5, wt = 0.5$, and $v_w = 5.4$. We find that $P_{line} = 74.8, P_{stripe} = 75.3$, and $P_{dist} = 71.5$. This means the distributed wave protocol scans the field faster than the other two protocols at the cost of extra energy consumption.

To compare the DP, ASD, and energy consumption of different wave protocols, we use the same set of parameters except P for one simulation scenario. In the line wave and stripe wave protocols, ad is slightly less than $2R$. Note that $L = 200$, and $P = \frac{2L}{v_w}$.

DP and ASD Results. In all our experiments on DP, we restrict that $t_a < P$ to make sure that DP varies between 0 and 100%. Figures 13 and 14 demonstrate that all three protocols have close DP and ASD results. However, the distributed wave protocol performs slightly different from the other two protocols, it has a higher DP and a lower ASD. When either v or t_a increases, DP increases too, which is shown in Fig. 13(a). Fig. 13(b) shows that a larger wt incurs a smaller DP. On the other hand, a larger v incurs a larger ASD, as shown in Fig. 14(a). Interestingly, the ASD increases linearly when node settling time wt increases, as we can observe from Fig. 14(b). This is because for a larger wt, it takes longer for an active line or stripe to scan the field than a smaller wt, thus the object can travel a longer distance.

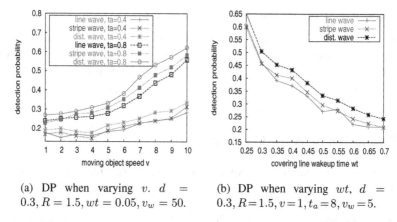

(a) DP when varying v. $d = 0.3, R = 1.5, wt = 0.05, v_w = 50$.

(b) DP when varying wt, $d = 0.3, R = 1.5, v = 1, t_a = 8, v_w = 5$.

Fig. 13. DP of the wave protocols when varying different parameters

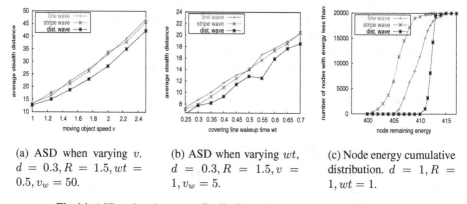

(a) ASD when varying v. $d = 0.3, R = 1.5, wt = 0.5, v_w = 50$.

(b) ASD when varying wt, $d = 0.3, R = 1.5, v = 1, v_w = 5$.

(c) Node energy cumulative distribution. $d = 1, R = 1, wt = 1$.

Fig. 14. ASD and node energy distribution when varying different parameters

Node Energy Distribution Result. All nodes have the same energy E at the beginning, and node energy consumption rate is $er = C \cdot R^3 \cdot \frac{2R}{v_w} / \frac{L}{2v_w} = \frac{4CR^4}{L}$, where C is a constant being dependent on hardware design of the sensor nodes. We set $C = 0.00625$. We draw the node energy cumulative distribution in Fig. 14(c) to further show energy variance among nodes. For any curve point in this figure, its x value represents the node remaining energy, and its y value represents the number of nodes with energy less than the value specified by the x-axis. We observe that the remaining energy of most nodes is around the average node energy of the network. On the other hand, the node energy distribution of the distributed wave protocol has a narrower range than those of the other two wave protocols.

5.2   Comparison of Line Wave, Random, and Synchronized Schedules

In [12], we have formally studied the random sensing schedule and the synchronized sensing schedule. In Figures 15 and 16, we compare the analytical DP and ASD results

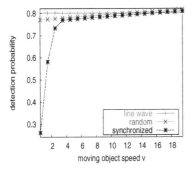

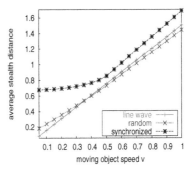

Fig. 15. DP when varying v. $d = 1, R = 0.75, t_a = 0.4, wt = 0.01, v_w = 2$

Fig. 16. ASD when varying v. $d = 1, R = 0.75, wt = 0.01, v_w = 2$

of the line wave schedule, the random schedule, and the synchronized schedule, respectively, when varying v and fixing all other parameters. We observe that, with a small v, the line wave schedule and the random schedule have a larger DP than the synchronized sensing schedule; however, as v increases, the synchronized schedule begins to catch up the line wave schedule, and eventually outperforms it. Similarly, the line wave schedule has a smaller ASD for small v. When v increases, the ASD of the line wave schedule exceeds that of the random sensing schedule, and the synchronized schedule has a larger ASD than the other two schedules.

6   Related Work

Extensive work has been conducted on object-tracking with different approaches: system design and deployment [6, 13], design of high tracking-precision protocols [2], design of energy-efficient tracking protocols [11], and design of protocols to exploit node collaborations [3, 7, 9, 10, 17]. In [11], Pattem *et al.* proposed duty-cycled activation and selective activation algorithms to balance tracking errors and energy expenditure. Gui *et al.* [5] studied the quality of surveillance of several sensing scheduling protocols. [13] is a work covering design of hardware, networking architecture, and control and management of remote data access. Increasing the degree of sensing coverage can usually improve object detection and tracking quality. Along this direction, many power-efficient sensing coverage maintenance protocols [1, 4, 14, 15] have been proposed in the literature. Yan *et al.* [15] presented an energy-efficient random reference point sensing protocol to achieve a targeted coverage degree. In [16], Zhang and Hou studied the system lifetime of a k-covered sensor network, and proved that it is upper bounded by k times node continuous working time.

7   Conclusion

In this paper, we designed three wave sensing scheduling protocols to provide bounded worst-case object detection quality for many surveillance applications. One of the protocols is distributed and can be implemented locally on each individual node. In these protocols, sensing field is scanned by a set of connected active sensors periodically. In addition, we characterized the interactions among network parameters, and analyzed average-case object detection quality and energy consumption of the protocols. We proved the correctness of the proposed protocols and evaluated their performances through extensive simulations.

Acknowledgment

We are grateful to the anonymous reviewers for their constructive comments and suggestions. We also acknowledge William L. Bynum for reading an earlier version of the paper. This work is partially supported by the U.S. National Science Foundation under grants CNS-0098055, CCF-0129883, and CNS-0405909.

References

1. Z. Abrams, A. Goel, and S. Plotkin. Set k-Cover algorithms for energy efficient monitoring in wireless sensor networks. In *Proceedings of IPSN'04*.
2. J. Aslam, Z. Butler, F. Constantin, V. Crespi, G. Cybenko, and D. Rus. Tracking a moving object with a binary sensor network. In *Proceedings of ACM SenSys'03*.
3. R. Brooks, P. Ramanathan, and A. Sayeed. Distributed target classification and tracking in sensor networks. In *International Journal of High Performance Computer Applications, special issue on Sensor Networks*, 16(3), 2002.
4. P. B. Godfrey and D. Ratajczak. Naps: scalable, robust topology management in wireless ad hoc networks. In *Proceedings of IPSN'04*.
5. C. Gui and P. Mohapatra. Power conservation and quality of surveillance in target tracking sensor networks. In *Proceedings of ACM MobiCom'04*.
6. T. He, S. Krishnamurthy, J. Stankovic, T. Abdelzaher, L. Luo, R. Stoleru, T. Yan, L. Gu, J. Hui, and B. Krogh. Energy-efficient surveillance system using wireless sensor networks. In *Proceedings of ACM/USENIX MobiSys'04*.
7. D. Li, K. Wong, Y. Hu, and A. Sayeed. Detection, classification and tracking of targets in distributed sensor networks. In *IEEE Signal Processing Magazine*, pages 17–29, March 2002.
8. Q. Li and D. Rus. Global clock synchronization in sensor networks. In *Proceedings of IEEE INFOCOM'04*.
9. J. Liu, M. Chu, J. Liu, J. Reich, and F. Zhao. Distributed state representation for tracking problems in sensor networks. In *Proceedings of IPSN'04*.
10. J. Liu, J. Liu, J. Reich, P. Cheung, and F. Zhao. Distributed group management for track initiation and maintenance in target localization applications. In *Proceedings of IPSN'03*.
11. S. Pattem, S. Poduri, and B. Krishnamachari. Energy-quality tradeoffs for target tracking in wireless sensor networks. In *Proceedings of IPSN'03*.
12. S. Ren, Q. Li, H. Wang, X. Chen, and X. Zhang. Analyzing object detection quality under probabilistic coverage in sensor networks. In *Proceedings of IWQoS'05*.
13. R. Szewczyk, A. Mainwaring, J. Polastre, and D. Culler. An analysis of a large scale habitat monitoring application. In *Proceedings of ACM SenSys'04*.
14. X. Wang, G. Xing, Y. Zhang, C. Lu, R. Pless, and C. Gill. Integrated coverage and connectivity configuration in wireless sensor networks. In *Proceedings of ACM SenSys'03*.

15. T. Yan, T. He, and J. Stankovic. Differentiated surveillance for sensor networks. In *Proceedings of ACM SenSys'03*.
16. H. Zhang and J. Hou. On deriving the upper bound of α-lifetime for large sensor networks. In *Proceedings of ACM MobiHoc'04*.
17. W. Zhang and G. Cao. Optimizing tree reconfiguration for mobile target tracking in sensor networks. In *Proceedings of IEEE INFOCOM'04*.

Multiple Controlled Mobile Elements (Data Mules) for Data Collection in Sensor Networks

David Jea, Arun Somasundara, and Mani Srivastava

Networked and Embedded Systems Laboratory,
Department of Electrical Engineering, UCLA
{dcjea, arun, mbs}@ee.ucla.edu

Abstract. Recent research has shown that using a mobile element to collect and carry data mechanically from a sensor network has many advantages over static multihop routing. We have an implementation as well employing a single mobile element. But the network scalability and traffic may make a single mobile element insufficient. In this paper we investigate the use of multiple mobile elements. In particular, we present load balancing algorithm which tries to balance the number of sensor nodes each mobile element services. We show by simulation the benefits of load balancing.

1 Introduction

Recently there has been an increased focus on the use of sensor networks to sense and measure the environment. Some practical deployments include NIMS [1], James Reserve [2]. Both these deployments focus mainly on the problem of habitat and environment monitoring. In most cases the sensors are battery-constrained which makes the problem of energy-efficiency of paramount importance.

There are multiple ways in which the sensor readings are transferred from the sensors to a central location. Usually, the readings taken by the sensor nodes are relayed to a base station for processing using the ad-hoc multi-hop network formed by the sensor nodes. While this is surely a feasible technique for data transfer, it creates a bottleneck in the network. The nodes near the base station relay the data from nodes that are farther away. This leads to a non-uniform depletion of network resources and the nodes near the base station are the first to run out of batteries. If these nodes die, then the network is for all practical purposes disconnected. Periodically replacing the battery of the nodes for the large scale deployments is also infeasible.

A number of researchers have proposed mobility as a solution to this problem of data gathering. Mobile elements traversing the network can collect data from sensor nodes when they come near it. Existing mobility in the environment can be used [3, 4, 5, 6] or mobile elements can be added to the system [7, 8, 9], which have the luxury to be recharged. This naturally avoids multi-hop and removes the relaying overhead of nodes near the base station. In addition, the sensor nodes no longer need to form a connected network (in a wireless sense). Thus a network can be deployed keeping only the sensing aspects in mind. One need not worry about adding nodes, just to make sure that data transfer remains feasible.

V. Prasanna et al. (Eds.): DCOSS 2005, LNCS 3560, pp. 244–257, 2005.

But this technique of using mobile elements comes at a cost of increased latency for data collection. As a result, this is more suitable for delay tolerant networks [10], for instance, habitat monitoring [1, 2] mentioned earlier.

In this paper we consider using multiple mobile elements for purposes of data collection. We first briefly review our prior work on single mobile element [7] in Sect. 3. Next we describe the necessity of using multiple mobile elements for scalability reasons in Sect. 4. When multiple mobile elements are used to collect data from sensor nodes (# sensor nodes $\gg$ # mobile elements), it is better to have the mobile elements serve more or less the same number of nodes. We describe these ideas of load balancing in Sect. 5. The operation with load balancing is described precisely in Sect. 6. We present simulation methodology and results in Sect. 7, and finally end with conclusions and some directions for future work in Sect. 8. We begin with presenting the related work.

2   Related Work

Various types of mobility have been considered for the mobile element. These can be broadly classified as random, predictable or controlled. An algorithm for routing data among randomly mobile users was suggested in [11] where data is forwarded to nodes which have recently encountered the intended destination node. Random motion of mobile entities was also used for communication in [4, 5], where the mobile entities were zebras and whales. An important difference in these two from others is that the sensor nodes themselves are mounted on the mobile entities (animals), and the goal is to track their movements. In [3], randomly moving humans and animals act as "data mules" and collect data opportunistically from sensor nodes when in range. However, in all cases of random mobility, the worst case latency of data transfer cannot be bounded.

Predictable mobility was used in [6]. A network access point was mounted on a public transportation bus moving with a periodic schedule. The sensor nodes learn the times at which they have connectivity with the bus, and wake up accordingly to transfer their data.

Controlled mobility was considered in our previous work [7], where a robot acts as a mobile base station. The speed of the mobile node was controlled to help improve network performance. This is briefly summarized in the next section. Controlled mobility was also used in [8], where a mobile node is used to route messages between nodes in sparse networks. However, all nodes are assumed to have short range mobility and can modify their locations to come within direct range of the mobile node which has long range mobility and is used for transferring data.

In [12], mobile nodes in a disconnected ad hoc network modify their trajectories to come within communication range, and [13] considered moving the intermediate nodes along a route, so that the distances between nodes are minimized and lower energy is used to transmit over a shorter range. This system also assumes that all nodes are mobile, which may be expensive or infeasible in many deployments where node locations depend on sensing or application requirements. A mobile base station was also used in [9] to increase network lifetime. A scheduling problem for the mobile node with buffer constraints on static nodes and variable sampling rates at each static node is studied in [14]. Connectivity through time protocols were proposed in [15] that exploit robot motions to buffer and carry data if there is no path to the destination.

Henceforth, we will use the term **data mule**, borrowed from [3] to denote a mobile element.

3   Single Data Mule

For sake of completeness and having continuity, we briefly describe the single controlled data mule approach in this section. Sensor nodes are deployed in an area, and are sampling the physical phenomenon. There is a data mule whose job is to collect data from these sensor nodes. The data mule moves in a straight line up and down. The operation can be divided into two parts: Network algorithms (specifies how sensor nodes interact with each other and the data mule) and Motion Control algorithms (specifies how the data mule moves)

3.1   Network Algorithms

The network may be such that some sensor nodes may never hear the data mule directly. In this situation, they transfer the data through other nodes, which can directly hear the data mule. The algorithm can be divided into three phases:

1. **Initialization:** This is used to find out the number of hops each node is from the path of the data mule (initialized to ∞). The data mule moves broadcasting the beacons (with hop count as 1). All nodes which hear it mark the hop count, and also rebroadcast it (after incrementing the hop count). A node which hears a beacon with hop count less than what has, updates itself (also noting the node from which it came). At the end of this phase, all nodes know if they are on path of data mule (at 1 hop). If they are not on path of the data mule, they know the parent through which to reach a node which is on path. Basically, this is tree building, with number of trees being equal to the number of nodes on path of the data mule. All nodes are members of exactly one tree.
2. **Local multihops:** Each of the trees formed above do a local multihop within themselves, with the root of the tree collecting data of its children nodes. Directed Diffusion [16] is suitable for this.
3. **Data Collection by Data Mule:** After one round of initialization phase, the data mule moves polling for data. The nodes which hear the data mule respond with the data (their own and that of their children). To prevent loss of data due to data mule going out of range, we can have a acknowledgement based scheme.

Once the initialization phase is over, the other two proceed in parallel.

3.2   Motion Control Algorithms

Motion can be controlled in two dimensions: space (where the data mule goes), and time (how or what speed the data mule moves). By fixing the path to be a straight line, we need to decide the speed. Two options are possible:

1. Fix the round trip time (RTT) of the data mule. With this, there are few approaches possible.

   – We can traverse the path at a fixed speed (at which we get maximum efficiency from the data mule). Suppose this takes time $T(< RTT)$. Then we have $RTT - T$ spare time. We can divide this time equally among all nodes. The data mule would stop for this time at each node. We do not assume that the data mule knows the node locations. So we stop when we first hear from a node.
   – Alternately, we can cover the trail at constant speed (Length of path / RTT), and not stop at any node.
   – Finally, we can also have an adaptive speed control algorithm, where the data mule would normally move at twice the speed above. This leaves $RTT/2$ time to service the nodes by stopping at them. This time can be divided among a subset of the nodes, from which the mule had collected less data than a threshold in the previous round. Thus unlike the first case, the sets of nodes at which the mule would stop would change with each round.

   We have not gone into details of the above algorithms. Our earlier paper [7] can be referred where we have implemented a version of these ideas. Although two nodes are one hop from the data mule, the time which the data mule stays in their contact may vary, depending on the distance of the node to the path. One way to take care of this is adaptive motion control mentioned above. The mule would have collected less data from a node which is far from the path, and in the next round would stop at it, giving it more time. Nevertheless, this approach only tries to maximize the amount of data collected. It does not guarantee that all the generated data is collected.

2. Give an equal amount of service time to each node (with the mule stopping for this amount of time at each node). The service time for a node can be set equal to Buffer size / Communication data rate. This would ensure that the data mule is able to collect all the data. The mule can collect data even when it is moving, and this can be considered as a bonus. As mentioned before, all nodes need not be one hop from the data mule. For this, the root of the trees (all roots are one hop from the data mule) can be given time equivalent to the number of nodes in its tree. The RTT would depend on number of nodes in the network.

4   Multiple Data Mules

The single data mule approach presented in the previous section does not scale well. Suppose the density of the network increases due to increasing number of nodes. Considering the approach of fixed round trip time for data mule, there are more nodes from which data has to be collected, in the same amount of time. This leads to loss of data due to buffer overflows at the nodes. If the second approach of stopping at each node is used, the data mule will take a longer time to complete a round. In this case, although at time of each service, the buffer of a node is cleared, it may not be possible for the data mule to return to this node before its buffer fills again. Again this leads to loss of data. Another issue arises if the network is deployed over a larger area. The distance over which the data mule moves increases. The battery capacity may not be sufficient for moving this length, requiring recharge on the path.

These problems can be addressed by using multiple data mules. A trivial solution would be dividing the area into equal parts and having one data mule in each. This solves the problem if the nodes are uniformly randomly deployed, so that each mule gets approximately same number of nodes to service. Each mule covers the same area. Now each mule can independently run the same single mule algorithms presented in the previous section. To analytically calculate the required number of data mules, let us define the following:

– num_nodes nodes are deployed in an area of $l \times l$ units.
– A data mule moves in a straight line from one end to another and back at speed s.
– Time to fill a node's buffer is $buffer\_fill\_time$. This can be calculated using the sampling period and buffer size. We assume that all nodes are sampling at the same rate.
– Time for the data mule to empty a node's buffer is $service\_time$ given by $buffer\_size/communication\_data\_rate$. Let us assume the second form of motion control (Sec. 3.2), where the data mule stops at each node for this amount of time.
– Round trip time (RTT) for the data mule will be $(l/s)+(num\_nodes\times service\_time)+(l/s)$.

Now, if $RTT \leq buffer\_fill\_time$, one data mule will suffice. Otherwise, $\lceil RTT/buffer\_fill\_time \rceil$ mules would be required.

Two things are to be noted with respect to the last calculation. Firstly, in the expression for round trip time, the first two terms denote the time it takes the data mule to move from one side of the area to the other along with time for data collection. The last term denotes the time to come back to the starting point. There is no data collection in the reverse path, so as to approximate the whole motion to a closed loop. Secondly, some nodes may not be able to communicate directly to the data mule. The data of these nodes will be available at root of the tree this node belongs to using local multihopping.

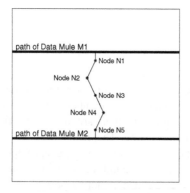

Fig. 1. Nodes between 2 mules

Table 1. Hop count at nodes in Fig. 1

Nodes	Data Mule M1	Data Mule M2
N1	1	5
N2	2	4
N3	3	3
N4	4	2
N5	5	1

(The root will be on path of data mule as mentioned in previous section). A related issue is the fact that a node will belong to more than one tree (one tree per data mule). In such cases, the node will send data towards the closer data mule. This is illustrated in Fig. 1, with the 2 data mules moving on straight line paths. Table 1 shows the hop count variable at each of the sensor nodes $N1 - N5$ due to the two data mules. As can be seen $N1$, $N2$ will be serviced by $M1$, and $N4$, $N5$ by $M2$. $N3$ is at equal hop count from both the data mules. Such ties can be broken randomly.

5   Load Balancing

The previous section made the case for using multiple data mules. If the nodes are uniformly randomly distributed, then the obvious thing to do is to divide the area into equal regions.

But in practice, at real deployments it may not be so trivial. Firstly, the nodes need not be uniformly deployed. They will be placed by the field experts such as the biologists. They would want to deploy nodes in areas where they suspect interesting activities to take place. This will naturally lead to non-uniform placement. In addition, in these environments it may not be feasible to have the data mule trails according to system designer's requirements.

5.1   Problem Description

We are given a set of nodes deployed in an area, and straight line paths for the data mules ($M1$, $M2$, ...) to move (which are not necessarily equally spaced). We assume for simplicity that each node is one hop away from atleast one data mule (and atmost two data mules). Fig. 2 describes the scenario with three data mules $M1$, $M2$, $M3$. The various regions are marked $A - H$. Table 2 describes the various regions, and the data mules they can talk to. The regions which are serviced by a single data mule have no choice. The nodes in these regions will be called non-shareable nodes. But the nodes in regions which are serviced by 2 data mules, can be attached to either of them. Such nodes are called shareable nodes. The goal is to find the data mule assignment for these shareable nodes, so that each data mule services approximately same number of nodes.

5.2   Why It Is Important

Consider a simple scenario with 50 nodes and 2 data mules $M1$, $M2$. Suppose $M1$ has 25 non-shareable nodes, $M2$ has 5 non-shareable nodes, and 20 nodes are shared between them. If these 20 nodes are equally divided between the two data mules, $M1$ will end up servicing 35 nodes and $M2$ 15 nodes. Consider $M1$. If we use the approach of fixed round trip time for the data mule, the time given to each node will be reduced. On the other hand, if we use the approach of stopping at each node (for amount of time required to empty its buffer), the round trip time of the data mule will increase, leading to possibility of buffer overflows, when the data mule returns to service them. Instead, if both the data mules serviced 25 nodes, the above mentioned problem will be solved. For this to happen, all 20 shareable nodes are to be serviced by $M2$.

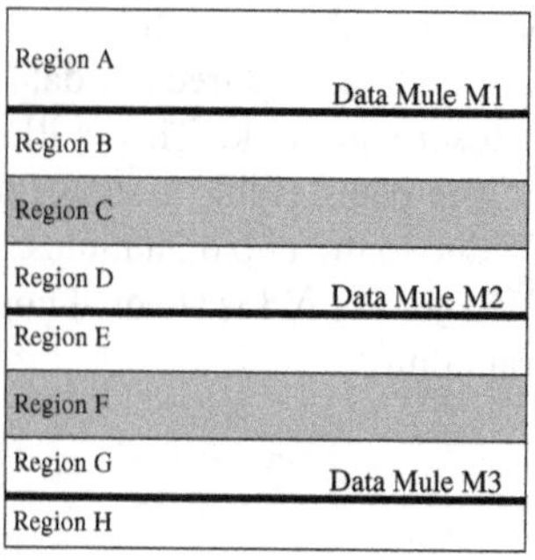

Fig. 2. Problem description

Table 2. Illustration of Fig. 2

Region	Visible Data Mule(s)	Region type
A	M1	non_shareable
B	M1	non_shareable
C	M1, M2	shareable
D	M2	non_shareable
E	M2	non_shareable
F	M2, M3	shareable
G	M3	non_shareable
H	M3	non_shareable

Load balancing is common concept in distributed systems [17]. In our case, the tasks are the servicing of sensor nodes, and the processing elements (PEs) are the data mules. In addition, there are constraints on the tasks, as to which PEs can process them. This refers to the fact that sensor nodes (if they are shareable) can only be serviced by 2 data mules.

6   Multiple Data Mules with Load Balancing

We now describe the multiple data mule approach with load balancing. This is one of the main contributions of this paper. This can be divided into five parts: initialization, leader election, load balancing, assignment, and data collection.

6.1   Initialization

The data mules make a round broadcasting the beacons. The nodes which can hear, reply back with their id's. The data mules note down the list of distinct node id's they got the response from. At the end of this round, each data mule has a list of nodes which are one hop from its path.

6.2   Leader Election

We assume that the data mules are equipped with powerful radios, and can communicate with each other. They elect a leader among themselves, and everyone sends the information gathered in the initialization round to the leader. The data mule with the smallest id becomes the leader in our case.

6.3   Load Balancing

The leader data mule has the information of all the data mules. For each data mule i, the leader can classify its nodes into 2 classes: shareable nodes and non_shareable nodes. The shareable nodes can further be classified as being shared with previous or next data mule. Let us define an array structure DM, of size equal to number of data mules (N). The structure $DM[i]$, denoting data mule i has the following members:

- $non\_shareable\_nodes$ denotes the set of nodes it is solely responsible.
- $shareable\_nodes\_neg$ denotes the set of nodes it shares with the previous data mule $i - 1$. For $DM[1]$, this is a null set.
- $shareable\_nodes\_pos$ denotes the set of nodes it shares with the next data mule $i+1$. For $DM[N]$, this is a null set.
- $non\_shareable\_load$ denotes the size of the set $non\_shareable\_nodes$.
- $shareable\_load\_neg$ denotes the size of the set $shareable\_nodes\_neg$.
- $shareable\_load\_pos$ denotes the size of the set $shareable\_nodes\_pos$.

The above variables are calculated by the leader, and form the input to the load balancing algorithm. It may be noted that $DM[i].shareable\_nodes\_neg$ & $DM[i - 1].shareable\_nodes\_pos$ are same. Similarly, $DM[i].shareable\_nodes\_pos$ is same as $DM[i + 1].shareable\_nodes\_neg$.

- $my\_shareable\_load\_neg$ denotes the number of nodes this data mule is responsible for out of $shareable\_load\_neg$.
- $my\_shareable\_load\_pos$ denotes the number of nodes this data mule is responsible for out of $shareable\_load\_pos$.
- my_total_load is the total number of nodes this data mule will be responsible for. It is the sum of $non\_shareable\_load$, and the above two variables.
These three variables evolve as the algorithm proceeds.

Initially all data mules are in the same single group, with first mule called the $start\_mule$, and last one called the end_mule. The idea is to make the load of (number of nodes serviced by) each data mule equal to the average load of that group. This may not always be possible. For instance, consider the simple scenario with 50 nodes and 2 data mules $M1, M2$. Suppose $M1$ has 35 non_shareable nodes, $M2$ has 5 non_shareable nodes, and 10 nodes are shared between them. The best possible result would be to assign all the 10 shareable nodes to $M2$, making it responsible for 15 nodes, and $M1$ for 35 nodes. In such a case, we divide the original group into two, and try to balance the load of each group recursively. The recursion is terminated when we reach the last mule of the group.

Table 3. Meaning of flag variables for a mule

Flag	If Flag is $TRUE$	If Flag is $FALSE$
$start\_flag$	$my\_shareable\_load\_neg = shareable\_load\_neg$	$my\_shareable\_load\_neg = 0$
end_flag	$my\_shareable\_load\_pos = shareable\_load\_pos$	$my\_shareable\_load\_pos = 0$

The group splitting happens when we reach a mule such that the *minimum* load it should take is more than the group average. We form two groups with this mule belonging to the first group. This mule becomes the end_mule of the first group. Also, the load this mule shared with the next mule is given completely to the next mule, which becomes the $start\_mule$ of the second group. A group can also split when we reach a mule such that the *maximum* load it can take is less than the group average. Here again we form two groups with this mule belonging to the first group. But now this mule takes

---

ALGORITHM: Load_Balance(start_mule, end_mule, start_flag, end_flag)

1. Initialize $group\_has\_split$ to $FALSE$
2. Calculate the average load of this group $group\_avg$, as shown in Fig. 4.
3. Reset the following variables:
   - $DM[start\_mule..end\_mule].my\_shareable\_load\_neg$
   - $DM[start\_mule..end\_mule].my\_shareable\_load\_pos$
   - $DM[start\_mule..end\_mule].my\_total\_load$
4. Repeat the following **for** $i = start\_mule..end\_mule$
   (a) if $group\_has\_split$ is $TRUE$, return.
      - This is to terminate recursion, when we come to next iteration of for loop during back tracking of recursion.
   (b) Calculate the minimum load that can be assigned to this mule.
      i. If $i = start\_mule$ AND $start\_flag = TRUE$
         A. $DM[i].my\_shareable\_load\_neg = DM[i].shareable\_load\_neg$
      ii. Else If $i \neq start\_mule$
         A. $DM[i].my\_shareable\_load\_neg =$
            $DM[i].shareable\_load\_neg - DM[i-1].my\_shareable\_load\_pos$
      iii. $DM[i].my\_total\_load =$
            $DM[i].non\_shareable\_load + DM[i].my\_shareable\_load\_neg$
   (c) If $i = end\_mule$
      i. If $end\_flag = TRUE$
         A. $DM[i].my\_shareable\_load\_pos = DM[i].shareable\_load\_pos$
         B. $DM[i].my\_total\_load+ = DM[i].my\_shareable\_load\_pos$
      ii. **Return**
   (d) If $DM[i].my\_total\_load > group\_avg$
      - The load on this mule is more than group average; we split into two groups.
      i. Set $group\_has\_split = TRUE$
      ii. Call **Load_Balance(start_mule, i, start_flag, FALSE)**
      iii. Call **Load_Balance(i+1, end_mule, TRUE, end_flag)**
         • All nodes shared between mules i and $i + 1$ is taken by mule $i + 1$.
   (e) Else (i.e. if $DM[i].my\_total\_load \leq group\_avg$)
      i. Calculate the extra load that can be given to this mule.
         - $extra\_load = group\_avg - DM[i].my\_total\_load$
         - $DM[i].my\_shareable\_load\_pos =$
            $\min(extra\_load, DM[i].shareable\_load\_pos)$
         - $DM[i].my\_total\_load+ = DM[i].my\_shareable\_load\_pos$
      ii. If $DM[i].my\_total\_load < group\_avg$
         - The maximum load this mule can have is less than the group average. Here also, we split into two groups.
         A. Set $group\_has\_split = TRUE$
         B. Call **Load_Balance(start_mule, i, start_flag, TRUE)**
         C. Call **Load_Balance(i+1, end_mule, FALSE, end_flag)**
            • All nodes shared between mules i and $i + 1$ is taken by mule i.
      iii. Else (i.e. if $DM[i].my\_total\_load = group\_avg$, **continue**
         - We continue to check mule $i + 1$

END

Fig. 3. Load Balancing Algorithm

ALGORITHM: Calculate group average

- Input parameters: $start\_mule, end\_mule, start\_flag, end\_flag$
- Procedure:
  1. Initialize $group\_load = \sum_{i=start\_mule}^{end\_mule} DM[i].non\_shareable\_load$
  2. Switch depending on $(start\_flag, end\_flag)$
     - $(TRUE, TRUE)$
       * $group\_load+ = \sum_{i=start\_mule}^{end\_mule} DM[i].shareable\_load\_neg+$
         $$DM[end\_mule].shareable\_load\_pos$$
     - $(TRUE, FALSE)$
       * $group\_load+ = \sum_{i=start\_mule}^{end\_mule} DM[i].shareable\_load\_neg$
     - $(FALSE, TRUE)$
       * $group\_load+ = \sum_{i=start\_mule}^{end\_mule} DM[i].shareable\_load\_pos$
     - $(FALSE, FALSE)$
       * $group\_load+ = \sum_{i=start\_mule+1}^{end\_mule} DM[i].shareable\_load\_neg$
  3. $group\_avg = \frac{group\_load}{end\_mule - start\_mule + 1}$
- Return $group\_avg$

END

Fig. 4. Algorithm for calculating $group\_avg$ in step 2 of Fig. 3

all the load it shares with the next mule. We use two flags $start\_flag$, and end_flag to denote these states. Mules are affected by these flags only if they are $start\_mule$, or end_mule respectively. Table 3 describes these variables. Initially, when there is only one group comprising all the mules, the values of these flags do not matter, as the $start\_mule$ (mule #1) does not have any predecessor, and end_mule (mule #N) does not have any successor.

The precise algorithm is given in Fig. 3. Initially, it is invoked with parameters $(1, N, FALSE, FALSE)$[1]. The algorithm has comments explaining each of the steps, and is also described below. One thing to be noted is the use of local boolean variable $group\_has\_split$. There are recursive calls inside the *for* loop in step 4. If it goes inside recursion, there is no more meaning for the current group. When control comes back to this *for* loop again during backtracking of recursion, we should not process the remaining mules in the *for* loop. Step 4.*a* achieves this.

We begin by calculating the group average in step 2, making sure not to count shareable nodes twice. The average depends on the two flag values, as shown in Fig. 4. We next run a loop for all nodes in the group. First, we calculate the minimum load the mule under consideration should take (step 4.*b*). In particular, if this mule is not the $start\_mule$, 4.*b.ii.A* calculates the part of of shareable_load which the previous mule did not take.

The group splitting can happen in two cases. Firstly, if the minimum load (which was calculated in step 4.*b*) that has to be assigned to the mule under consideration is more than the group average. When this happens we break into two groups, and the

---

[1] as mentioned previously, the two flags do not matter initially.

mule under consideration becomes part of the first group. This is shown in $4.d$ in the algorithm. We recursively call the algorithm for the two groups. If the above does not happen, we try to assign some shareable_load , which this shares with the next mule, as shown in $4.e.i$. Now the other reason of splitting can arise. If the maximum load that can be assigned to this mule is less than the group average, we split into two groups, as shown in $4.e.ii$.

The current recursion ends when we reach end_mule in step $4.c$. In addition, if the end_flag is $TRUE$, we add some more load, as shown in $4.c.i.A$. This is in accordance to Table 3.

The worst case complexity of this load balancing algorithm is $O(n^2)$, where n is the number of data mules. This occurs when the group splitting always happens such that one of the resultant groups has only one mule, and the other group has all the remaining mules.

6.4   Assignment

The load balancing algorithm of the previous subsection outputs three counts for each data mule: $my\_shareable\_load\_neg$, $my\_shareable\_load\_pos$, my_total_load. Now we have to find the corresponding sets i.e.$my\_shareable\_nodes\_neg$ and $my\_shareable\_nodes\_pos$. Two data mules would be sharing some nodes. These nodes are ordered by node id. The idea is to assign the first part of this ordered set to the first mule (resulting in $DM[i].my\_shareable\_nodes\_pos$), and the second part to the second mule (resulting in $DM[i + 1].my\_shareable\_nodes\_neg$). The size of the two parts depends on the counts mentioned above. Finally, each data mule is responsible for nodes in the three sets: $non\_shareable\_nodes$, $my\_shareable\_nodes\_neg$, and $my\_shareable\_nodes\_pos$. After the assignment has taken place, the leader can inform all the data mules, the set of nodes they have to service.

6.5   Data Collection

With the assignment done, the data mules traverse their paths, polling for data. The shareable sensor nodes do not know which of the two data mules they belong to. The nodes respond for data when they hear the poll packet. The data mule will send back an acknowledgement only if it is responsible for servicing that node. The sensor node marks the data mule from which it hears an acknowledgement, and does not respond to poll packets from the other data mule in future.

7   Simulation Methodology and Results

We now present our simulation methodology. We implemented our algorithms in TinyOS [18]. The simulator used is TOSSIM [19]. The advantage of this combination is that, the same TinyOS code can be put on real sensor nodes. To simulate mobility, we use tython [20]. We consider three schemes for sharing the shareable load between data mules.

Table 4. Simulation Results

	Data Mule $M1$	Data Mule $M2$	Data Mule $M3$	Data Mule $M4$
non_shareable_load	13	5	5	9
shareable_load_neg	0	3	3	2
shareable_load_pos	3	3	2	0
FCFS	16	8	5	11
Equal Sharing	15	8	7	10
Load Balancing	13	9	9	9

1. First Come First Serve (FCFS): The shareable sensor node will get attached to the data mule from which it hears the beacon packet first in the Initialization round of data mules (Sect. 6.1).
2. Equal sharing: Each adjacent data mules have a set of shareable nodes between them. Here, half the shareable nodes are assigned to one data mule, and the other half to the other mule.
3. Load balancing: Result of applying the load balancing algorithm of Sect. 6.3.

For the simulation topology, we had 40 sensor nodes, and 4 data mules, $M1 - M4$. The nodes are randomly distributed inside the 100x100 grid region from $(24, 30)$ to $(76, 70)$. $M1$ moves from $(32, 10)$ to $(32, 90)$, $M2$ moves from $(44, 90)$ to $(44, 10)$, $M3$ moves from $(56, 10)$ to $(56, 90)$ and $M4$ moves from $(68, 90)$ to $(68, 10)$. To have effect of closed-loop path, whenever a data mule reaches its end point, we immediately place it at the starting point, from where it starts moving again. The nodes are placed inside a smaller region than whole grid to avoid edge-effects.

The result after the leader mule ($M1$) gets information from other data mules is shown in upper half of Table 4. This result depends on topology. Now we execute the three schemes of load sharing presented above, resulting in *my_total_load* values shown in lower half of Table 4. As can be seen, the result of load balancing does not necessarily result in equal distribution of load, because non_shareable_load of $M1$ is more than the group average. This results in group splitting, and we end up balancing the load of the second group.

With these three node assignments to mules, we ran the experiment for 5 rounds, with the first strategy mentioned in Sect. 3.2 of fixed round trip time. The RTT was set to 120 time units, and it took each mule 40 time units to complete a round. This gave 80 time units for stopping at the nodes, which was divided equally among the nodes to be serviced. Fig. 5 shows the average number of packets received per node per round, at each of the data mules. Results are shown for the three assignments. It is evident that load balancing leads to more uniformity. Although data mules $2, 3, 4$ service the same number of nodes (9) in the load balanced case, we see a minor variation. This is due to fact that we are collecting data even when moving, in addition to when being stopped. So if the nodes assigned to $M4$ are closer to its path (when compared to nodes assigned to $M3$ and its path), we end up collecting more data at $M4$.

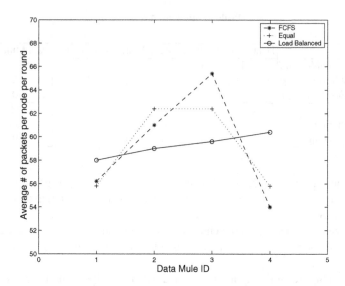

Fig. 5. Simulation Results

8   Conclusions and Future Work

Deployments of sensor networks are taking place. Using a controlled mobile element is a promising approach to collect data from these sensor nodes. We showed that as the network scales, using a single mobile element may not be sufficient, and would require multiple of them. The sensor nodes and (or) the mobile elements may not be uniformly placed in practice, necessitating the use of load balancing, so that each mobile element as far as possible, serves the same number of sensor nodes. We gave a load balancing algorithm, and described the mechanism these multiple mobile elements can be used. Finally we presented simulation results justifying our approach.

The work presented here can be extended in many directions. For load balancing, we can remove the assumption that each sensor node can talk to at least one mobile element. This will lead to case similar to Figure 1. Now, when doing load balancing, we also need to consider the cost of doing multihop, to reach either of the mobile elements. We can also extend to cases where mobile elements can be added or removed once the system is in operation. Node dynamics can be handled by running the initialization and load balancing periodically.

References

1. Kaiser, W.J., Pottie, G.J., Srivastava, M., Sukhatme, G.S., Villasenor, J., Estrin, D.: Networked Infomechanical Systems (NIMS) for Ambient Intelligence. Technical Report 31, Center for Embedded Networked Sensing, UCLA (2003)
2. Cerpa, A., Elson, J., Estrin, D., Girod, L., Hamilton, M., Zhao, J.: Habitat Monitoring: Application Driver for Wireless Communications Technology. In: SIGCOMM Workshop on Data Communications in Latin America and the Caribbean. (2001)

3. Shah, R.C., Roy, S., Jain, S., Brunette, W.: Data MULEs: Modeling a Three-tier Architecture for Sparse Sensor Networks. In: IEEE Workshop on Sensor Network Protocols and Applications (SNPA). (2003)

4. Juang, P., Oki, H., Wang, Y., Martonosi, M., Peh, L., Rubenstein, D.: Energy-efficient Computing for Wildlife Tracking: Design Tradeoffs and Early Experiences with Zebranet. In: ACM ASPLOS. (2002)

5. Small, T., Haas, Z.: The Shared Wireless Infostation Model-A New Ad Hoc Networking Paradigm (or Where there is a Whale, there is a Way). In: ACM MobiHoc. (2003)

6. Chakrabarti, A., Sabharwal, A., Aazhang, B.: Using Predictable Observer Mobility for Power Efficient Design of Sensor Networks. In: IPSN. (2003)

7. Kansal, A., Somasundara, A., Jea, D., Srivastava, M., Estrin, D.: Intelligent Fluid Infrastructure for Embedded Networks. In: ACM MobiSys. (2004)

8. Zhao, W., Ammar, M., Zegura, E.: A Message Ferrying Approach for Data Delivery in Sparse Mobile Ad Hoc Networks. In: ACM MobiHoc. (2004)

9. Luo, J., Hubaux, J.P.: Joint Mobility and Routing for Lifetime Elongation in Wireless Sensor Networks. In: IEEE INFOCOM. (2005)

10. Fall, K.: A Delay-Tolerant Network Architecture for Challenged Internets. In: ACM SIGCOMM. (2003)

11. Dubois-Ferriere, H., Grossglauser, M., Vetterli, M.: Age Matters: Efficient Route Discovery in Mobile Ad Hoc Networks Using Encounter Ages. In: ACM MobiHoc. (2003)

12. Li, Q., Rus, D.: Sending Messages to Mobile Users in Disconnected Ad-hoc Wireless Networks. In: ACM MobiCom. (2000)

13. Goldenberg, D., Lin, J., Morse, A.S., Rosen, B., Yang, Y.R.: Towards Mobility as a Network Control Primitive. In: ACM MobiHoc. (2004)

14. Somasundara, A., Ramamoorthy, A., Srivastava, M.: Mobile Element Scheduling for Efficient Data Collection in Wireless Sensor Networks with Dynamic Deadlines. In: IEEE RTSS. (2004)

15. Rao, N., Wu, Q., Iyengar, S., Manickam, A.: Connectivity-through-time Protocols for Dynamic Wireless Networks to Support Mobile Robot Teams. In: IEEE ICRA. (2003)

16. Intanagonwiwat, C., Govindan, R., Estrin, D.: Directed Diffusion: A Scalable and Robust Communication Paradigm for Sensor Networks. In: ACM MobiCom. (2000)

17. Shirazi, B.A., Hurson, A.R., Kavi, K.M.: Scheduling and Load Balancing in Parallel and Distributed Systems. IEEE Computer Society Press (1995)

18. Hill, J., Szewczyk, R., Woo, A., Hollar, S., Culler, D., Pister, K.: System architecture directions for network sensors. In: ACM ASPLOS. (2000)

19. Levis, P., Lee, N., Welsh, M., Culler, D.: TOSSIM: accurate and scalable simulation of entire tinyOS applications. In: ACM Sensys. (2003)

20. Demmer, M., Levis, P.: Tython scripting for TOSSIM (2004) Network Embedded Systems Technology Winter 2004 Retreat.

Analysis of Gradient-Based Routing Protocols in Sensor Networks*

Jabed Faruque, Konstantinos Psounis, and Ahmed Helmy

Department of Electrical Engineering,
University of Southern California, Los Angeles, CA 90089
{faruque, kpsounis, helmy}@usc.edu

Abstract. Every physical event results in a natural information gradient in the proximity of the phenomenon. Moreover, many physical phenomena follow the diffusion laws. This natural information gradient can be used to design efficient information-driven routing protocols for sensor networks. Information-driven routing protocols based on the natural information gradient, may be categorized into two major approaches: (*i*) the single-path approach and (*ii*) the multiple-path approach. In this paper, using a regular grid topology, we develop analytical models for the query success rate and the overhead of both approaches for ideal and lossy wireless link conditions. We validate our analytical models using simulations. Also, both the analytical and the simulation models are used to characterize each approach in terms of overhead, query success rate and increase in path length.

1 Introduction

Sensor networks are envisioned to be widely used for habitat and environmental monitoring where the attached tiny sensors sample various physical phenomena. More specifically, advances in the MEMS technology make it possible to develop sensors to detect and/or measure a large variety of physical phenomena like temperature, light, sound, radiation, humidity, chemical contamination, nitrate level in the water, etc. Every physical event leaves some fingerprints in the environment in terms of the event's effect; e.g., fire increases the temperature, chemical spilling increases the contamination, nuclear leakage increases the radiation so on. Moreover, most of the physical phenomena follow the diffusion law[13][14] with distance, i.e., $f(d) \propto \frac{1}{d^\alpha}$, where d is the distance from the point having the maximum effect of the event, $f(d)$ is the magnitude of the event's effect, and α is the exponent of the diffusion function that depends on the type of effect and the medium; e.g., for light $\alpha = 2$, and for heat $\alpha = 1$. Hence, routing protocols used in sensor networks for habitat and environmental monitoring applications can exploit this natural information gradient to efficiently forward the queries toward the source of the event. Throughout this document the term *"source"* is used to refer to the source of the event; e.g., the contaminant, the epicenter of an earthquake, etc.

---

* This research has been partially supported by NSF award 0435505.

V. Prasanna et al. (Eds.): DCOSS 2005, LNCS 3560, pp. 258–275, 2005.

Traditional data-centric routing protocols for sensor networks are based mostly on flooding [1] or random-walks [3], [4]. However, these approaches do not utilize the domain-specific knowledge, i.e., the information gradient about the monitored phenomenon. Here, we keep our focus on the routing protocols that exploit the information gradient to route a query efficiently in sensor networks from the sink to the source.

In real life, sensors are not always perfect and are subject to malfunction due to obstacles or failures. Also, the characteristics of the sensor nodes, e.g., limited battery life, the energy expensive wireless communication, and the unstructured nature of the sensor network, make the data-centric routing protocols based on the information gradient a challenging problem. Several routing protocols have been proposed to exploit the information gradients. These query routing protocols use greedy forwarding and can be broadly classified in two categories: (1) *Single-path approach*[5][6][8], where the query reaches the source from the sink through a single path, and (2) *Multiple-path approach*[7], where the query uses multiple paths to reach the source.

In this paper, we do not aim to design new routing protocols per se. Rather, *the objective of the research is the evaluation and the analysis of the general approaches to route a query using the natural information gradient in the sensor networks.* In particular, we use probabilistic modeling methods to derive analytical expressions for the energy overhead and the query success rate of each approach. Also, we compare the performance of the query routing approaches using carefully selected performance metrics. Our analysis is validated through extensive simulations. For the analysis and the simulations, we only consider sensor networks with static nodes, which is usually the case for environmental monitoring, and we assume that the queries are triggered from a sink to identify the origin (i.e., source) of the event, after the event's occurrence. To keep the analysis simple, we ignore potential packet collisions, which can be (and usually is) effectively reduced by inserting a random delay time before forwarding the query packet. However, wireless link loss is considered in both the analysis and the simulations. The main contributions and findings of this paper include:

- Analytical models for the query success rate and the overhead of the single-path and the multiple-path approaches to route a query using the information gradients in sensor networks. Validation of the analytical models using extensive simulations.
- Performance analysis of both routing approaches using the analytical models and simulations in ideal and lossy link conditions.
- In the ideal wireless link case, it is found that the multiple-path approaches are more energy efficient than the single-path approaches when the source is relatively close to the sink. Also, the multiple-path approaches yield shorter paths than the single-path approaches. Further, as the number of malfunctioning nodes in the network increases, the query success rate of the multiple-path approaches degrades a lot slower than that of the single-path approaches.
- In the lossy wireless link case, the query success rate of the single-path approaches drops drastically while the multiple-path approaches are quite resilient.

2   Related Work

Several approaches have been proposed for routing in sensor networks. The major difference between the information-gradient based approach[5][6] and the flooding[1][2] and the random-walks[3][4][9] based approaches is that the former uses the sensors measurements about the event's effect for routing decisions. In [9], Servetto and Barrenechea have shown that multiple random-walks improve load balancing and minimize latency with increased communication cost. Also, they analyzed the random-walk approach in regular/irregular and static/dynamic grid topology, but they did not consider the existence of information gradients. We now briefly summarize previously proposed gradient based routing protocols.

Chu, Hausseker and Zhao propose CADR (Constrained Anisotropic Diffusion Routing) mechanism[5], especially designed for localization and target tracking. CADR uses a proactive sensor selection strategy for correlated information based on a criterion that combines information gain and communication cost. CADR is a single path greedy algorithm that routes a query to its optimal destination using the local gradients to maximize the information gain through the sensor network.

Later work by Liu, Zhao and Petrovic [6] proposed the min-hop routing algorithm to overcome the limitation of CADR to handle local maxima and minima. The algorithm uses a multiple step look-ahead approach with single path query forwarding. Here, the initial network discovery phase determines the minimum look-ahead horizon (in hops) so that the path planning phase can avoid network irregularities. The algorithm improves the success rate of routing message with additional search cost. Also, the increase in the neighborhood size causes more communications between the cluster leaders and their neighbors.

Some recent studies on information-driven routing protocols also use the single-path approach [8]. It is important to note that all the above information-driven protocols based on the single-path approach, use a proactive phase to prepare the gradient information repository. In our study we analyze the performance of the query routing mechanisms without considering the cost of the proactive phase.

In [7], a multiple path exploration mechanism is proposed to discover a route or an event. It is a reactive and distributed routing mechanism to effectively exploit the natural information gradient repository. The protocol controls the instantiation of multiple paths using a probabilistic function based on the simulated annealing concept. It uses flooding to forward the query when no information gradient is available. The efficiency of the protocol depends on properly selecting the parameters of the probabilistic function.

According to [7], the diffusion information in the environment consists of a flat (i.e. zero) and a gradient information region. However, the single-path protocols are unable to forward a query in the flat information region. In this work, we only consider the performance of various routing approaches in the gradient information region, while malfunctioning sensors are uniformly distributed.

In this work, our interest is primarily focused on the systematically analyzing the performance e.g., the query success rate and the overhead, of the single-path and the multiple-path approaches to design data centric routing protocols in the presence of a natural information gradient. In the analysis, we use a probabilistic framework to develop simple analytical models for the success rate and the overhead for both approaches in ideal and lossy wireless link conditions. We also simulate these protocols with more realistic scenarios.

3   Query Routing Approaches

According to the discussion of Section 2, two major approaches, (i) the single-path approach and (ii) the multiple-path approach, are used for query routing protocols to exploit the information gradients. To properly describe these routing approaches, we need to define two terms: (1) *Active node*, a node which is currently holding the query, and (2) *Candidate node*, a node which has never received the query. Now, a brief description of both routing approaches is given below:

Single-Path Approach: The query follows a single path to reach the source from the sink. At each step of the query forwarding, the active node uses a look-ahead parameter r, $r \geq 1$, to collect information from all candidate nodes within r-hops. For $r > 1$, all nodes within $r - 1$ hops need to transmit the request of the active node to gather information about the event. Note that for $r = 0$, the single-path approach becomes a random-walk and is unable to utilize the gradient information repository.

Single-path approach based protocols can be designed in several ways using different selection policies for the next active node. In our study, we consider the following two policies:

a) *Basic single-path approach*: In this policy, the protocol always selects the node with the maximum information among all candidate nodes within r-hops of the active node, when the node's information is higher than that of the active node. This selection policy is sensitive to local maxima and arbitrarily high readings of the malfunctioning nodes that cause these local maxima. The resilience of the protocols based on this approach can be improved by using filters to avoid such arbitrarily high readings.

b) *Improved single-path approach*: In this policy, the active node forwards the query to a node having the maximum information among all candidate nodes within r-hops of the active node. So, the information content of all candidate nodes can be less than that of the active node. Here, the query forwarding ends either at the source node or at an active node having no candidate nodes within r-hops.

Multiple-path Approach: This approach forwards the query through multiple paths towards the source without any look-ahead phase. These paths may not be disjoint paths. Usually the active nodes forward the query greedily when information level improves. In the presence of malfunctioning nodes having wrong information, the protocols based on this approach can use probabilistic forwarding. For example, the protocol proposed in [7] uses a diffusion function for probabilistic forwarding. It creates some extra paths but the protocol can adaptively change the forwarding probability to control the instantiation of these extra

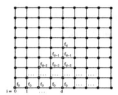

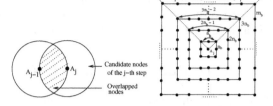

Fig. 1. A regular grid topology. Here, f_j indicates the magnitude of information and $f_0 < f_1 < \cdots < f_d$. The triangular pattern represented by white dots is present •••• ••• •• in the grid. Information magnitude near the source is f_d and gradually reduces towards the edge

(a) A_{j-1} and A_j are the active nodes of steps $(j-1)$ and j.

(b) Active node A_j with r-hop neighbors.

Fig. 2. Single path approach with look-ahead r

paths. To capture this in the analysis, the forwarding probability is considered different at each step of the query forwarding.

All query routing protocols considered in this paper use unique query IDs to suppress duplicates and to avoid loops.

4   Analytical Model

In this section, we derive models to describe the characteristics of the approaches used to design information-driven routing protocols for sensor networks.

4.1   Assumptions and Metrics

Let a sensor network consist of N nodes and the nodes be deployed as a regular grid as shown in Fig.1. Assume that only one event occurs and the effect of the event follows the diffusion law as previously described. Assume also that the information gradient is available in the whole network, i.e., there is no flat information region. Further, consider that the malfunctioning nodes have arbitrary information and that these nodes are uniformly distributed in the network which may cause failure during the route discovery. Let p_f be the probability that a node is malfunctioning. The stored information in the malfunctioning node can be arbitrarily high or low and this is equally likely. Finally, assume each node is able to communicate via broadcast with its eight neighbors on the grid.

Suppose that the querier, i.e. the sink, is located d hops away from the source node. The query is forwarded step by step, where the term *"step"* is defined as follows:

1) Single-path approach: The active node collects information from all candidate nodes within r-hops. Then it forwards the query to the next active node which is r-hops away.

2) Multiple-path approach: The active nodes forward the query either greedily or probabilistically via broadcasts. Then the query reaches the candidate nodes which are 1-hop away.

Due to greedy forwarding based on the information gradient, after each step of the query forwarding, the query reaches one step closer to the source with some probability.

Here, we are interested to develop analytical models for the following two metrics:

1) *Query Success rate*, i.e. success probability, is the probability that the query initiated from the sink reaches the source.
2) *Overhead* in terms of energy dissipation, calculated as the number of transmissions required to forward the query to the source and to get the reply back from the source using the reverse path.

4.2   Single-Path Approach

Let n_b be the number of nodes that are one hop away from the active node. Overlap of the sensor nodes radio coverage causes some nodes to receive the same query multiple times. The query ID is used to suppress duplicate queries. If we consider the radio coverage of a node to be circular and the radius to be the same for all nodes, then using simple geometry, it can be shown that the overlap is one-third. For all except the first step of the query forwarding, let $n_c = \frac{2}{3} n_b$ denote the number of candidate nodes within one-hop of the active node. Now, it is easy to show that the total number of neighbors and candidate nodes within r-hops of the active node equal $n_B = \frac{r(r+1)}{2}$ and $n_C = \frac{2}{3} n_B$ respectively. However, for the first step of the query forwarding, $n_C = n_B$.

Within one-hop of the active node, let n_h and n_l be the number of candidate nodes having high and low information respectively according to the diffusion pattern of the event's effect in the grid, where $n_c = n_l + n_h$. Thus, except for the first step of the query forwarding, it can be shown from Fig.2 that the total number of high information candidate nodes within r-hops of the active node equals $n_H = \frac{r(r+1)}{2} n_h - \frac{r(r-1)}{2}$. Finally, $n_L = n_C - n_H$ denotes the number of low information candidate nodes within r-hops, since $n_C = n_L + n_H$.

Basic Single-path Approach: At each step of the query forwarding, the protocol selects the node that has the highest information among all nodes (including the active node) within r-hops and forwards the query to that node. When all candidate nodes function properly, the query forwarding ends at the source. However, the query forwarding also halts at a local maxima. A malfunctioning node contains arbitrarily high information with probability $\frac{p_f}{2}$. Let $f(i)$ be the diffusion function value when the source is i-hops away and R_{max} be the maximum reading of the sensor attached to the source node. The arbitrarily high reading of a malfunctioning node can be between 0 and R_{max}. But at the j-th step, the query forwarding halts for any candidate node having value between $f\left(\left(\lceil \frac{d}{r} \rceil - j\right) r\right)$ and R_{max}. Therefore, the probability of success equals

$$P_{single_B} = \prod_{j=1}^{\lceil \frac{d}{r} \rceil} \left[1 - \frac{p_f}{2} \cdot \frac{R_{max} - f\left(\left(\lceil \frac{d}{r} \rceil - j\right)r\right)}{R_{max}} \right]^{n_C}. \tag{1}$$

Here, the product term is the probability that at each step of the query forwarding, there is no malfunctioning candidate node having arbitrarily high information when the source is i-hops away and a total of $\lceil \frac{d}{r} \rceil$ such steps are required to reach the source. In the expression, we assume that after each forward of the query, the next active node is r-hops away. This may not be possible if all high information candidate nodes that are r-hops away are malfunctioning. However, a protocol can still select a node that is r-hops away as the next active node and notify the maximum information collected from all of the nodes within r-hops.

At each step of the query forwarding, to gather information for the active node, each node within $(r-1)$-hops from the active node sends the request to its n_b neighbors and finally all nodes within r-hops send the reply to the active node through their one-hop away neighbors. Thus, the number of transmissions required to forward the query to the next active node is $\left(1 + n_b \sum_{i=1}^{r-1} i\right) + n_b \sum_{i=0}^{r-1} (r-i) + r$, that equals $1 + r^2 n_b + r$. Now, the total number of transmissions require to forward the query from the sink to the source and get the reply equals

$$T_{single_B} = \left\lceil \frac{d}{r} \right\rceil (1 + r^2 n_b + r) + d. \tag{2}$$

Here, $1 + r^2 n_b + r$ is the required number of transmissions at each step of the query forwarding and total $\lceil \frac{d}{r} \rceil$ such steps are required. Also, it requires d transmissions to reply to the sink through the reverse path. If we consider that nodes in the overlapping regions respond only one time, then the total number of transmissions equals

$$T_{single_{B_O}} = (1 + r^2 n_b + r) + \left(\left\lceil \frac{d}{r} \right\rceil - 1\right) \left[1 + \frac{2r^2 n_b}{3} + r\right] + d, \tag{3}$$

since except for the first step of the query forward, each remaining step requires $1 + \frac{2r^2 n_b}{3} + r$ transmissions for non-overlapping nodes.

Improved Single-Path Approach: In this policy, at each step of the query forwarding, the protocol selects the node with the maximum information among all candidate nodes within r-hops of the active node and forwards the query to that node. For protocols based on this approach, the query success rate and the overhead depend on the length of the path followed by the protocol. Though the sink node is d-hops away from the source, in the presence of arbitrary information in the malfunctioning nodes, the query may follow a path other than the shortest path. Let l_j denote the length of the path after the j-th forward of the query. If all sensor nodes in the network were perfect, the query should follow the shortest path and $l_j - l_{j-1} = r$. However, due to malfunctioning nodes, some low information candidate nodes may contain arbitrarily high information with probability $\frac{p_f}{2}$. The probability of selecting such a node as the next active node is $\left(\frac{p_f}{2}\right) \frac{n_L}{n_C}$ and the path length difference per step equals $l_j - l_{j-1} = r + L_{err}$, where L_{err} is the average path length increase per step. For simplicity of the analysis, if we consider that each step of the query forwarding is independent, then the length of the path after the j-th step can be expressed as

$$l_j = \begin{cases} \frac{\frac{p_f}{2}.n_L}{n_C}\left(l_{j-1}+r+L_{err}\right) + \left(1 - \frac{\frac{p_f}{2}.n_L}{n_C}\right)\left(l_{j-1}+r\right), & \text{for } j > 1 \\ r, & \text{for } j = 1. \end{cases}$$

Thus, l_d denotes the length of the path followed by the protocol while the actual distance of the source is d.

Here, the query forwarding halts either at the source node or at an active node with no candidate nodes within r-hops. In the gradient information repository, there are always some candidate nodes as the query forwarding proceeds from low to high information nodes. However, due to malfunction, with probability $\frac{p_f}{2}$, high information candidate node(s) may contain arbitrarily low information which may be lower than that of some low information candidate nodes. When all high information candidate nodes are malfunctioning and containing arbitrarily low information, the query forwarding proceeds through a low information candidate node. Such low information candidate nodes are unable to find any candidate node as its all neighbors may already have received the query and the query fails to reach the source. Therefore, the probability of query success equals

$$P_{single_I} = \left[1 - \left(\frac{p_f}{2}\right)^{n_H}\right]^{\left\lceil \frac{l_d}{r} \right\rceil}. \tag{4}$$

Here, $1 - \left(\frac{p_f}{2}\right)^{n_H}$ is the probability that not all high information candidate nodes are malfunctioning at each step and a total of $\left\lceil \frac{l_d}{r} \right\rceil$ such steps are required.

Similar to the Equation(2), here the total number of transmissions required to forward the query from the sink to the source for path length l_d and get the reply equals

$$T_{single_I} = \left\lceil \frac{l_d}{r} \right\rceil \left(1 + r^2 n_b + r\right) + l_d. \tag{5}$$

Here, l_d transmissions are required to reply to the sink using the reverse path.

Further, considering that nodes in the overlapping regions respond only one time, the total number of transmissions equals

$$T_{single_{IO}} = \left(1 + r^2 n_b + r\right) + \left(\left\lceil \frac{l_d}{r} \right\rceil - 1\right)\left[1 + \frac{2r^2 n_b}{3} + r\right] + l_d, \tag{6}$$

since except for the first step of the query forward, each remaining step requires $1 + \frac{2r^2 n_b}{3} + r$ transmissions for the non-overlapping nodes.

4.3   Multiple-path Approach

In this routing approach, except for the first step of the query forwarding, multiple active nodes may forward the query to the candidate nodes without any look-ahead phase. The active nodes with lower information forward the query probabilistically. Let p_j denote the probability of forwarding the query probabilistically at the j-th step of the query forwarding. So, at the j-th step of

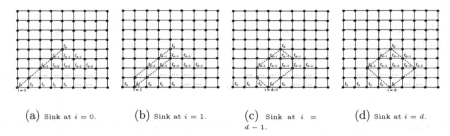

(a) Sink at $i = 0$.   (b) Sink at $i = 1$.   (c) Sink at $i = d - 1$.   (d) Sink at $i = d$.

Fig. 3. Query forwarding pattern using the multiple path approach. Depending on the position of the sink patterns are different. Here, the white dots indicate the participating nodes to forward the query towards the source and d is the distance between the source and the sink

the query forwarding, a high information candidate node fails to forward the query with probability $q_j = \left(\frac{p.f.}{2}\right)(1 - p_j)$, where $\left(\frac{p.f.}{2}\right)$ is the probability that the high information candidate node is malfunctioning and containing low information. For simplicity of the analysis, we assume that the query forwarding steps are independent. This simplified model still captures the characteristics of the multiple-path approach, while the analysis is kept tractable.

Let $P_{multiple}$ and $T_{multiple}$ denote the query success rate and the overhead of the multiple-path approach. Consider that i denotes the position of the querier, i.e. the sink, in the last row of the grid as shown in Fig.1 and Fig.3. The query forwarding patterns, i.e. the number of participating nodes, are different for different values of i as shown in Fig.3. Also, it is easy to show that according to the diffusion pattern in the grid, the average number of low information candidate nodes is four at each step of the query forwarding. These nodes forward the query probabilistically.

According to the query forwarding patterns as shown in Fig.3, for an even value of i, $0 \leq i \leq d$, the query success probability equals

$$P_{multiple_e}(i) = (1 - q_1) \cdot$$
$$\left[\prod_{m=\frac{i}{2}}^{d-\frac{i}{2}} \left(1 - q_{m+1}^{i+1}(1 - p_{m+1})^4\right) \right] \cdot \prod_{m=1}^{\frac{i}{2}-1} \left(1 - q_{m+1}^{2m+1}(1 - p_{m+1})^4\right)\left(1 - q_{d-m+1}^{2m+1}(1 - p_{d-m+1})^4\right),$$

and for an odd value of i, $0 \leq i \leq d$, it equals $P_{multiple_o}(i)$, where the only difference with the above expression is the limits of the products (i.e., $\lceil \frac{i}{2} \rceil \leq m \leq d - \lceil \frac{i}{2} \rceil$ and $1 \leq m \leq \lfloor \frac{i}{2} \rfloor$ in the first and the second products respectively). Here, the terms having the form q_y^x, $1 \leq y \leq d$, in the above equation expresses the probability of not forwarding the query at the y-th step by all high information candidate nodes, x, as they are malfunctioning and containing less information. Also, the surrounding four low information candidate nodes fail to forward the query with probability $(1 - p_y)^4$. Thus, at the y-th step, $1 - q_y^x(1 - p_y)^4$ is the probability of forwarding the query towards the source. Detailed derivation of these equations and all the equations of the remaining document are presented in [15].

Similarly, to compute the energy dissipation, we also consider the different forwarding patterns of the query as shown in Fig.3. The total number of transmissions required to forward the query to the source for an even value of i, $0 \leq i \leq d$, equals

$$T_{multiple_e}(i) = \left[\frac{i^2}{2} - 1 + (i+1)(d-i+1)\right] - \left[\sum_{m=1}^{\frac{i}{2}} (2m-1)q_m\right.$$

$$\left. + \sum_{m=1}^{\frac{i}{2}-1} (2m+1)q_{d-m+1} + (i+1)\sum_{m=1}^{d-i+1} q_{\frac{i}{2}+m}\right] + 4\sum_{m=1}^{d} p_{m+1}, \quad (7)$$

and for an odd value of i, $0 \leq i \leq d$, it equals $T_{multiple_o}(i)$, where the differences with Equation(7) are the first term ($\left[\frac{(i+1)^2}{2} - 1 + (i+1)(d-i)\right]$) and the limits of the second term (i.e., $1 \leq m \leq \frac{i+1}{2}$, $1 \leq m \leq \frac{i+1}{2} - 1$ and $1 \leq m \leq d - i$ respectively). Here, the first term of the above equation computes the number of transmissions due to high information candidate nodes, if all such nodes are working properly. However, some of these nodes are malfunctioning and unable to forward the query with probability q_y, $1 \leq y \leq d$, at the y-th step of the query forwarding. This reduces the overhead. The second term of the above equation computes this reduction. Finally, the third term computes the overhead due to the probabilistic forwarding of the four low information candidate nodes.

Now, if each value of i, $0 \leq i \leq d$, is equally likely, then the average probability of success equals

$$P_{multiple} = \frac{1}{d+1}\left[\sum_{k=0}^{\lfloor \frac{d}{2}\rfloor} P_{multiple_e}(2k) + \sum_{k=1}^{\lceil \frac{d}{2}\rceil} P_{multiple_o}(2k-1)\right]. \quad (8)$$

Similarly, the average number of transmissions required to forward the query from the sink to the source and get the reply using the reverse path equals

$$T_{multiple} = \frac{1}{d+1}\left[\sum_{k=0}^{\lfloor \frac{d}{2}\rfloor} T_{multiple_e}(2k) + \sum_{k=1}^{\lceil \frac{d}{2}\rceil} T_{multiple_o}(2k-1)\right] + d. \quad (9)$$

4.4   Wireless Link Loss

Recent experimental studies on wireless sensor networks [11][12] have shown that in practice, the wireless links of the sensor networks can be extremely unreliable and deviate from the idealized perfect-reception-range models at a large extent. Due to the lossy links, the transmissions of a node may not reach to some of its neighbor nodes. This affects the performance of the routing protocols. In this section, we consider the wireless link loss in the derivation of analytical models of the query success rate and the overhead for both routing approaches. Since the *basic single-path* approach shows poor performance even for the ideal wireless link condition (see Fig.4), only the *improved single-path* and the *multiple-path* approaches are considered in this section.

Let, p_c be the probability of a link loss, and assume that the lossy links are uniformly distributed in the network. Assume also that no automatic repeat request (ARQ) is used to broadcast or to forward the query towards the source, which is usually the case in sensor networks for energy conservation. However, notice that the ARQ mechanism is used to send the reply (if the query is successful) to the sink using the reverse path.

Improved Single-Path Approach: Due to the lossy links, at each step of the query forwarding, the broadcast of the active node may not reach to all candidate nodes within r-hops. Thus, at each step of the query forwarding, $n_H(1 - p_c)$ high information candidate nodes receive the broadcast. Similarly, the active node receives responses from $n_H(1 - p_c)^2$ high information candidate nodes. Also, the probability to forward the query to the next active node, which is r-hops away, is $(1 - p_c)^r$. Thus, for the improved-single path approach, the probability of success equals

$$P_{single_{Ic}} = \left[\left(1 - \left(\frac{p_f}{2} \right)^{n_H(1-p_c)^2} \right) (1 - p_c)^r \right]^{\left\lceil \frac{l_d}{r} \right\rceil}. \tag{10}$$

Here, $\left(1 - \left(\frac{p_f}{2} \right)^{n_H(1-p_c)^2} \right) (1 - p_c)^r$ is the probability of success at each step of the query forwarding and a total of $\left\lceil \frac{l_d}{r} \right\rceil$ such steps are required.

In this routing approach, the first step of the query forwarding requires $1 + \frac{r(r-1)}{2}n_b(1 - p_c) + \frac{r(r+1)}{2}n_b(1 - p_c) + r(1 - p_c)$ transmissions, which equals $1 + r(1 - p_c)(rn_b + 1)$. With probability p_c, the nodes of the overlapped region can be candidate nodes of the current active node as they failed to receive the broadcast of the previous active node due to the lossy links. Further, consider that the overlapped region nodes respond only one time. So, each remaining step of the query forwarding requires $1 + \frac{1}{3} \left[\frac{r(r-1)}{2}n_b(1 - p_c) + \frac{r(r+1)}{2}n_b(1 - p_c) \right] p_c + \frac{2}{3} \left[\frac{r(r-1)}{2}n_b(1 - p_c) + \frac{r(r+1)}{2}n_b(1 - p_c) \right] + \frac{r}{1-p_c}$ transmissions, which equals $1 + \frac{1}{3}r^2(1 - p_c) \left[n_b(p_c + 2) + \frac{3}{r} \right]$. Thus the total number of transmissions equals

$$T_{single_{IOc}} = 1 + r(1 - p_c)(rn_b + 1) + \left(\left\lceil \frac{l_d}{r} \right\rceil - 1 \right) \cdot \left[1 + \frac{1}{3}r^2(1 - p_c) \left(n_b(p_c + 2) + \frac{3}{r} \right) \right] + \frac{l_d}{1 - p_c}. \tag{11}$$

Since except for the first step of the query forwarding, we need to consider the overlapping region nodes for the remaining steps. Also, note that to reply to the sink through the reverse path requires $\frac{l_d}{1-p_c}$ transmissions.

Multiple-path Approach: In this approach, multiple active nodes forward the same query to the candidate nodes and reduce the possibility of the failure due to the wireless link loss. Since, the probability of lossy links, p_c is small, so in the analytical models, we consider that $1 - p_c^n \approx 1, n \geq 2$, where n is the number of copies of the same query received by a candidate node. Results with this assumption are later compared to (and validated with) simulations in Section 5.5. Now, according to the query forwarding patterns as shown in Fig.3, for an even value of i, $0 \leq i \leq d$, the success probability equals

$$P_{multiple_{e_c}}(i) = (1 - q_1) \left[\prod_{m=\frac{i}{2}}^{d-\frac{i}{2}} \left(1 - q_{m+1}^{i+1-p_c} (1 - p_{m+1})^{4+3p_c} \right) \right] \cdot$$

$$\prod_{m=1}^{\frac{i}{2}-1} \left(1 - q_{m+1}^{2m+1-2p_c} (1 - p_{m+1})^{4-4p_c} \right) \left(1 - q_{d-m+1}^{2m+1} (1 - p_{d-m+1})^{4-2p_c} \right),$$

and for an odd value of i, $0 \le i \le d$, it equals $P_{multiple_{o_c}}(i)$, where the only difference with the above expression is the limits of the products (i.e., $\lceil \frac{i}{2} \rceil \le m \le d - \lceil \frac{i}{2} \rceil$ and $1 \le m \le \lceil \frac{i}{2} \rceil$ for the first and the second products respectively) Here, the terms having the form $1 - q_y^x (1 - p_y)^z$ in the above equation expresses the probability of forwarding the query at the y-th step of the query forwarding, where x and z are the number of nodes that perform greedy and probabilistic forwarding respectively.

Similarly, to compute the overhead, we consider the different forwarding patterns of the query as shown in Fig.3. Due to the lossy links, the query may fail to reach some nodes. So the overhead reduces due to the less number of participating nodes. Thus, using the Equations (7), the total number of transmissions required to forward the query to the source for an even value of i, $0 \le i \le d$, equals

$$T_{multiple_{e_c}}(i) = T_{multiple_e}(i)$$

$$- p_c \left[2 \sum_{m=2}^{\frac{i}{2}} (1 - q_m) + \sum_{m=1}^{d-i+1} \left(1 - q_{\frac{i}{2}+m} \right) + 4 \sum_{m=2}^{\frac{i}{2}} p_m + 3 \sum_{m=1}^{d-i+1} p_{\frac{i}{2}+m} + 2 \sum_{m=1}^{\frac{i}{2}-1} p_{d-m+1} \right],$$

and for an odd value of i, $0 \le i \le d$, it equals $T_{multiple_{o_c}}(i)$, where the differences expression are the first term ($T_{multiple_o}(i)$) and the limits of the second term (i.e., $2 \le m \le \frac{i}{2}$, $1 \le m \le d-i+1$, $2 \le m \le \frac{i}{2}$, $1 \le m \le d-i+1$, and $1 \le m \le \frac{i}{2}-1$ respectively). Here, the second term of the above equation computes the overhead reduction due to the lossy links.

Now, similar to the Equation (8) and (9), the average success probability and the average number of transmissions can be computed. Here, the reply to the sink through the reverse path requires $\frac{d}{1-p_c}$ transmissions.

5   Simulations and Results

In this section, we validate our analytical models by conducting extensive simulations. The objective of these simulations is to compare the simulation results about the performance of the routing approaches with the analytical models. Following performance metrics are considered in the simulations:

1) *Success probability*, is the ratio of the total number of queries that reached the source over the total number of queries sent.
2) *Overhead* in terms of energy dissipation is the average number of transmissions required to forward a query successfully to the source and to get the reply using the reverse path.

3) *Path quality* in terms of path length increase factor is the ratio of the average
   length of the path discovered by a routing approach over the length of the
   shortest path from a set of sinks to the source. Here, the shortest path length
   between the source and any sink node in the set is the same. This metric is
   important for long-lived continuous queries.

5.1 Simulation Model

In our simulations, we use a $100m \times 100m$ grid with 10^4 sensor nodes placed
at distance $1m$ from each other. Except for the border nodes, each sensor node
is able to communicate with eight neighbors. For all simulations, the exponent
of the phenomenon diffusion function, i.e., the parameter α, is set to 0.8. To be
consistent with the analytical models, the information gradient is available in
the whole network and the malfunctioning nodes are uniformly distributed with
some arbitrary values.

The querier, i.e., the sink, and the source are different and can be any node.
We use a flooding technique to find the set of sink nodes that are specific shortest distance away from the source. In the simulations, we use only single-value
queries, that search for a specific value and have a single response.

The simulated protocol based on the single-path approach uses a look-ahead
parameter $r = 1$. For $r = 1$, it can be easily shown from Fig.1 that $n_B = 8$, $L_{err} \approx 2$
and $n_H \approx 2.5$. So, using the expressions of Section 4.2, we get $n_C = \frac{2}{3} n_B \approx 5$ and
$n_L \approx 2.5$. These parameter values are used in the analytic models of the single-
path approach to compare the analytical results with the simulation results.

The simulated protocol based on the multiple-path approach uses a probabilistic diffusion function with exponent β as specified in [7] for probabilistic
forwarding. Thus $p_j = f(j) = \frac{1}{j^\beta}$, where j is the hop count in the information
gradient region and $\beta < \alpha$.

5.2 Query Success Rate i.e., Probability of Success

The basic single-path approach is not resilient to local maxima. The query success rate of the protocol based on this approach is shown in Fig. 4, where the
analytical result closely matches with the simulation result. With the increase of
malfunctioning nodes, local maxima increases and the query success rate drops
close to zero. However, using a filter to avoid the nodes having arbitrarily high
information, the query success rate of this approach can be improved. Detailed
derivation for the filter and corresponding simulation results are presented in
[15]. Also, The remaining results of the performance analysis of this approach
are detailed in [15].

For the improved single-path approach, the query success rate of the routing
protocols depends on the availability of high information candidate nodes. From
Fig.5, it is obvious that the analytically results are more or less in line with
the simulation results. The number of high information candidate nodes reduces
with the exploration of more nodes, especially for large d and causes some minor differences between the analytical and the simulation results. The improved
single-path approach is resilient to local maxima due to its selection policy for

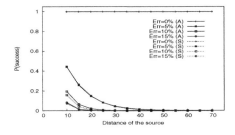

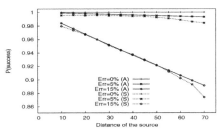

Fig. 4. Query success rate of the basic single-path approach. 'A' and 'S' indicate the analytical and the simulation results respectively

Fig. 5. Query success rate of the improved single-path approach

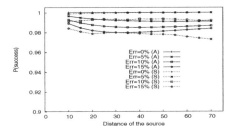

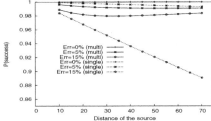

Fig. 6. Probability of success for the multiple path approach. The exponent of the probabilistic function $\beta = 0.65$.

Fig. 7. Comparison of the query success rate of the improved single-path and the multiple-path routing approaches using analytical results. (Simulation results yield very similar plots)

the next active node. So, the query success rate of this approach is significantly higher than that of the basic single-path approach.

In the analytical model for the multiple-path approach, we consider that each step of the query is independent, and that low information candidate nodes forward the query probabilistically. However, due to correlation with previous steps of the query forwarding, some extra nodes may also forward the query and create few more extra paths, which actually improve the query success rate when less number of nodes are malfunctioning. Also, with the increase of malfunctioning nodes, active nodes use more probabilistic forwarding that results less number of paths and the query success rate drops. For these reasons, we notice some minor difference between the analytical and the simulation results in Fig.6.

The use of multiple paths and the probabilistic forwarding in the presence of malfunctioning nodes improves the query success rate of the multiple-path approach with compare to that of the single-path approach as shown in Fig.7. For the single-path approach, it is important to notice that the query success rate drops fast as the number of the malfunctioning nodes in the network increase.

5.3   Overhead i.e., Energy Dissipation

Fig.8 shows the overhead in terms of the average energy dissipation of the improved single-path approach. In both models, the overlapped region nodes respond only one time. The analytical and the simulation results are very similar. The minor discrepancy is due to following reason. With the exploration of more nodes, the number of high information candidate nodes reduces and the path length increases due to choosing the malfunctioning nodes to forward the query.

Both analytical and simulation results for the overhead of the multiple-path approach are given in Fig.9. The analytical results are quite close to the simulation results with a small discrepancy. In the analytical model, we assume that the sink is located at the last row of the grid i.e., an edge node as shown in Fig.1 and 3. However, in our simulation scenarios, the sink is not always at an edge node and the initial high value of the probabilistic function causes the exploration of some nodes in the low information gradient region. Also, as we have already explained in Section 5.2, due to correlations between the query forwarding steps, this routing approach creates some extra paths and increases the overhead which is not considered in the analytical model. Thus, simulation results show sightly more overhead than the analytical results.

In Fig.10, the overhead of both approaches is compared using the analytical results. It is obvious that in our model, if the source is less than 22 hops away from the sink then the multiple-path approach is more energy efficient; otherwise, the single path approach is preferable when energy dissipation is only considered. The overhead of the multiple-path approach increases more due to the extra paths created by probabilistic forwarding.

Using analytical models, the percentage of energy savings of the multiple-path approach over the single-path approach is shown in Fig.11. As the number of malfunctioning nodes increase, the overhead of the single-path approach increases. Since, the length of the path followed by this approach increases. On the other hand, with the increase of malfunctioning nodes, the multiple-path approach uses more probabilistic forwarding. This creates less number of paths, and the overhead reduces.

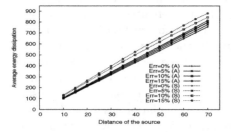

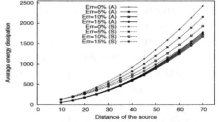

Fig. 8. Overhead for the single path approach. 'A' and 'S' indicate the analytical and the simulation results respectively

Fig. 9. Overhead for the multiple path approach. The exponent of the probabilistic function is $\beta = 0.65$

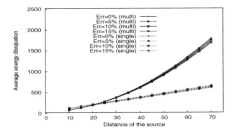

Fig. 10. Comparison of the overhead of the improved single-path and the multiple-path approaches using analytical results

Fig. 11. Percentage of energy saving of the multiple path approach over the improved single-path approach for $d \leq 25$ using analytical models

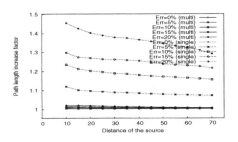

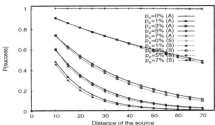

Fig. 12. Path length increase factor for the improved single-path and the multiple-path approaches. The exponent of the probabilistic function $\beta = 0.60$

Fig. 13. Query success rate of the improved single-path approach with the varying lossy link conditions. The probability of malfunctioning nodes is $p_f = 0.05$

5.4   Path Quality

Fig.12 shows the path length increase factor of both routing approaches. The multiple-path approach results the shorter paths which are very close to the shortest path length. We notice that the path length for the single-path approach increases with the increase of the malfunctioning nodes. As expected, in the presence of malfunctioning nodes, the single path approach fails to follow the shortest path towards the source. On the other hand, the instantiation of multiple paths and probabilistic forwarding help the multiple path approach to alleviate the problem for malfunctioning nodes.

5.5   Wireless Link Loss Effect

So far, all the results presented in Section 5 consider ideal wireless links. However, the wireless links are lossy and likely affect the query success rate of the routing protocols significantly. Fig.14 shows the query success rate of both approaches in the presence of lossy links with probability $p_c = 0.05$. Here, the analytical (Fig.14(a)) and simulation (Fig.14(b)) results are identical. In both cases, the query success rate of the multiple-path approach drops quite slowly and it is

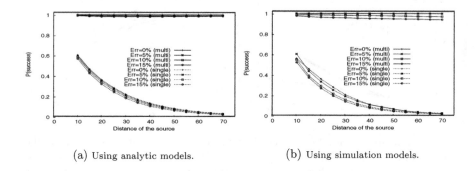

(a) Using analytic models.　　　　(b) Using simulation models.

Fig. 14. Comparison of the query success rate of the improved single-path and the multiple-path approaches in the presence of link loss, $p_c = 0.05$. The exponent of the probabilistic function $\beta = 0.65$

more than 93% even at the presence of 15% malfunctioning nodes in the sensor network. The ability to send the same query from multiple active nodes towards a candidate node improves the resilience of this approach significantly.

For the improved single-path approach, the query success rate drops drastically with the increase of the distance between the source and the sink. Also, Fig.13 shows that the query success rate drops further with the increase of loss probability of the lossy links, i.e., p_c. From Equation (10), it is obvious that the term $(1 - p_c)^r$, which corresponds to forwarding the query from the current active node to the next active node in the presence of lossy links, is responsible for the low success rate of this approach. Using ARQ, the query success rate can be improved significantly [15].

6   Summary and Conclusions

In this paper, we have presented a detailed performance analysis of information-driven routing approaches in ideal and lossy wireless link conditions using analytical models and simulations. We consider the effect of (various kinds of) noise, malfunctioning nodes and node failures in our analysis.

From our study, it is found that the query success rate of the single-path approach drops quite fast as the number of malfunctioning nodes in the network increase while the multiple-path approach retains very high query success rate. Also, it is found that the multiple-path approach is more energy efficient when the source is less than 22 hops away from the sink; otherwise, the single-path approach is more energy efficient. For example, in Fig. 11, for 5% malfunctioning nodes and a source 15 hops away from the querier, the overhead of the multiple-path approach is only 75% of that of the improved single-path approach. Further, the multiple-path approach results in shorter paths which are close to the shortest path. Finally, in the lossy link case, the query success rate of the single path approach drops drastically with the increase of the link loss probability and the distance between the source and the querier. On the other hand, the multiple-path approach achieves over 93% success rate even at the presence of 15% malfunctioning nodes in the sensor network.

The analytical models of both routing approaches can be used to determine the performance and the bottleneck of a protocol for a large sensor network without simulations or when simulations are not possible due to resource constraints. Further, the performance of a new protocol, based on either of these two approaches, can be determined using our models.

From the analytical models, it is obvious that more efficient information-driven routing protocols can be designed based on these two approaches by tuning the parameters of the models, which can be one possible future research direction.

References

1. C. Intanagonwiwat, R. Govindan and D. Estrin, "Directed Diffusion: A Scalable and Robust Communication Paradigm for Sensor Networks", MobiCom 2000.
2. C. Intanagonwiwat, D. Estrin, R. Govindan, and J. Heidemann, "Impact of Network Density on Data Aggregation inWireless Sensor Networks", ICDCS'02.
3. D. Braginsky and D. Estrin, "Rumor Routing Algorithm for Sensor Networks", WSNA 2002.
4. N. Sadagopan, B. Krishnamachari, and A. Helmy, "Active Query Forwarding in Sensor Networks (ACQUIRE)", SNPA 2003.
5. M. Chu, H. Haussecker, and F. Zhao, "Scalable Information-Driven Sensor Querying and Routing for ad hoc Heterogeneous Sensor Networks", Int'l J. High Performance Computing Applications, 16(3):90-110, Fall 2002.
6. J. Liu, F. Zhao, and D. Petrovic, "Information-Directed Routing in Ad Hoc Sensor Networks", WSNA 2003.
7. J. Faruque, A. Helmy, "RUGGED: RoUting on finGerprint Gradients in sEnsor Networks", IEEE International Conference on Pervasive Services (ICPS), 2004.
8. Q. Li, M.D. Rosa and D. Rus, "Distributed Algorithms for Guiding Navigation across a Sensor Network", MobiCom 2003.
9. S. D. Servetto and G. Barrenechea, "Constrained Random Walks on Random Graphs: Routing Algorithms for Large Scale Wireless Sensor Networks", WSNA 2002.
10. Fan Ye, Gary Zhong, Songwu Lu and Lixia Zhang, "GRAdient Broadcast: A Robust Data Delivery Protocol for Large Scale Sensor Networks", ACM WINET (Wireless Networks).
11. A. Woo, T. Tong, and D. Culler. "Taming the Underlying Issues for Reliable Multhop Routing in Sensor Networks". ACM SenSys, November 2003.
12. J. Zhao and R. Govindan. "Understanding Packet Delivery Performance in Dense Wireless Sensor Networks". ACM Sensys, November 2003.
13. D.R. Askeland, The Science and Engineering of Materials, PWS Publishing Co., 1994.
14. J.F. Shackelford, Intro to Materials Science For Engineers, 5th Ed., Prentice Hall, 2000.
15. J. Faruque, K. Psounis, A. Helmy, "Analysis of Gradient-based Routing Protocols in Sensor Networks", CS Technical report, USC 2005.

Analysis of Target Detection Performance
for Wireless Sensor Networks

Qing Cao, Ting Yan, John Stankovic, and Tarek Abdelzaher

Department of Computer Science, University of Virginia, Charlottesville VA 22904, USA
{qingcao, ty4k, stankovic, zaher}@cs.virginia.edu

Abstract. In surveillance and tracking applications, wireless sensor nodes collectively monitor the existence of intruding targets. In this paper, we derive closed form results for predicting surveillance performance attributes, represented by detection probability and average detection delay of intruding targets, based on tunable system parameters, represented by node density and sleep duty cycle. The results apply to both stationary and mobile targets, and shed light on the fundamental connection between aspects of sensing quality and deployment choices. We demonstrate that our results are robust to realistic sensing models, which are proposed based on experimental measurements of passive infrared sensors. We also validate the correctness of our results through extensive simulations.

1   Introduction

A broad range of current sensor network applications involve surveillance. One common goal for such applications is reliable detection of targets with minimal energy consumption. Although maintaining full sensing coverage guarantees immediate response to intruding targets, sometimes it is not favorable due to its high energy consumption. Therefore, designers are willing to sacrifice surveillance quality in exchange for prolonged system lifetime. The challenge is to obtain the analytical relationship that depicts the exact tradeoff between surveillance quality and system parameters in large-scale sensor networks. In this paper, we characterize surveillance quality by average detection delay and detection probability of intruding targets. For system parameters, we are mainly concerned with duty cycle and node density. This knowledge answers the question of whether a system with a set of parameters is capable of achieving its surveillance goals.

In this paper, we establish the relationship between system parameters and surveillance attributes. Our closed-form results apply to both stationary and moving targets, and are verified through extensive simulations. Throughout this paper, we adopt the model of unsynchronized duty-cycle scheduling for individual nodes. In this model, nodes sleep and wake-up periodically. Nodes agree on the length of the duty cycle period and the percentage of time they are awake within each duty cycle. However, the wakeup times are not synchronized among nodes. We call this model, *random duty-cycle scheduling*. There are two reasons for this choice. First, random scheduling is probably the easiest to implement in sensor networks since it requires no coordination among nodes. Coordination among nodes takes additional energy and may be severely impaired by clock drifts. Second, many surveillance scenarios pose the requirement of

V. Prasanna et al. (Eds.): DCOSS 2005, LNCS 3560, pp. 276–292, 2005.

stealthiness (i.e., that nodes minimize their electronic signatures). Coordinating node schedules in an ad hoc network inevitably involves some message exchange which creates a detectable electronic signature. In contrast, random scheduling does not require communication. Each node simply sets its own duty-cycle schedule according to the agreed-upon wakeup ratio, and starts surveillance. No extra messages need to be exchanged before a potential target appears. Based on these two considerations, we focus this paper only on random duty-cycle scheduling.

We make two major contributions in this paper. First, for the first time, we obtain closed-form results to quantify the relationship between surveillance attributes and system parameters. The advantage is that we can now answer a variety of important questions without the need for simulation. For example, to decrease the average detection delay for a potential target by half, how much should we increase the node density? In this paper, we demonstrate that this problem, along with many others, can be answered analytically based on our results. Second, we propose a realistic (irregular) model of sensing based on empirical measurements. This model is then incorporated into our simulation to prove the robustness of our analytical predictions. The results show that even under the irregular sensing model, the analytical closed-form results are still quite accurate.

The rest of this paper is organized as follows. Section 2 presents the notations and assumptions of the paper. Section 3 considers stationary target detection. Section 4 addresses mobile target detection. Section 5 presents the irregular sensing model based on experiments. Section 6 discusses related work and Section 7 concludes the paper.

2   Notations and Assumptions

In this section, we present the notations and assumptions for our derivations. First of all, we assume that nodes are independently and identically distributed conforming to a uniform distribution. Density d is defined as the total number of nodes divided by area in which the system is deployed, A_{system}. In our analysis, we adopt a simple sensing model in which a point is covered by a node if and only if the distance between them is less than or equal to the sensing range R of the node. For simplicity, we first assume that all nodes have the same sensing range in all directions. We shall address sensing irregularity in Section 5. Each node is assumed to have a scheduling period T and a duty cycle ratio β, $0 \leq \beta \leq 1$, that defines the percentage of time the node is awake. Each node chooses its wakeup point t_{start} independently and uniformly within $[0, T)$, wakes up for a period of time βT, and then goes back to sleep until $T + t_{start}$. Finally, we assume that (for all practical purposes) the entire system area is covered when all nodes are awake.

Each target in the area moves in a straight line at a constant speed v in our analysis. We assume that all targets are point targets so that their physical sizes can be neglected. Observe, however, that larger targets can still be analyzed by increasing the sensing range used in the analysis by the diameter of the target to account for the larger sensory signature.

3   Stationary Target Sensing Analysis

In this section, we analyze the expected value and probability distribution of detection delay t_d for a stationary, persistent target (e.g., a localized fire).

First, according to our assumptions, since nodes are deployed with a uniform distribution, the number of nodes within an area of πR^2 conforms to a binomial distribution $B(A_{system}d, \pi R^2/A_{system})$. For a sensor network with a reasonable scale, $A_{system}d$ is a large number, and $(A_{system}d)(\pi R^2/A_{system}) = \pi R^2 d$ is a constant, denoted as λ. With these conditions, the binomial distribution can be approximated by a Poisson distribution with parameter λ. Observe that $\lambda = \pi R^2 d$ is the average number of nodes within a sensing range. The probability that there are n nodes covering an arbitrary geometric point is $\frac{\lambda^n}{n!}e^{-\lambda}$, $n = 0, 1, 2, \ldots$. The probability that no node covers the point (i.e., $n = 0$) is $e^{-\lambda}$. To make this probability smaller than 0.01, λ should be at least $-\ln 0.01 \approx 4.6$. Observe that the detection delay for points that are not covered is infinite, which makes the expected detection delay for the whole area infinitely large. To avoid this problem, we take into account only those points that are covered by at least one node. Our calculations above indicate that as long as there are more than 4.6 nodes on average within a sensing range, the probability that a point is not covered is less than 0.01, which is negligible.

In the rest of the paper, we extensively use two general results from theory of probability. First, if the probability of an event A occurring in a single experiment is p, and if the *number of experiments* conforms to a Poisson Distribution with parameter λ, the probability of event A occurring at least once in the series of experiments is:

$$P = 1 - e^{-p\lambda} \tag{1}$$

In the analysis, we first calculate the probability of detection when the target is covered by one node in a certain area. Since the number of nodes in a certain area conforms to a Poisson distribution, we get the probability that at least one node covers the target.

Another important result relating to our derivation is Proposition 11.6 in [16]: Let X be a nonnegative random variable, then $E(X) = \int_0^\infty P(X > x)dx$.

Since the cumulative distribution function (CDF) of X $cdf(x) = P(X \leq x)$, we have

$$E(X) = \int_0^\infty (1 - cdf(x))dx \tag{2}$$

We will use this fact to derive the average detection delay for both stationary and mobile target detection, after we get the probability that the target is detected within a certain period of time.

Now consider an arbitrary point in the area to be monitored. Suppose node O is the only node covering this point. The duty cycle of node O is shown in Figure 1. Denote the random variable corresponding to the detection delay at this point as t_d. Since the node has a probability β of being awake, we have $P(t_d = 0) = \beta$.

The probability density function (PDF) of t_d $f(\tau)$, where $0 < \tau \leq T - \beta T$, conforms to a uniform distribution, so $f(\tau) = \frac{1}{T}$.
as long as the target can arrive uniformly anywhere within the duty-cycle. Therefore, when there is only one node covering the point, the cumulative probability distribution for the detection delay is:

$$F_{t_d}(\tau) = P(t_d \leq \tau) = \beta + \frac{\tau}{T}, \tag{3}$$

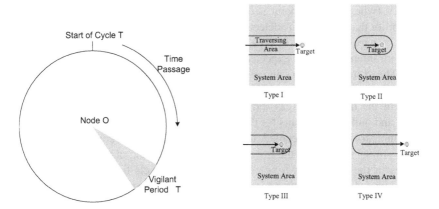

Fig. 1. Duty Cycle of a Particular Node O **Fig. 2.** Target Detection Scenarios

where $0 \leq \tau \leq T - \beta T$. When there are n nodes covering the point, we consider the detection of each node as an experiment. Let event A correspond to the fact that the target is detected within an interval time no larger than t_d. Substituting for the single event probability p in Equation (1) from Equation (3) we get the probability of event A:

$$P(A) = 1 - e^{(\frac{-\lambda \tau}{T} - \beta \lambda)} \tag{4}$$

Since we focus only on those points that are covered, the CDF of detection delay for such points is $P(A)/1 - e^{-\lambda}$, or:

$$F_{t_d}(\tau) = \frac{1 - e^{(\frac{-\lambda \tau}{T} - \beta \lambda)}}{1 - e^{-\lambda}}, 0 \leq \tau \leq T - \beta T \tag{5}$$

where $e^{-\lambda}$ is the probability of voids.

Second, according to Equation (2), the expected value of the delay t_d is:

$$E(t_d) = \int_0^\infty P(t_d > t)dt = \int_0^\infty \frac{e^{-\lambda(\frac{t}{T} + \beta)}}{1 - e^{-\lambda}} dt = \frac{T}{\lambda} e^{-\beta \lambda}/(1 - e^{-\lambda}) \tag{6}$$

In particular, when the duty cycle β approaches 0 and $1 - e^{-\lambda}$ approaches 1, Equation (6) turns into T/λ. Therefore, we conclude that when the duty cycle is sufficiently small and the density is sufficiently large, the average detection delay is inversely proportional to node density.

4   Mobile Target Sensing Analysis

In this section, we analyze the detection delay for mobile targets. We consider a target that moves at velocity v along a straight path, and consider four detection scenarios, as shown in Figure 2. We categorize these scenarios based on whether or not the start or end point of the target path is inside the system area. For target scenarios of type I and type IV (where the end-point is outside the monitored area), we are interested in the probability that the target is detected by at least one node before it leaves the area. For target scenarios of type II or type III (where the end-point is inside the monitored area), we are interested in the expected time and distance the target travels before it is detected. The system area is assumed to be rectangular, in which nodes are randomly deployed. The target is detected within a radius R (or diameter $2R$) which constitutes the width of the traversing areas shown in Figure 2. We focus on type I and type II scenarios, and outline the results for the other two types.

We demonstrate a more detailed model for type II target detection in Figure 3 (type I model is similar). Consider the traversing area S consisting of the rectangle and two half-circles in the figure. For a potential target that travels from point A to point B (where distance $AB = L$), only nodes located within this area can detect the target, for example, node M, which has an intersection length of l with the target's moving track. Therefore, we call this area the detection area. Since all nodes are deployed conforming to a uniform distribution, the number of nodes in the detection area also conforms to Poisson Distribution approximately. According to Equation (1) mentioned above, in order to find the probability of detection, we only need to find the detection probability when there is only one node within this area.

Now consider node M. Since it has an intersection length with the target track, the potential target takes time l/v to pass its sensing area. Therefore, the target appears to be a temporary event with a lifetime of l/v to node M, which means it has a probability of $\min(\beta + \frac{l}{vT}, 1)$ to be detected by node M. Considering that node M could be anywhere within the detection area, we now analyze the average detection probability for node M to detect the target.

Notice that $\beta + \frac{l}{vT}$ can be at most 1, and if it is, the target will definitely be detected when it passes. This fact categorizes potential targets into two types, fast and slow. For fast targets, the expression $\beta + \frac{l}{vT}$ is always smaller than or equal to 1. Therefore, we can obtain the expectation of detection delay directly by integrating over the detection area. On the other hand, if v is small enough, $\beta + \frac{l}{vT}$ can become larger than 1, thus, we have to partition the detection area first before the integral. Observe that the maximal intersection length between the target path and a sensor's range is $2R$. Therefore, if target velocity $v \geq \frac{2R}{(1-\beta)T}$, then $\beta + \frac{l}{vT}$ is always smaller than or equal to 1. We use $\frac{2R}{(1-\beta)T}$ as the threshold between fast and slow targets.

4.1   Detection Analysis for Fast Targets

For a fast target we can express the probability that the target is detected given there is only one node in the detection area, S, as follows:

$$P = \int_S (\beta + \frac{l}{vT}) ds / S = \beta + \frac{\int_S l ds}{vTS} \tag{7}$$

Thus, we only need to calculate $\int_S l ds$ and S.

Because of symmetry, we only need to calculate the integral $\int_S l ds$ within the first quadrant, as shown in Figure 4.

For a type I target, we need to do the integral over area A and B, while for a type II target, we need to do the integral over A, B and C.

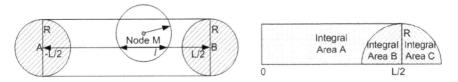

Fig. 3. Target Detection Example	Fig. 4. Fast Target Detection

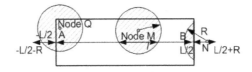

Fig. 5. Fast Target Detection Illustration

Fast Type I Target Analysis. For a type I target, the area under consideration consists of area A and B. Therefore, we have:

$$\int_{S_{A+B}} lds/S_{A+B} = \frac{\phi_A + \phi_B}{LR/2} \tag{8}$$

while

$$\phi_{A+B} = \int_0^R dy \int_0^{\frac{L}{2}} 2\sqrt{R^2 - y^2}dx = \frac{\pi R^2 L}{4} \tag{9}$$

Thus, we get the overall probability as:

$$P = \beta + \frac{\pi R^2 L}{2RLvT} = \beta + \frac{\pi R}{2vT} \tag{10}$$

Fast Type II Target Analysis. Similar to the type I target analysis, we have:

$$\int_{S_{A+B+C}} lds/S_{A+B+C} = \frac{\phi_A + \phi_B + \phi_C}{LR/2 + \pi R^2/4} \tag{11}$$

Therefore,

$$\phi_A = \int_0^R dy \int_0^{\frac{L}{2}-\sqrt{R^2-y^2}} 2\sqrt{R^2 - y^2}dx = \frac{\pi R^2 L}{4} - \frac{4R^3}{3} \tag{12}$$

$$\phi_B = \int_0^R dy \int_{\frac{L}{2}-\sqrt{R^2-y^2}}^{\frac{L}{2}} (\frac{L}{2} + \sqrt{R^2 - y^2} - x)dx = R^3 \tag{13}$$

$$\phi_C = \int_0^R dy \int_{\frac{L}{2}}^{\frac{L}{2}+\sqrt{R^2-y^2}} (\frac{L}{2} + \sqrt{R^2 - y^2} - x)dx = \frac{R^3}{3} \tag{14}$$

Thus, we get the overall probability as:

$$P = \beta + \frac{\pi R^2 L}{(2RL + \pi R^2)vT} = \beta + \frac{\pi RL}{(2L + \pi R)vT} \tag{15}$$

Note that for the case $L < 2r$, the derivation is a little different. However, the result remains the same and for simplicity we omit the details of derivation.

We also have a more intuitive explanation for the calculation of $\int_S l\,ds$. This expression is an integral of a line segment of length l on the target's locus for each point p in the traversing area S. Note that a point p' belongs to the line segment if and only if its distance from p is smaller or equal to R, the sensing range. Therefore, the integral $\int_S l\,ds$ is equivalent to the integral $\int_L s\,dl$, where L is the locus of the target, and s is the area in which each point has a shorter distance than R to dl. For type II targets, s is simply πR^2 and therefore $\int_S l\,ds$ is simply $\pi R^2 L$, which verifies Equation (15). It is a little more complicated for type I targets since s can be smaller than πR^2. For example, for point N in Figure 5, s is a half-lens. The expression then becomes the integral of the overlapping area of the circular disk with radius R and the traversing area over L. We can think it as a circular disk virtually moving from $-L/2 - R$ to $L/2 + R$ and calculate the accumulation of overlapping areas. To make the calculation simpler, equivalently, we can also imagine a fixed circular disk and virtually move the traversing area and calculate the accumulation overlapping areas. This also gives a result of $\pi R^2 L$, which verifies Equation (10).

4.2   Detection Analysis for Slow Targets

Now consider $v < \frac{2R}{(1-\beta)T}$. The main difference in this case is that it is possible that for certain node positions (x, y), $l(x, y) > (1 - \beta)vT$, therefore $p(x, y)$ is 1 instead of $\beta + l(x, y)/vT$. Suppose there are two partitions, U and V. In U, $p(x, y) = \beta + l(x, y)/vT$, and in V, $p(x, y) = 1$. Therefore, we have:

$$P = \frac{\int_{S_U}(\beta + \frac{l}{vT})ds + \int_{S_V} 1\,ds}{S_{A+B}} = \frac{\int_{S_{U+V}}(\beta + \frac{l}{vT})ds + \int_{S_V}(1 - \beta - \frac{l}{vT})ds}{S_{U+V}}$$

$$= \beta + \frac{\int_{S_{U+V}} l\,ds}{(S_{U+V})vT} + \frac{S_V(1 - \beta) - \int_{S_V} l\,ds/vT}{S_{U+V}} \tag{16}$$

Obviously, the term $\beta + \frac{\int_{S_{U+V}} l\,ds}{(S_{A+B})vT}$ is exactly what we have obtained for fast object analysis. So now we need to calculate $S_V(1 - \beta)$ and $\int_{S_V} l$. We classify the analysis into two cases according to type I and type II targets.

Slow Type I Target Analysis. In this type, the partition where $p(x, y) = 1$ is the rectangle area A in Figure 6. We define another variable a such that $\beta + 2a/vT = 1$. The height of the rectangle is $\sqrt{R^2 - a^2}$.

Observing that $\frac{S_V(1-\beta) - \int_{S_V} l\,ds/vT}{S_{U+V}} = \frac{\int_{S_V}[(1-\beta)vT - l(x,y)]dxdy}{S_{U+V}vT}$. Therefore, we have:

$$\int_V [(1 - \beta)vT - l(x, y)]dxdy = \int_A [2a - l(x, y)]dxdy \tag{17}$$

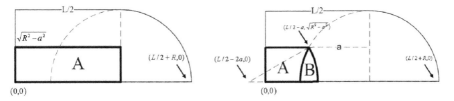

Fig. 6. Type I target Detection   **Fig. 7.** Regions A and B

in which

$$\iint_A [2a - l(x,y)]dxdy = \int_0^{\sqrt{R^2-a^2}} \int_0^{\frac{L}{2}} (2a - 2\sqrt{R^2 - y^2})dxdy$$

$$= \int_0^{\sqrt{R^2-a^2}} [aL - L\sqrt{R^2 - y^2}]dy = aL\sqrt{R^2 - a^2} - L(\frac{a\sqrt{R^2 - a^2}}{2} + \frac{R^2}{2}sin^{-1}\frac{\sqrt{R^2 - a^2}}{R})$$

$$= [aL\sqrt{R^2 - a^2} - LR^2 sin^{-1}\frac{\sqrt{R^2 - a^2}}{R}]/2$$

Therefore we obtain:

$$\iint_A [2a - l(x,y)]dxdy = Lk(R,a)/4 \tag{18}$$

where

$$k(R,a) = 2a\sqrt{R^2 - a^2} - 2R^2 cos^{-1}(\frac{a}{R}) \tag{19}$$

Finally, we get when $L \geq 2R$, $v < \frac{2R}{(1-\beta)T}$, for a type I target,

$$P = \beta + \frac{\pi R^2 + k(R,a)}{2RvT} \tag{20}$$

Slow Type II Target Analysis. In this type, the partition where $p(x,y) = 1$ is less regular. We still define a such that $\beta + 2a/vT = 1$. Observe that Figure 7 plots the region where $p(x,y) = 1$, which includes A and B, whose boundaries are formed by line $y = \sqrt{r^2 - a^2}$ and two circles which centered at $(L/2 - 2a, 0)$ and $(L/2, 0)$, respectively. For $(x,y) \in A \cup B$, $p(x,y) = 1$.

Similarly, we have:

$$\int_V [(1 - \beta)vT - l(x,y)]dxdy = \int_{A+B} [2a - l(x,y)]dxdy \tag{21}$$

in which

$$\iint_A [2a - l(x,y)]dxdy = \int_0^{\sqrt{R^2-a^2}} \int_0^{\frac{L}{2} - \sqrt{R^2-y^2}} (2a - 2\sqrt{R^2 - y^2})dxdy$$

$$= \int_0^{\sqrt{R^2-a^2}} (\frac{L}{2} - \sqrt{R^2 - y^2})(2a - 2\sqrt{R^2 - y^2})dy$$

$$\iint_B [2a - l(x,y)]dxdy = \int_0^{\sqrt{R^2-a^2}} \int_{\frac{L}{2}-\sqrt{R^2-y^2}}^{\frac{L}{2}-2a+\sqrt{R^2-y^2}} [2a - (\sqrt{R^2-y^2} - \frac{L}{2} - x)]dxdy$$

$$= \int_0^{\sqrt{R^2-a^2}} [(2a - \sqrt{R^2-y^2} - \frac{L}{2})(2\sqrt{R^2-y^2} - 2a) + 2a^2 - aL - 2a\sqrt{R^2-y^2} + L\sqrt{x^2-y^2}]dy$$

Therefore we can obtain:

$$\iint_{A+B} [2a - l(x,y)]dxdy = (L - 2a)k(R,a)/4$$

where

$$k(R,a) = 2a\sqrt{R^2-a^2} - 2R^2cos^{-1}(\frac{a}{R}) \tag{22}$$

Finally, we get when $L \geq 2R$, $v < \frac{2R}{(1-\beta)T}$, for a type II target,

$$P = \beta + \frac{\pi R^2 L + (L-2a)k(R,a)}{(2RL + \pi R^2)vT} \tag{23}$$

Indeed, in Figure 7, we only plotted the case where $2a > R$. For $2a < R$, the derivation is similar, and the results are the same.

When $a = 0$, $k(R,a) = -\pi R^2$. When $a = R$, $k(R,a) = 0$. We also have

$$\frac{\partial k(R,a)}{\partial a} = 4\sqrt{R^2-a^2} \geq 0, \tag{24}$$

so $k(R,a)$ is a monotonically non-decreasing function of a. Interestingly, the area of B is exactly $-k(R,a)/2$.

Denote $m(R,\beta)$ as the function substituting a with $(1-\beta)vT/2$ in $k(R,a)$, then

$$m(R,\beta) = (1-\beta)vT\sqrt{R^2 - \frac{(1-\beta)^2v^2T^2}{4}} - 2R^2\cos^{-1}[\frac{(1-\beta)vT}{2R}] \tag{25}$$

$m(R,\beta)$ is a monotonically non-increasing function of β. When $\beta = 1-\frac{2R}{vT}$, $m(R,\beta) = 0$. When $\beta = 1$, $m(R,\beta) = -\pi R^2$.

Additionally, for a slow type II target, if the travel distance L is less than $(1-\beta)vT$, the detection probability for a node located within the detection area cannot be larger than 1. Therefore, we should treat this period of time in the same way as a fast target. Therefore, we can revise the final probability as:

$$P = \beta + \frac{\pi R^2 L + min((L-2a)k(R,a),0)}{(2RL + \pi R^2)vT} \tag{26}$$

Average Detection Delay and Probability. So far, we have finished the derivation of $P(L,v)$ for both slow and fast objects. Next we find the detection probability for a type I target, and average detection delay for a type II target.

First, consider fast targets. According to Equation 1, for a fast type I target, with a deployment width of L, we obtain the detection probability as follows:

$$P_{detection}(v) = 1 - e^{-2RLdP} = 1 - e^{-2RLd(\beta+\frac{\pi R}{2vT})} \tag{27}$$

while for slow targets, the result is:

$$P_{detection}(v) = 1 - e^{-2RLdP} = 1 - e^{-2RLd(\beta + \frac{\pi R^2 + k(R,a)}{2RvT})} \tag{28}$$

For fast type II targets, the cumulative distribution function of the detection delay t_d is:

$$F_{t_d}(t) = P(t_d \leq t) = 1 - e^{-d(2Rvt + \pi R^2)(\beta + \frac{\pi Rvt}{(2vt + \pi R)vT})} \tag{29}$$

While for slow targets, the function is:

$$F_{t_d}(t) = P(t_d \leq t) = 1 - e^{-d(2Rvt + \pi R^2)(\beta + \frac{\pi R^2 vt + min(0,(vt - 2a)k(R,a))}{(2Rvt + \pi R^2)vT})} \tag{30}$$

For fast targets, the expected detection delay is

$$E(T_d) = \int_0^\infty e^{-\beta \pi R^2 d - vt(2R\beta + \frac{\pi R^2}{vT})d} dt = \frac{e^{-\beta \pi R^2 d}}{(2R\beta v + \frac{\pi R^2}{T})d} \tag{31}$$

Similarly, when $v < \frac{2R}{(1-\beta)T}$, the expected detection delay is[1]:

$$E(T_d) = \int_0^{2a} e^{-\beta \pi R^2 d - vt(2R\beta + \frac{\pi R^2}{vT})d} dt + \int_{2a}^\infty e^{-\beta \pi R^2 d + \frac{2ak(R,a)d}{vT} - vt(2R\beta + \frac{\pi R^2 + k(R,a)}{vT})d} dt$$

$$= \frac{e^{-\beta \pi R^2 d}}{(2R\beta v + \frac{\pi R^2}{T})d}[1 - \frac{m(R,\beta)e^{-(2R\beta vT + \pi R^2)(1-\beta)d}}{2R\beta vT + \pi R^2 + m(R,\beta)}] \tag{32}$$

4.3   Summary

So far, we have explained in detail the derivation for type I and type II targets in this paper. The authors have also finished the derivation of the detection delay and probability for type III and type IV targets, using a similar derivation approach. Due to space limitations, we only outline our final results for type III and type IV targets, as follows:

Expected detection delay for fast type III targets:

$$E(T_d) = \frac{e^{-\beta \pi R^2 d/2}}{(2R\beta v + \frac{\pi R^2}{T})d} \tag{33}$$

Expected detection delay for slow type III targets:

$$E(T_d) = \frac{e^{-\beta \pi R^2 d/2}}{(2R\beta v + \frac{\pi R^2}{T})d}[1 - \frac{m(R,\beta)e^{-(2R\beta vT + \pi R^2)(1-\beta)d/2}}{2R\beta vT + \pi R^2 + m(R,\beta)}] \tag{34}$$

Detection probability for fast type IV targets:

$$P = 1 - e^{-(2RL + \pi R^2/2)d(\beta + \frac{\pi RL}{(2L + \pi R/2)vT})} \tag{35}$$

Detection probability for slow type IV targets:

$$P = 1 - e^{-(2RL + \pi R^2/2)d(\beta + \frac{\pi R^2 L + min((L-a)k(R,a),0)}{(2RL + \pi R^2/2)vT})} \tag{36}$$

---

[1] Observe that if $v = 0$ (this is equivalent to a stationary target), we obtain the expected delay to be ∞. This is because we do not exclude voids in the mobile target model, and the existence of voids leads to infinitely large expected detection delay in stationary target detection.

4.4   Discussion

We now discuss the implications of our analytical results. First, we assume that almost all of the area is covered, that is, $1 - e^{-\lambda} \approx 1$, thus, $\lambda \geq 4.6$ according to our earlier discussion. Second, in order to save energy, we assume that β approaches 0, that is, the waking period is sufficiently small compared with the total scheduling period. Based on these two assumptions, we have several interesting observations.

First, for fast type I target detection, Equation (27) can be simplified to $1 - e^{-\frac{\lambda}{vT}L}$. Therefore, in order to obtain 99% detection probability, we have $L \geq \frac{4.6vT}{\lambda}$. Assume $\lambda \approx 4.6$, $L \geq vT$. The result is quite intuitive: in order to almost certainly catch an intruding target, the deployment width can be no smaller than the product of the target velocity and the scheduling period.

Second, for a fast type II target, Equation (29) can be simplified to $1 - e^{-\frac{\lambda t}{T}}$. This means that for a target that starts from within the system area, the probability of being caught increases exponentially. Also, to make this probability larger than 99%, $t \approx T$, given that λ is 4.6. This result is also confirmed in our simulations.

Third, for a type II target, the detection delay can be simplified as T/λ if β approaches 0. This result is the same as the stationary detection delay, which means that as β approaches 0, regardless of the movement pattern of the target, the detection delay is approximately constant, determined only by the scheduling period and node density.

4.5   Simulation-Based Verification

We now demonstrate that the derivation results are consistent with simulation results under perfect circular range assumptions. In the first set of simulations of stationary targets, locations of nodes are generated conforming to a uniform random distribution over a unit area with size $100m \times 100m$, without loss of generality. The period T is chosen to be $1s$. The waking points of the nodes are generated according to a uniform distribution over $[0s, 1s]$. The sensing range for circular model is $10m$. We choose a point $(50, 50)$ and generate $10,000$ sets of random target locations to run the simulations. The parameter $\lambda = \pi r^2 d$ is set to 5. We record the number of experiments where the detection delay is smaller than or equal to $0s$, $0.05s$, $0.10s$, ..., $0.90s$, and $0.95s$. We then compare these frequencies with the cumulative distribution functions (CDF's) obtained in the analysis section. The simulation results are shown in Figure 8. From the figure we can see that our analysis and simulation results match well.

In the second set of simulations, we choose various λ values (5, 7, 9 and 11) and β values (0, 0.1, 0.2, ..., 1), and run simulations to gather the average detection delays. Other settings are the same as the previous set. We compare the average detection delays with the theoretical expected detection delays obtained in the analysis section and plot Figure 9. The figure shows that the average delays are very close to the analytical expected delays.

The settings for simulation verification with mobile target tracking are the same as the stationary setting, except that now the target has a velocity. We only consider type II target detection in this section. Other types can be similarly verified. The simulation result is shown in Figure 10. Observe that velocity $2.5m/s$ means a slow target, according to our threshold. In particular, its cumulative detection probability has a relatively long tail, compared to the distribution of targets with higher velocity, implying that for slower

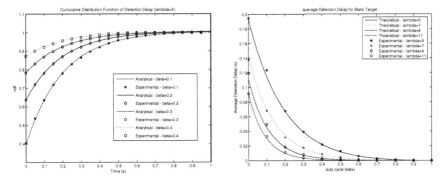

Fig. 8. Stationary Detection Delay Distribution   **Fig. 9.** Average Stationary Detection Delay

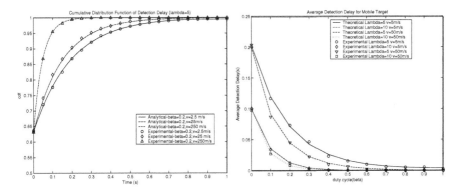

Fig. 10. Mobile Detection Delay Distribution   **Fig. 11.** Average Mobile Detection Delay

targets, there is a higher chance for the target to remain undetected after the scheduling cycle T. Again, we observe that our analysis and the simulation results match well.

In the second set of simulations, we choose various λ values (5 and 10), velocity ($5m/s$ and $50m/s$), and β values (0, 0.1, 0.2, ..., 1), and run simulations to gather the average detection delays. Other settings are the same as in the previous set. We compare the average detection delays with the theoretical expected detection delays obtained in the analysis section and plot Figure 11. The figure shows that the average delays are very close to the analytical expected delays.

The simulation results demonstrate the viability of using our results to predict the system performance. For example, suppose that we have a deployment of $\lambda = 5$ and $T = 1s$, in order to make sure that the expected detection delay to be no larger than $0.1s$, we can calculate that β must be at least 0.14. The simulation results confirm this calculation (Figure 9).

5   Practical Application

An example application of random duty-cycle sensor networks is a surveillance system that tracks trespassers. A person is detected by motion sensors. Once such an event is triggered, the detecting node sends a wakeup command to its neighbors to track the intruder. These commands constitute messages with a preamble that is longer than the duty-cycle period T. Thus, all neighbors eventually wake up and receive the message. It alerts them to remain awake and track, possibly waking up their neighbors as well. Subsequent detections from multiple sensor nodes eventually reconstruct the target's path. Other types of sensors, such as magnetic ones, can be applied to tell if the person carries a weapon.

5.1   Application Analysis and Sensing Irregularity

We simulate the aforementioned application and check whether simulation results match analytical predictions. However, our discussions so far have assumed a perfectly circular sensing model. This assumption makes the analysis tractable, and leads to closed-form results. Real sensing devices used in this application do not have such a perfect sensing area. In experiments, we use passive infrared sensors (PIR sensors) as motion sensors, and identify the causes of irregular sensing range. We then propose a realistic sensing model for PIR sensors. We integrate this model into our simulations and show that the predictions based on our analysis are quite robust under realistic sensing irregularities.

We have PIR sensors installed on the ExScal XSM motes from OSU and Cross-Bow [6]. At present, four motion sensors are integrated on a single ExScal Mote to provide a 360 degree surveillance range (each individual motion sensor can only handle a 90 degree area). We carried a series of experiments on ExScal nodes. Specifications of ExScal nodes can be found at [1].

5.2   Experimental Results

In principle, PIR sensors can detect anything that generates an infrared field disturbance, such as vehicles, persons, etc. We are interested in the ability of PIR sensors to detect walking persons, as demanded by the application. The central metric is sensing range of PIRs. In individual tests, we let a person walk by at different distances and angles relative to the sensing node, and measure the maximal range of detection. All experiments are repeated in an open parking lot. We program the node with a simple frequency analysis procedure to report events. The procedure relies on an adaptively adjusted threshold to compare the current readings. The main purpose of this adaptive threshold is to avoid false alarms introduced by weather changes, since this change can produce infrared noise that may trigger the sensor. More specifically, our procedure monitors the average readings of the filtered signal within a moving window. If the average reading observed is much smaller than the current threshold, it decreases the

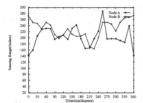

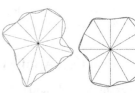

Fig. 12. Range with Directions   **Fig. 13.** Ranges of Two Nodes   **Fig. 14.** Range Approximations

threshold by taking a weighted average. On the other hand, if the average energy is close to or larger than the threshold for a certain period, the sensor decides that the weather is noisy, and increases the threshold. Practically we are able to filter out almost all false alarms using this technique. We note that this technique also filters out those slight disturbances that may, in fact, be caused by the target. Thus, the results proposed below present effective sensing range in slightly noisy environments, which are different from precisely controlled environments.

We measured the sensing range from 24 equally divided directions. At each direction, a person moves with different distances from the sensor node. If the fluctuation of the filtered sensor reading exceeds a predefined noise threshold due to the nearby motion, the measured point is within the sensing range. If the motion does not cause a fluctuation greater than the threshold, the measured point is out of the sensing range. In the experiments we found out that the sensitivity of the sensors changes dramatically at the edge of the sensing range. The motion can always be detected in the sensing range and the motion 10 inches beyond the sensing range never triggers the sensors. Therefore, the precision of the range measurement is always within 10 inches (the range itself being hundreds of inches). The experimental results for two representative nodes are shown in Figure 12. Obviously, with four sensors, the range is far from being circular. A more intuitive illustration is shown in Figure 13, where the sensing range in each direction is plotted to scale.

Based on the experimental results of multiple tests, we have the following observations regarding the PIR sensing capability. First, the sensing range of one node is not isotropic, that is, the node exhibits different ranges in different directions. Second, the boundary of the sensing range is delineated quite sharply, with predictably no detection when the range is exceeded by about 3-7%. Third, we can consider the variation of ranges relatively continuous. We do observe sudden changes in the sensitivity of some nodes, but this is very uncommon. Therefore, a model may consider connecting sensitivity ranges in different directions using continuous curves. Fourth, the sensing range distribution in different directions roughly conforms to a Normal Distribution. This conclusion is based on statistical analysis of our experimental data using a Kolmogorov-Smirnov Test [24]. We concluded from our experimental data set that the sensing range in one direction can be approximated with a Normal Distribution with an expectation of 217 inches and a standard deviation of 32 inches.

5.3   Realistic Sensing Model for PIR sensors

In this section, we present a realistic sensing model for PIR nodes. This model is designed to reflect the three key observations: non-isotropic range, continuity and normality. First, the model determines sensing ranges for a set of equally-spaced directions. Each range can either be specified based on actual measurements or obtained from a representative distribution, for example, the normal distribution $N(217, 32^2)$. Next, sensing ranges in all other possible directions are determined based on an interpolation method. For simplicity, we use linear interpolation to specify the boundary of the sensing area. As an example, the approximation of the two representative nodes based on this model are shown in Figure 14. Clearly, our simplified model pretty accurately reflects the fluctuations of sensing ranges in different directions.

5.4   Robustness of Theoretical Predictions to Realistic Sensing Model

We now incorporate the realistic sensing range model into simulations to test the robustness of the performance predictions to sensing irregularity. We scaled the normal distribution $N(217, 32^2)$ to $N(10, 1.47^2)$ to fit the simulation setting. We use the same formulas as the previous idealistic experiments. The CDF's of detection delay under different parameter settings are shown in Figure 15. To illustrate the difference, we also plot the error relative to theoretical predictions in Figure 16.

We have two observations regarding these results. First, for our simulation settings, sensing irregularity has a very small effect. One primary reason is that even though the detection ranges do vary with different directions, the overall degree of coverage for the area remains almost the same, approximated by λ. Therefore, the overall detection delay distribution is almost not affected. Second, we observe that the maximal error relative to theoretical predictions is no more than 1.2%. As an example to show its implications, suppose that we have a set of system parameters that guarantees that 99% of intruding targets are detected within a certain time. Then, the actual detection rate should be no less than 97% with the existence of irregularity. Based on this observation, we conclude that our model is quite robust to realistic sensing conditions.

At last, we acknowledge that so far, our modeling of the realistic sensing model is only concerned with PIR sensors. Other types of sensors may well exhibit different characteristics, therefore, may have varied effect on the detection performance. However, we envision that with a relatively large system with considerable density, the effect of sensing irregularity will be considerably limited, leading to improvements in the accuracy of the aforementioned results.

6   Related Work

Research on minimizing energy consumption [5, 7, 8, 15, 18, 25, 28] has been one central topic in the sensor network community in recent years. Various effective techniques have been proposed, evaluated and implemented. Protocols to schedule nodes while maintaining full coverage have been proposed by [27, 14, 21, 10, 23]. The full sensing coverage model is very suitable for areas where continuous vigilance is required, but

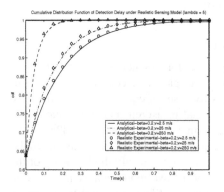

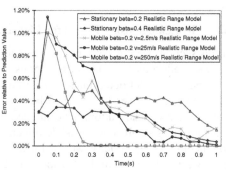

Fig. 15. Realistic Sensing Effect   **Fig. 16.** Robustness of Theoretical Predictions

consumes considerable energy. There exists a lower bound on the minimal energy consumption if the requirement of full sensing coverage is to be fulfilled. To guarantee a longer sensor network lifetime, in certain situations, the designer may wish to keep partial sensing coverage in exchange of longer product lifetime.

Another related research direction is tracking and surveillance [2, 3, 4, 12, 13, 19, 20, 22, 26]. In these efforts, various tracking approaches are examined to optimize the overall surveillance quality. In particular, recent research efforts [9, 17, 11] have turned attention to the study of target tracking in the context of partial sensing coverage. In [9], the authors proposed the metric of quality of surveillance, which is defined as the average traveled distance of the target before it is detected. This is consistent with the model in our paper, where the average detection delay rather than distance is used. However, the authors didn't address the precise determination of this metric. Instead, they used an approximation model based on coverage process theory. Their attention is mainly focused on evaluating various sleep models with respect to the average travel length, which lead to considerably different results compared with ours. The work in [17] considers an equivalent problem to the type II target detection in our paper. However, we are more interested in deriving closed-form expressions for all four scenarios, whereas [17], while valuable, only considers fast targets and does not present closed-form results for detection delay. Another interesting work is [11], which considered the effect of different random and coordinated scheduling approaches, but provides no closed-form results. Therefore, we believe our work is a necessary complement towards thorough understanding of detection delay performance.

7   Conclusion and Future Work

This paper is the first to derive closed-form formulas for the distribution and expectation of detection delay for both stationary and mobile targets, and is also the first to propose and evaluate a realistic sensing model for sensor networks, to the best of the authors' knowledge. Extensive simulations are conducted and the results show the validity of the analysis. The analytical results are highly important for designers of energy efficient sensor networks for monitoring and tracking applications. Designers can apply these formulas to predict the detection performance without costly deployment and testing. Based on these formulas, they can make decisions on key system or protocol parameters, such as the network density and the duty cycle, according to the detection requirements of the system. Therefore, this work is a major contribution towards a thorough understanding of the relationship between system and protocol parameters and achievable detection performance metrics.

In the future, we intend to apply our results in large scale experiments. We also intend to expand our analysis to more irregular movement models, such as random waypoint movement. We hope the results can give us further insight on the fundamental tradeoffs in sensor network detection performance.

Acknowledgements

This work is supported in part by NSF grant CCR-0098269, ITR EIA-0205327 and CCR-0325197, the MURI award N00014-01-1-0576 from ONR, and the DARPA IXO offices under the NEST project (grant number F336615-01-C-1905).

References

1. In *http://www.cast.cse.ohio-state.edu/exscal/*.
2. J. Aslam, *et al.* Tracking a moving object with a binary sensor network. In *ACM Sensys*, 2003.
3. M. Batalin, *et al.* Call and response: Experiments in sampling the environment. In *ACM Sensys*, 2004.
4. K. Chakrabarty, *et al.* Grid coverage for surveillance and target location in distributed sensor networks. In *IEEE Transaction on Computers, 51(12)*, 2002.
5. B. J. Chen, *et al.* Span: An energy-efficient coordination algorithm for topology maintenance in ad hoc wireless networks. In *ACM Mobicom*, 2002.
6. CrossBow. In *http://www.xbow.com*.
7. D. Ganesan, *et al.* Power-efficient sensor placement and transmission structure for data gathering under distortion constraints. In *Proceedings of IPSN*, 2004.
8. L. Gu and J. Stankovic. Radio-triggered wake-up capability for sensor networks. In *IEEE RTAS*, 2004.
9. C. Gui and P. Mohapatra. Power conservation and quality of surveillance in target tracking sensor networks. In *ACM Mobicom*, 2004.
10. T. He, *et al.* An energy-efficient surveillance system using wireless sensor networks. In *ACM Mobisys*, 2004.
11. C. Hsin and M. Y. Liu. Network coverage using low duty-cycled sensors: Random and coordinated sleep algorithms. In *Proceedings of IPSN*, 2004.
12. J. J. Liu, *et al.* Distributed group management for track initiation and maintenance in target localization applications. In *Proceedings of IPSN*, 2003.
13. S. Megerian, *et al.* Exposure in wireless sensor networks. In *ACM Mobicom*, 2001.
14. S. Meguerdichian, *et al.* Coverage problems in wireless ad-hoc sensor networks. In *IEEE Infocom*, 2001.
15. S. Pattem, *et al.* Energy-quality tradeoff for target tracking in wireless sensor networks. In *Proceedings of IPSN*, 2003.
16. S. Port. *Theoretical Probability for Applications*. John Wiley and Sons,Inc, 1994.
17. S. Ren, *et al.* Analyzing object tracking quality under probabilistic coverage in sensor networks. In *ACM Mobile Computing and Communications Review*, 2005.
18. V. Shnayder, *et al.* Simulating the power consumption of large-scale sensor network applications. In *ACM Sensys*, 2004.
19. G. Simon, *et al.* Sensor network-based countersniper system. In *ACM Sensys*, 2004.
20. R. Szewczyk, *et al.* An analysis of a large scale habitat monitoring application. In *ACM Sensys*, 2004.
21. D. Tian, *et al.* A node scheduling scheme for energy conservation in large wireless sensor networks. In *Wireless Communications and Mobile Computing Journal*, 2003.
22. G. Veltri, *et al.* Minimal and maximal exposure path algorithms for wireless embedded sensor networks. In *ACM Sensys*, 2003.
23. X. R. Wang, *et al.* Integrated coverage and connectivity configuration in wireless sensor networks. In *ACM Sensys*, 2003.
24. E. W. Weisstein. Kolmogorov-smirnov test. In *MathWorld at http://mathworld.wolfram.com/Kolmogorov-SmirnovTest.html*.
25. J. Wu and M. Gao. On calculating power-aware connected dominating sets for efficient routing in ad hoc wireless networks. In IEEE ICPP, 2001.
26. N. Xu, *et al.* A wireless sensor network for structural monitoring. In *ACM Sensys*, 2004.
27. T. Yan, *et al.* Differentiated surveillance for sensor networks. In *ACM Sensys*, 2003.
28. F. Ye, *et al.* Peas:a robust energy conserving protocol for long-lived sensor networks. In *IEEE International Conference on Distributed Computing Systems(ICDCS)*, 2003.

Collaborative Sensing Using Sensors of Uncoordinated Mobility*

Kuang-Ching Wang[1] and Parmesh Ramanathan[2]

[1] Clemson University, Clemson SC 29634, USA
kwang@clemson.edu
[2] University of Wisconsin, Madison WI 53705, USA
parmesh@ece.wisc.edu

Abstract. Wireless sensor networks are useful for monitoring physical parameters and detecting objects or substances in an area. Most ongoing research consider the use of stationary sensors or controlled mobile sensors, which incur substantial equipment costs and coordination efforts. Alternatively, this paper considers using •••••••••••• • •••••• •••••, who is not directed for any specific sensing activity. Each node independently observes a ••••• ••••••• of the field along its own path. The limited observation can be extended via information exchange among nodes coming across each other. For this model, the inherently noisy mobile measurements, incomplete individual observations, different sensing objectives, and collaboration policies must be addressed. The paper proposes a design framework for uncoordinated mobile sensing and one sensing approach based on ••••••• ••••• ••••• for target detection, field estimation, and edge detection. With simulations, we study its strengths and trade-offs with stationary and controlled mobile approaches.

1 Introduction

Wireless sensor networks are envisioned useful in a wide range of applications for monitoring physical attributes or detecting objects or substances in an area of interest. So far, there have been examples in military applications such as target detection, classification, tracking [1], and in civilian applications such as environmental monitoring [2], infrastructure protection [3], and ecological studies [4]. Typically, sensor nodes make measurements of their local surroundings and store them on-board for later retrieval or relay towards remote sinks over an ad hoc network.

Most research in literature have focused on stationary sensor networks with utmost interest on their sensing coverage and network connectivity. On the one

---

* The work reported here is supported in part by DARPA and Air Force Research Laboratory, Air Force Material Command, USAF, under agreement number F30602-00-2-0555, and by U.S. Army Research grant DAAD19-01-1-0504 under a subrecipient agreement S01-24 from the Pennsylvania State University. The views and conclusions contained herein are those of the authors and should not be interpreted as necessarily representing the views of funding agencies.

V. Prasanna et al. (Eds.): DCOSS 2005, LNCS 3560, pp. 293–306, 2005.

hand, it is always desirable to deploy sensors well covering the entire area with measurements reachable from all locations. On the other hand, it has also been shown that given a limited sensor communication range and a relatively large area of interest, full connectivity is achievable only with a prohibitively large number of stationary sensors [5]. Similar density constraints exist due to sensors' limited sensing range [6]. With mobile sensors, such constraints are relaxed if each mobile node can make measurements at multiple locations [7]. The apparent tradeoff is the latency needed to complete the measurements sequentially. It is also challenging to have a mobile node come to the right place at the right time to detect an occurring event. In these studies, motion control algorithms are proposed to direct mobile sensors towards specific locations such that a higher detection probability is achieved. Even so, steering unmanned mobile platforms or having dedicated personnel to carry out such sensing tasks remains expensive and by no means trivial.

In this paper, we study a different model of mobile sensing utilizing mobile sensors with *uncoordinated mobility* (UM). In this model, no efforts are made to coordinate the motion of each node and each node makes measurements independently along its arbitrary path. Consider UM nodes as vehicles mounted with sensors or individuals carrying sensors. In a military setting, vehicles and personnel can be equipped with sensors even if they are not dedicated to performing an explicit sensing task. The sensors operate in the background and make measurements from time to time on the move or at places they visit. In a civilian setting, individuals may have sensors embedded in daily gadgets such as cell phones or wrist watches that are carried around. These people may not be keeping track of the sensor readings and they are never directed to go to specific places for sensing. Nevertheless, the collected measurements contain information that can be used to estimate attributes of the area. Decisions made about the measurements may be used to raise alerts of abnormal events, to provide auxiliary information for individuals, or to establish a profile of the monitored field. As UM nodes measure areas they actually visit, the sensing model essentially provides coverage of places where future visits are highly possible.

Sensing with UM nodes faces a number of challenges. First of all, by nature a UM node makes measurements along its path and therefore observes only a "cross section" of the field. Unless a node is only interested in traveling back and forth along a single path, or its path winds through a significant portion of the field, the scope of its estimation is of limited use for itself. Second, as a node measures the field on the move, the measurements are inherently noisy; a node does not usually allow sufficient time to acquire multiple samples at one location to filter out the noise. Third, measurements may be stored of different spatial-temporal ranges and resolutions according to individual preferences, storage constraints, and task-specific requirements at each node, and a node can always adapt its strategy as time goes on. For instance, a newly arriving node may be interested in fine details about its immediate vicinity. As a node explores a wider range over time and observes the field to have slow variations, it may choose to lower its resolution to fit in estimations of a larger area.

Via information exchange, UM nodes coming across each other can aggregate their estimations of a larger area. If multiple nodes have had measurements at the same location, their estimations can also be robustly fused. Such information exchange is opportunistic, as it occurs only among nodes coming across the same vicinity. It is also voluntary, as each node decides the scope of information it is willing to share subject to its own constraints on resources, security policies, or privacy concerns. With information exchange, the sensing model is flexible to incorporate more nodes and larger coverage without limitation. The model can also be facilitated with data exchange infrastructure such as a number of publicly recognized data fusion centers, which may be placed at locations where large flows of people from different areas come across, e.g. airports, train stations, cargo fleet posts, military bases, etc. In a fully distributed manner, the aggregated estimations account for all areas that have ever been visited by at least one of a large group of UM nodes. In contrast to a centralized effort to deploy a large number of stationary or controlled mobile sensors, this model based on voluntary participation is much more cost efficient for long-term monitoring of our surroundings. With homeland security in mind, the model is a practical candidate for precautionary monitoring and detection over a geographic area too large to be deployed with centralized infrastructure.

With UM sensing, different sensing objectives are achieved by making, sharing, and processing measurements of different physical phenomena using different procedures. To facilitate modular design and extensible incorporation of different UM sensing tasks, we propose a structured framework with three components, namely the *sensing methods*, *information exchange*, and *decision making* of a UM node. Given a specific query, a UM node with the corresponding sensing modality defines its sensing schedule and storage format, its exchange schedule and format, exchange policy, and the processing algorithms to derive its sensing decisions. In this paper, we examine one such example, which is based on *adaptive profile estimation* and is suitable for a large class of applications. We demonstrate its use in response to queries of point target detection, field estimation, and edge detection. Simulation studies are conducted to assess its performance and design tradeoffs versus solutions with stationary sensors and controlled mobile sensors.

The rest of this paper is organized as follows. In Section 2, we describe the UM sensing design framework. In Section 3, we present the adaptive profile estimation method and the corresponding target detection, field estimation, and boundary detection algorithms. Simulation studies are presented in Section 4. The paper concludes in Section 5.

2   Design Framework

Given a specified query, the framework defines three components:

- *Sensing methods.*
  According to its sensing modalities, a UM node produces certain types of measurements, e.g., binary, multi-modal, or continuous in range. To fulfill

the query, a node specifies its sensing schedule (frequency of periodic measurements or events of interrupt-driven measurements), storage formats (preprocessing procedures and data resolution), and memory management policies. For example, a UM node may decide to make periodic measurements once per hour indefinitely, compute the mean of the measurements, store results in a data structure, and keep only the latest results in its memory.

– *Information exchange.*
A node decides its information exchange policy regarding when, how, and what information it chooses to exchange with other nodes. An exchange can be sender-initiated or receiver-initiated, and senders and receivers negotiate the type, scope, and resolution of information to be exchanged. A sender-initiated exchange allows a node to pro-actively disseminate critical information of common interests, while a receiver-initiated exchange allows a node to request specific information it wishes to receive. A sender can deny a request if it does not possess the matching data or the request is against its information exchange policy.

– *Decision making.*
Based on its available information, a node makes an independent decision in response to the query. In an extended UM sensing model with central data fusion centers, the fusion centers make aggregate decisions based on information collected from a much larger set of UM nodes. The component defines all data handling procedures and inference algorithms needed to derive a sensing decision in the form of a literal response, a triggered action, or dissemination of critical information to other nodes.

We follow the prescribed methodology to design the adaptive profile estimation approach for target detection, field estimation, and edge detection.

3   Adaptive Profile Estimation

3.1   Sensing Methods

Sensing of UM nodes are intrinsically *localized* in space and *instantaneous* in time. While moving along its path, a node collects a series of measurements, each of which is made at a different location different time. Over time, the set of measurements characterizes a cross section of the field. A measurement is associated with its measured time and location, and a measurement of node u is stored as a three-tuple,

$$\mathbf{m}_i^u = [\mu_i^u, \tau_i^u, \rho_i^u] \ \forall \ i = 1, ..., n_u, \tag{1}$$

where μ_i^u is the measured value, τ_i^u is the measured time, ρ_i^u is the coordinates of the measured location, and n_u is the number of measurements u currently possesses. The set $\{\mathbf{m}_i^u\}$ is a sequential observation of the physical field along u's path. If a slow varying field remains stationary over a time period $[t_0, t_1]$, $\{\mathbf{m}_i^u \mid t_0 \leq \tau_i^u \leq t_1\}$ is effectively a cross section snapshot of the field.

With time, the number of measurements will grow to approach the storage capacity if all of them are stored. To extract and store only useful information from the measurements in a structured form, we define a *profile* as a set of grids associated with estimations at corresponding locations. The *grids* are defined with respect to a *chosen origin* to map locations in a geographic coordinate system onto cells of a *chosen resolution*. While one or more measurements at nearby locations may be mapped to the same grid, a single *estimation* is associated with each grid. A profile is then the collection of all grids with a valid estimation. Each node potentially maintains a profile covering a different area from others. A node acquires valid grid estimations based on either its own measurements or exchanged estimations of other nodes. Estimation rules are set based on an intuitive reasoning of the field energy model. If a measurement $\mathbf{m}_i$ with value μ_i can be modeled as,

$$\mu_i = e_i + n_i \tag{2}$$

where e_i is the field energy to be estimated and n_i is an i.i.d. additive white Gaussian noise energy perceived at the node, multiple measurements made at *the same location* by *different sensors* must consist of the same field energy but potentially different noise; therefore, the minimum measurement contains the least noise energy. Hence, the profile estimation rules are:

- The minimum of all measurements mapped to the same grid is the associated estimation of the grid;
- An estimation received from another node is added to the profile if the grid is previously unknown;
- An estimation received from another node overwrites the matching grid if it is smaller than the current estimation.
- An estimation is valid only for a limited duration specified by a node.

The valid duration is specified according to the temporal variations of a field. A longer duration is specified if fields exhibit slower variations. As each node may have chosen a different origin, resolution, and valid duration for its profile, conversion rules are defined for information exchange among nodes.

3.2   Information Exchange

Nodes exchange profiles to expand their knowledge of the field beyond their immediate vicinity as well as to improve their estimation accuracy. While such exchange can be sender- or receiver-initiated, we consider the latter where a node requests others' profile on demand. As the extent of profile exchange constitutes a trade-off between communication costs and actual needs, a node decides the need of exchange based on the *coverage* and *confidence* of its current profile. A profile's coverage is defined as the summed area of valid grids over the query-specified area of interest. The higher the coverage, the less necessary another profile exchange will be. A profile's confidence is defined as the reciprocal of its expected rate of adjustment, which is the average difference per updated grid in recent profile exchanges. The higher the confidence, the more possible the current profile accurately reflects the field of interest, and the less necessary another profile exchange will be. The information exchange rules are:

- When coverage is below threshold Θ_v, a node requests a profile from an available neighbor.
- When confidence is below threshold Θ_f, a node requests a profile from an available neighbor.

In both cases, a request indicates the requested area and resolution of interest. The requested node responds with a matching partial profile. If the requested node has a finer resolution than the requester, the profile is down-sampled before it is delivered. Down-sampling multiple grids into one larger grid is done by retaining the minimum of the multiple estimations for the larger grid. If the requested resolution is finer than the requested node's profile, or the requested area is entirely beyond the requested node's profile, no profile but an error indication is returned. If two nodes have had profiles based on different origins, the received profile is first aligned according to the node's preference, e.g., towards the nearest grid, and then aggregated into the current profile. In practice, a common origin selection guideline can be defined for consistent profile maintenance.

3.3   Decision Making

Based on a node's profile, the decision making components respond to the respective queries of point target detection, field estimation, and field edge detection.

Point Target Detection. The query of target detection is as follows. Given a specified target type and the *expected* target energy, each node independently determines if a target exists in an area of interest.

The query decision is made with a Bayesian energy detector algorithm. Given a measured value μ of a phenomenon H, the classical Bayesian detection theory formulates a binary hypothesis testing problem [8]. Detection of H is determined based on the likelihood ratio $\Delta(\mu)$ of the probability that H is present (positive hypothesis, H_1) to the probability that it is not present (negative hypothesis, H_0). Given a specified threshold ϵ, the likelihood ratio is

$$\Delta(\mu) = \frac{P(\mu|H_1)}{P(\mu|H_0)} \begin{cases} > \epsilon, \text{ then } H \text{ is present.} \\ < \epsilon, \text{ then } H \text{ is not present.} \\ = \epsilon, \text{ then } H \text{ is present with probability } p. \end{cases} \quad (3)$$

Note that the classical theory is formulated based on one single measurement (single-point) assuming the target and the sensor being in the same vicinity at the same time. The detector fails when a target is far from a sensor and the minute energy measured always results in claims of no detection. This assumption is not particularly useful for a UM node, as it often possesses measurements spanning a range much larger than a single vicinity. A proper inference for such measurements must incorporate the location of the measurements. Accordingly, we propose a *multi-point location-based* Bayesian detector. Given a set of measurements, the multi-point location-based likelihood ratio is defined as the product of the single-point likelihood ratios of all individual measurements. Given a target at location g and a measurement $\mathbf{m}_i$, the H_1 probability is dependent on the value μ_i, measured location ρ_i, and target location g. The H_0 probability does not depend on locations, since there is no target at all. The location-based single-point likelihood ratio is

$$\Delta(\mu_i, \rho_i, g) = \frac{P(\mu_i, \rho_i | H_1, g)}{P(\mu_i | H_0)}. \tag{4}$$

Given a set of measurements, $M = \{\mathbf{m}_i\}$, the multi-point likelihood ratio is

$$\Delta(M, g) = \prod_M \frac{P(\mu_i, \rho_i | H_1, g)}{P(\mu_i | H_0)} \begin{cases} > \epsilon, \text{ then } H \text{ is present.} \\ < \epsilon, \text{ then } H \text{ is not present.} \\ = \epsilon, \text{ then } H \text{ is present with probability } p. \end{cases} \tag{5}$$

For a UM node, M is the set of all valid estimations in its profile.

To determine the likelihood ratio without prior knowledge of g, a set of possible target locations is searched. Given the area of interest as specified in a query, the area is divided into G evenly spaced grids $(g_1, \ldots, g_G)$, among which the grid, $g_{\max}$, with the highest multi-point likelihood ratio is identified. A target is claimed to exist close to $g_{\max}$ if its likelihood ratio exceeds ϵ. The multi-point location-based Bayesian detector is then

$$D(M) = 1 \left(\max_{j \in G} \prod_{i \in M} \frac{P(\mu_i, \rho_i | H_1, g_j)}{P(\mu_i | H_0)} \geq \epsilon \right) \tag{6}$$

where $1(\cdot)$ is the binary indicator function. Note that when the ratio equals ϵ, a detection is declared. It is also interesting to note that ϵ controls the detection sensitivity; the smaller is the ϵ, the more sensitive is the detector.

The H_1, H_0 probabilities are estimated based on the target energy model and the random noise process. Given a single point energy source with a distance-squared decay, the signal energy e at a distance d from the source is modeled as $e = e_{\max}/d^2$, with $e_{\max}$ being the unit-distance target energy. The energy of an i.i.d. additive white Gaussian noise is an i.i.d. chi-squared random variable of degree one, whose distribution $P(n)$ is known and permits efficient lookup. Hence, for each measurement $\mathbf{m}_i$

$$P(\mu_i, \rho_i | H_1, g) = P(n_i = \mu_i - e_i) \tag{7}$$

where $e_i = e_{\max}/(|g - \rho_i|)^2$. Similarly, $e_i = 0$ when there is no target and

$$P(\mu_i | H_0) = P(n_i = \mu_i). \tag{8}$$

Note that an accurate estimation of the probabilities requires a fair, if not precise, assumption of the peak target energy $e_{\max}$ which thereby determines e_i. If e_i is much larger than the actual energy, $P(\mu_i, \rho_i | H_1, g)$ is much larger than it should be and may cause the detector to claim false detections. On the other hand, if e_i is much less than the actual energy, $P(\mu_i, \rho_i | H_1, g)$ is less distinct from $P(\mu_i|H_0)$ and may cause the detector to miss detecting a target. Moreover, for targets with larger dimensions, their energy model may not be adequately captured by a single point energy source. Nevertheless, as long as the assumed e_{max} is not far-off, the detector remains robust as observed in our simulations.

Target detection is assessed using two metrics. Given that there is a target in the monitored region, the fraction of instances a node claims a positive detection is defined as the *probability of detection*. Given that there is no target in the monitored region, the fraction of instances a node incorrectly claims a positive detection is defined as the *false alarm rate*. A good detection performance must have a high probability of detection and a low false alarm rate. For example, if a node detects a target 90% of the time when there is a target, but it also falsely reports detections 50% of the time when there is actually no target, the approach is not likely to be useful.

Field Estimation. The query of field estimation is as follows. Given a specified type of field, each node independently determines the energy profile in an area of interest.

The proposed sensing method readily produces a profile in response to the query. The profile maintained by each node accounts for the field in areas it traversed and areas traversed by nodes it has exchanged profiles with. Due to the uncoordinated nature of the nodes' motion, the resulting profile coverage is not necessarily comprehensive and the confidence of the estimations may not be satisfactory. If a node has traversed a limited area and had limited profile exchange opportunities, its coverage is to be limited. If a node has rapidly come across a region and made a limited number of measurements along its way, its measurements can be noisy and its confidence will be low. A natural enhancement of coverage and confidence is by way of increasing chances for nodes to aggregate their profiles. An aggressive enhancement is possible if publicly recognized data fusion centers are available to opportunistically collect and disseminate the profiles of a large set of nodes.

Prior to posing a field estimation query to a UM node, it is important to note the fact that UM nodes are only to traverse areas of their own interest. Such a query should not expect nodes to provide estimations of areas hardly visited by individuals. The profile resolution also depends on nodes' mobility properties. If nodes make measurements at fixed intervals, a slow moving node can estimate the field in a finer resolution than a fast moving node. If a fixed resolution is requested, a node's sampling rate must be proportionally adjusted according to its speed. Finally, while the proposed approach intrinsically inhibits noise, random noise remains in the estimations. Beyond the scope of this paper, additional noise filtering are readily applicable to the estimated profile [9].

Field Edge Detection. The query of edge detection is as follows. Given a specified type of field with an inhomogeneous distribution, each node independently determines the existence and geometry of boundaries between its inhomogeneous segments.

Given the profile of each node, different edge detection techniques can be readily applied. Three classes of solutions have been considered in literature.

In the one class, *thresholds* separating segments of high and low intensities are identified to classify the measured locations as being *inside* or *outside* a high energy segment [10]. The classification is then presented to human as a binary map and the edges are implicitly perceived. In the second class, *intensity gradients* are computed at selected locations in the field. Locations exhibiting large gradients are identified as *edge points*, which are then properly linked to reveal the entire estimated boundary. A profile is analogous to a digital image, for which numerous derivative operators have been widely adopted to effectively compute its grid gradients [9]. Similarly, grids of large gradients are identified and linked to produce complete boundaries. Note that the two classes of solutions both rely on the choice of a good threshold to classify either the intensities or the gradients. The third class of solutions utilize *contouring* to identify continuous line segments connecting locations of the same constant intensities.

In this paper, we demonstrate the use of contouring with simple pre-processing operations to reveal edges in a profile. Specifically, the profile is first linearly scaled to increase the dynamic range of intensities to within $[0, 256]$. A two-dimensional median filter is then applied to suppress its noisy components. In addition to noise effects, the profile can also be *sparse* with lots of coverage holes. While segments of entirely undiscovered areas can be excluded from processing, segments with sparse valid estimations always exist and must be properly handled. As we initialize all undiscovered grids with a zero value, the median filter effectively interpolates a profile among the sparse estimations.

4   Simulation Studies

The proposed profile based target detection, field estimation, and edge detection algorithms are evaluated with simulations in MATLAB. Presented in three parts, the simulation results contrast their performance measures in a UM network with respect to those of a stationary network and a set of controlled mobile nodes.

4.1   Network Model and Parameters

We consider a network of 50 UM sensor nodes in a 500 units by 500 units field. Each node can communicate with any other nodes within a radio range of 100 units. Nodes are initially placed at 10 random base locations following a uniform distribution and then start moving according to a *multi-homed* random way point mobility model. At each simulated time instance, a currently stationary node probabilistically determines to remain at the same location or start moving towards a randomly chosen destination at unit speed (1 unit distance per unit time). Once the destination is reached, a node probabilistically determines to either return to its base or move towards a new randomly chosen destination. The mobility model is adopted with the intention to emulate practical situations where people stay idle at a number of gathering places, such as offices or shops, and occasionally proceed to random places for their individual errands.

Be it mobile or stationary, a node periodically makes measurements at its present location and constructs its profile accordingly. A node also periodically discovers any neighbors within its communication range. The information exchange rules dictate neighboring nodes to exchange and update their profiles at periodic intervals whenever their coverage or confidence is below a specified threshold. In the simulations, we consider both thresholds to be high and nodes always exchange their profiles at specified exchange intervals. In response to each query, each node makes an independent decision based on its profile at the end of every decision interval. Table 1 summarizes all relevant parameters.

Table 1. Network configuration and algorithm parameters

• •••• ••••	• • •• •	• •••• ••••	• • •• •
Nodes	50	Warmup time	100
Time	2000	Sampling interval	1
Area	500 by 500	Exchange interval	100
Radio range	100	Decision interval	100
Speculated e_{max}	6	Node speed	1
Coverage threshold Θ_v	80%	Node memory	100 measurements
Confidence threshold Θ_f	1	Field radius r	30
Bayesian detector ϵ	$\{0, ..., 15\}$	Field energy δ_e	1, 0.1

The same algorithms are also applied to (i) a network of 50 stationary sensors, and (ii) 50 controlled mobile sensors. In the stationary scenario, 50 sensors are placed at uniformly random locations within the area of interest. The controlled mobile sensors are dispatched simultaneously along fixed paths to traverse disjoint rectangular sections of the area.

4.2   Target Detection

A target detection query specifies an area of interest and a speculated target energy level. The query states: "In the square area $[0 \leq x \leq 500, 0 \leq y \leq 500]$, detect any target with an approximate peak energy e_{max}." Recall that the detector is defined based on the assumption of a point energy source of e_{max}. Nevertheless, we are interested in its performance in general use where a target does not necessarily produce a point source energy profile. For example, substances spilled over an extended area result in a planar rather than point energy source. In our simulations, we consider a circular field of radius $r = 30$ located at the center of the area of interest. Within the field, each grid is modeled as a point source of energy δ_e contributing to a cumulative energy field. An additive white Gaussian noise of mean zero and variance one is considered. In practice, the detector may speculate e_{max} based on general knowledge about a specific target type of interest. In our simulations, we speculate $e_{max} = 6$ while the peak energy seen in the cumulative field is about 5.

Figure 1 shows the receiver operating characteristics (ROC) curves, i.e., the average detection probabilities vs. false alarm probabilities, obtained with UM nodes, stationary nodes, and controlled mobile nodes using different detector

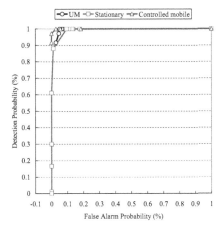

Fig. 1. Target detection ROC curves

thresholds, ϵ. On the one hand, slight gaps between the curves can be observed when false alarm probabilities are low (higher detector ϵ), where controlled mobile nodes out-performs UM nodes, and static nodes degrade the most. This is perceivable as the controlled mobile nodes traverse paths covering the entire area, UM nodes randomly traverse parts of the area, and static nodes are confined to observe only its local vicinity. On the other hand, the performance differences among the three are marginal. In this particular setting, the noise level has been significant relative to the field of interest. As we observe simulation runs with higher and higher source energy, the gaps become even smaller, suggesting the feasibility of using UM nodes as an alternative approach for target detection at low costs and little degradation in performance.

4.3   Field Estimation

A field estimation query specifies an area of interest and each node independently infers the energy profile of the area. With time, a node's coverage and confidence of a node are expected to have a steady growth. Figure 2(a) shows the average per-node coverage with time. As expected, controlled mobile nodes steadily explore the entire area, UM nodes expand their coverage at a smaller rate, and stationary nodes cover a fixed vicinity around themselves. To observe confidence with time, Figure 2(b) shows the per node per grid update adjustment rate, i.e., the reciprocal of confidence. In this case, stationary nodes show the least change and thereby the most confident estimations. Both UM and controlled mobile nodes perceive less changes as time goes by, while controlled nodes show more change as they steadily explore new areas not estimated before.

4.4   Edge Detection

An edge detection query specifies an area of interest and each node independently detects the presence and location of edges. With time, each node's profile expands

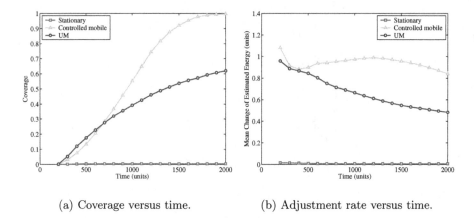

(a) Coverage versus time.   (b) Adjustment rate versus time.

Fig. 2. Average UM profile coverage and adjustment rate versus time

and edge detection is conducted periodically. At any instance, the present profile of a node is scaled (into the range of $[0, 256]$), filtered (with a 3 by 3 median filter), and its contours are identified. Figure 3 shows contour plots obtained by a UM node at time 500 with and without the scaling and filtering pre-processing operations. It is clearly observed that the original profile is sparse and creates lots of "islets" in its contour. The narrow dynamic range ($[0, 9]$) also limits the perceivable resolution of the plot. Scaled and filtered, the processed profile better distinguishes its contours and greatly suppressed most of the noise components.

In the case where a primary edge must be extracted, the profile is also readily applicable to intensity or gradient based edge detecting algorithms. The use of

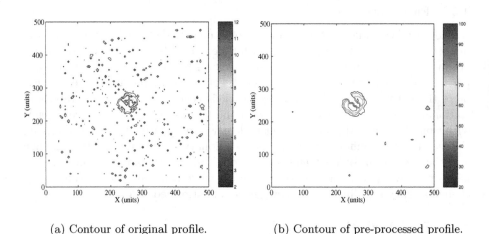

(a) Contour of original profile.   (b) Contour of pre-processed profile.

Fig. 3. Contour plots of a UM node

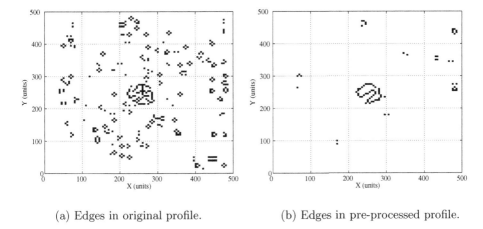

(a) Edges in original profile.   (b) Edges in pre-processed profile.

Fig. 4. Edges detected with a Sobel operator

a gradient based edge detector using a first order Sobel derivative operator is illustrated in Figure 4. Similarly, the pre-processing operations greatly enhance the edge detector's performance in excluding most false edges due to noise. It is worth noticing that, in these simulations, the simulated field has been sufficiently weak in strength relative to the additive noise. The algorithms have remained robust and correctly extracted the edges in the field.

Envisioning a comprehensive sensing infrastructure for the future, the story does not end here. Aside from UM nodes, there can always be stationary nodes deployed at strategic locations and scheduled mobile surveillance nodes. The simulations characterize the behavior of *fully uncoordinated nodes* that do not coordinate under any circumstances. This is useful for long-term monitoring *before a target is known to exist* or *before any preliminary knowledge of a field is available*. Once a target is detected or a field profile is estimated, nodes may benefit in coordinated movements for safety purposes or public policies, while controlled nodes may be dispatched for systematic exploitation of the area.

5   Conclusion

Sensing with UM nodes requires no planning overheads and is useful for precautionary surveillance in surroundings of people's daily activities. We presented a modular framework for designing collaborative UM sensing and proposed the adaptive profile estimation approach for target detection, boundary estimation, and edge detection queries. Due to the uncoordinated nature, the proposed solutions do not intend to out-perform other types of networks. Instead, the paper intends to demonstrate a low-cost sensing alternative useful for precautionary sensing. With simulations, we assessed the performance of the proposed solutions with UM networks as well as stationary networks and fully controlled mobile nodes to illustrate their relative strengths and performance tradeoffs.

References

1. Brooks, R., Ramanathan, P., Sayeed, A.: Distributed target classification and tracking in sensor networks. In: Proceedings of IEEE. Volume 91. (2003) 1163–1171
2. Biagioni, E., Bridges, K.: The application of remote sensor technology to assist the recovery of rare and endangered species. International Journal of High Performance Computing Applications Special Issue on Distributed Sensor Networks 16 (2002)
3. Kottapalli, V.A., Kiremidjian, A.S., Lynch, J.P., Carryer, E., Kenny, T.W., Law, K.H., Lei, Y.: Two-tiered wireless sensor network architecture for structural health monitoring. In: Proceedings of SPIE's 10th Annual International Symposium on Smart Structures and Materials. (2003)
4. Mainwaring, A., Polastre, J., Szewczyk, R., Culler, D., Anderson, J.: Wireless sensor network for habitat monitoring. In: Proceedings of ACM International Conference of Wireless Sensor Networks and Applications. (2002)
5. Santi, P., Blough, D.M.: The critical transmitting range for connectivity in sparse wireless ad hoc networks. IEEE Transactions on Mobile Computing 2 (2003) 25–39
6. Clouqueur, T., Phipatanasuphorn, V., Ramanathan, P., Saluja, K.: Sensor deployment strategy for target detection. In: Proceeding of The First ACM International Workshop on Wireless Sensor Networks and Applications. (2003)
7. Ogren, P., Fiorelli, E., Leonard, N.E.: Cooperative control of mobile sensor networks:adaptive gradient climbing in a distributed environment. IEEE Transactions on Automatic Control 49 (2004) 1292–1302
8. Scharf, L.: Statistical Signal Processing: Detection, Estimation, and Time Series Analysis. Addison-Wesley Publishing Company (1991)
9. Gonzalez, R., Woods, R.: Digital Image Processing. Prentice Hall (2002)
10. Chintalapudi, K., Govindan, R.: Localized edge detection in sensor fields. Ad-hoc Networks Journal (2003)

Multi-query Optimization for Sensor Networks*

Niki Trigoni[1], Yong Yao[2], Alan Demers[2], Johannes Gehrke[2],
and Rajmohan Rajaraman[3]

[1] Department of Computer Science and Inf. Systems, Birkbeck College,
University of London
[2] Department of Computer Science, Cornell University
[3] College of Computer and Information Science, Northeastern University

Abstract. The widespread dissemination of small-scale sensor nodes has sparked interest in a powerful new database abstraction for sensor networks: Clients "program" the sensors through queries in a high-level declarative language permitting the system to perform the low-level optimizations necessary for energy-efficient query processing. In this paper we consider multi-query optimization for aggregate queries on sensor networks. We develop a set of distributed algorithms for processing multiple queries that incur minimum communication while observing the computational limitations of the sensor nodes. Our algorithms support incremental changes to the set of active queries and allow for local repairs to routes in response to node failures. A thorough experimental analysis shows that our approach results in significant energy savings, compared to previous work.

1 Introduction

Wireless sensor networks consisting of small nodes with sensing, computation and communication capabilities will soon be ubiquitous. Such networks have resource constraints on communication, computation, and energy consumption. First, the bandwidth of wireless links connecting sensor nodes is usually limited, on the order of a few hundred Kbps, and the wireless network that connects the sensors provides only limited quality of service, with variable latency and dropped packets. Second, sensor nodes have limited computing power and memory sizes that restrict the types of data processing algorithms that can be deployed. Third, wireless sensors have limited supply of energy, and thus energy conservation is a major system design consideration. Recently, a database approach to programming sensor networks has gained interest [1, 2, 3, 4, 5, 6, 7], where the sensors are programmed through declarative queries in a variant of SQL. Since energy is a highly valuable resource and communication consumes most of the available power of a sensor network, recent research has focused on devising query processing strategies that reduce the amount of data propagated in the network.

Our Model and Assumptions. We are assuming a network of small-scale sensor nodes, similar to the Berkeley Motes connected through a multi-hop wireless network. The latest generation of Mica Motes, the MICAz mote, has a 2.4GHz 802.15.4 radio with

* This work was partially supported by NSF grants CCR-9983901 and IIS-0330201.

V. Prasanna et al. (Eds.): DCOSS 2005, LNCS 3560, pp. 307–321, 2005.

a data rate of 250 kbps, running TinyOS. We assume that nodes are stationary and battery-powered, and thus severely energy constrained. Users inject queries into a special type of node, referred to as a *gateway*. The sensor network is programmed through declarative queries posed in a variant of SQL or an event-based language [1, 2, 3, 4, 5]. We concentrate on aggregation queries, and the sensor network performs in-network aggregation while routing data from source sensors through intermediate nodes to the gateway.

Existing work has focussed on the execution of a single long-running aggregation query. In our new usage model, we allow *multiple* users to pose both long-running and snapshot queries (i.e. queries executed once). As new queries occur, they are not sent immediately to the network for evaluation, but are gathered at the gateway node into batches and are dispatched for evaluation once every *epoch*. The query optimizer groups together queries with the same aggregate operator and optimizes each group separately. Hence, in our presentation of our optimization techniques, we assume that queries use the same aggregate operator. Each epoch consists of a *query preparation* (QP) and a *result propagation* (RP) phase. In the QP phase, all queries gathered during the previous epoch are sent to the network together for evaluation. In the RP phase, query answers are forwarded back to the gateway.

Our model is general enough to include queries with different result frequencies and different lifespans. Although the algorithms proposed in this paper apply to this general usage model, for ease of presentation we will restrict our discussion to a simpler scenario: all queries asked in a QP phase have the same result frequency, and their computation spans a single (and entire) RP phase. The duration of an RP phase is a tunable application-specific parameter. Typically, an RP phase includes multiple *rounds* of query results. To summarize, an *epoch* has a QP and an RP phase, and an RP phase has many *rounds* in which query results are returned to the gateway.

The Intelligent National Park. To give an example of a scenario with multiple queries, consider a sensor network deployed in a national park. Visitors of the park are provided with mobile devices that allow them to access a variety of information about the surrounding habitat by issuing queries to the network through a special purpose *gateway*. For instance, visitors may wish to know counts of certain animal species in different regions of the park. The region boundaries will vary depending on the location of the visitors. The queries also change with time, as visitors move to different sections of the park, and certain queries are more popular than others. In addition, the sensor readings change probabilistically as animals move around the park, and there might be different update rates during the day than at night.

Our Contributions. This paper addresses the problem of processing multiple aggregate queries in large sensor networks, and makes the following contributions:

- **Multi-Query Optimization In Sensor Networks: Concepts and Complexity.** We formally introduce the concept of *result sharing* for efficient processing of multiple aggregate queries. We also address the problem of irregular sensor updates by developing *result encoding* techniques that send only a minimum amount of data that is sufficient to evaluate the updated queries. Our result sharing and encoding techniques achieve optimal communication cost for *sum* and related queries (such as *count* and *avg*). While some of our techniques also extend to *max* and *min* queries, we show that the problem of minimizing communication cost is NP-hard for these queries [8].

- **Distributed Deployment of Multi-Query Optimization.** We refine our multi-query optimization algorithms to account for computational and memory limitations of sensor nodes, and present fully distributed implementations of our algorithms. Besides a communication-optimal algorithm, we propose a near-optimal algorithm that significantly decreases the computational effort. We show how to tune our algorithms to take into account the node computational capabilities, and the relative energy expended for communication and for computation. In [8], we show how to adapt our algorithms to link failures that change the structure of the dissemination tree.

- **Implementation Results Validating our Techniques.** We present results from an empirical study of our multi-query optimization techniques with several synthetic data sets and realistic multi-query workloads. Our results clearly demonstrate the benefits of effective result sharing and result encoding. We also present a prototype implementation on real sensor nodes and demonstrate the time and memory requirements of running our code with different query workloads.

Relationship to Traditional Approaches for Multi-Query Optimization (MQO).
The problem considered in this paper is significantly different from the traditional MQO problems. The difficulty in devising efficient MQO algorithms for sensor networks is not only in finding common subexpressions, but in dealing with the challenges of distribution and resource constraints at the nodes. This paper is, to the best of our knowledge, the first piece of work to i) formulate this important problem, and ii) give efficient algorithms with provable performance guarantees that are shown to work well in practice.

2 Optimization Problems and Complexity

We now formally present the multi-query optimization, and study its complexity, focusing on algorithms that aim to minimize the communication cost of query evaluation ignoring any computation limitations or issues of distributed implementation. In Sect. 3 we will develop fully distributed algorithms that take into consideration the computation and memory constraints in sensor networks.

We consider a set of aggregate queries $Q = \{q_1, \ldots, q_m\}$ over a set of k distinct sensor data sources. A set of sensor readings is a vector $x = \langle x_1, \ldots, x_k \rangle \in \Re^k$. Each query q_i requests an aggregate value of some subset of the data sources at some desired frequency. This allows each query q_i to be expressed as a k-bit vector: element j of the vector is 1 if x_j contributes to the value of q_i, and 0 otherwise. The *value* of query q_i on sensor readings x is expressed as the dot product $q_i \cdot x$.

In our multi-query optimization problem, we are given a dissemination tree connecting the k sensor nodes and the gateway, over which the aggregations are executed. Note that our solutions apply to any given tree. The goal is to devise an execution plan for evaluating queries, that minimizes total communication cost. The communication cost includes the cost of query propagation in the QP phase and the cost of result propagation in each round of the RP phase. While we discuss the implementation of the QP phase in detail in Sect. 3.2, we ignore the query propagation cost in the following analysis, since it is negligible compared to the total result propagation costs, whenever the RP phase of an epoch consists of a sufficiently large number of rounds. We consider two classes of aggregation: (i) *min* queries and (ii) *sum* queries. Clearly our results for *min* queries also apply to *max*, and our results for *sum* queries can be extended to *count*, *average*, moments and linear combinations in the usual way. For *min* queries, we establish the NP-hardness of the multi-query optimization problem using a straightforrwad reduction from the Set Basis problem [8].

Complexity of *sum* Queries. For *sum* queries the underlying mathematical structure is a field. We can exploit this fact, using techniques from linear algebra to optimize the number of data values that must be communicated. Let N be an arbitrary node in the tree. Let $P(N)$ denote the parent of N and let T_N denote the subtree rooted at N. We denote as $x(N)$ the vector of sensor values in the subtree T_N and $Q_{x(N)}$ the set of query vectors projected only onto sensors in T_N.

We present a simple method to minimize the amount of data that N sends to $P(N)$ in each round. Let $B(Q_{x(N)}) = \{b_1, \ldots, b_n\}$ be a basis of the subspace of $\Re^k$ spanned by $Q_{x(N)}$. Then any query $q \in Q_{x(N)}$ can be expressed as a linear combination of the basis vectors $q = \sum_j \alpha_j \cdot b_j$, where $\alpha_j \in \Re$, $j = 1, \ldots, n$. By linearity of inner product we get, for sensors $x(N)$ (in the subtree T_N)

$$q \cdot x(N) = \left(\sum_j \alpha_j \cdot b_j\right) \cdot x(N) = \sum_j \alpha_j \cdot (b_j \cdot x(N))$$

That is, to evaluate the answers of queries in $Q_{x(N)}$ it suffices to know the answers for any basis of the query space spanned by $Q_{x(N)}$. Any maximal linearly independent subset of $Q_{x(N)}$ is a (not necessarily orthogonal or normal) basis of the space and every such basis has the same cardinality. So we can use any maximal linearly independent subset of $Q_{x(N)}$ as our basis, and N can forward the answers of the queries in this basis to $P(N)$. The parent $P(N)$, using the same set of basis vectors, can easily interpret the reduced results that it receives from N. We assume that N and $P(N)$ use the same algorithm in order to identify the basis vectors of $Q_{x(N)}$, and the factors α_j. We refer to this procedure as *linear reduction*.

Theorem 1. *The size of the query result message sent by the above algorithm in each round is optimal.*

3   Multi-query Optimization

The linear reduction technique outlined in Sect. 2 provides an elegant solution for minimizing the cost of processing multiple *sum* queries. However, a number of system considerations have to be taken into account to apply to a real sensor network. In this section, we develop fully distributed multi-query optimization algorithms for *sum* queries. Due to space constraints, we do not consider the impact of failures on our algorithms. We refer the reader to the full paper for a discussion on failures and detailed experiments that measure the tradeoff between communication and computation cost in the presence of failures [8]. We start our discussion by introducing the notion of *equivalence class*, which is central to the algorithms proposed in the remainder of the section.

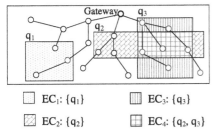

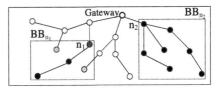

EC₁: {q₁}   EC₃: {q₃}

EC₂: {q₂}   EC₄: {q₂, q₃}

Fig. 2. The bounding box of a subtree is the minimum rectangle that covers all sensors in the subtree. Grey nodes represent sensors that belong to a bounding box of a subtree without belonging to the subtree

Fig. 1. Equivalence classes formed by three queries

3.1 Queries and Equivalence Classes

Rectangular Queries. We have represented each query as a k-bit vector, where k is the number of sensors. Expressing queries in this form requires that the user have complete knowledge of the sensor topology. It is more natural, and generally more compact, to represent queries spatially. We focus our attention on queries that aggregate sensor values within a rectangular region, and represent such a query as a pair of points $((x_0, y_0), (x_1, y_1))$ at opposite corners of the rectangle. Since queries do no longer enumerate nodes specifically, we can even evaluate queries in an acquisitional manner [9], e.g. by selecting a sample of sensor values generated within a query rectangle.

Equivalence Classes (ECs). To deal efficiently with rectangular queries and distribution, we introduce the notion of *Equivalence Class* (*EC*). An equivalence class is the union of all regions covered by the same set of queries. For example, Fig. 1 shows that queries $\{q_1, q_2, q_3\}$ form four ECs $\{EC_1, EC_2, EC_3, EC_4\}$, each one of which corresponds to a different set of queries. For instance, EC_4 is covered only by queries q_2 and q_3, and can be represented by the column bit-vector $[0, 1, 1]^T$; EC_1 is represented by $[1, 0, 0]^T$, EC_2 by $[0, 1, 0]^T$ and EC_3 by $[0, 0, 1]^T$. Notice that an equivalence class is not necessarily a connected region (see EC_2). An equivalence class may contain no sensors. Equivalence classes are identified based solely on spatial query information; they are independent of the node locations in the network or of the dissemination tree that connects nodes. We can, however, speak of the *value* of an equivalence class – this is the aggregate of the data values of sensors located in the EC region. The value of an EC can be obtained by a subset of sensors located in the EC region, if an acquisitional processing style is adopted.

Query Vectors and Query Values. We can now express queries in terms of ECs as follows. We number equivalence classes (instead of sensors) from 1 to ℓ, where ℓ is the number of equivalence classes. Let x denote the column vector in $\Re^\ell$ representing the values of the equivalence classes; thus, x_i denotes the sum of (all or a sample of) sensor values in EC_i. Each query q is a linear combination of the set of equivalence classes and can be captured by a row vector in $\{0, 1\}^\ell$. For example, query q_3 in Fig. 1 can be represented as the vector $[0, 0, 1, 1]$, since it only covers EC_3 and EC_4. The value of a query q given the EC values x is simply the product $q \cdot x$. Given the above representation of the queries and EC values, it is natural to represent a set of m queries as an $m \times \ell$ (bit) matrix Q, in which the (i, j) element is 1 if the ith query in Q covers the jth equivalence class. The value of the query set Q given the EC values x is again given by the product $Q \cdot x$, which is a column vector in $\Re^m$. We often refer to the rows of a query-EC matrix as *query vectors*, and to the columns as *EC vectors*.

Bounding Boxes (BBs). Expressing queries in terms of ECs brings out the dependencies among queries. In order to exploit these dependencies fully, each node needs to view queries in the context of its subtree, rather than the entire network. Therefore, in our algorithms, a node expresses queries in terms of ECs intersecting with its subtree; an EC intersects with a subtree if any of the sensor nodes in the subtree lies within the EC region. A node N can accurately determine which ECs intersect with subtree T_N, if it either knows the locations of all nodes in T_N or receives from its children a list of all ECs intersecting with their subtrees. Both approaches are prohibitive in terms of the communication involved. An approximation of the set of equivalence classes intersecting T_N can be obtained if we consider the minimum rectangle that contains all sensors in the subtree. This rectangle is hereafter called the *bounding box* of T_N and is denoted as BB_N. Figure 2 depicts the bounding boxes of the subtrees rooted at nodes n_1 and n_2. Note that a bounding box may contain nodes that do not belong in the subtree (grey nodes in Fig. 2).

Queries and ECs Projected to the Bounding Box of a Subtree. Let X_N denote the set of equivalence classes that intersect with the bounding box of the subtree T_N. It is easy to see that X_N is a superset of the equivalence classes that actually intersect with the subtree T_N. For given query set Q, we let Q_N denote the projection of Q on to X_N; that is, we obtain Q_N by setting all entries of Q that appear in columns not in X_N to be zero. Duplicate and zero rows are removed. We extend the notation to let x_N denote the vector of projected EC values onto the subtree T_N (*not* onto BB_N since a node N can only receive values generated by its descendants). The ith entry corresponds to the sum of sensor values lying in the intersection of EC_i and the subtree T_N. The entries of ECs that do not intersect with T_N are set to 0. If we denote the values of queries Q that are contributed from sensors in the subtree T_N as $V(Q, N)$, then the vector of values of Q_N contributed from subtree T_N is $V(Q_N, N) = Q_N \cdot x_N$.

3.2   The Query Preparation (QP) Phase

The query preparation phase consists of three steps: a *bounding-box calculation* step, a *query propagation* step, and an *EC evaluation step*. Some of our algorithms for the RP phase do not require the evaluation of ECs, in which case the last step is omitted.

Bounding-box calculation: A dissemination tree is first created using a simple flooding algorithm. Given the dissemination tree, each node N computes the bounding box BB_N of its subtree T_N from the bounding boxes of the subtrees of its children (if any) as follows. If x, (resp., x') and y (resp, y') are the smallest (resp., largest) x- and y-coordinates of the child bounding boxes, then (x, y) and (x', y') form two opposite corner points of the bounding box of N.

Query propagation: The next step is to send query information down the dissemination tree. We distinguish query propagation schemes based on whether bounding boxes are used to reduce the query propagation cost: (i) AllQueries: flood all queries to the entire network; (ii) BBQueries: each node propagates down only queries that have a non-empty intersection with its bounding box. This is performed using semantic routing information, discussed in detail in [9]. Once a node receives query information, it computes for each round in the epoch the set of queries that are active in the round.

EC computation: Given a set of m query rectangles, we can compute all the ECs formed by the m queries using a two-dimensional sweep algorithm in $O(m^3)$ time using $O(m^3)$ space. Due to space constraints, we defer the algorithm description and its analysis to the full paper. Using this algorithm, each node locally computes the ECs intersecting with its bounding box.

3.3   The Result Propagation (RP) Phase

Each RP phase consists of a number of *rounds*, in which aggregation results are forwarded through the tree paths from the leaves to the gateway. Consider a result message sent by a node N to its parent $P(N)$. The forwarded data should be sufficient to evaluate $V(Q_N, N)$, i.e. the *contribution* of sensors in T_N to the values of the projected queries Q_N. A result message   consists of a pair $\langle$RESULTCODE, RESULTDATA$\rangle$; RESULTDATA includes updated values, and RESULTCODE encodes what has been updated, showing how to interpret the values in RESULTDATA.

We now propose a series of result propagation algorithms, all of which use the above message format. These algorithms can be classified according to four dimensions. The first two dimensions are the methods employed for computing the RESULTCODE and RESULTDATA components. The third dimension is whether the linear reduction technique of Sect. 2 is applied. The last dimension is whether these choices are identical for all nodes, yielding a *pure* algorithm, or these choices may differ across nodes, yielding a *hybrid* algorithm.

Pure Algorithms Without Reduction. We consider two methods for determining the RESULTCODE component of a result message. In *Query-encoding*, a node sends to its parent information about *which queries have been updated* since the last round. Formally, let $UpdRows(Q_N)$ be the matrix derived from Q_N after removing all queries (row vectors) that are not affected by the current sensor updates in T_N. Both N and $P(N)$ agree on unique labels for the queries in Q_N from the integer interval $[1, |Q_N|]$. Then, RESULTCODE consists of a set of $\lg |Q_N|$-bit labels listing the queries in $UpdRows(Q_N)$. We note that Query-encoding does not require computation of equivalence classes. In *EC-encoding*, a node sends to its parent information about *which equivalence classes have been updated* since the last round. Let $UpdCols(Q_N)$ be the matrix derived from Q_N after removing all ECs (column vectors) that do not include any updated sensors in T_N (and after removing duplicate and zero rows). Since both N and $P(N)$ can compute X_N (i.e. the set of equivalence classes that intersect with BB_N) they can agree on a unique label in the range $[1, |X_N|]$ for each equivalence class in X_N. In EC-encoding, RESULTCODE includes the identifiers of ECs (columns) of $UpdCols(Q_N)$.

We also consider two methods for populating the RESULTDATA component of a result message that a node sends to its parent. In the *Query-data* approach, RESULTDATA is the set of values of updated queries. In the *EC-data* approach, RESULTDATA is the set of values of updated EC values.

One can combine the two dimensions above to obtain four different algorithms for the RP phase: QueryQuery, QueryEC, ECQuery and ECEC, respectively, where the first part of the name refers to the encoding, and the second part to the data. EC-encoding results in messages with smaller RESULTDATA components than Query-encoding, independent of whether the Query-data or EC-data policy is used. This is because both (row and column) dimensions of $UpdCols(Q_N)$ are smaller than those of $UpdRows(Q_N)$. Therefore, if the computational capabilities of the sensor nodes allow EC-computations, then we only consider ECQuery and ECEC. On the other hand, if the computational limitations of the sensor nodes do not allow them to compute the ECs, then QueryQuery is the only algorithm of interest. Consequently, we focus our attention on three of these four algorithms, namely, QueryQuery, ECQuery, and ECEC.

- **ECQuery:** In the RESULTCODE component, each node N sends to $P(N)$ the identifiers of the updated ECs in the subtree rooted at N. In the RESULTDATA component, node N includes delta values only of the distinct row vectors of matrix $UpdCols(Q_N)$. That is, query vectors are projected only onto the updated ECs (columns), and one value is sent for each distinct projected query vector.

- **QueryQuery:** In the RESULTCODE component of the message that N sends to $P(N)$, it includes the identifiers of updated queries. In RESULTDATA, node N includes delta values of the distinct row vectors of matrix $UpdRows(Q_N)$. Since the number of distinct query (row) vectors in $UpdRows(Q_N)$ is larger or equal to their number in $UpdCols(Q_N)$, the size of RESULTDATA in QueryQuery is larger or equal to its counterpart in ECQuery.

- **ECEC:** The RESULTCODE here is identical to that of ECQuery. Unlike ECQuery, ECEC sends up EC values in the RESULTDATA component of the message. For each updated EC in the subtree, it sends up the aggregate value of all sensors in the intersection of the EC and the subtree T_N.

An Optimal Pure Algorithm Using Linear Reduction. Both ECQuery and ECEC decrease the communication cost of result propagation by explicitly encoding irregular updates. Additional communication savings can be achieved by carefully applying the linear reduction technique (introduced in Sect. 2) in a distributed manner to reduce the size of propagated irregular updates. We now present the algorithm **ECReduced** which uses EC-encoding, and is provably optimal with respect to the amount of result data that is communicated. The RP phase of ECReduced at each node consists of two steps: i) a basis evaluation step and ii) a result evaluation step. Detailed pseudocode for both steps is presented in [8]. The basis evaluation step is executed whenever the set of active queries changes or the set of updated ECs changes. Thus, if every query has the same frequency and all sensors are updated regularly (D-scenario), then the basis evaluation step is executed only once at the beginning of the RP phase. This step is the most computationally demanding part of our algorithm since it involves matrix linear reduction; the complexity of reducing a matrix with m rows and n columns is $O(mn^2)$.

Basis evaluation step: Consider a node N with ch children nodes. Node N initially performs ch row-based linear reductions on matrices $UpdCols(Q_{N_k})$, $k = 1, \ldots, ch$, in order to interpret the results received from its children $N_1, \ldots, N_{ch}$. It derives a coefficient matrix A_{N_k} for each child k, such that the product of A_{N_k} and the basis vectors $B(UpdCols(Q_{N_k}))$ yields the original projected queries $UpdCols(Q_{N_k})$. Node N then reduces its own query-EC matrix $UpdCols(Q_N)$ into a set of linearly independent query vectors. Overall, N performs $ch + 1$ matrix reductions.

Result evaluation step: This step is executed once per round of the RP phase, and it includes simple operations wrt time and memory space. Relying on the output of the basis evaluation step, node N combines the incoming (delta) values received from its children and forwards a minimum number of values to its parent $P(N)$. Details of this step are given in [8]. In summary, for each child N_k, node N evaluates the values of queries (row vectors of) $UpdCols(Q_{N_k})$, based on the values of the basis vectors (received from child k) and matrix A_{N_k} (from previous step). Combining the values of $UpdCols(Q_{N_k})$ (and the node's own sensor value), node N proceeds to evaluate the results of queries $UpdCols(Q_N)$ for the entire subtree T_N. It is sufficient to evaluate only the values of queries that belong to the basis $B(UpdCols(Q_N))$. Only those values are finally forwarded to the parent node $P(N)$. Notice that if all nodes use the same algorithm to linearly reduce a query matrix, there is no need to communicate the selected basis vectors; a node only forwards up the values of these vectors to its parent. The following result is derived from Theorem 1.

Theorem 2. *The size of the* RESULTDATA *component in the ECReduced algorithm is optimal; it is a lower bound on the size of the optimal result message.*

Hybrid Algorithms with no Reduction: The algorithms introduced so far are executed in an identical manner at all nodes. We now consider two hybrid algorithms that perform differently across nodes, depending on the load of results contributed by the underlying subtrees. The first algorithm, referred to as HybridBasic, attempts to approximate the optimal cost achieved by the ECReduced algorithm, while avoiding the high computational requirements for linear reduction.

- **HybridBasic:** Consider the bounding box of a node and the set of queries and ECs intersecting with the bounding box. For a given sensor update rate, when the number of (projected) queries is small, the number of (projected) ECs is greater than the number of queries. In this case, the ECQuery algorithm is expected to outperform the ECEC algorithm. However, for a large number of queries the equivalence classes might be fewer than the queries. In this case, the ECEC algorithm is expected to outperform the ECQuery algorithm. The point where the two algorithms cross depends on the sensor update frequency. The HybridBasic algorithm combines the ECEC and ECQuery approaches. A node selects the approach that locally yields the least cost, and sends an additional bit to denote its choice. The only constraint is that if a child uses the ECQuery approach, it only provides information about the values of updated queries; hence, its parent can only implement the ECQuery approach. On the contrary, a parent of a node that implements ECEC can implement either of the two approaches.

Surprisingly, HybridBasic performs extremely well in terms of communication; as will be shown in Sect. 4, it closely approximates the cost of the ECReduced algorithm, without requiring a linear reduction task. In fact, HybridBasic can be viewed as an approximate application of linear reduction in the following sense: the rank of a matrix is always smaller or equal to the smallest dimension of the matrix; given a query-EC matrix, HybridBasic effectively chooses to propagate values of row vectors (queries), or of column vectors (ECs) depending on which ones are fewer. In practice, this policy works well, since the cardinality of the smallest matrix dimension often coincides with the matrix rank.

HybridBasic assumes that each node is able to evaluate equivalence classes within the bounding box of its subtree. The following algorithm, named HybridWithThreshold, lifts this requirement for nodes close to the gateway, whose bounding boxes overlap with many queries.

- **HybridWithThreshold:** If the input query workload is light, EC evaluation for the entire network is easy to perform locally at each node. Otherwise, nodes close to the leaves may opt for local EC computation, i.e. computation of ECs within the context of the bounding box of the node's subtree. As we approach the gateway, the bounding box of a node's subtree increases, and so do the number of query rectangles that intersect with the bounding box. The computational cost of evaluating ECs may become prohibitively expensive for nodes close to the gateway. The HybridWithThreshold algorithm behaves like the HybridBasic algorithm at nodes that are able to perform EC computation. When the effort for EC computation exceeds a certain threshold at a node (its computational capability), the node switches to Query-encoding and sends up one result per updated query.

4   Experimental Evaluation

In this section we measure the communication cost of the proposed algorithms using a home-grown simulator. We also present our feasibility test of the linear reduction technique, which we performed on the Mica2 mote. In the full version of our paper [8] we evaluate how the proposed algorithms trade communication for computation; we also show the benefits of our techniques by drawing data from a real sensor network infrastructure deployed in the Intel Berkeley Research Lab.

4.1   Synthetic Experimental Setup

We deploy 400 sensors in a square region of 400 m^2 and randomly select their x and y coordinates to be any real numbers in $[0, 20]$. We ensure that with a communication range of $2m$ the random deployment of nodes results in a (100%) connected network (otherwise the random deployment is repeated). A flooding algorithm is used to generate a minimum spanning tree that connects all nodes to the gateway. Each node selects as its parent a randomly chosen neighbor that lies on a shortest path to the root. The queries considered in our framework are *sum* queries that cover all sensors in a rectangular area. In our experiments we test a number of different query workloads, each defined as a set of tuples of the form (numberOfQueries, minQueryWidth, maxQuery-Width, minQueryHeight, maxQueryHeight). We assume that all the queries in a workload have the same frequency. We set the minimum values of the query dimensions (minQueryWidth, minQueryHeight) to $1m$ and the maximum values to $20m$. Given

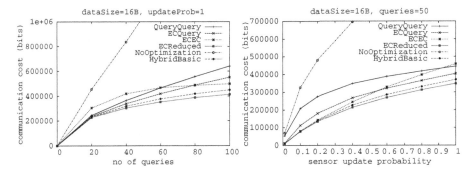

Fig. 3. Comm. cost of algorithms (D-scenario)   **Fig. 4.** Comm. cost of algorithms (I-scenario)

query input patterns, a random workload generator generates specific *instances* in each epoch that satisfy the patterns. The sensor update workload defines the probability that a sensor is updated at the end of a round. Given a sensor update input *pattern*, a random workload generator selects a specific set of sensors to be updated in a round.

For simplicity, we assume long-running queries that are propagated once at the beginning of an epoch (in the QP phase) and are evaluated at every round of the RP phase until the end of the epoch. Since the query propagation cost occurs once per epoch, it is negligible compared to the result propagation cost and is not accounted for. In our evaluation, we measure the result (communication) cost *per round*, averaged over 200 rounds (10 epochs of 20 rounds each).

4.2   Communication Cost

To make the measurements of communication cost realistic, we consider a packet size of 34 bytes (similar to the size of TOS_Msg used for Mica motes) that consists of a 5-byte header and a 29-byte payload. If the number of bits in a message is x, then the communication cost is $\lceil x/29 \rceil \times 34$; that is, we account for a fixed header cost (5 bytes) and only consider fixed size packets. The size of each query result is set to 16 bytes.

Deterministic Sensor Updates. In Fig. 3, we compare the performance of different algorithms as we increase the number of queries sent together for evaluation at the beginning of an epoch. In this initial experiment, we assume that all sensors are updated in each round with probability 1 (D-scenario). We first compare our techniques with the existing approach, namely an extension of the TAG algorithm [3] to process multiple queries. Since this algorithm, which we refer to as NoOptimization, performs in-network aggregation independently for each query, the average (per round) result propagation cost increases linearly in the number of queries. The performance advantage of our proposed techniques is apparent even for light query workloads.

Figure 3 validates our analysis of Sect. 3.3 that EC-encoding outperforms Query-encoding, if we restrict our attention to communication cost. Between ECQuery and ECEC, Fig. 3 shows that ECQuery outperforms ECEC for query workloads with less than 80 queries, but as we increase the number of queries, the number of ECs became smaller than the number of queries and ECEC wins.

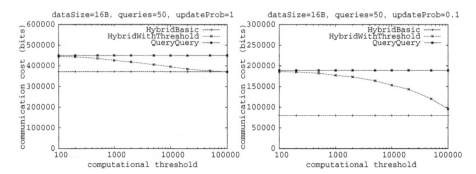

Fig. 5. HybridWithThreshold (D-scenario)   **Fig. 6.** HybridWithThreshold (I-scenario)

We now consider the cost and benefit of the reduction technique in the D-scenario. Figure 3 shows that the proposed ECReduced algorithm performs better than all the other algorithms, thus validating Theorem 2. An interesting observation is that the HybridBasic algorithm performs almost as well as the ECReduced algorithm, without requiring any computational cost for linear reduction (Fig. 3). This shows that a very simple distributed algorithm, which can easily be implemented on constrained sensor nodes, gives a very good approximation of the optimal solution.

In addition to our simulations, we implemented the linear reduction technique on the Berkeley Mica2 motes (4MHz ATMEL processor128kB flash, 4kB RAM, 4kB ROM) using the NesC programming language. We measured the time in seconds required for reducing an $m \times m$ matrix of floats as a function of m. The observed time grows as $\Theta(m^3)$, which is consistent with the complexity of the reduction algorithm. The code for matrix reduction was compiled with "make mica" to 12604 bytes in ROM and 428 bytes in RAM. For matrices of dimension from 5 to 15, the linear reduction algorithm takes 0.07 to 1 seconds, but the algorithm time increases rapidly for larger matrices.

Probabilistic Sensor Updates. In Fig. 4, we compare algorithms as we vary the probability that a sensor generates an updated reading in a given round. We set the number of queries to 50. Recall that in the D-scenario (Fig. 3), which corresponds to the I-scenario with probability 1, ECQuery is preferred to ECEC for workloads of less than 80 queries. As we decrease the sensor update probability to less than 0.6, however, Fig. 4 shows that it becomes beneficial for nodes to send up EC values instead of query values in RESULTDATA. For an update rate of 10%, ECEC is 30% cheaper than ECQuery. The ECReduced algorithm, which applies the linear reduction technique in a distributed manner, outperforms all other algorithms (Fig. 4). Moreover, the HybridBasic algorithm has a very good performance, approaching closely the cost of ECReduced.

In Figs. 5 and 6, we consider the limited computational power of sensor nodes. From the two algorithms that do not require EC computation (NoOptimization and QueryQuery), we only consider QueryQuery because it has smaller communication cost. Among the algorithms that do not perform reduction but require EC computation, we only consider HybridBasic because it has similar computational cost with the others yet smaller communication cost. We omit ECReduced because it requires matrix reduction without yielding noteworthy cost savings compared to HybridBasic. In

Figs. 5 and 6, we set the sensor update probability to 1 and 0.1 respectively. In both figures, QueryQuery has a higher cost than HybridBasic. The former algorithm does not require knowledge of ECs, whereas the latter assumes knowledge of ECs independent of the nodes' computational capabilities. We study the performance of the HybridWithThreshold algorithm, where the significance of the threshold value is as follows: if the effort of computing ECs at a node N (measured as m^3, where m is the number of distinct projected queries onto the local bounding box BB_N) exceeds the threshold value at N, then EC computation cannot be performed, and the node switches to using the QueryQuery algorithm. Figure 5 shows that as we increase the threshold value (plotted on a logarithmic scale), more nodes are able to compute ECs, and the cost of HybridWithThreshold approaches the cost of HybridBasic.

5   Related work

Query Processing in Sensor Networks. Several research groups have focused on in-network query processing as a means of reducing energy consumption. The TinyDB Project at Berkeley investigates query processing techniques for sensor networks including an implementation of the system on the Berkeley motes and aggregation queries [1, 2, 3, 4, 5]. An acquisitional approach to query processing is proposed in [9], in which the frequency and timing of data sampling is discussed. The sensor network project at USC/ISI group [10, 11] proposes an energy-efficient aggregation tree using data-centric reinforcement strategies (directed diffusion). A two-tier approach (TTDD) for data dissemination to multiple mobile sinks is discussed in [12]. An approximation algorithm for finding an aggregation tree that simultaneously applies to a large class of aggregation functions is proposed in [13]. Duplicate insensitive skethches for approximate aggregate queries are discussed in [14, 15]. Our study differs from previous work in that we consider multi-query optimization for sensor networks.

Communication Protocols for Sensor Networks. The data dissemination algorithms that we study in this paper are all aimed at minimizing energy consumption, a primary objective in communication protocols designed for sensor (and ad hoc) networks. A number of MAC and routing protocols have been proposed to reduce energy consumption in sensor networks [16, 17, 18, 19, 20, 21, 22, 23] While these studies consider MAC and routing protocols for arbitrary communication patterns, our study focuses on multi-query optimization to minimize the amount of data.

6   Conclusions and Future Work

Our work addresses several issues in the area of Sensor Databases. We have introduced two major extensions to the standard model of executing a single long-running query: A workload of multiple aggregate queries and a workload of sensor data updates. We have given efficient algorithms for multi-query optimization, and tested their performance in several scenarios. To the best of our knowledge this is the first work to *formally* examine the problem of multi-query optimization in sensor networks.

The main conclusions drawn in this paper are the following: First, the notion of equivalence class (EC) is important for distributed query evaluation: encoding sensor updates in terms of ECs enables better compression of the result messages. Second, the result data size is minimized for a certain class of aggregate queries (sum, count and avg) by applying the linear reduction technique in a distributed manner. Third, in applications where the computationally expensive task of linear reduction is infeasible for the nodes, a very good approximation of the optimal can be obtained by having each node select an appropriate local data encoding strategy. This encoding strategy can itself be defined in terms of a threshold that specifies the computational limitation.

There are a number of directions for further research. First, we would like to extend our ideas to a wider class of aggregation functions. Second, our paper has focused on accurate query evaluation. It would be worthwhile to study approximate query processing and obtain error-energy tradeoffs. We would also like to adapt our techniques to multi-path aggregation methods that provide more fault-tolerance.

References

1. Hellerstein, J., Hong, W., Madden, S., Stanek, K.: Beyond average: Towards sophisticated sensing with queries. In IPSN. (2003)
2. Madden, S., Franklin, M.: Fjording the stream: An architecture for queries over streaming sensor data. In ICDE. (2002)
3. Madden, S., Franklin, M., Hellerstein, J., Hong, W.: Tag: A tiny aggregation service for ad-hoc sensor networks. In OSDI. (2002)
4. Madden, S., Hellerstein, J.: Distributing queries over low-power wireless sensor networks. In SIGMOD. (2002)
5. Madden, S., Szewczyk, R., Franklin, M., Culler, D.: Supporting aggregate queries over ad-hoc sensor networks. In WMCSA. (2002)
6. Yao, Y., Gehrke, J.: The Cougar approach to in-network query processing in sensor networks. Sigmod Record **31** (2002)
7. Yao, Y., Gehrke, J.: Query processing in sensor networks. In CIDR. (2003)
8. Trigoni, N., Yao, Y., Demers, A., Gehrke, J., Rajaraman, R.: Multi-query optimization for sensor networks. In TR2005-1989, Cornell Univ. (2005)
9. Madden, S., Franklin, M., Hellerstein, J., Hong, W.: The design of an acquisitional query processor for sensor networks. In SIGMOD. (2003)
10. Heidemann, J., Silva, F., Intanagonwiwat, C., Govindan, R., Estrin, D., Ganesan, D.: Building efficient wireless sensor networks with low-level naming. In SOSP. (2001) 146–159
11. Heidemann, J., Silva, F., Yu, Y., Estrin, D., Haldar, P.: Diffusion filters as a flexible architecture for event notification in wireless sensor networks. In ISI-TR-556, USC/ISI. (2002)
12. Ye, F., Luo, H., Cheng, J., Lu, S., Zhang, L.: A two-tier data dissemination model for large-scale wireless sensor networks. In MOBICOM. (2002)
13. Goel, A., Estrin, D.: Simultaneous optimization for concave costs: Single sink aggregation or single source buy-at-bulk. In SODA. (2003)
14. Considine, J., Li, F., Kollios, G., Byers, J.: Approximate aggregation techniques for sensor databases. In ICDE. (2004)
15. Nath, S., Gibbons, P.: Synopsis diffusion for robust aggregation in sensor networks. In SENSYS. (2004)
16. Heinzelman, W., Wendi, R., Kulik, J., Balakrishnan, H.: Adaptive protocols for information dissemination in wireless sensor networks. In MOBICOM. (1999)

17. Johnson, D., Maltz, D.: Dynamic source routing in ad hoc wireless networks. In: Mobile Computing. Volume 353 of The Kluwer International Series in Engineering and Computer Science. (1996)
18. Perkins, C.: Ad hoc on demand distance vector (aodv) routing. (Internet Draft 1999, http://www.ietf.org/internet-drafts/draft-ietf-manet-aodv-04.txt)
19. Perkins, C., Bhagwat, P.: Highly dynamic destination-sequenced distance-vector routing (DSDV) for mobile computers. In SIGCOMM. (1994) 234–244
20. Xu, Y., Bien, S., Mori, Y., Heidemann, J., Estrin, D.: Topology control protocols to conserve energy inwireless ad hoc networks. In TR6, UCLA/CENS (2003)
21. Xu, Y., Heidemann, J., Estrin, D.: Geography-informed energy conservation for ad hoc routing. In MOBICOM. (2001) 70–84
22. Ye, W., Heidemann, J., Estrin, D.: An energy-efficient MAC protocol for wireless sensor networks. In INFOCOM. (2002) 1567–1576
23. Ye, W., Heidemann, J., Estrin, D.: Medium access control with coordinated, adaptive sleeping for wireless sensor networks. In ISI-TR-567, USC/ISI. (2003)

Distributed Energy-Efficient Hierarchical Clustering for Wireless Sensor Networks

Ping Ding, JoAnne Holliday, and Aslihan Celik

Santa Clara University
{pding, jholliday, acelik}@scu.edu

Abstract. Since nodes in a sensor network have limited energy, prolonging the network lifetime and improving scalability become important. In this paper, we propose a distributed weight-based energy-efficient hierarchical clustering protocol (DWEHC). Each node first locates its neighbors (in its enclosure region), then calculates its weight which is based on its residual energy and distance to its neighbors. The largest weight node in a neighborhood may become a clusterhead. Neighboring nodes will then join the clusterhead hierarchy. The clustering process terminates in O(1) iterations, and does not depend on network topology or size. Simulations show that DWEHC clusters have good performance characteristics.

1 Introduction

Sensor nodes are relatively inexpensive and low-power. They have less mobility and are more densely deployed than mobile ad-hoc networks. Since sensor nodes are always left unattended in sometimes hostile environments, it is difficult or impossible to re-charge them. Therefore, energy use is a key issue in designing sensor networks.

Energy consumption in a sensor network can be due to either *useful* or *wasteful* work. Useful energy consumption results from transmitting/receiving data, querying requests, and forwarding data. Wasteful energy consumption is due to collisions and resulting retransmissions, idle listening to the channel, and overhead of each packet header (even when the data packet is short). Energy consumption reduces network lifetime, which is defined as the time elapsed until the first node (or a certain percentage of nodes [1]) use up their energy.

To reduce energy consumption, clustering techniques have been suggested [3-20]. These techniques organize the nodes into clusters where some nodes work as clusterheads and collect the data from other nodes in the clusters. Then, the heads can consolidate the data and send it to the data center as a single packet, thus reducing the overhead from data packet headers. Clustering has advantages for: 1) reducing useful energy consumption by improving bandwidth utilization (i.e., reducing collisions caused by contention for the channel); 2) reducing wasteful energy consumption by reducing overhead.

In a clustered network, the communication is divided into intra and inter cluster communication. The intra-cluster communication is from the nodes inside a cluster to the head. The inter-cluster communication is from the heads to the data center (sink node). The energy efficiency of a clustered sensor network depends on the selection of the heads. Heinzelman et al. [3] propose a low-energy adaptive clustering hierarchy (LEACH), which generates clusters based on the size of the sensor network.

V. Prasanna et al. (Eds.): DCOSS 2005, LNCS 3560, pp. 322–339, 2005.

However, this approach needs a priori knowledge of the network topology. Younis and Fahmy [4] propose a Hybrid Energy-Efficient Distributed clustering (HEED), which creates distributed clusters without the size and density of the sensor network being known. However, the cluster topology fails to achieve minimum energy consumption in intra-cluster communication. Also, as we show in Section Simulation, the clusters generated by HEED are not well balanced.

In this paper, our goal is to achieve better cluster size balance and obtain clusters such that each has the minimum energy topology. We propose a distributed weight-based energy-efficient hierarchical clustering protocol (DWEHC). DWEHC makes no assumptions on the size and the density of the network. Every node implements DWEHC individually. DWEHC ends after seven iterations that are implemented in a distributed manner. Once DWEHC is over, the resulting clusters have a hierarchical structure. Each node in the network is either a clusterhead, or a child (first level, second level, etc). The number of levels depends on the cluster range and the minimum energy path to the head. Within a cluster, TDMA (Time Division Multiple Access) is used. Each node responds to their nearest parent's polling with their data, then that parent is polled by the next parent until the data gets to the clusterhead. For inter-cluster communciation, the heads contend for the channel using IEEE 802.11 to send data to the data center.

The paper is organized as follows: We review the literature in Section 2, and propose DWEHC in Section 3. In Section 4, we provide an analysis of correctness and energy-effiency, complexity and scalability of DWEHC. In Section 5, we present our performance studies. Section 6 concludes the paper.

2 Related Work

Many clustering algorithms have been proposed [3-20]. Most of these algorithms are heuristic in nature and their aim is to generate the minimum number of clusters. In the Linked Cluster Algorithm [6], a node becomes the clusterhead if it has the highest id among all nodes within two hops. In updated LCA [7], those nodes with the smallest id become cluster head. All the other nodes which are 1-hop to the heads become children of the heads. In [8], those nodes with highest degree among their 1-hop neighbors become cluster heads. In [10], the authors propose two load balancing heuristics for mobile ad hoc networks, where one is similar to LCA and the other is degree-based algorithm. The Weighted Clustering Algorithm (WCA) [11] elects clusterheads based on the number of surrounding nodes, transmission power, battery-life and mobility rate of the node. WCA also restricts the number of nodes in a cluster so that the performance of the MAC protocol is not degraded. The Distributed Clustering Algorithm (DCA) uses weights to elect clusterheads [12]. These weights are based on the application and the highest weight node among its one hop neighbors is elected as the clusterhead. All of the above algorithms generate 1-hop clusters, require synchronized clocks and have a complexity of $O(n)$, where n is the number of sensor nodes. This makes them suitable only for networks with a small number of nodes.

The Max-Min d-cluster algorithm [5] generates d-hop clusters with a complexity of $O(d)$, which achieves better performance than the LCA without clock synchronization. In [13], the authors aim at maximizing the lifetime of a sensor network by determining optimal cluster size and assignment of clusterheads. They assume that both the number and the location of the clusterheads are known, which is generally not possible in all scenarios. McDonald et al. [14] propose a distributed clustering algorithm for mobile ad hoc networks that ensures that the probability of mutual reachability between any two nodes in a cluster is bounded over time. In [15], the authors generate a 2-level hierarchical telecommunication network in which the nodes at each level are distributed according to two independent homogeneous Poisson point processes and a node is connected to the closest node lying on the another level. Baccelli et al. extend previous study to hierarchical telecommunication networks with more than two levels in [16]. They use point processes and stochastic geometry to determine the average cost of connecting nodes for assigning them to multiple levels in the networks.

Heinzelman et al.[3] propose a low-energy adaptive clustering hierarchy (LEACH) for microsensor networks. LEACH uses probability to elect clusterheads. The remaining nodes join the clusterhead that requires minimum communication energy, thus forming a 1-hop cluster. LEACH also calculates the optimal number of clusterheads that minimizes the energy used in the 1-level network. Bandyopadhyay et al.[17] propose a hierarchical clustering algorithm to minimize the energy used in the network. They generate a hierarchical structure, which is up to 5-levels in intra-cluster communication. They assume all nodes transmit at the same power levels and hence have the same radio ranges. Based on the size of the network, they calculate the optimal number of clusterheads in a network and the optimal number of hops from the nodes to the clusterheads.

All the previous protocols require either knowledge of the network density or homogeneity of node dispersion in the field. Younis and Fahmy.[4] propose Hybrid Energy Efficient Distributed clustering (HEED). HEED does not make any assumptions about the network, such as, density and size. Every node runs HEED individually. At the end of the process, each node either becomes a clusterhead or a child of a clusterhead. Residual energy of a node is the first parameter in the election of a clusterhead, and the proximity to its neighbors or node degree is the second. HEED generates a 1-level hierarchical clustering structure for intra-cluster communication. In DWEHC, we do not make any assumptions about the network similar to HEED. DWEHC creates a multi-level structure for intra-cluster communication, which uses the minimum energy topologies.

3   A Distributed, Weighted, Energy-Efficient Hierarchical Clustering Algorithm

3.1   Network Model

We assume that a sensor network can be composed of thousands of nodes. Also:

1) Nodes are dispersed in a 2-dimensional space and cannot be recharged after deployment.
2) Nodes are quasi-stationary.
3) Nodes transmit at the same fixed power levels, which is dependent on the transmission distance.
4) Nodes base decisions on local information.

5) Nodes are location-aware, which can be defined using GPS, signal strength or direction.

6) The energy consumption among nodes is not uniform.

We do not make any assumptions about:

1) the size and density of the network;
2) the distribution of the nodes;
3) the distribution of energy consumption among nodes;
4) the probability of a node becoming a clusterhead;
5) the synchronization of the network.

We believe this model and these assumptions are appropriate for many real networks. In a sensor network, sensor nodes collect their local information and send them to the data center. Frequently, the information is location-dependent, so the nodes know their own position via GPS or by other means. On the other hand, density is not uniform or known.

3.2 Clustering Structure

After running DWEHC, a clustered network has the features:

1) A node is either a clusterhead or a child in the cluster. The level of the node depends on the cluster range and the minimum energy path to the head.
2) Heads are well distributed over the sensor field.
3) Each cluster has a minimum energy topology.
4) A parent node has a limited number of children.

3.3 Related Concepts

We will introduce some concepts used in DWEHC. As shown by Li and Wan [2], given a set of wireless nodes, and a directed weighted transmission graph, there exists the minimum power topology. This topology is the smallest subgraph of the transmission graph and contains the shortest paths between all pairs of nodes. Li and Wan [2] propose a distributed protocol to contruct an enclosure graph which is an approximation of the minimum power topology for the *entire network*. In DWEHC, we generate a minimum power topology for each *cluster*, which is an enclosure graph, using their protocol. As shown in [2], enclosure graph is a planar graph and the average number of edges incident to a node is at most 6.

1. *Relay.* In this paper, we assume that all mobile devices have similar antenna heights as [2], so we will only concentrate on path loss that is distance-dependent. With this assumption, the power required by distance r is r^{α}, which is called the transmitter power of a node s. For example, in Figure 1, the node s tries to send a packet to d. Node s can send the packet directly with transmission power $\|sd\|^{\alpha} + c$, where α is equal to 2 or 4 and c is a constant. Or, s sends the packet through r with transmission   power $\|sr\|^{\alpha} + c + \|rd\|^{\alpha} + c$ which   is   called   relay.   If $\|sd\|^{\alpha} + c > \|sr\|^{\alpha} + c + \|rd\|^{\alpha} + c$,   then   relaying   through   node r consumes   less transmission energy than directly transmitting from s to d.

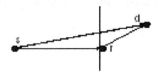

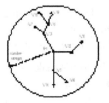

Fig. 1. Relaying through node r   **Fig. 2.** Levels in DWEHC

2. *Relay Region* [2]. Given sender node *s* and relay node *r*, the nodes in the relay region can be reached with the least energy by relaying through *r*. And, the region $R_{\alpha,c}(s,r)$, is called the relay region of *s* with respect to *r*.

$$R_{\alpha,c}(s,r) = \{x | \text{such that } \|sx\|^{\alpha} > \|sr\|^{\alpha} + \|rx\|^{\alpha} + c\} \tag{1}$$

3. *Enclosure Region* [2]. Enclosure region of *s* with respect to a node *r*, $E_{\alpha,c}(s,r)$, is the complement of $R_{\alpha,c}(s,r)$. The enclosure region $E_{\alpha,c}(s)$ of a node *s* is defined in Equation 2, where *T(s)* is the set of nodes lying inside the transmission range of node *s*.

$$E_{\alpha,c}(s) = \bigcap_{r \in T(s)} E_{\alpha,c}(s,r) \tag{2}$$

4. *Neighbors* [2]. Equation 3 defines the set of neighbors $N_{\alpha,c}(s)$ of a node *s*. These are the nodes which do not need relaying when *s* transmits to them,

$$N_{\alpha,c}(s) = \{u | \text{such that } u \in T(s), u \in E_{\alpha,c}(s)\} \tag{3}$$

5. *Cluster range (cluster radius)*, R. This parameter specifies the radius of a cluster, i.e., the farthest a node inside a cluster can be from the clusterhead. The cluster radius is a system parameter and is fixed for the entire network.

6. *Weight used in clusterhead election.* The weight will be the only locally calculated parameter used in clusterhead election, and is represented by *my_weight* in DWEHC. We define it in Equation 4, where *R* is the cluster range and *d* is the distance from node *s* to neighboring node *u*, $E_{residual}(s)$ is the residual energy in node *s*, and $E_{initial}(s)$ is the initial energy in node *s*, which is the same for all nodes.

$$W_{weight}(s) = \left(\sum_{u \in N_{\alpha,c}(s)} \frac{(R-d)}{6R} \right) \times \frac{E_{residual}(s)}{E_{initial}(s)} \tag{4}$$

Our primary goal is to improve energy efficiency and prolong the lifetime of the network. Since a clusterhead needs to forward all data from the nodes inside its cluster to the data center, the clusterhead will consume much more energy than its child nodes. Therefore, the residual energy is a key measurement in electing clusterheads. On the other hand, the more nodes inside a cluster, the more energy the clusterhead will consume. We build the clusters by considering the neighboring nodes, of which there are at most 6 [2]. Each parent node has a limited number of child nodes. Additionally, intra-cluster communication will consume less energy if the cluster is a minimum energy topology. Therefore, the distances from a node to all its neighbors inside its enclosure region becomes another key measurement in electing clusterheads.

Notations:

$N(s)$: the neighbors of node s;

$T(s)$: the nodes inside cluster range of node s;

Algorithm:

1. $N(s) = \phi$; Q= $T(s)$

2. while $(Q \neq \phi)\{$

   2.a. Let $v \in$ Q and be the nearest node to s

   2.b. $N(s) = N(s) \cup \{v\}$

   2.c..Eliminate all nodes x from Q, such that

$$\|sv\|^{\alpha} + \|vx\|^{\alpha} + c < \|sx\|^{\alpha}\}$$

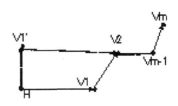

Fig. 3. Finding Neighbors   **Fig. 4.** HV1V2 is a minimum energy path

7. Levels in a cluster. Each cluster is multi-level. There is no optimal number of levels. This is because we make no assumptions about the size and topology of the network. The number of levels in a cluster depends on the cluster range and the minimum energy path to the clusterhead, which is represented by *my_level* in DWEHC. Figure 2 shows a multi-level cluster generated by DWEHC, where H is the clusterhead, first level children are v_0, v_1, and v_2, second level children are v_3, v_4, v_5, v_6 and v_7, and the third level children are v_8 and v_9. A parent node and its child nodes are neighbors. For example, H's neighbors are v_0, v_1, and v_2.

8.my_range and *my_dis.*In a cluster, each child node should be inside the cluster range of its clusterhead. *my_range* is the distance (in Figure 2) from v_3 to H and calculated $\sqrt{(x_H - x_{v_3})^2 + (y_H - y_{v_3})^2}$, where (x_H, y_H) and (x_{v_3}, y_{v_3}) are the (x, y) coordinates for H and v_3 respectively. *my_dis* denotes the minimum energy path to the clusterhead. In Figure 2, *my_dis* of v_3 is calculated by

$$(x_H - x_{v_0})^2 + (y_H - y_{v_0})^2 + (x_{v_0} - x_{v_3})^2 + (y_{v_0} - y_{v_3})^2 .$$

3.4 The DWEHC Algorithm

Before the communications may start, the clusters need to be generated. It takes $T_{generating}$ to generate the clusters. Then, two types of communication may occur in a clustered network: intra-cluster and inter-cluster. The time commited for these is $T_{cluster}$. $T_{cluster}$ should be much longer than $T_{generating}$ to guarantee good performance. To prevent a clusterhead from dying due to energy loss, the DWEHC algorithm, runs periodically, every $T_{cluster}+T_{generating}$. During $T_{generating}$, each node runs DWEHC to generate the clusters. During initialization, a node broadcasts its (x, y) coordinates and then uses the algorithm in Figure 3 to find its neighbors[2].The complexity of a node u finding its neighbors is $O(min(d_{Gt}(u)d_{Ge}(u), d_{Gt}(u)logd_{Gt}(u)))$, where $d_{Gt}(u)$ is the degree of node u in its cluster range and $d_{Ge}(u)$ is the degree of node u in its enclosure region [2]. The DWEHC algorithm is Figure 4.After running the algorithm, each node will have $N(s)$, the set of neighbors inside its enclosure region, and these neighbors will be its first level children if it becomes a clusterhead. All the other nodes inside its cluster range will be reached through at least one relay using one of these neighbors. All nodes set their *my_level* to -1 in the beginning, which indicates that they have not joined any cluster yet. Next, we explain how a node either becomes a clusterhead or a child in a cluster.

Notations:

my_id, my_x, my_y: the id, x and y coordinates of a node;

my_level, my_weight: the level and the weight of a node; my_range: the distance to the head;

my_dis: the minimum enegy distance to the head;

my_temp_head: 1 (if a node is a temporary head); 0 (o.w

my_head_num: the temporary head id in the neighborhood;

my_per: percentage of neighbors choosing the node as their temporary head;

temp_head_id: the id of a temporary head;

head_x, head_y: the x and y coordinate of a head

my_dir_parent: the id of my directly parent node;

MAX: the maximum number of neighbors (6) ;

Algorithm:

1. Initialization

1.1. broadcast x and y coordinates;

1.2. collect broadcasts inside cluster range;

1.3. find neighbors inside own enclosure region using Fig 3;

1.4. calculate my_weight and broadcast it;

2. Cluster Generation

FOR (i=0; i<MAX; i++) {

2.1. IF my_level = -1 {

2.1.a. IF my_weight is lagest among my neighbors then {
        my_temp_head = 1;
        my_head_num = my_id; }ELSE {
        my_temp_head = 0;}
    my_head_num = temp_head_id; } //use neighbsor with the lagest weight

2.1.b. broadcast my_head_num;

2.1.c. IF ((my_per$\geq \dfrac{MAX-i}{MAX}$ AND my_temp_head = 1) OR

(no neighbors) OR (all neighbòs my_level > -1)) {
    my_level = my_dis =0; head_x = my_x; head_y = my_y; //become a real cluster head
    broadcast my_level, my_dis,head_x, head_y;} //end 2.1.c

} //end 2.1

2.2. IF(my_level <>0) {

2.2.a. receives broadcast message from a neighbor;

2.2.b. my_range$=\sqrt{(head\_x- my\_x)^2 + (head\_y- my\_y)^2}$;

my_dis_new=neighbds my_dis+(distance to the neighbor^2

2.2.c. IF ((my_dis_new<my_dis AND my_level>0) OR (my_range<=Cluster Range AND my_level=-1)) {
    my_level = neighbòs my_level + 1; my_dis = my_dis_new; my_dir_parent = neighbserid;
    head_x = neighbòs head_x; head_y = neighbßs head_y;
    broadcast my_level, my_dis,head_x, head_y;}}//end 2.2

} //end 2

3. Finalization

Repeat cluster generation one more time.

Fig. 5. DWEHC

Clusterhead: A node that has the largest weight of all its neighbors will become a temporary clusterhead (*temp_head*=1). A node can become a real clusterhead only if a given percentage of its neighbors elect it as their temporary clusterhead. In the first iteration (i=0), this percentage is 100%. In subsequent iterations (i++), it is decreased to $(6-i)/6$. When a node becomes a real clusterhead, it sets *my_level=0*. Those nodes with *my_level* = -1 could still become clusterheads during the following iterations.

Those who become real clusterheads broadcast the information including their *my_level*, x (*head_x*) , and y (*head_y*) coordinates.

Non Clusterhead: A node will become a child node in the following three cases. The first case occurs when a node's *my_level* is equal to -1. The node receives a broadcast message from its neigbors, which includes *my_level* of the neighbor and the x and y coordinates of a clusterhead (*head_x* and *head_y*). If the distance from the clusterhead to the node is less than or equal to the cluster range, then the node chooses the clusterhead as its clusterhead and sets its *my_level* to *my_level* from the broadcast message plus one, and its *my_dis* to *my_dis* from the broadcast message plus its distance to the neighbor. The second case is when a node's *my_level* is not equal to -1. This indicates that it has already chosen its clusterhead. If the neighbor who sends the broadcast has a different clusterhead, and the distance from the node to the neighbor's clusterhead is inside the cluster range, and the new calculated *my_dis* is less than the previous *my_dis* from the node to its current clusterhead; then the node will choose the neighbor's clusterhead as its clusterhead and its *my_level* will be changed to current neighbor's level plus one. The third case occurs when a node's *my_level* is not equal to -1. The node receives a broadcast from its neighbor. If the neighbor has the same clusterhead as the node, and the *my_dis* from the node to this neighbor is less than the previous *my_dis*, then the node will choose the neighbor as its parent and reset its *my_level* and *my_dis* as in the second case.

Iteration: The cluster generating process runs at most seven times (including finalization) since each node has at most six neighbors[2]. After running DWEHC, a node either becomes a clusterhead or becomes a child in a cluster, and its level in the cluster is represented by *my_level*.

4   Analysis of DWEHC

4.1   Intra-cluster Communication

Intra-cluster communication is contentionless using TDMA. Each parent node polls its direct children and forwards the data to its parent node until the data reaches the clusterhead. The parent node may combine several data packets from its children together with its own data into one packet.

Correctness and Energy Efficiency: DWEHC is completely distributed on the whole network. Each node is either a clusterhead or a child node in a cluster. Each cluster contains the minimum-power topology, which is locally optimal. Lemmas 1-4 prove these statements.

Lemma 1: After running DWEHC, a node is either a clusterhead or a child in a cluster.

Proof. Assume a node *A* is neither a clusterhead nor a child after running DWEHC. Node *A* can only have two conditions before running DWEHC: 1) *A* does not have neighbors; 2) *A* has neighbors.

In the first case, since *A* does not have any neighbor, it will become a clusterhead in step 2.1.c of cluster generation, which contradicts the assumption. In the second case, since *A* is neither a clusterhead nor a child, this indicates that none of the *A*'s neighbors have become clusterheads. If one of them were a clusterhead, A would have to be a first level child. If A and some of its neighbors do not join any clusters, then, one of them will become a clusterhead, which will satisfy one of the conditions to become a clusterhead in 2.1.c of DWEHC. A will then become a child node. Therefore, all A's neighbors should join other clusters before finalization of DWEHC. A is the only one which does not join any clusters at most after DWEHC's six iterations of cluster generation. In the finalization of DWEHC, A will become a clusterhead, which contradicts the assumption.

Lemma 2: A node is covered by only one clusterhead.

Proof. A node A will set its *my_level* to only one value based on the distance to the clusterhead (inside the clusterhead's cluster range) and its *my_dis* variable. Therefore, A will only belong to one clusterhead.

Lemma 3: DWEHC distributes the clusterheads well, i.e., when two nodes are within each other's cluster range, the probability of both of them becoming clusterheads is very small.

Proof. Omitted because of space limitations.

Lemma 4: DWEHC generates each cluster with the minimum energy topology.

Proof. Let Figure 5 show a cluster generated by DWEHC. Let H be the clusterhead. V_1 is the child of H, V_2 is the child of V_1, and V_m is the child of V_{m-1}. Path HV_1 should be on the minimum energy path from H to V_1 since V_1 is a neighbor of H (using the definition of *neighbor* from Section 3.3 **Related** Concepts). Similarly, the path V_1V_2 should be on the minimum energy path from V_1 to V_2 (V_2 is a neighbor of V_1).

Let us assume that HV_1V_2 is not a minimum energy path, which indicates that there should exist another path $HV_1'V_2$, where V_1' is a neighbor of H, V_2 is a neighbor of V_1' and $\|HV_1'\|^\alpha + \|V_1'V_2\|^\alpha < \|HV_1\|^\alpha + \|V_1V_2\|^\alpha$. If such a node V_1' existed, then V_2 would have chosen V_1' to be its parent in step 2.2.c of DWEHC cluster generation. Therefore, V_1' does not exist. This contradicts the assumption, thus, HV_1V_2 is a minimum energy path. The same argument can be made to prove that $HV_1V_2...V_{m-1}V_m$ is the minimum energy path. Therefore, DWEHC provides the minimum energy topology inside each cluster.

Complexity: DWEHC generates clusters in at most seven iterations. Each node sends only $O(1)$ broadcast messages. Lemmas 5-7 prove these statements.

Lemma 5. The complexity of broadcast message exchange is $O(1)$ for each node.

Proof. In $T_{generating}$, during initialization, each node broadcasts two messages: 1) a broadcast message of its coordinates; 2) a broadcast message including its weight. During cluster generation, each node will announce information in two broadcast messages: 1) *my_head_num*; and 2) *my_level, my_dis, head_x,* and *head_y* variables after joining a cluster. A node which becomes a clusterhead only needs to send two messages. As for the non clusterhead nodes, the minimum number of broadcast messages exchanged is two and the maximum is twelve (changing its *my_level* in each iteration). During finalization, a node with *my_level*=-1 broadcasts the same variables as in cluster generation. Therefore, the minimum number of exchanged broadcast messages is four, and the maximum is fourteen. Therefore, the complexity of broadcast message exchange is $O(1)$ for each node.

Lemma 6. The complexity of a node becoming a clusterhead is $O(1)$.

Proof. We will show that the complexity of a node becoming a clusterhead is $O(1)$. Note that, only a temporary clusterhead will become a clusterhead. Suppose A is a temporary clusterhead and nodes from V_1 to V_5 are its neighbors as shown in Figure 6. We will discuss two cases: A has neighbors which choose it as their temporary clusterhead (we call this set neighbor1) and neighbors that do not choose it (we call this set neighbor2). If at least one node is in neighbor1 and remains there, then A becomes a clusterhead by the sixth iteration. Since some neighbors in neighbor2 may join other clusters, the percentage of remaining nodes choosing A increases and A will become a clusterhead even sooner.

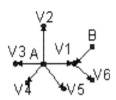

Fig. 6. Node A and its neighbors

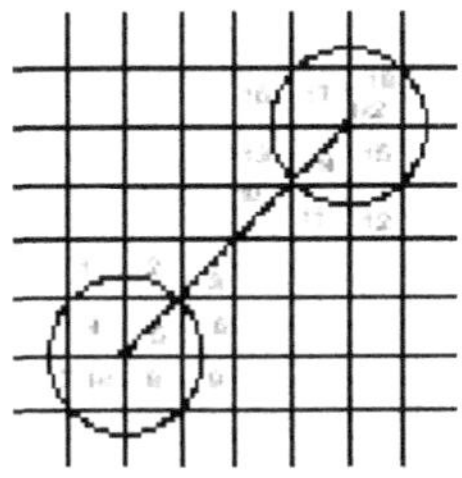

Fig. 7. The longest distance between two clusterheads

In the second case, A only has neighbors in neighbor2. In this case, all the nodes in neighbor2 may choose other nodes as their temporary clusterhead. For example, V_1 may choose B to be its temporary clusterhead. B could be a temporary head or not. If B is already a temporary head, then B could become a head for the same reason as A becoming head in the previous case. If B is not a temporary head and B does not join any clusters, there are two possibilities. 1) B will become a temporary head with neighbor V_1. If B becomes a head, then V_1 joins it or not. 2) B never becomes a temporary head with V_1 or not. Then there must exist another temporary head which will become a head and B joins that cluster. The same is valid for the other nodes. Therefore, if A does not join any clusters, then A itself will become a head by the time DWEHC is finalized.

Scalability: Each parent node has a limited number of child nodes. This is important in terms of scalability. DWEHC achieves good load balance per node, prolonging the life time of a clusterhead.

4.2 Inter-cluster Communication

Inter-cluster communication is contention based. The clusterheads poll their first level children, include their own data and transmit to the data center. The clusterheads consolidate several data packets into one data packet thus reducing overhead. Next, we prove that DWEHC provides end-to-end connectivity in the network.

To find out the conditions under which the clusters generated by DWEHC are *asymptotically almost surely* (a.a.s) connected, let us consider a sufficiently large network $R = [0,L]^2$, which is divided into square cells, with side $R_c/\sqrt{2}$, where R_c is the cluster range. This is because all the nodes within the same cluster should be reachable from each other and the highest two nodes should be at the end of the diagonal. We will show DWEHC ensures connectivity a.a.s when the clusterheads transmitting range, R_t, is greater than or equal to $4R_c$.

Lemma 7: Assume that n nodes, each with transmitting range R_c, are distributed uniformly, independently, and randomly in $R = [0,L]^2$ where R is a 2 dimensional plane, and assume that the area is divided into square cells of size $(R_c/\sqrt{2}) \times (R_c/\sqrt{2})$. If $R_c^2 N = kL^2(\ln L)$, for some $k>0$, then each cell contains at least one node a.a.s.

Proof. Our proof is similar to [1], so, we omit it.

Lemma 8: When Lemma 7 holds, two clusterheads, *H1* and *H2*, can communicate if $R_t \geq 4R_c$, where R_t is the minimum transmitting range.

Proof. Our proof is similar to that in [18]. In Figure 7, we show a boundary case, where H_1 and H_2 are heads. DWEHC generates clusters with the largest radius, R_c (R_c is the cluster range). With lemma 3 holding, there is no overlap between the clusters generated by DWEHC. With lemma 7 holding, there is at least one node in each square cell size $R_c/\sqrt{2}$. To cover cell 3, H_1 can be any position of cell 5. The farthest position for covering cell 3 is the H_1 position shown in Figure 7. To cover cell 10, H_2 can be in any position in cell 14, and the H_2 position shown is the farthest position. Thus, the distance between H_1 and H_2 is the farthest distance between any two clusterheads. The distance between clusterheads H_1 and H_2 is $4R_c$, which is the minimum transmission range for H_1 to reach H_2.

Lemma 9: DWEHC generates multi-hop clusters a.a.s.

Proof. Omitted because of space limitations.

5 Simulation

In this section, we will evaluate the performance of DWEHC via simulation. We ran simulations with 300 and 1000 sensor nodes, which are randomly dispersed into a field with dimensions 1000 by 1000 meters and 2000 by 2000 meters. Simulations with 300 nodes are run for cluster ranges of 75, 100, 150, 200, 250, 300 meters. Those for 1000 nodes are run for cluster ranges of 100, 150, 200, 250, 300, 350, 400 meters. In each simulation, we randomly initialize the nodes' residual energy and generate the topology. Each result represents the average of 20 simulation runs with the same parameters.

Wireless transmission laws dictate that power attenuation be proportional to the square of the covered distance (assuming fixed transmission power). If the distances are small (up to hundreds of meters), then the power attenuation can be assumed to be linear [19]. Other factors may also affect the received power, such as combined noise or physical obstacles. For simplicity, we ignore all these factors in our simulations[4], therefore, we consider the distance between two nodes as the only requirement to decide the transmission power.

In the simulations, we will compare DWEHC with HEED-AMRP[4]. HEED-AMRP (average minimum reachability power) considers the average minimum power levels required by the nodes within the cluster range of a node as the second parameter besides using residual energy as the first parameter to elect clusterheads. Both DWEHC and HEED-AMRP consider minimum power levels in the protocol design with no assumption about the size and density of the network. The simulation results will be discussed in two parts. In the first part, we compare DWEHC with HEED-AMRP with respect to cluster characteristics. In the second part, we will compare their energy usage and network lifetime.

5.1 Cluster Characteristics

In this section, we will compare DWEHC with HEED-AMRP in number of iterations to terminate, and the features of chusterheads, such as the number of clusterheads, the number of single node clusterheads, the maximum number of nodes in a cluster, and the distribution of nodes in clusters. The clustering code is written in C and all nodes run DWEHC synchronously. (Synchronization improves performance but is not necessary for correctness.)

a) Iterations to Terminate

In HEED-AMRP, we initialize both CH_{prob} and C_{prob} to be 0.05 (same as the simulations in [4]), where CH_{prob} is used to decide the probability to be a clusterhead, and C_{prob} is used to initialize the residual energy. In each iteration, CH_{prob} is multiplied by two. A node terminates when CH_{prob} reaches 1. So, HEED-AMRP takes six iterations to terminate. In DWEHC, we use weight as the parameter to elect clusterheads (see details about weight calculation in section 3.3). Theoretically, DWEHC will take at most seven iterations (see proof in lemma 4). In fact, all simulations took at most three iterations to terminate. So, we can see DWEHC will use less time, $T_{generating}$ in generating clusters .

b) Clusterhead Characteristics

1) With 300 nodes. In this section, all the simulations are done on a sensor network with 300 nodes. Table 1 shows the number of clusters for various cluster ranges. We see that HEED-AMRP generates significantly more clusters than DWEHC when the cluster range varies from 75 to 150. Since more clusters means more interference between the clusterheads and more overhead, HEED-AMRP will consume more energy than DWEHC. From 200 to 300, HEED-AMRP produces slightly more clusterheads. The performance also depends on the number of nodes inside a cluster, which we will show in Table 4 and 5.

Table 2 shows the number of single node clusters. Single node clusters are not desirable. When the cluster range varies from 75 to 100m, HEED-AMRP generates more than twice as many single node clusters as DWEHC. Starting at 150, DWEHC does not generate single node clusters, but HEED-AMRP has 4 and 1 when the cluster range is 150 and 200, respectively. Single node clusters disrupt the load balance. So, we expect the clusters generated by DWEHC to have better load balance than the clusters generated by HEED-AMRP.

Table 3 shows the maximum number of nodes in a cluster. The maximum number of nodes in a cluster in DWEHC is less than that of HEED-AMRP. This is because

Table 1. # of clusterheads (300)

Cluster range	75	100	150	200	250	300
DWEHC	57	30	19	16	12	8
HEED-AMRP	81	54	29	18	13	11

Table 2. # of single node cluster (300)

Cluster range	75	100	150	200	250	300
DWEHC	7	3	0	0	0	0
HEED-AMRP	14	8	4	1	0	0

Table 3. # of nodes in clusters(300)

Cluster range	75	100	150	200	250	300
DWEHC	8	12	23	28	39	53
HEED-AMRP	11	15	24	40	43	78

Table 4. # of clusters with size 75 (300)

Number of nodes	<=5	<=10	>10
DWEHC	42	15	0
HEED-AMRP	65	15	1

each parent node has a limited number of child nodes and a node could join a cluster only when at least one of its neighbors are inside the cluster. Therefore, DWEHC can achieve load balance. HEED-AMRP may cause many nodes to join a cluster which causes the clusterhead to deplete its energy quickly. For example, using DWEHC, the maximum number of nodes is 53 when the cluster range is 300. However, HEED-AMRP generates a 78-node cluster at this cluster range. Table 4 shows the number of clusters (distribution of nodes) with different cluster size (the cluster range is 75m). DWEHC has 42 clusters which have 5 or less, and 15 clusters which have 10 or fewer nodes. However, HEED-AMRP generates more clusters (65) which have 5 or less nodes. Note that, if child nodes send data to the clusterhead frequently, the intra-cluster communication will consist of many short data packets. Thus, HEED-AMRP will cause more overhead.

Table 5. # of clusters with size 300(300)

# of nodes	10	15	20	25	30	35	40	45	50	55	=78
DWEHC	2	2	0	1	1	0	0	1	0	1	0
HEED-AMRP	3	2	0	1	1	0	1	0	2	0	1

Table 6. # of clusters (1000)

Cluster range	100	150	200	250	300	350	400
DWEHC	125	71	60	58	50	45	22
HEED-AMRP	203	111	66	43	32	27	19

Table 5 shows the number of clusters with different cluster size (the cluster range is 300m). For example, when the number of nodes is less than or equal to 10 in a cluster, DWEHC has two such clusters and HEED-AMRP has 3 such clusters. HEED-AMRP also generates a cluster with 78 nodes but DWEHC generates one cluster with the maximum number of nodes to 53 (see Table 3). From the previous tables, we can see DWEHC realizes better load balance.

2) With 1000 nodes

In this section, all the simulations are done on a sensor network with 1000 nodes. Table 6 shows the number of clusterheads when cluster range is between 100 and 400m. DWEHC generates fewer clusters than HEED-AMRP when the cluster range varies from 100 to 200. This is because DWEHC realizes better clusterhead distribution than HEED-AMRP does and has fewer single node clusters. Starting at cluster range 250, HEED-AMRP generates fewer clusters than DWEHC. The reason is each parent node in DWEHC has a limited number of nodes (see lemma 7) in a

Table 7. # of single node clusters(1000)

Cluster range	100	150	200	250	300	350	400
DWEHC	10	2	0	0	0	0	0
HEED-AMRP	22	10	2	0	1	0	0

Table 8. # **MAX** of node in clusters (1000)

Cluster range	100	150	200	250	300	350	400
DWEHC	12	20	30	42	42	49	71
HEED-AMRP	15	30	48	85	97	97	100

Table 9. # of clusters with size 100(1000)

Number of nodes	<=5	<=10	<=15
DWEHC	74	46	5
HEED-AMRP	121	72	10

Table 10. # Of clusters with size 400(1000)

Number of nodes	<=40	<=60	<=80	<=100
DWEHC	9	9	4	0
HEED-AMRP	8	5	4	2

cluster which is determined by the cluster range and the my_dis (see details in section 3.3) variable of a node. In HEED-AMRP, all nodes inside a clusterhead's range can join the cluster without the limited number of nodes. Therefore, DWEHC has better load balance than HEED_AMRP. Table 7 shows the number of single node clusters when the cluster range varies from 100 to 400. HEED-AMRP generates many more single node clusters than DWEHC. Table 8 shows the maximum number of nodes in clusters. Similar to the 300-node network, HEED-AMRP always generates more nodes in a cluster, which will make the clusterhead fail sooner. Table 9 shows the number of clusters with different cluster sizes (the cluster range is 100). For example, DWEHC generates 74 clusters with the number of nodes less than or equal to 5, however, HEED-AMRP generates 121 such clusters, where 22 of those clusters are single node clusters (as shown in Table 7). Therefore, DWEHC achieves better load balance than HEED-AMRP. Table 10 shows the number of clusters with different cluster size (cluster range 400). Here, DWEHC achieves better load balance than HEED-AMRP. Since HEED-AMRP has two clusters which have 100 nodes, the clusterheads will consume more energy in transmissions than other clusters which have fewer nodes.

5.2 Simulation of Clustering Applications

The simulation results are implemented by using ns[20]. Table 11 shows the parameters used in simulations. We have implemented two sets of simulations, one with 300 nodes in one square kilometer, and the other with 1000 nodes in a four square kilometer area. The sink node (i.e., the data center) is placed in the middle.

The following parameters are the same as those in [4]. In the simple radio model that we use, energy expenditure is due to: 1) digital electronics, E_{elec}, (actuation, sensing, signal emission/reception), and 2) communication, E_{amp}. In our model, E_{amp} varies according to the distance d between a sender and a receiver where $E_{amp} = \varepsilon_{fs}$ assuming a free space model when $d < d_0$, while $E_{amp} = \varepsilon_{mp}$, assuming a multipath model when $d \geq d_0$, where d_0 is a constant distance (i.e., Threshold) that depends on the environment. To transmit n_b bits for a distance d, the radio expends $n_b(E_{elec} + E_{amp} \times d^n)$ J, where $n=2$ for $d < d_0$, and $n=4$ for $d \geq d_0$. To receive n_b bits at the receiver, the radio expends $n_b \times E_{elec}$ J. This energy model assumes a continuous function for energy consumption.

In DWEHC, a data packet size is 100 bytes. A parent non clusterhead node polls its children. The children respond to the polling with a data packet of 100 bytes. The parent node responds to the polling from its parent node with a data packet and the size can be up to 800 bytes. The clusterheads send data packets to the data center with the data packet size up to 800 bytes also. Clusterheads in both HEED-AMRP and DWEHC send 800 byte-long data packets. In our simulations, a node always responds to a polling. The clusters are regenerated every 300 TDMA frames ($T_{cluster}$=300) in the 300-node network. In the 1000-node network, the clusters are regenerated every 1000 TDMA frames ($T_{cluster}$=1000). In a real application, $T_{cluster}$ should be high enough to avoid re-clustering too often.

Next, we show the energy consumption in intra-cluster, inter-cluster communications in a $T_{cluster}$, and the number of rounds until the first node dies. A node is considered dead when 99% of its energy is used up.

Table 11. Simulation Parameters

Parameter	Simulation1	Simulation2
Number of Nodes	300	1000
Network grid	(0,0), (1000,1000)	(0,0), (2000,2000)
Sink (data center)	(500, 500)	(1000, 1000)
Threshold distance	75m	75m
Cluster range	75,100, ..., 300	100,150, ..., 400
E_{elec}	50 nJ/bit	50 nJ/bit
ε_{fs}	10pJ/bit/m^2	10pJ/bit/m^2
ε_{mp}	0.0013 pJ/bit/m^4	0.0013 pJ/bit/m^4
E_{fusion}	5nJ/bit/signal	5nJ/bit/signal
Data packet size	100 bytes	100 bytes
Broadcast packet	25 bytes	25 bytes
Packet header size	25 bytes	25 bytes
Round ($T_{cluster}$)	300 TDMA frames	1000 TDMA frames
Initial energy	10 J/battery	10 J/battery

a) Intra-cluster Communication. Figure 8 shows the average energy spent per node for intra-cluster communication in the network with 300 nodes in time $T_{cluster}$. HEED-AMRP consumes more energy than DWEHC, and especially as the cluster range is increased. Even at small cluster range, DWEHC saves energy. Since in HEED-AMRP the nodes communicate directly with the clusterhead, when cluster range is increased, so will the distance from the senders to the clusterheads. This will cause the senders to consume more energy. In DWEHC, senders relay messages through their parent, achieving optimal energy consumption within a cluster. For example, DWEHC consumes 82.2% and 65.7% of the energy consumed by HEED-AMRP when the cluster range is 75 and 100 respectively.

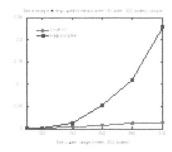
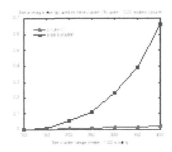

Fig. 8. Average energy used for intracluster(300)   **Fig. 9.** Average energy used for intra-cluster (1000)

Figure 9 shows the average energy spent per node in intra-cluster communication with 1000 nodes in time $T_{cluster}$. The results are similar to the 300-node network. HEED-AMRP consumes significantly more energy than DWEHC as the cluster range is increased. For example, DWEHC consumes *68.4%* and *39.1%* of the energy consumed by HEED-AMRP when the cluster range is 100 and 150 respectively.

b) Inter-cluster Communication. Figure 10 shows the energy consumed in inter-cluster communication in time $T_{cluster}$ (simulation1) when the total number of nodes is 300. The x_axis is the cluster range and y_axis is the energy consumed for inter-cluster communication in time $T_{cluster}$. HEED-AMRP consumes more energy than DWEHC does. This is because DWEHC achieves better distribution than HEED-AMRP. As shown in Table 1 and Table 2, HEED-AMRP generates more clusterheads especially single node clusterheads than DWEHC. More clusterheads will cause more contention between clusterheads during transmissions from clusterheads to data center, and more overhead because fewer data packets can be consolidated by the clusterheads. When the cluster range is 75, both because they both have the maximum number of clusterheads. With the increase of cluster range, the number of clusterheads decreases. The interferences caused by the contentions between clusterheads will decrease. But, the energy consumption is also related to the distances from the clusterheads to the data center. So, in DWEHC, energy consumed with cluster range of 200 is larger than the energy consumed with DWEHC and HEED-AMRP reach the highest in their energy consumption. This is cluster range of 150. The same condition occurs at HEED-AMRP. Figure 11 shows the energy

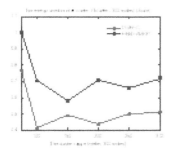
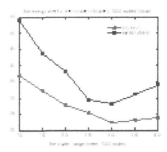

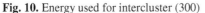

Fig. 10. Energy used for intercluster (300)   **Fig. 11.** Energy used for inter-cluster (1000)

consumed in inter-cluster communication in time $T_{cluster}$ (simulation2) when the total number of nodes is 1000. The results are similar to the previous one. DWEHC consumes less energy in inter-cluster communication. Both of them consume the maximum energy when cluster range is 100 since both of them generate the largest number of clusterheads at cluster range 100.

c) The number of rounds until the first node dies. Figure 12 shows the number of rounds of $T_{cluster}$ until the first node dies when the total number of nodes is 300 in simulation1. The x_axis is the cluster range and y_axis is number of rounds until the first node dies. DWEHC lasts much longer than HEED-AMRP, except it is close to HEED-AMRP when the cluster range is 75 and 300 where it is still *20.51%* and *28.25%* longer respectively. Although both DWEHC and

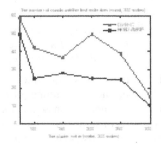   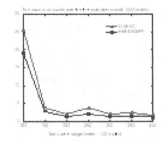

Fig. 12. # of rounds until the first node dies (300)   **Fig. 13.** # of rounds until the first node dies(1000)

HEED-AMRP consume the most energy at cluster range of 75 (see Figure 10), both last the longest at this cluster range. This is because they contain many single node clusters, which do not have much data to send. Figure 13 shows the number of rounds of $T_{cluster}$ until the first node dies when the total number of nodes is 1000 in simulation2. The results are similar to the 300 nodes network. DWEHC lasts much longer than HEED-AMRP does. They are closest when the cluster range is 100, where DWEHC lasts *23.4%* longer than HEED-AMRP.

6   Conclusions

In this paper, we proposed a distributed, weight-based Energy-Efficient Hierarchical Clustering algorithm (DWEHC). DWEHC operates with only realistic assumptions. The algorithm constructs multilevel clusters and the nodes in each cluster reach the clusterhead by relaying through other nodes. DWEHC is well-distributed, and runs in *O(1)* time which is a major advantage in a power-constrained sensor network. Our simulations demonstrated that DWEHC generates well balanced clusters. Both intra-cluster and inter-cluster energy consumption is greatly improved over clusters generated by the HEED-AMRP algorithm.

References

[1] D. M. Blough and P. Santi, "Investigating Upper Bounds on Network Lifetime Extension for Cell-Based Energy Conservation Techniques in Stationary Ad Hoc Networks", in Proceedings of the ACM/IEEE International COnference on Mobile Computing and Networking (MOBICOM), 2002.

[2] X.-Y. Li, and P.-J. Wan, "Constructing Minimum Energy Mobile Wireless Netwroks", in Proceedings of the ACM/IEEE International Conference on Mobile Computing and Networking (MOBICOM), Rome, Italy, July 2001.

[3] W. R. Heinzelman, A. Chandrakasan and H. Balakrishnan, "Energy-Efficient Communication Protocol for Wireless Microsensor Networkings", in Proceedings of IEEE HICSS, Jan 2000.

[4] S. Younis, S. Fahmy, "Distributed Clustering in Ad-hoc Sensor Networks: A Hybrid, Energy-Efficient Approach", in Proceedings of IEEE INFOCOM, March, Hong Kong, China, 2004

[5] A. D. Amis, R. Prakash, T. H. P. Vuong, and D. T. Huynh, "Max-Min D-cluster Formation in Wireless Ad Hoc Networks", in Proceedings of IEEE INFOCOM, March 2000.

[6] D. J. Baker and A. Ephremides, "The Architectural Organization of a Mobile Radio Netwrok via a Distributed Algorithm", IEEE Transactions on Communications, Vol. 29, No. 11, pp. 1694-1701, Nov. 1981.

[7] A. Ephremides, J. E. Wieselthier and D. J. Baker, "A Design concept for Reliable Mobile Radio Networks with Frequency Hopping Signaling", Proceedings of IEEE, vol. 75, No. 1, pp. 56-73, 1987.

[8] A. K. Parekh, "Selecting Routers in Ad-Hoc Wireless Networks", in Proceedings of ITS, 1994.

[9] C. R. Lin and M. Gerla, "Adaptive Clustering for Mobile Wireless Networks", Journal on Selected Areas in Communication, Vol. 15, pp. 1265-1275, Sep. 1997.

[10] A. D. Amis, and R. Prakash, "Load-Balancing Clusters in Wireless Ad Hoc Networks", in Proceedings of ASSET 2000, Richardson, Texas, Mar. 2000.

[11] M. Chatterjee, S. K. Das, and D. Turgut, "WCA: A Weighted Clustering Algorithm for Mobile Ad Hoc Networks", Cluster Computing, pp. 193-204, 2002

[12] S. Basagni, "Distrbuted Clustering for Ad Hoc Networks", in Proceedings of International Symposium on Parallel Architectures, Algorithms and Networks, pp. 310-315, June. 1999.

[13] C. F. Chiasserini, I. Chlamtac, P. Monti and A. Nucci, "Energy Efficient design of Wireless Ad Hoc Networks", in Proceedings of European Wireless, Feb. 2002.

[14] A. B. McDonald, and T. Znati, "A Mobility Based Framework for Adaptive Clustering in Wireless Ad-Hoc Networks", IEEE Journal on selected Areas in Communications, vol. 17, no. 8, pp. 1466-1487, Aug. 1999.

[15] S. G. Foss and S. A. Zuyev, "On a Voronoi Aggregative Process Related to a Bivariate Poisson Process", Advances in Applied Probability, vol. 28, no. 4, pp. 965-981, 1996.

[16] F. Baccelli and S. Zuyev, "Poisson Voronoi Spanning Trees with Applications to the optimization of Communication Networks", Operations Research, vol. 34, no. 1, pp. 619-631, 1999.

[17] S. Bandyopadhyay and E. Coyle, "An Energy-Efficient Hierarchical Clustering Algorithm for Wireless Sensor Networks", in Proceedings of IEEE INFOCOM, April. 2003.

[18] F. Ye, G. Zhong, S. Lu, and L. Zhang, "PEAS: A Robust Energy Conserving Protocol for Long-lived Sensor Netwroks", in International Conference on Distributed Computering Systems (ICDCS), 2003.

[19] W. C. Y. Lee, Mobile Cellular Telecommunications. McGraw Hill, 1995.

[20] The CMU Monarch Project. The CMU Monarch Project's Wireless and Mobility Extensions to NS.

A Distributed Greedy Algorithm for Connected Sensor Cover in Dense Sensor Networks

Amitabha Ghosh and Sajal K. Das

Center for Research in Wireless Mobility and Networking (CReWMaN),
Department of Computer Science and Engineering,
The University of Texas at Arlington, Arlington, TX 76019
{aghosh, das}@cse.uta.edu

Abstract. Achieving optimal battery usage and prolonged network lifetime are two of the most fundamental issues in wireless sensor networks. By exploiting node and data redundancy in dense networks, and by scheduling nodes efficiently, minimum battery drainage is possible. In this paper, we focus on the problem of • •• •• •• • • • • • •• ••• •••• • • •••• • • •••• (MCSC), an NP-hard problem, and describe a distributed greedy algorithm to generate sub-optimal connected sensor covers for homogeneous dense static sensor networks. Our greedy algorithm is based on the notions of maximal independent sets on random geometric graphs, and on the structure of Voronoi diagram. We provide complexity analysis and bounds on the cardinalities of maximal independent sets (MIS) for our problem scenario, and derive an analytical expression for the size of the sub-optimal minimum connected sensor cover. We verify the bounds on the MIS using simulation.

1 Introduction

Wireless sensor networks are distributed, self-organizing, pervasive systems that perform the tasks of sensing and collaborative data processing to provide useful information about some physical phenomenon, which is typically stochastic in nature. Sensor nodes are severely energy constrained due to limited battery power, and are not likely to be replenished during their lifetime. Therefore, it is very important to make use of their energy as optimally as possible and enhance the overall network lifetime. In dense sensor networks, the data sensed by geographically neighboring nodes exhibit a high degree of spatio-temporal correlation, and thus, many such data maybe redundant. Being able to exploit this redundancy is one of the ways to optimize energy usage and extend network lifetime.

Some of the existing works on conserving battery power have focussed on node scheduling algorithms based on the concept of sponsored sectors. Tian and Georganas [1] proposed a node self-scheduling scheme based on sponsorship criteria, by which each node decides whether to turn itself off or on using only local information. Gao, et. al [2] analyzed the problem of estimating redundant

V. Prasanna et al. (Eds.): DCOSS 2005, LNCS 3560, pp. 340–353, 2005.
© Springer-Verlag Berlin Heidelberg 2005

sensing area and described observations concerning minimum and maximum number of neighbors required for complete redundancy of a particular node.

In literature some methods have been proposed ([3], [4]) to determine the optimal number of nodes and their locations to provide complete coverage of a given sensing region, while maintaining connectivity. Zhang and Hou [5] proved that if the communication radius of a node is at least twice the sensing radius, then complete coverage of a convex region guarantees a connected network. Although, on one hand, the problem of determining the optimal number of nodes for complete coverage is important; deploying redundant nodes in the region on the other hand, contributes to network robustness and can overcome degradation in signal propagation or loss of nodes. However, when there are redundant nodes, keeping them active all the time will lead to faster drainage of energy; and thus, will reduce network lifetime. Now, since the data sensed by geographically neighboring nodes are spatio-temporarily correlated, it is often sufficient to turn on only an optimal number of nodes, which can provide the required data reliability, while putting others to sleep. However, the optimal number of nodes selected should be able to guarantee the same quality of data, which would have been provided if all the nodes were kept active. Since two of the factors which determine data quality are coverage and connectivity [6], the problem boils down to finding an optimal number of nodes that will provide the same quality of coverage and network connectivity.

This leads to the problem of finding the minimum connected sensor cover (MCSC), which is proved to be NP-hard for random deployment of nodes, as the less general problem of covering points using line segments is already known to be NP-hard [7]. Gupta, et. al [8] described an algorithm to construct a connected sensor cover for a network topology with fixed sensing radius, within an $O(logN)$ factor of the optimal, where N is the number of nodes in the network. Zhou, et. al [9] approached the MCSC problem, where each node can vary its sensing and transmission radii and provided a Voronoi diagram based localized algorithm and a greedy algorithm to construct sub-optimal network topologies within a factor of $O(logN)$ of the optimal size. Funke, et. al [10] proposed improved approximation algorithms for connected sensor covers and gave worst case approximation factors of 6π and 12 for grid placement and fine grid algorithms, respectively. They also described a greedy algorithm that provides complete coverage with an approximation factor of $\Omega(logN)$ from the optimal connected sensor cover.

In this paper, we propose a distributed greedy algorithm that constructs a sub-optimal MCSC using only local neighborhood information. Our basic idea is to greedily select a large set of nodes in the first phase, such that none of their sensing circles overlap with each other. Then we make those nodes select an optimal number of neighbor nodes using local information, such that the whole query region gets covered. We use the notions of maximal independent sets (MIS) on *random geometric graphs* and the structure of *Voronoi diagram* to find out sub-optimal MCSC. We also do complexity analysis of our algorithm and provide bounds on the cardinality of MIS for our problem scenario. These bounds are directly related to the size of the sup-optimal MCSC, as will be seen later.

The rest of the paper is organized as follows. In section 2, we formally introduce the problem of MCSC and discuss the notions of *Independent Set* (IS) and *Voronoi diagram*. In section 3, we describe a distributed greedy algorithm to find a sub-optimal MCSC and illustrate it on a grid and random deployment of nodes. In section 4, we show some preliminary simulation results and analyze the algorithm in terms of time complexity and provide bounds on the size of MIS. We conclude the paper in section 5.

2   Preliminaries

In this section, we give a formal description of the problem and introduce some of the basic concepts related to independent sets in graphs and the structure of Voronoi diagrams that we will use in our algorithm to find a sub-optimal connected cover set.

2.1   Problem Formulation

We consider homogeneous static (dense) sensor networks, where all the nodes have the same sensing radius, R_s and the same communication radius, R_c. We assume that the communication radius is α times the sensing radius, i.e., $R_c \geq \alpha R_s$, where $\alpha \geq 2$, and that the sensing range is a circular region of radius R_s with the node at the center. Throughout this paper, we also assume that a node is aware of its own location and its one-hop neighbors' locations. Such localization can be achieved by triangulation methods using received signal strengths or proximity measurements [11].

Let us define the *induced communication graph* on the network as the undirected graph $G_C = (V, E_{R_c})$, where the nodes act as vertices and an edge exists between any two nodes if the Euclidean distance[1] between them is less than the communication radius.

Definition 1. Connected Sensor Cover: *Let a set* $S = \{s_1, s_2, ..., s_N\}$ *of* N *nodes, be deployed in a sensing field of area* A *and let the sensing region covered by node* s_i *be denoted by* A_i. *Given a query* Q *over a region* A_Q *in the sensing field, where* $A_Q \subseteq A$, *a set* $\Gamma = \{s_{i_1}, s_{i_2}, ..., s_{i_m}\}$ *of* m *nodes is called a connected sensor cover if the following two conditions hold.*

1. $A_Q \subseteq A_{i_1} \cup A_{i_2} \cup ... \cup A_{i_m}$
2. *The induced communication graph* G_C *is connected, i.e., any pair of nodes in the connected sensor cover can communicate with each other, either directly or indirectly over a multi-hop communication path.*

The *Minimum Connected Sensor Cover* problem is to find the Γ with minimum number of nodes, such that the above two conditions hold. As mentioned

---

[1] We will denote $d(.)$ as the Euclidean distance function

earlier, MCSC is an NP-hard problem. In this paper, we propose a distributed greedy algorithm to find a sub-optimal MCSC, and derive an analytical expression for the cardinality of this sub-optimal set.

Under our particular problem scenario, where a communication link exists between a pair of nodes only if they are less than a certain distance (R_c) away, the structure of *Random Geometric Graphs* (RGG) provides the closest resemblance for modelling such networks. It is a more realistic model compared to the classical random graph models of Erdos and Renyi. We define the *induced sensing graph* over the set of nodes as a *random geometric graph*, $G_S = (V, E_{R_s})$, where the nodes act as vertices and an edge exists between two nodes s_i and s_j if the Euclidean distance between them is less than twice the sensing radius, i.e., $e(s_i, s_j) \in E_{R_s}$ if $d(s_i, s_j) < 2R_s$. The rationaile for defining the *induced sensing graph* in this way is to make sure that an edge exists between any pair of nodes, only when their sensing circles intersect with each other. We will see in latter sections how we select a large number of nodes, whose sensing circles do not overlap with each other, and hence, cover a maximum area in the sensing field. Next, we define the notion of *independent set* over the *induced sensing graph*.

Definition 2. Maximal and Maximum Independent Sets: *An independent set (IS) of a graph G is a subset of vertices, such that no two vertices in the subset has an edge in G. A maximal independent set (MIS) is an independent set that is not a proper subset of any other independent set, i.e., it is a largest set with respect to set containment. A maximum independent set is an independent set that has the largest possible number of vertices.*

Note that, a maximum independent set is always maximal, but the converse is not always true. In Fig. 1.(a), the set of vertices colored black forms a maximal independent set. In this particular case, this is also a maximum IS. However, the set of white vertices also forms a maximal independent set, but it it not the maximum. It is well known that finding the maximum IS for a general graph is NP-hard [12]. It is also NP-hard to approximate the size of the maximum IS. In the case of sensor networks, we are interested in finding an MIS on the induced sensing graph, which is basically a maximal set of nodes none of whose sensing circles overlap with each other (see Fig. 1.(b)).

2.2   Voronoi Diagrams

In 2-dimension, the Voronoi diagram [13] for a set of discrete points divides the plane into a set of convex polygons according to the nearest neighbor rule: all points inside a polygon are closest to only one point. In other words, if S is the set of points in the 2-D plane, then the convex polygon $V(p_i)$ for any point $p_i \in S$, is defined as the set of all points in the plane that are closest p_i than any other point p_j, for $i \neq j$. Mathematically,

$$V(p_i) = \left\{ x \in \Re^2 \mid d(x, p_i) \leq d(x, p_j) \right\}, \quad p_j \in S, \ i \neq j. \tag{1}$$

$V(p_i)$ is called the Voronoi polygon for p_i, and the edges that constitute the polygon are called its Voronoi edges. The Voronoi diagram VD(S) for the set of

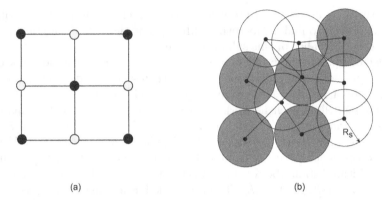

Fig. 1. (a) Set of black vertices forms a maximum IS; set of white vertices forms a MIS but not a maximum. (b) Induced sensing graph of 10 nodes. The set of 5 darkened nodes forms a MIS

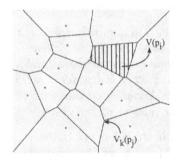

Fig. 2. Voronoi diagram for a set of randomply deployed points in 2-D. $V_k(p_j)$ denotes a Voronoi vertex

points S is the union of such polygons for all the points in S (see Fig. 2). A pair of points p_i and p_j are called Voronoi neighbors if their polygons share a common edge. The number of edges of $V(p_i)$ is equal to the number of neighbors of p_i, and each Voronoi edge is basically the perpendicular bisector between the two points which share the common edge. We will use these properties of Voronoi diagram in developing the proposed greedy algorithm for connected sensor cover.

3 Sub-optimal MCSC Algorithm

In this section, we describe the distributed greedy algorithm that generates a *sub-optimal MCSC* for homogeneous dense static sensor networks. Our algorithm runs in two phases, and always inherently maintains connectivity by selecting the best node at each step, only when it is connected to one or more already selected set of nodes.

Table 1. Notations and their meanings

$N_{s_i}(R_x)$	Set of nodes lying within radius R_x with s_i at the center
$N_{s_i}(R_x - R_y)$	Set of nodes lying in the annular region between the two circular regions of radii R_x and R_y with s_i at the center and $R_x > R_y$
Γ	Set that will contain the sub-optimal MCSC
$V(s_i)$	Voronoi polygon of s_i
$V_k(s_i)$	A Voronoi vertex of $V(s_i)$
$A_{holes}(s_i)$	Total area of the holes lying within $V(s_i)$
A_Γ	Area covered by the nodes in the sub-optimal set Γ

The basic idea behind the first phase is to select a maximum number of nodes, such that none of their sensing circles overlap with each other, and they form a connected network. This essentially boils down to finding a *connected MIS* on the induced sensing graph. In case of certain special graphs, it is possible to find out the maximum IS in polynomial time; however, in case of RGG, upon which we base our network model, finding the maximum IS is NP-hard. It is well known and has been experimentally evaluated in [14], that for *random graphs $G(N, p)$* [2], the standard randomized and greedy algorithms can generate an IS of size $log_{1/(1-p)}N$, with high probability for fixed p. However, in our case of induced sensing graph, where the sensing region is bounded and the degree of a node is directly proportional to its number of neighbors, we need to consider inter-node distances and maintain network connectivity while choosing the next node to be included in the *connected MIS*. Thus, the sensing radius, rather than the node density, plays a bigger role in determining the cardinality of the *connected MIS*, as derived analytically and verified by simulation later, for our RGG model. Here, we present a greedy algorithm that generates a large MIS on the induced sensing graph, and derive lower and upper bounds on the cardinality of the MIS in section 4.

In the second phase, the nodes that form the *connected MIS* in the first phase construct a localized Voronoi diagram. Each of the nodes then finds out coverage holes (regions in the sensing field that are not covered by any sensor) [15] within its Voronoi polygon and chooses the best nodes that can optimally cover those holes. Therefore, if each node follows the principle of optimally covering all the holes within its own polygon, then at the end of phase 2 the whole sensing field will get covered, and the selected set of nodes will form a sub-optimal MCSC. In the next subsections, we describe in detail the two phases of the algorithm and illustrate how they run on a grid and random deployment of nodes. We introduce some notations in Table 1 that are used in our discussion.

---

[2] N is the number of vertices and p is the probability that an edge exists between a pair of nodes.

3.1   Phase 1

The *connected MIS* finding algorithm starts with the node that has comparatively fewer neighbors, i.e., one of the perimeter nodes, because intuitively they
will have fewer neighbors than the ones which are towards the center of the field.
The successive steps in this phase differ from that of the standard greedy algorithms for random graphs, in the sense that we follow a distributed approach
that makes it practically difficult at every step to choose the minimum degree
node from the remaining eligible set of nodes. Also, we want the network to
be connected at every step, and because of which our eligible set of nodes is
constrained within one-hop neighbors of the last selected node. At every step,
the last selected node chooses a new eligible node which is closest to itself, and
includes it in Γ. A node s_j is called *eligible* to another node s_i, if it satisfies the
following three criteria:

1. s_j has not yet been included in the *connected MIS*, Γ
2. s_j is a one-hop neighbor of s_i, i.e., $s_j \in N_{s_i}(R_s)$
3. s_j's sensing circle does not overlap with any of the already selected node's
   sensing circles.

Every node also informs the chosen closest eligible node about the area covered so far. However, if there exists no eligible node for s_i, then it passes over the
responsibility to its farthest one-hop neighbor and requests it to choose the next
node within that node's one-hop neighborhood, such that none of the sensing circles overlap with each other. This process continues until none of the nodes has
any more eligible nodes left to choose from, at which point phase 1 terminates
with Γ containing a *connected MIS* that covers a total sensing area of $|\Gamma|\pi R_s^2$
(because none of the sensing circles overlap with each other). Formalizing the
afore-mentioned rules we state the *best eligibility criteria* as follows.

Definition 3. Best Eligibility Criteria*: If the the last selected node in Γ is s_i,
then the best eligibility criteria for it to choose the next node s_j are the following:*

1. $s_j \in N_{s_i}(R_c - 2R_s) \setminus \left(N_{s_i}(R_c - 2R_s) \cap \left(\bigcup_{k=0}^{i-1} N_{s_k}(2R_s) \right) \right)$
2. $d(s_i, s_j) = \min \{d(s_i, s_k), \forall s_k \in N_{s_i}(R_c - 2R_s)\}$

The first condition makes sure that s_j's sensing circle does not overlap with
the sensing circle of any other already selected node, while the second condition
chooses the nearest eligible node of s_i. Note that, there can be more than one
node satisfying the best eligibility criteria, in which case, one of them is chosen at
random to break the tie. The stepwise description of phase 1 is given in Algorithm
1. Next, we illustrate the algorithm running on a grid and on randomly deployed
set of nodes.

Grid Network: Let the nodes are deployed on the intersection points of a
grid as shown in Fig. 3.(a). Without loss of generality, we assume that $R_c = 2R_s$. Phase 1 begins by choosing the first node s_0 having minimum node degree
(though not unique in this particular case) that lies on the intersection of the

first row and first column. In the next step, s_0 finds that there are two nodes s_1 and s_3, which tie on the best eligibility criteria. Let s_0 choose s_1 at random to break the tie. Next, s_1 chooses s_2 because that is the only node that satisfies the best eligibility criteria, and in a similar way, s_2 chooses s_3. At this point, phase 1 ends because no more nodes can be chosen without their sensing circles getting overlapped. Note that, these four nodes construct a MIS on the grid, which in this case is also the maximum IS. For special graphs like this, it is always possible to find the maximum IS in polynomial time.

Random Deployment: We illustrate few steps of phase 1 of the algorithm in case of random deployment as shown in Fig. 3.(b). Let s_0 be the first node chosen. In the second step, since s_1 is the nearest one-hop neighbor of s_0, which falls in the annular region and belongs to the set $N_{s_0}(R_c - 2R_s)$, it is included in Γ. In the fourth step, while it is s_2's responsibility to choose the next best eligible node, it chooses s_3 that satisfies the best eligibility criteria. That is, s_3 is the closest one-hop neighbor of s_2 that falls outside the hashed region in the diagram. This hashed region is where the set of nodes $N_{s_2}(R_c - 2R_s) \cap \left(\bigcup_{k=0}^{2} N_{s_k}(2R_s) \right)$ fall.

Algorithm 1. Phase 1: Distributed greedy algorithm to find a *connected MIS*

1: **Initialization:**
2: $\Gamma \leftarrow \phi$;
3: Choose the first node s_0 and include it in Γ; $s_b \leftarrow s_0$;
4: **Steps at each s_b:**
5: $N_{s_b}(R_c - 2R_s) \leftarrow \phi$;
6: **for all** $s_k \in N_{s_b}(R_c)$ **do**
7:    **if** $2R_s \leq d(s_b, s_k) \leq R_c$ **then**
8:       $N_{s_b}(R_c - 2R_s) \leftarrow N_{s_b}(R_c - 2R_s) \bigcup s_k$;
9:    **end if**
10: **end for**
11: **if** $N_{s_b}(R_c - 2R_s) \neq \phi$ **then**
12:    Find that $s_k \in N_{s_b}(R_c - 2R_s) \setminus \left(N_{s_b}(R_c - 2R_s) \cap \left(\bigcup_{s_j \in \Gamma, s_j \neq s_b} N_{s_j}(2R_s) \right) \right)$,
       such that $d(s_b, s_k)$ is minimum;
13:    $\Gamma \leftarrow \Gamma \cup s_k$;
14: **else if** $N_{s_b}(R_c - 2R_s) == \phi$ **then**
15:    $s_k \leftarrow s_q$, such that $d(s_b, s_q) = \max \{d(s_b, s_i), \forall s_i \in N_{s_b}(R_c)\}$;
16: **end if**
17: s_k becomes the next s_b to execute the same steps $5 - 16$.

3.2   Phase 2

In this phase, the nodes that were chosen in phase 1 construct a Voronoi diagram using neighborhood information. By properties of Voronoi diagram, the number of edges of the Voronoi polygon for node s_i is equal to the number of its one-hop neighbors. Now, since the nodes selected in phase 1 do not have their sensing circles overlapped with each other, there will definitely exist holes in each of the

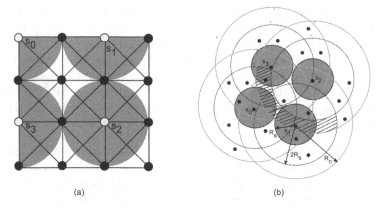

(a)   (b)

Fig. 3. (a) The induced sensing graph and the total coverage (shaded area) achieved by the four nodes (s_0, s_1, s_2, s_3), selected in phase 1. (b) Best Eligibility Criteria for a set of nodes deployed randomly

Voronoi polygons. In the first step of phase 2, one of these nodes $s_i \in \Gamma$ is selected at random, which then finds out holes existing within its polygon by the method described in [15]. It basically splits the convex Voronoi polygon into a set of disjoint and mutually exhaustive triangles, and using simple techniques of line and curve intersections finds out the area of coverage holes lying within its polygon. Next, the best one-hop neighbor is chosen, such that it covers maximum amount of hole within its polygon. The criteria for choosing the best node is as follows. Node s_i determines a set of optimal points $P_{s_i} = \{p_{s_i}^k, k = 1, ..., |N_{s_i}(R_c)|\}$, in the neighborhood of each of its Voronoi vertex $V_k(s_i)$. Each optimal point satisfies the following rules [15]:

1. Rule1: $p_{s_i}^k$ should lie on the angle bisector of the Voronoi vertex,
2. Rule2: $d(s_i, p_{s_i}^k) = \min \{2R_s, d(s_i, V(s_i))\}$.

These criteria ensure that if a node is placed at $p_{s_i}^k$, then the amount of coverage hole lying in the vicinity of $V_k(s_i)$ will get eliminated maximally. From these optimal locations, node s_i also finds out approximate estimates $Cov(p_{s_i}^k)$ of the amount of coverage holes that will get eliminated if nodes were placed at these optimal locations. This is illustrated in Fig. 4. Nodes A, B, C, D form Voronoi diagram, and let A be the first node to eliminate holes within its polygon. P, Q, R are the intersection points of the Voronoi edges with the lines that connect the nodes. From the properties of Voronoi diagram, $Area(\Delta AVP) = Area(\Delta BVP)$ and $Area(\Delta AVR) = Area(\Delta BVR)$, and node A knows that if it chooses the best node (s_b) close to vertex V, then that node will also eliminate almost an equal amount of holes in the polygons of B and C combined, as it will do within its own polygon. The point which corresponds to maximum such elimination of holes is chosen as the best location for the next best node.

Once the new best node s_b is selected, node s_i rechecks whether there still exist holes within its polygon. If so, it recalculates the area of the holes lying

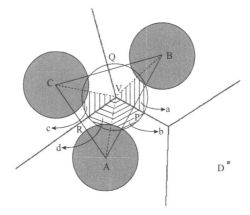

Fig. 4. Optimal position to choose a best node using Voronoi diagram. The hashed areas are equal: $a = b, c = d$

in the vicinity of the Voronoi vertices, except the one where s_b lies and repeats the same steps to choose another best node. This process continues until all the holes get eliminated from s_i's polygon, at which point, one of the neighbor nodes of s_i that was part of the *connected MIS* in phase 1 gets the chance to fill the holes within its polygon and so on. Thus, at the end of this phase, when each node chosen from the first phase have guaranteed the existence of no more holes within their polygons, the monitoring region gets completely covered with a connected set of nodes that form the sub-optimal MCSC. Phase 2 of the algorithm is described in Algorithm 2.

4   Analysis and Simulation

In this section, we present our analysis and preliminary simulation results of the distributed greedy algorithm to calculate a sub-optimal MCSC. We derive lower and upper bounds on the cardinality of the MIS generated in phase 1, using which we calculate an approximate size of the sub-optimal MCSC. We also provide a brief discussion on time complexity analysis for both the phases of the algorithm.

4.1   Analysis

Let the total number of nodes deployed in the square sensing field of area A, following a uniform random distribution be N, and let the cardinality of the *connected MIS* generated in phase 1 be ς. At any step during the first phase, when a node gets included in Γ, all nodes that lie within a distance of $2R_s$ from that node become ineligible to be included in Γ at a latter step. For instance, if the first node chosen falls atleast $2R_s$ distance away from any edge of the field, then the number of nodes that become ineligible is $4\rho\pi R_s^2 - 1$, where $\rho = N/A$ denotes uniform node density (here 1 is subtracted for the node which gets

Algorithm 2. Phase 2 of the sub-optimal MCSC algorithm

1: **Notations:**
2:     S_{MIS}: Set of •••••••• • •• that were selected in Phase 1;
3: **Initialization:**
4:     $S_{MIS} \leftarrow \Gamma$, $A_\Gamma \leftarrow \pi |S_{MIS}| R_s^2$;
5:     Nodes $\in S_{MIS}$ construct a localized Voronoi diagram;
6:     Randomly choose one of the nodes $s_i \in S_{MIS}$ as the starting node;
7: **Steps at each $s_i \in S_{MIS}$:**
8: **if** $A_\Gamma < A_Q$ **then**
9:     Calculate $A_{holes}(s_i)$;
10:    **while** $A_{holes}(s_i) \neq 0$ **do**
11:      Find optimal points $\{p_{s_i}^k\}$ that satisfy Rules 1, 2 near all unmarked $V_k(s_i)$;
12:      Calculate $Cov(p_{s_i}^k)$ for each optimal point;
13:      Choose the point for which $Cov(p_{s_i}^k)$ is maximum, call it $p_{s_i}^b$;
14:      s_i chooses the node (s_b) closest to $p_{s_i}^b$ and includes it in Γ;
15:      Update $A_{holes}(s_i)$, A_Γ and mark the vertex near $p_{s_i}^b$;
16:      s_i informs its neighbors of the amount of holes it eliminated, so that they can update their calculations;
17:    **end while**
18: **end if**

---

included). Note that, this is the maximum number of ineligible nodes at any step, because there could be some nodes which are ineligible to more than one node, and hence, should be counted only once. Similarly, the minimum number of ineligible nodes at any step is zero. Therefore, let the number of ineligible nodes at any step be $\beta(4\rho\pi R_s^2 - 1)$, where $0 \le \beta \le 1$.

Now, since the ζ nodes do not have their sensing circles overlapped with each other, the total number of nodes that lie within those ζ sensing circles is $\zeta\rho\pi R_s^2$. The remaining $(N - \zeta\rho\pi R_s^2)$ nodes must have become ineligible at some step while finding the *connected MIS*. Therefore, we have:

$$\zeta\beta(4\rho\pi R_s^2 - 1) = (N - \zeta\rho\pi R_s^2) \tag{2}$$

or

$$\zeta = \frac{N}{\rho\pi R_s^2(1 + 4\beta - \beta/\rho\pi R_s^2)} \tag{3}$$

Neglecting the term $\beta/\rho\pi R_s^2$ because it is very small, we get

$$\zeta = \frac{N}{\rho\pi R_s^2(1 + 4\beta)}. \tag{4}$$

Now, substituting $\beta = 0$ and $\beta = 1$ we get the upper and lower bounds, respectively, for the cardinality of the *connected MIS*, which gives us

$$\frac{N}{5\rho\pi R_s^2} \le \zeta \le \frac{N}{\rho\pi R_s^2}. \tag{5}$$

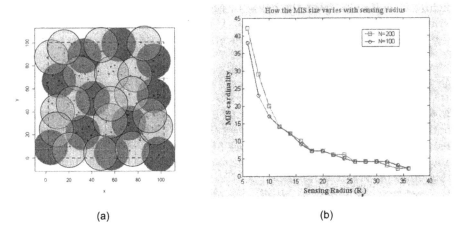

(a)   (b)

Fig. 5. Simulation: Dark circles represent sensing ranges of nodes belonging to *connected MIS*: (a) $N = 150$, $R = 15m$, $A_Q = 10,000m^2$. (b) Variation of *connected MIS* cardinality with sensing radius R_s

Using Eq. (4) and assuming that each of the ζ nodes on the average selects η nodes in phase 2 to optimally cover the holes within its Voronoi polygon, we can estimate the cardinality of the sub-optimal MCSC Γ as,

$$|\Gamma| = \zeta(1 + \eta) = \frac{N(1 + \eta)}{\rho \pi R_s^2 (1 + 4\beta)}. \tag{6}$$

The time complexity of the first phase of the algorithm is output sensitive, i.e., it depends on the cardinality of the *connected MIS* generated. Now, since the "for loop" in Algorithm 1, in the worst case, runs over all the one-hop neighbors of a node while choosing the best eligible node, the time complexity of the first phase is $O(\zeta N)$. For deriving the complexity of the second phase, note that the localized Voronoi diagram is constructed by the nodes in the *connected MIS*. This can be performed in $O(\zeta log \zeta)$ time. Next, each node checks for coverage holes in the vicinity of each of its Voronoi vertices, the time complexity of which is bounded by its number of one-hop neighbors. Hence, time complexity of the second phase is $O(\zeta log \zeta)$.

4.2   Simulation

We considered a sensing field of size 100 x 100 m^2 (the query region A_Q is also assumed to be of the same size) and performed simulation using Matlab and GNU R, a software package for statistical computing and graphics. We deployed the nodes uniformly randomly and varied their number as well as their sensing radii, and assumed that the communication radius is thrice the sensing radius. In Fig. 5.(a), we show a sample simulation run for $A_Q = 10,000m^2, N = 150, R = 15m$. The 11 darker nodes form the MIS in the first phase, which then construct Voronoi diagram to select another 18 nodes (the ones with ligher shade in the figure) to optimally cover the holes. In Fig. 5.(b), we show the variation of the *connected MIS* size with respect to sensing radius. Note that, the cardinality of the *connected MIS* satisfies the bounds given by Eq. (5).

5 Conclusions

In this paper, we described a distributed greedy algorithm for generating a sub-optimal minimum connected sensor cover (MCSC) for homogeneous dense sensor networks. Possibility to extend network lifetime by exploiting node redundancy and guaranteeing 100% coverage of the query region has been the motivation to this problem. We used the concepts of independent sets and Voronoi diagrams to construct a sub-optimal MCSC. We also provided upper and lower bounds on the cardinality of the MIS for random geometric graph model used to model the sensor network. The exact determination of the heuristic parameter, η, that we use to find the size of the sub-optimal MCSC is part of our future work. We also plan to do extensive simulations and derive tighter bounds on the size of the MIS. Instead of constructing the connected cover set greedily in a single phase, like in [8], where the sensing ranges of nodes could overlap right from the first step, we follow a different strategy of finding a large set (the *connected MIS*) of nodes in the first phase where none of the sensing circles overlap with each other. This gives us the maximum coverage that can be achieved by the set of nodes in phase 1, which is $\zeta \pi R_s^2$ (where, ζ is given by Eq. (4). In phase 2, since these set of non overlapping nodes optimally cover the holes within their Voronoi polygons, the sensing field can be covered with a number of nodes, that is close to optimal. The exact derivation of this sub-optimal bound is part of our future work.

Acknowledgment

This work is supported by NSF ITR grant under Award No. IIS-0326505.

References

1. Di Tian and Nicolas D. Georganas, "A coverage-preserving node scheduling scheme for large wireless sensor networks", •• •••••••••••• •• •• ••• ••• • • • •••••••••••••• • •••••••• •• • •••••• •••••• • ••• •••• ••• • ••••••••••• •• • ••• • •••, pp. 32–41, Atlanta, Georgia, USA, Sep 2002.
2. Yong Gao, Kui Wu, and Fulu Li, "Analysis on the redundancy of wireless sensor networks", •• • ••••••••• •• •• ••• ••• • • • •••••••••••••••••••••••• •• • ••••••• •••••• • ••• •••• ••• • ••••••••••• • •• ••• • •••, pp. 108–114, San Diego, California, USA, Sep 2003. ACM Press.
3. Chi-Fu Huang and Yu-Chee Tseng, "The coverage problem in a wireless sensor network", •• • ••••••••• •• •• ••• ••• • • • •••••••••••••••••••••• •• • ••••• •••••• • ••• •••• ••• • ••••••••••• • •• ••• •••, pp. 115–121, San Diego, CA, Sep 2003.

4. Xiaorui Wang, Guoliang Xing, Yuanfang Zhang, Chenyang Lu, Robert Pless, and Christopher Gill, "Integrated coverage and connectivity configuration in wireless sensor networks", •• • ••••••••••• •• •••• ••• •• •••••••••••••• •• •• •• ••••• • ••• •••••• •••••• •••••• • • ••••••••••, pp. 28–39, Los Angeles, CA, Nov 2003.

5. Honghai Zhang and Jennifer C Hou, "Maintaining sensing coverage and connectivity in large sensor networks", •• • ••••••••••• •• •• ••• •••••••• ••••••• • ••••••• •• • ••••••••••••• • •••••••• •• •• •••••••••• ••• • •••••••• ••• • •••••••• ••• • ••• ••••• •• •• ••••••••••••, Florida, USA, Feb 2004.

6. Amitabha Ghosh and Sajal K. Das, • •••••••• ••• • ••••••••••• ••••••• •• • ••••••• •••••• • •••• •••••• • ••••••••• • •••• ••• ••••••• • ••• ••••, John Wiley & Sons, 2005.

7. V. S. Anil Kumar, Sunil Arya, and H. Ramesh, "Hardness of set cover with intersection 1", •• • ••••••••• •• •• ••• •••• ••••••••••••• •••••••••• •• • •••• •••• •••••••••• ••• • ••••••• • ••• ••• ••• • ••••, pp. 624–635, Geneva, Switzerland, July 2000. Springer-Verlag.

8. Himanshu Gupta, Samir R. Das, and Quinyi Gu, "Connected sensor cover: self-organization of sensor networks for efficient query execution", •• • ••••••••• •• •••• ••• • • • ••••••••••••••••••• ••••••• •• • ••••• •• • •• •• ••• •••••• • • ••• ••••• •••••, pp. 189–200, Annapolis, Maryland, USA, Jun 2003. ACM Press.

9. Zongheng Zhou, Samir Das, and Himanshu Gupta, "Variable radii connected sensor cover in sensor networks", •• • ••••••••••• •• ••••• •• • • •••••••••••• •• ••• •• •• •••••• ••• • •••••••••• •••• • ••• •••• • •••• • ••• •••, pp. 387–396, Santa Clara, California USA, Oct 2004.

10. Stefac Funke, Alex Kesselman, Zvi Lotker, and Michael Segal, "Improved approximation algorithms for connected sensor cover", •• • ••••••• •• •• ••••• •••••••••••• • ••••••••• •• •• • • ••• • ••• ••••• • • •••••• • ••••••• •• • •••, pp. 56–69, Vancouver, British Columbia, Canada, July 2004.

11. Neal Patwari and III Alfred O. Hero, "Using proximity and quantized rss for sensor localization in wireless networks", •• • •••••••••• •• • ••• ••• • •• • ••••••••••••• • ••••••••• •• • •••••• •••••• • ••• •••• •• • •••••••••• •• • ••• •••, pp. 20–29, San Diego, CA, Sep 2003.

12. Michael R. Garey and Davis S. Johnson, • ••• •••••• ••• •••••••••••••••• • •••• •• ••• • •••••• ••• • •• •• ••••••••, W. H. Freeman and Company, 1991.

13. Franz Aurenhammer, "Voronoi diagrams a survey of a fundamental geometric data structure", • • • • •• •• •••• ••••••, vol. 23, pp. 345–405, 1991.

14. Mark K. Goldberg, D. Hollinger, and M. Magdon-Ismail, "Experimental evaluation of the greedy and random algorithms for finding independent sets in random graphs.", •• • •••••••••• •• •• ••• ••• ••••••••••• • •••••• •• • • ••••• ••• • ••• ••••• •••••• ••••••• ••• • •• •••, Santorini Island, Greece, May 2005 (to appear).

15. Amitabha Ghosh, "Estimating coverage holes and enhancing coverage in mixed sensor networks", •• • •••••••••• •• ••• •••• •••••••••• • •••••••• •• •••• • ••• ••••• • ••• ••••••• •• •••, pp. 68–76, Tampa, Florida, Nov 2004.

Infrastructure-Establishment from Scratch in Wireless Sensor Networks

Stefan Funke and Nikola Milosavljevic*

Computer Science Department,
Stanford University, Stanford, CA 94305, U.S.A
{sfunke, nikolam}@stanford.edu

Abstract. We present a distributed, localized and integrated approach for establishing both low-level (i.e. exploration of 1-hop neighbors, interference avoidance) and high-level (a subgraph of the unit-disk graph) infrastructure in wireless sensor networks. More concretely, our proposed scheme constructs a subgraph of the unit-disk graph which is connected, planar and has power stretch factor of 1 (the well-known Gabriel graph intersected with the unit disk-graph) and – most importantly – deals *explicitly* with the problem of interference between nearby stations. Due to our interleaved approach of constructing low- and high-level infrastructure simultaneously, this results in considerable improvements in running time when applied in dense wireless networks.

To substantiate the advantages of our approach, we introduce a novel distribution model inspired by actual sensing applications and analyze our new approach in that framework.

1 Introduction

Different from wired networks, wireless networks typically consist of a set of nodes that initially have no information about how to communicate with each other. Hence before running any application-specific algorithms and procedures on such a network, a basic communication infrastructure must be established. Low-level communication infrastructure essentially comprises learning about which other stations are within direct communication range and providing for interference-free communication between such neighboring stations. The high-level communication infrastructure is typically built by selecting a subset of the links identified during the low-level infrastructure establishment such that only those links will be used for communication within the network.

Assuming disk radii correspond to transmission ranges of individual stations, the information about which stations can receive messages from which other stations is captured in the so-called disk graph DG, or unit-disk graph UDG in case of uniform transmission ranges (the former a directed, the latter a bidirectional or undirected graph). Common high-level structures on top of the UDG/DG are for example the Gabriel graph

---

* Both authors were supported by the Max Planck Center for Visual Computing and Communication (MPC-VCC) funded by the Federal Ministry of Education and Research of the Federal Republic of Germany (FKZ 01IMC01).

V. Prasanna et al. (Eds.): DCOSS 2005, LNCS 3560, pp. 354–367, 2005.

$GG(UDG)$, the relative neighborhood graph $RNG(UDG)$, the Yao graph $Y(UDG)$, or the (localized) Delaunay graph $LD(UDG)$. Regarding interference-free communication there are basically two approaches. First, there is the possibility to resolve interference on the MAC-layer on a per-communication basis. Secondly, there is the concept of a so-called D2-coloring of the nodes of the disk graph. Here one color corresponds to one time slot/frequency and makes sure that no nodes at hop-distance less than or equal to 2 in the UDG transmit at the same time; hence no interference – not even at hidden terminals – can occur.

Typically low-and high-level infrastructure are constructed *sequentially*, in a sense that first the low-level infrastructure is built, and then the algorithms constructing the high-level infrastructure build upon the existing low-level infrastructure. As we will exhibit in Section 2, this inherently induces some inefficiencies, especially in scenarios where inter-station distance is small compared to their transmission range.

Related Work

There is a plethora of proposed algorithms for identifying a high-level infrastructure from a given unit-disk graph, such as [2], [9], [10], [14]. Common to all of these approaches is that they rely on a pre-existing low-level infrastructure, in particular they assume knowledge of all $1-$hop neighbors in the UDG and interference-free communication between adjacent stations in the UDG. [5] is an exception as it explicitly deals with the problem of interference and solves it by constructing a D2-coloring of all nodes in the UDG. Still, their algorithm assumes knowledge of all 1-hop neighbors in the UDG.

Most related to our work are the very recent papers by Kuhn et al. [6],[8], where the authors develop very interesting protocols for structuring a set of newly deployed wireless nodes without any assumption about a preexisting infrastructure. In [1] and [7] high-level infrastructures are examined with respect to their interference-inducing properties. Both papers introduce formal definitions of interference and propose new algorithms for which the resulting infrastructures exhibit low interference.

Our Results

In this paper we present an integrated approach for establishing both low- and high-level infrastructure for a set of deployed wireless stations. Our distributed protocol assumes no knowledge about 1-hop neighbors or interference-free communication between neighboring stations. It constructs the Gabriel graph $GG(UDG)^1$ consisting of all Gabriel edges of UDG, respective transmission ranges for each node of the network such that all nodes connected by an edge in $GG(UDG)$ can communicate with each other, as well as a valid coloring of the nodes such that interference-free communication between nodes in $GG(UDG)$ is ensured. Choosing $GG(UDG)$ as high-level structure guarantees that multi-hop communication between any two stations can be performed using the minimal amount of energy.

---

[1] In fact we chose the Gabriel graph as an example; other high-level structures should be constructible in a similar fashion.

Our scheme requires network nodes to have variable transmission power upper bounded by 1, to be able to detect if one or more neighboring stations, including itself, are sending a signal (the message is actually received only if exactly one station is transmitting), and to have access to a global clock and positioning device like GPS.

Furthermore, upon deployment each node is provided with a value ε, which can be set as some lower bound for the minimum inter-node distance (e.g. as derived from the physical size of the nodes), and a parameter $\widetilde{\Delta}$, which denotes the maximum number of stations nearby a node, where 'nearby' is defined for each node relative to the transmission radius assigned after the construction of the high-level infrastructure, and *not* relative to the maximum transmission radius.

The running time of our distributed protocol is $O(\widetilde{\Delta}^{3/2} \log \frac{1}{\varepsilon} \log^2 n)$ and all the above properties are fulfilled with high probability of $1 - \frac{1}{n^\phi}$ for any desired value ϕ.

To substantiate the usefulness of our protocol we introduce a novel distribution model for node deployment which is inspired by actual sensing applications. We show that in this model the value $\widetilde{\Delta}$ is a constant, hence making our algorithm extremely fast in such scenarios. We emphasize though, that by choosing $\widetilde{\Delta} = n$, our algorithm works for any node distribution.

In Section 2 we recap basic procedures of low- and high-level infrastructure establishment and give an intuitive derivation of the parameter $\widetilde{\Delta}$. Section 3 describes our algorithm in detail and analyzes its performance. In Section 4 we present our new distribution model and prove that in this model the parameter $\widetilde{\Delta}$ is a constant. Finally we conclude with some remarks and open problems.

2   Disk Intersection Graphs, Interference, and Power Efficiency

In this section we will introduce the basic concepts necessary for understanding our protocol, presented in the next section. Throughout we assume the standard communication model based on the *disk graphs* (*DG*), where the transmission ranges are represented by disks (of possibly different radii), centered at node locations. A disk graph corresponding to equal communication radii for all nodes is called a *unit-disk graph* (*UDG*). We remark, as this will be important later, that *DG*s are in general directed graphs, while *UDG*s are bidirectional (equivalently, undirected).

2.1   Interference

We assume that all station use the same frequency, so due to interference, not all communication suggested by the unit-disk graph is guaranteed to succeed. Suppose two stations transmit at the same time. If they are within each other's communication range, they both experience *direct interference*. Also, any station in the intersection of their communication ranges experiences *indirect interference*. The latter is also called *hidden terminal* problem.

The problem of interference can be dealt with on per-communication basis in the MAC-layer, typically using a handshake mechanism. Clearly, if n nodes are distributed densely enough so that the communication graph is complete, it might take $\Omega(n)$ time for a handshake to succeed. Thus in the following we focus on a different approach.

Resolution by D2-coloring. A natural way to avoid collisions or interference in case of uniform transmission ranges is to make sure that stations within 2-hop distance in UDG never transmit at the same time. Abstractly this corresponds to a coloring of the vertices of UDG such that two vertices with graph distance less or equal to 2 always have different colors.

Computing a D2-coloring with the minimum number of colors is NP-hard, but several approximation algorithms are known, both centralized and distributed. See [11], [12], [13]. The distributed algorithm in [5] computes with high probability (w.h.p.) a valid D2-coloring of a given UDG, in time $O(\delta_2 \log^2 n)$, where δ_2 is an upper bound on the size of any 2-hop neighborhood [2]. A variant of this algorithm will be used in our approach as a subroutine.

The concept of D2-coloring also extends to directed disk graphs DG derived from stations with non-uniform transmission radii. A necessary and sufficient condition for interference-free communication between nodes in DG is the following: for any node u, and incoming edges $(v_1, u), (v_2, u), \ldots$, each pair of nodes v_i and v_j, $i \neq j$ must be assigned different colors/time slots (also different from u). Otherwise they might transmit at the same time to u, resulting in a collision. Hence, if we denote by δ the maximum indegree of a node in DG, $\delta + 1$ is a *lower bound* on the number of colors/time slots required for interference-free communication. Still, to our knowledge there is no result stating that $O(\delta)$ colors are enough to color this *directed* graph such that the above condition is satisfied.

2.2   Power Efficiency — The Gabriel Graph [4]

Wireless sensor networks typically employ indirect, multi-hop communication, so having a good routing algorithm is crucial. Among other properties, *power efficiency* is often required: total energy needed to transmit a message along a route produced by the algorithm has to be close to the minimum over all possible routes for given source and destination. If we assume that required energy grows at least quadratically with distance to be spanned, and that total energy of a path is the sum of the energies of the individual links traversed, then a simple argument shows that power-optimal paths within UDG only contain *Gabriel edges*, i.e. edges whose diametral balls are empty of other stations. Hence restricting to the Gabriel edges in UDG (we will call this graph $GG(UDG)$) ensures that power-optimal routing is still possible. Schemes for computing the Gabriel graph and variations of it have been proposed in the literature, but all of them assume knowledge of the 1-hop neighborhood as well as a solution to the interference problem.

2.3   A Typical Scenario in Sensor Networks

Suppose that many wireless sensors are dispersed over an area for the purpose of monitoring physical quantities such as temperature, humidity, exposure to light etc. The sensors have to self-organize into a network, in order to efficiently answer queries injected from the outside world. One way to achieve such an organization is as follows:

---

[2] In fact, they state the algorithm in terms of the maximum degree (maximum number δ_1 of 1-hop neighbors). But in the proof they derive δ_2, which is $O(\delta_1)$ for unit-disk graphs.

1. establish a *low-level infrastructure*, i.e.
   - Let every node discover its 1-hop neighborhood in the communication graph w.r.t. its maximal transmission radius.
   - Compute a D2-coloring of this communication graph to provide for conflict-free communication.
2. establish the *high-level infrastructure* and adjust the low-level infrastructure
   - Construct a suitable routing substructure, e.g. the Gabriel graph $GG(UDG)$, using the existing low-level infrastructure.
   - Adjust low-level infrastructure to the new adjacency relationships by decreasing the transmission radii where possible, and and reducing the number of colors in the D2-coloring.

See Figure 1 for a graphical illustration of the process. Observe that in the initial low-level infrastructure as many as n colors are necessary, since the nodes are densely placed. Next, the Gabriel edges are identified, and each node adjusts its transmission range to just be able to reach its furthest neighbor in $GG(UDG)$. The last phase is D2-coloring of the resulting disk graph DG, corresponding to the reduced radii.

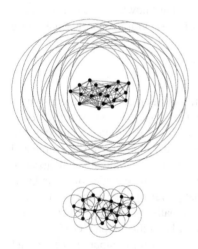

We expect the number of required colors/time slots to drop considerably due to the decreased transmission radii. In particular, for dense node distributions, after construction of the high-level infrastructure, nodes communicate only with stations at a very short distance. The amount of interference that the coloring has to cope with is drastically reduced. Still, to construct the high-level infrastructure, most known algorithms require collision-free communication between stations within distance 1, and hence a coloring w.r.t. the maximum transmission ranges has to be computed before, even though it will not be needed in the final version of the communication graph.

The goal of this paper is to avoid this potentially wasteful procedure and use time slots only up to an amount as roughly required by the final assignment of transmission powers anyway.

Fig. 1. The communication graph for a dense node distribution (top) and the adjusted transmission radii after computing a high-level infrastructure (bottom)

2.4   The Local Vicinity Size Δ_k

Before we can present our remedy for the problem sketched above, we need to have a closer look at some properties of the resulting communication graph DG after adjustment of the transmission powers. In particular, we are interested in a parameter that estimates how many colors are necessary in a D2-coloring of the final structure.

As mentioned above, if δ is the maximum indegree of a node in a DG, then $\delta + 1$ is a lower bound on the number of colors/time slots required for interference-free communication. For the analysis of our algorithm we need to get a handle on a simple and *geometric* parameter that relates the node distribution and this quantity δ. Hence let us introduce a parameter Δ_k, which provides us with an upper bound on δ (and later it will also be an upper bound on the number colors needed for a valid coloring).

Definition 1. *For a node v, let $\Delta_k(v)$ be the number of nodes contained in a disk of radius $k \cdot |vu|$, where (v, u) is the longest edge in $GG(UDG)$ adjacent to v. The quantity $\Delta_k = \max_v \Delta_k(v)$ is called the **local vicinity size**.*

For not too contrived node distributions we expect $\Delta_k = O(\delta)$, but again, showing a tight bound seems hard. See Figure 2 for an example of the local vicinity size. It is now easy to see that for large enough k, Δ_k yields an upper bound on δ.

Lemma 1. *For $k \geq 2$ we have $\delta + 1 \leq \Delta_k$.*

Proof. Consider the node v which maximizes the indegree in DG. Let (u, v) be the longest incoming edge of v. Clearly u has a Gabriel edge adjacent with length at least $|uv|$. Considering the ball B of radius $2|uv|$ centered at u, B must contain all w with $(w, v) \in DG$. Hence $\Delta_2 \geq \delta + 1$.

Our algorithm in the next section will use the local node density $\widetilde{\Delta} = \Delta_k$ for some constant $k \geq 2$ as an estimate for the number of time slots necessary to achieve interference-free communication. The intuition behind this parameter is that the number of colors required should be somewhat proportional to the maximum number of nearby stations of some node. Here 'nearby' has to be seen relative to the longest Gabriel edge adjacent to that node. In non-contrived node distributions we expect $\widetilde{\Delta}$ to be quite small or even constant, but certainly smaller than the maximum degree of the original unit-disk graph corresponding to the assignment of maximum transmission ranges.

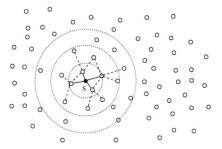

Fig. 2. Local vicinity sizes $\Delta_1(s) = 5, \Delta_2(s) = 12, \Delta_3(s) = 20$ (only Gabriel edges of S and its 1-hop neighbors are drawn)

3   An Integrated Approach to Infrastructure Establishment

As we have seen, in case of very dense sensor distributions, sequential infrastructure establishment introduces an inherent overhead. In the following we will remedy this problem by providing an integrated approach for constructing both low- and high-level infrastructure.

The basic idea of our approach is very straightforward: locally, at every node, we start exploring its neighborhood gradually by increasing its transmission range. For each transmission radius, one then constructs a local D2-coloring using a variation of the algorithm in [5]. As we will then see, one can guarantee that w.h.p. this approach actually finds exactly the edges of $GG(UDG)$.

- $r \leftarrow \varepsilon$
- while $(r < 1)$
  1. **ExploreDirectNeighborhood**$(s, r, \widetilde{\Delta})$
  2. **LocalD2Color**$(s, r, \widetilde{\Delta})$
  3. **AnnounceNApproveEdges**(s)
  4. $r \leftarrow r \cdot 2$
- **FinalD2Color**$(s, |l_s|, \widetilde{\Delta})$

Fig. 3. Infrastructure-establishment protocol

Provided with two parameters, ε and $\widetilde{\Delta}$, every node s locally executes the protocol in Figure 3. Here ε is a parameter essentially capturing the minimum distance between any two nodes, that can be for example derived just from the physical dimensions of the wireless nodes. So each node executes for $\log \frac{1}{\varepsilon}$ rounds the exploration and announcement protocol to learn about its neighborhood and construct the high-level substructure $GG(UDG)$. At the end, a valid D2-coloring of the final communication graph is computed. Let us now describe the subroutines in more detail.

3.1   Exploration of the r-Neighborhood – ExploreDirectNeighborhood$(s, r, \widetilde{\Delta})$

This routine makes sure that a node s w.h.p. announces its presence to all nearby nodes that need to know about it. More precisely, we will show that for $r \leq \varepsilon \cdot 2^{\lceil \log_2(l_s/\varepsilon) \rceil}$, where l_s denotes the length of the longest edge adjacent to s in $GG(UDG)$, w.h.p. all nodes 'nearby' s in the unit-disk graph of disks of radius r get to know about their 1-hop neighborhood. **Note:** We do not guarantee this property for larger values of r, since it turns out that this information is not necessary for constructing all edges of $GG(UDG)$.

This exploration/announcement procedure consists of $\kappa \cdot \log n$ rounds, each of which lasts for $2\widetilde{\Delta}$ time steps. In each round a node chooses a random number $1 \leq t \leq 2\widetilde{\Delta}$ and transmits its ID together with its position at time step t in this round within radius r. Let us first convince ourselves that all neighbors of s get to know about s's presence w.h.p.

Lemma 2. *After* $2\kappa\widetilde{\Delta}\log n$ *time steps, with high probability of* $1 - \frac{1}{n^{\kappa-1}}$, *all neighboring nodes (w.r.t. transmission radius r) of a node s with $r \leq \varepsilon \cdot 2^{\lceil \log_2(l_s/\varepsilon) \rceil}$ know about s's presence if $\widetilde{\Delta} = \Delta_k$ for $k \geq 4$.*

Proof. For a given node, the probability that its announcement was overheard by its neighbors is at most $1/2$, since there are at most $\widetilde{\Delta}$ other nodes which could interfere with its announcement (since we have $r \leq 2l_s$ and Δ_k for $k \geq 4$ counts the number of all 2-hop neighbors of s w.r.t. transmission radius r). The probability that in none of the $\kappa \cdot \log n$ rounds its announcement was successful is bounded by $(\frac{1}{2})^{\kappa \cdot \log n} = \frac{1}{n^\kappa}$. Hence the probability that all nodes with $r \leq \varepsilon \cdot 2^{\lceil \log_2(l_s/\varepsilon) \rceil}$ have successfully announced their presence to their neighbors is at least $1 - \frac{1}{n^{\kappa-1}}$.

In the same setting, we can also show that all h-hop neighbors of s get to know about their 1-hop neighbors w.h.p. This will be needed later in the proofs.

Lemma 3. *After* $2\kappa\widetilde{\Delta}\log n$ *time steps, with high probability of* $1 - \frac{1}{n^{\kappa-2}}$, *all nodes within h-hop distance from some station s get to know about their neighborhood (w.r.t. transmission radius r) if $r \leq \varepsilon \cdot 2^{\lceil \log_2(l_s/\varepsilon) \rceil}$ and $\widetilde{\Delta} = \Delta_k$ for $k \geq 2(h+1)$.*

Proof. Consider some node t within h-hop distance of s. For t to get to know about the set N_1 of its 1-hop neighbors, each of the nodes in N_1 should in some round have announced its presence without interfering with another node in N_1. Hence if we have $\widetilde{\Delta} \geq |N_1|$, the probability for one announcement to get through is $\geq 1/2$ in each round. Using the same argumentation as in Lemma 2, it follows that the probability of each node getting at least one announcement through in $\kappa \log n$ rounds is at least $1 - \frac{1}{|n|^{\kappa-1}}$. Furthermore, *all* nodes within the h-hop neighborhood then get to know about their neighbors with probability $\geq 1 - \frac{1}{|n|^{\kappa-2}}$. We now need to relate that to $|l_s|$, i.e. the longest adjacent edge of s in $GG(UDG)$. This can be easily achieved by requiring that $\widetilde{\Delta}$ bounds the number of nodes contained within distance $(h+1) \cdot r$. Hence the Lemma holds for $\widetilde{\Delta} = \Delta_k$ for $k \geq 2(h+1)$.

Correctness of our whole algorithm will later be established by focusing on one particular node s and show that when $r = \varepsilon \cdot 2^{\lceil \log_2(l_s/\varepsilon) \rceil}$, all edges adjacent to s in $GG(UDG)$ are found w.h.p. and that wrong edges are never created. Towards this goal, Lemma 3 states that for the appropriate choice of $\widetilde{\Delta}$, **ExploreDirectNeighborhood**$(s, r, \widetilde{\Delta})$ with high probability ensures that all h-hop neighbors of s get to know about their immediate neighbors in the unit-disk graph induced by transmission radius r.

3.2   Local D2-Coloring – LocalD2Color$(s, r, \widetilde{\Delta})$

This subroutine ensures that for a node s and transmission radius r with $r \leq \varepsilon \cdot 2^{\lceil \log_2(l_s/\varepsilon) \rceil}$ all 1- and 2-hop neighbors and s have distinct colors with high probability. Again, we do not guarantee anything for larger values of r, since this is not necessary for constructing all edges of $GG(UDG)$. Our procedure is a slight variation of the one proposed in [5]. Next, we sketch the latter and briefly discuss the changes needed for our algorithm.

The Algorithm by Parthasarathy/Gandhi for D2-Coloring. Parathasarathy/Gandhi in [5] have presented an algorithm for D2-coloring of a unit-disk graph. Their algorithm assumes that each node knows its 1-hop neighbors and a bound on the number of 2-hop neighbors.

Each node maintains a list of c potential colors it can choose from. The algorithm proceeds in rounds, in each of which typically some so-far-uncolored nodes choose a color from their list, and if this color has not been chosen by any 1- or 2-hop neighbors, these color assignments become permanent, and all 1- and 2-hop neighbors remove the respective colors from their color lists. The algorithm terminates after t rounds after which with high probability all nodes have been assigned colors which are a valid D2-coloring provided a suitable choice of the parameters c and t.

One round consists of the following 4 phases: TRIAL, TRIAL-REPORT, SUCCESS, and SUCCESS-REPORT.

The **TRIAL** phase consists of c time slots. An uncolored node u decides with probability $1/2$ to wake up, choose a random color from its list, and transmit a TRIAL message $\{ID(u), color(u)\}$ at the time slot corresponding to the chosen color.

The **TRIAL-REPORT** phase consists of b blocks of c time slots each. At the beginning of this phase each node composes a TRIAL-REPORT containing all TRIAL messages that node received in the previous phase. Then in every block it chooses a random time slot among the c slots available in a block, and sends its TRIAL-REPORT.

The **SUCCESS** phase consists of c time slots. At the beginning of this phase, every awake node u determines if the color it had chosen in the first phase is safe, i.e. if TRIAL-REPORTs have been received from all neighbors, each contains u, and none contains the same color chosen by another node. If all these conditions are met, u sends a SUCCESS message $\{ID(u), color(u)\}$ in the time slot corresponding to $color(u)$. u will not participate in future TRIAL and SUCCESS phases.

The **SUCCESS-REPORT** phase is similar to the TRIAL-REPORT phase. Nodes prepare SUCCESS-REPORTs and send them in random time slots over r rounds. But additionally, at the end of this phase, all uncolored nodes remove the colors used in SUCCESS-REPORTs that they have received.

Parthasarathy and Gandhi prove that for $c = O(\max\# \text{ D2-neighbors})^3$, $t = O(\log n)$, and $b = O(\log n)$, their algorithm computes a valid D2-coloring of the unit-disk graph with high probability $\geq 1 - \frac{1}{n^\chi}$ for arbitrary $\chi > 0$ (the constant factors for c, t, b depend on the desired success probability). Clearly, the overall running time is $O(c \log^2 n)$. They establish correctness of their algorithm by proving the following key facts (we refer to [5] for a detailed exposition of the proofs):

1. the probability that after completion of the procedure a node is still uncolored can be bounded by $\frac{1}{n^\psi}$ for arbitrary $\psi \geq 1$
2. the probability that any two nodes u, v at 1- or 2-hop distance end up with the same color can be bounded by $\frac{1}{n^\psi}$ for arbitrary $\psi \geq 1$

Can we apply their algorithm directly in our setting, where for a fixed transmission range r we want to compute a valid D2-coloring of the induced disk graph? — Unfortunately not! In fact, in this graph we do not have a bound on the number of 1- or 2-hop neighbors that is valid for *all* nodes. For a choice of $\widetilde{\Delta} = \Delta_k$ with $k \geq 4$ we can only bound the number of D2-neighbors of nodes s with $r \leq \varepsilon \cdot 2^{\lceil \log_2(l_s/\varepsilon) \rceil}$.

What we actually want is that each node s with $r \leq \varepsilon \cdot 2^{\lceil \log_2(l_s/\varepsilon) \rceil}$ w.h.p. gets colored differently from its arbitrary 1- or 2-hop neighbor t. The details of the proofs in [5] reveal that the probability of t not being colored at all depends on the number of 2-hop neighbors of t, and *not* of s (essentially, if t has too many 2-hop neighbors, chances are that each trial of t is doomed as there is some 2-hop neighbor of t which has chosen the same color). Similarly, if both s and t have been colored, they only might have the same color if a SUCCESS-REPORT of their common neighbor was lost due to interference. This probability can be kept low if the number of D2-neighbors of this common neighbor is bounded.

But D2-neighborhoods of all these nodes can be easily bounded by requiring $\widetilde{\Delta} = \Delta_k$ with $k \geq 8$, that is by bounding the number of points in a larger disk around s which also contains all 2-hop neighbors of 2-hop neighbors of s for the current choice r.

So we can prove the two key facts under this new condition. We emphasize again, though, that our Lemmas hold only for nodes s with $r \leq \varepsilon \cdot 2^{\lceil \log_2(l_s/\varepsilon) \rceil}$ and not for larger values of r.

---

[3] In fact they show $c = O(\max \text{ 1-hop-neighbors})$, but in case of unit-disk graphs, this is the same quantity up to a constant factor.

Lemma 4. *Let s be a node with $r \le \varepsilon \cdot 2^{\lceil \log_2(l_s/\varepsilon) \rceil}$. Then with high probability of $1 - \frac{1}{n^\psi}$, s and all its 1- and 2-hop neighbors are colored after completion of* **LocalD2Color**$(s, r, \widetilde{\Delta})$ *for $\widetilde{\Delta} = \Delta_k$ and $k \ge 8$.*

Proof. All 2-hop neighbors of 2-hop neighbors of s in the unit-disk graph w.r.t. transmission radius r are contained in a ball of radius $4r$ around s. Since $r \le 2|l_s|$ we obtain a bound on the number of 2-hop neighbors of all 2-hop neighbors of s and can apply the proofs from [5].

By similar arguments, one can show that the coloring is valid w.h.p.

Lemma 5. *Let s be a node with $r \le \varepsilon \cdot 2^{\lceil \log_2(l_s/\varepsilon) \rceil}$, and let u be its fixed 1- or 2-hop neighbor, and assume both have been assigned a color during the procedure. Then with high probability of $1 - \frac{1}{n^\psi}$, s and u have been assigned distinct colors for $\widetilde{\Delta} = \delta_k$ and $k \ge 8$.*

Using these Lemmas and Lemma 3 with $h = 3$ it is easy to conclude the following.

Lemma 6. **LocalD2Color**$(s, r, \widetilde{\Delta})$ *computes in time $O(\widetilde{\Delta} \log^2 n)$ a coloring of the nodes such that for each node s with $r \le \varepsilon \cdot 2^{\lceil \log_2(l_s/\varepsilon) \rceil}$, s and its D2-neighbors in $UDG(r)$ are assigned different colors with probability $1 - \frac{1}{n^\psi}$ for arbitrary $\psi \ge 1$ if $\widetilde{\Delta} = \Delta_k$, $k \ge 8$.*

3.3   AnnounceNApproveEdges(s)

In this procedure each node first locally computes a list of adjacent Gabriel edges based upon its knowledge of the 1-hop neighbors, and then announces them one by one. Announcements take $O(\widetilde{\Delta}^{1/2})$ rounds of $2\widetilde{\Delta}$ time steps each. In each round, a node s announces in time slot $2 \cdot \text{color}(s)$ a Gabriel edge $e = (s, t)$ from its list that has not been announced by the other end-node t of e. If no neighbor of s transmits at the same time slot and no VETOs or collisions are encountered in time slot $2 \cdot \text{color}(s) + 1$, s and t regard the edge e as an edge of the final graph $GG(UDG)$. Furthermore, s listens to all announcements of other nodes and sends a VETO message in the following time slot if it either experiences collision (two or more neighboring nodes announce at the same time slot) or its own position contradicts the creation of the announced edge (i.e. s lies in the diametral circle of the announced edge).

Let us first show that non-Gabriel edges always get vetoed.

Lemma 7. **AnnounceNApproveEdges**(s) *never creates an edge that is not a Gabriel edge (for arbitrary transmission radius r).*

Proof. Assume otherwise, i.e. a node v created an edge e whose diametral circle contains a 1-hop neighbor u of v. Then e must have been announced at some point, and no VETO message, in particular none from u, was received and no collision was experienced at v. But u should have sent a VETO regardless of whether it received the announcement or experienced a collision. Thus we have a contradiction.

Next we show that w.h.p. all Gabriel edges of a fixed node s are constructed in the round where $r = \varepsilon \cdot 2^{\lceil \log_2(l_s/\varepsilon) \rceil}$. With Lemma 7, this implies that the computed $GG(UDG)$ is correct.

Lemma 8. *Let s be a node with $\varepsilon \cdot 2^{\lfloor \log_2(l_s/\varepsilon) \rfloor} \leq r \leq \varepsilon \cdot 2^{\lceil \log_2(l_s/\varepsilon) \rceil}$. Then with high probability all Gabriel edges adjacent to s are constructed if $\widetilde{\Delta} = \Delta_k$, $k \geq 8$.*

Proof. By Lemma 3, only correct edges are announced by s or its 1-hop neighbors. By Lemma 6, no collisions appear during the execution of **AnnounceNApproveEdges**(s). It remains to prove that $O(\widetilde{\Delta}^{1/2})$ rounds suffice to have all Gabriel edges of s announced (Note that there could be $\Theta(\widetilde{\Delta})$ many, and without the help of the other endnodes of these edges it would take $\Theta(\widetilde{\Delta})$ rounds or $\Theta(\widetilde{\Delta}^2)$ time slots for s to announce them all!).

Let us call a node *active* if not all of its adjacent Gabriel edges have been announced. Let a_i be the number of active neighbors of s in the beginning of round i. Then at every round, at least $a_i/2$ edges *in the 2-hop neighborhood* of s get announced, since each such announcement can deactivate at most two neighbors of s. The 2-hop neighborhood initially contains $O(\widetilde{\Delta})$ unannounced edges, so there are $O(\widetilde{\Delta}^{1/2})$ rounds with $a_i = \Omega(\widetilde{\Delta}^{1/2})$. Once it becomes $a_i = O(\widetilde{\Delta}^{1/2})$, it is clear that all the remaining Gabriel edges adjacent to s will be announced by s itself in additional $O(\widetilde{\Delta}^{1/2})$ rounds.

3.4   FinalD2Color$(s, |l_s|, \widetilde{\Delta})$

At this point, each node s w.h.p. knows about the topology of the *directed* disk graph DG, that is about the nodes that lie within its own transmission range, as well as the nodes in whose transmission range it lies. These correspond to the *outgoing* and *incoming* edges of s in DG, respectively. Indeed, s can learn about its outgoing edges at the time when its longest Gabriel edge is created, and about its incoming edges after running another 'announcement phase' in the spirit of **ExploreDirectNeighborhood**, with the only difference that each node actually transmits within the range determined by its longest adjacent Gabriel edge. Then with high probability all nodes get to know about their incoming edges (since again, interference is limited by the parameter $\widetilde{\Delta}$).

It remains to make the final assignment of time slots that will be used during the lifetime of the network. As explained previously, the purpose of this step is to save on the number of colors required by taking into account the nodes' effective radii of communication.

However, the algorithm of [5] cannot be directly applied to construct the final coloring. It may happen that nodes u and v propose the same color in TRIAL phase, and their messages collide *only* at node w, but w is unable to communicate that fact to u and v because its communication radius is too small. More precisely, the problem is that the DG induced by $GG(UDG)$ is *directed*, because the communication radii are different in general. The D2-coloring scheme requires that the DG be undirected, so that the nodes can communicate in "propose–respond" fashion.

Fortunately, there is a simple modification that gets around this problem without affecting the asymptotic performance. The TRIAL and SUCCESS phases do not change, while in the TRIAL-REPORT and SUCCESS-REPORT phase each node transmits within the radius equal to the length of its longest incoming edge. By keeping the TRIAL and SUCCESS unchanged, we make sure that the collisions generated during rounds of coloring are exactly those that would occur in the final communication scheme. On the other hand, modification to the REPORT phases makes sure that the coloring is correct, but may hurt the running time; increasing the radii for reporting purpose may create a large number of new D2-neighbors, and therefore many collisions among the REPORT messages. However, we show that the new neighborhood size has already been accounted for in the preceding analysis. The actions for fixing a color after reception the TRIAL-REPORT as well as for removal of colors from the list after reception of a SUCCESS-REPORT of course then restrict to reports received from nodes within the transmission range of a node.

Lemma 9. *Let G be the DG induced by $GG(UDG)$, and let H be the undirected version of G obtained by setting the radius of each node to the length of its longest incoming edge in G. Then any node in H has a D2-neighborhood of size at most Δ_k, for any $k \geq 5$.*

Proof. Consider a fixed node u and its D2-neighborhood N_u in H. Clearly, the edges in G that connect nodes in N_u are also present in H. Let (v, w) be the longest incoming edge of the nodes in N_u (w is in N_u). Then the whole N_u at most $5|vw|$ away from v. Also, (v, w) emanates from v in G, therefore $|vw| \leq l_v$. Thus, $B(v, 5l_v)$ contains all nodes in N_u, which implies $N_u \leq \Delta_5(v) \leq \Delta_5 \leq \Delta_k$, for any $k \leq 5$. This completes the proof.

In other words, the increased transmission radii (to the length of the longest *incoming* edge) create only limited potential interferences, so w.h.p. the TRIAL-REPORT and TRIAL-SUCCESS phases succeed.

Lemma 10. *In time $O(\widetilde{\Delta} \log^2 n)$, **FinalD2Color**$(s, |l_s|, \widetilde{\Delta})$ w.h.p. computes a valid D2-coloring of the final communication graph DG.*

3.5   Summary and Further Remarks

We summarize the statements of Lemmas 3, 6, 7, and 8, 10 and give our main theorem.

Theorem 1. *Our algorithm runs in time $O(\widetilde{\Delta}^{3/2} \log \frac{1}{\varepsilon} \log^2 n)$ and w.h.p. computes the $GG(UDG)$, as well as a coloring of its nodes which ensures interference-free communication between nodes adjacent in $GG(UDG)$.*

Here $\widetilde{\Delta} = \Delta_8$ denotes the maximum number of points within the distance of $8 \cdot |l_s|$ from s, where $|l_s|$ is the length of the longest Gabriel edge adjacent to s. Thus, we can relate the running time of the algorithm to the number of 'nearby' stations relative to its final assigned transmission range, instead of the number of stations within its maximum transmission range. ε can be set to the minimum inter-node distance (as e.g. derived from the physical size of the nodes) or even larger, as long as the number of nodes contained in any disk of radius 8ε is bounded by $\widetilde{\Delta}$.

We want to remark that k can be decreased from 8 to a value arbitrarily close to 4, by instead of doubling the radius r, multiplying it by factor smaller than 2.

4   Lipschitz-Type Node Distributions

Consider an application of monitoring physical quantities (such as temperature, humidity, exposure to light etc.) over an area A of a nature preserve. Suppose that the area contains a set H of interesting 'hotspots' (e.g. breeding areas of a bird species) which should be monitored more accurately. Scientists want to deploy more wireless sensors close to the hotspots, and fewer sensors further away.

Formally, they might assume that there is an 'interest function' $f : A \to \mathbb{R}$ which is small in regions of high interest and vice versa. An example is $f(x) = d(x, H)$, the distance to the closest hotspot. Then, if the deployment of the set of sensors S satisfies the condition $\forall x \in A : \exists s \in S : d(x, s) \le \varepsilon \cdot f(x)$ for some small $\varepsilon \in (0, 1)$, enough data is gathered. On the other hand, sensor nodes are expensive, so only the necessary number should be used; in terms of the interest function f, the required condition is $\forall x \in A : |\{s \in S : d(x, s) \le \varepsilon \cdot f(x)\}| \le \beta$ for some constant β.

If the distribution of the deployed sensors adheres to the above conditions, our algorithm performs very well, since the value $\tilde{\Delta}$, which has to be provided to every node upon deployment, is a reasonably small constant. Roughly speaking, our algorithm performs well if the node density has *bounded variation* as a function of the spatial coordinates. Below we give a more formal statement of this fact.

We want to remark that this property of gradually changing node-densities naturally also arises in other settings. For example, consider the application of tracking a set of point-sized *slowly moving* objects. If we are interested in minimizing the effort needed to adapt the sensor distribution to possible movement of objects, it is reasonable to have a gradually increasing sensor density as the distance to an object (in its current position) becomes smaller. Let us formalize the situation described above.

Definition 2. *Let S be a set of wireless nodes placed within some area of interest A. We call S a Lipschitz-type node distribution, if there exists an α-Lipschitz function $\varphi : A \to \mathbb{R}$ with $\alpha > 0$ and constants $\varepsilon > 0$, $\beta \ge 1$ such that for all $x \in A$, $1 \le |S \cap B(x, \varepsilon \varphi(x))| \le \beta$.*

In our example above we had $\varphi(x) = f(x)$ with $\alpha = 1$ (as f was defined to be the distance to a set of points, it is 1-Lipschitz). Observe though that this definition of Lipschitz-type node distributions also allows for 'oversampling' of the domain of interest, as long as the oversampling happens in a smooth manner.

Lemma 11. *Let S be a Lipschitz-type node distribution, with parameters $\alpha, \beta, \varepsilon$, and $\varepsilon < \frac{1}{2\alpha(k+1)}$, then we have $\Delta_k = O(1)$.*

Proof. Consider a fixed node u, and let (u, v) be its longest adjacent edge in $GG(UDG)$, $|uv| = l_u$. Let $x \in A$ be the midpoint of (u, v). Clearly, $l_u/2 \le \varepsilon \varphi(x)$. Let us define $R = (k + 1)l_u$, and hence $\Delta_k \le |B(x, R) \cap S|$. So in the following we will concentrate on bounding the number of nodes within $B(x, R)$.

Using the inequalities above we get $R \le (k + 1) \cdot 2 \cdot \varepsilon \varphi(x)$. As φ is α-Lipschitz, we have for any $y \in B(x, R)$ that $\varphi(y) \ge \varphi(x) - \alpha R \ge \varphi(x)(1 - 2(k+1)\alpha\varepsilon)$, that is, any ball of radius at most $r = \varepsilon \varphi(x)(1 - 2(k+1)\alpha\varepsilon)$ centered within $B(x, R)$ contains less than β nodes. It remains to bound the number of balls of radius r to cover $B(x, R)$. But that number is $O((\frac{k+1}{1 - 2(k+1)\varepsilon\alpha})^2)$, which for constant values of k, ε with $\varepsilon < \frac{1}{2\alpha(k+1)}$ remains a constant. That is, at $\Delta_k \le |B(x, R) \cap S| = O(\beta) = O(1)$.

It follows that for the above example with $\alpha = 1$, the condition of Definition 2 is satisfied with $\varepsilon < 1/18$ in order to guarantee $\tilde{\Delta} \le \Delta_8 = O(1)$, so the algorithm runs in $O(\log \frac{1}{\varepsilon} \log^2 n)$. We emphasize, however, that this theoretical analysis is very pessimistic and we expect the algorithm to perform very well for many practical node distributions.

5   Conclusions

In this paper we have presented a distributed protocol for constructing both the low- and the high-level infrastructure for a set of newly deployed wireless stations. Our *integrated approach* remedies some of the inherent problems incurred by a sequential construction of first low- and then high-level infrastructure, which is particularly apparent for dense and varying node distributions. We believe that other high-level infrastructures can be similarly constructed in this manner. It might be interesting to see whether our approach can also be adapted to work under the less restrictive model of node capabilities as used in [6] and [8].

References

1. M. Burkhart, P. von Rickenbach, R. Wattenhofer, and A. Zollinger, "Does topology control reduce interference?" *5th ACM Interational Symposium on Mobile Ad Hoc Networking and Computing (MobiHoc)*, 2004.
2. X. Cheng, X. Huang, D. Li and D.-Z. Du, "Polynomial-time approximation scheme for minimum connected dominating set in ad hoc wireless networks," *Networks*, to appear.
3. B. Clark, C. Colbourn and D. Johnson, "Unit Disk Graphs," *Discrete Mathematics*, Vol. 86, pp. 165-177, 1990.
4. K.R. Gabriel and R.R. Sokal, "A new statistical approach to geographic variation analysis", *Systematic Zoology*, vol 18, pp. 259–278, 1969
5. R. Gandhi and S. Parthasarathy, "Fast Distributed Well Connected Dominating Sets for Ad Hoc Networks," *CS-TR-4559*, UM Computer Science Department.
6. F. Kuhn, T. Moscibroda, and R. Wattenhofer. "Initializing Newly Deployed Ad Hoc and Sensor Networks" *10th Ann. Int. Conf. on Mobile Computing and Networking (MOBICOM)*, Philadelphia, USA, Sept. 2004.
7. F. Meyer auf der Heide, C. Schindelhauer, K. Volbert, and M. Grünewald, "Congestion, Dilation, and Energy in Radio Networks" *Theory of Computing Systems 37 (3), 2004*
8. T. Moscibroda and R. Wattenhofer "Efficient Computation of Maximal Independent Sets in Unstructured Multi-Hop Radio Networks" *1st IEEE Int. Conf. on Mobile Ad-hoc and Sensor Systems (MASS)*, Fort Lauderdale, USA, Oct. 2004.
9. X.-Y.Li, W.-Z.Song, Y.Wang. "Localized Topology Control for Heterogenous Wireless Ad Hoc Networks", *1st IEEE Int. Conf. on Mobile Ad hoc and Sensor Systems (MASS)*, 2004.
10. X.-Y. Li, P.-J. Wan, W. Yu, O. Frieder. "Sparse power efficient topology for wireless networks", *IEEE Hawaiian Int. Conf. on System Sciences (HICSS)*, 2002.
11. S.Ramanathan. "A unified framework and algorithm for channel assignment in wireless networks", *Wireless Networks*, 5(2):81-94, 1999
12. A.Sen and M.L.Huson. "A new model for scheduling packet radio networks", *Wireless Networks*, 3(1):71-82, 1997
13. A. Sen and E. Melesinska. "On approximation algorithms for radio network scheduling", *Proc. 35th Allerton Conf. on Communication, Control, Computing*, pp.573–582, 1997.
14. P. J. Wan, K. Alzoubi and O. Frieder, "Distributed Construction of Connected Dominating Set in Wireless ad hoc networks," *Proc. of INFOCOM 2002*.

A Local Facility Location Algorithm
for Sensor Networks

Denis Krivitski[1], Assaf Schuster[1], and Ran Wolff[2]

[1] Computer Science Dept., Technion – Israel Institute of Technology
{denisk, assaf}@cs.technion.ac.il
[2] Computer Science Dept., University of Maryland, Baltimore County
ranw@cs.umbc.edu

Abstract. In this paper we address a well-known facility location problem (FLP) in a sensor network environment. The problem deals with finding the optimal way to provide service to a (possibly) very large number of clients. We show that a variation of the problem can be solved using a *local* algorithm. Local algorithms are extremely useful in a sensor network scenario. This is because they allow the communication range of the sensor to be restricted to the minimum, they can operate in routerless networks, and they allow complex problems to be solved on the basis of very little information, gathered from nearby sensors. The local facility location algorithm we describe is entirely asynchronous, seamlessly supports failures and changes in the data during calculation, poses modest memory and computational requirements, and can provide an anytime solution which is guaranteed to converge to the exact same one that would be computed by a centralized algorithm given the entire data.

1   Introduction

Determining the location of facilities which provide system related services is a major issue for any large distributed system. The resource limited scenario of a sensor network makes the problem far more acute. A well-placed resource (cache server, relay, or high-powered sensor, etc.) can tremendously increase the lifespan and the productivity of dozens or even hundreds of battery operated sensors. In many cases, however, the optimal location of such resources depends on dynamic characteristics of the sensors (e.g., their remaining battery power), the environment (e.g., level of radio frequency white noise), or the phenomena they monitor (e.g., frequency of changes). Thus, optimal placement cannot be computed a priori, independently of the system's state.

One example of a facility location problem may occur in sensor networks that, in addition to regular sensors, use a few dozen relays. Regular sensors are low-power motion sensing devices, which are distributed from the air, covering the area randomly. Instead, relays are equipped with large batteries and long range transceivers and are placed at strategic points by ground transportation. The purpose of the relays is to collect data from the sensors and transmit it to a command station whenever it is requested. However, the question remains how to best utilize the relays which in themselves have but limited resources. It would make sense to shut down a relay if there was only mild activity in its nearby surroundings and report that activity via other relays. Note, however,

V. Prasanna et al. (Eds.): DCOSS 2005, LNCS 3560, pp. 368–375, 2005.
© Springer-Verlag Berlin Heidelberg 2005

that the amount of activity, the available resources of the relays and the motion sensors, as well as the environmental conditions in which they all operate may influence the decision, and these factors may vary over time. The optimal solution, thus, has to be regularly adjusted.

The facility location problem (FLP) has been extensively studied in the last decade. Like many other optimization problems, optimal facility location is NP-Hard [13]. Thus, the problem is usually solved using either a hill-climbing heuristic [14, 8, 4] or linear programming [11, 19, 12]. These approaches achieve constant factor approximation of the globally optimal solution [1]. FLP also has several versions, primarily divided according to whether facilities have finite or infinite capacity (i.e., *capacitated* vs. *uncapacitated* FLP).

To the best of our knowledge FLP has never been studied specifically in a distributed setting. Nevertheless, it is easy to see how distributed formulation of related hill-climbing algorithms such as k-means and k-median clustering [5, 7, 6] can be adapted to solve distributed FLP. We note, however, that all previous work on distributed clustering assumes tight cooperation and synchronization between the processors containing the data, and a central processor that collects the sufficient statistics needed in each step of the hill-climbing heuristic. Such central control is not practical in wireless networks, because of the energy required and because it is prone to errors even in the case of single failures. Even more importantly, central control is unscalable in the presence of dynamically changing data: any such change must be reported to the center, for fear it might alter the result.

Thus, it is clear that other features are required to qualify an algorithm for sensor networks. The most important of these are the following: the ability to perform in a routerless network (i.e., to be driven by data rather than by address), the ability to calculate the result in-network rather than collect all of the data to a central processor (which would quickly exhaust bandwidth [9]), and the ability to locally prune redundant or duplicate computations. These three features typify *local* algorithms.

A local algorithm is one in which the complexity of computing the result does not directly depend on the number of participants. Instead, each processor usually computes the result using information gathered from just a few nearby neighbors. Because communication is restricted to neighbors, a local algorithm does not require message routing, performs all computation in-network, and in many cases is able to locally overcome failures and minor changes in the input (provided that these do not change its output). Local algorithms have been mainly studied in the context of graph related problems [2, 15, 16, 17, 18]. Most recently, [20] demonstrated that local algorithms can be devised for complex data analysis tasks, specifically, data mining of association rules in distributed transactional databases. The algorithm presented in [20] features local pruning of false propositions (candidates), in-network mining, asynchronous execution, and resilience to changes in the data and to partial failure during execution.

In this work we develop a local algorithm that solves a specific version of FLP, one in which uncapacitated resources can be placed in any k out of m possible locations. Initiating our algorithm from a fixed resource location, say, in the first k locations, we show that the computation required to reach agreement on a single hill-climbing step—moving one resource to a free location—can be reduced to a group of majority votes. We then use a variation of the local majority voting algorithm presented in [20] to develop an algorithm which locally computes the exact same solution a hill-climbing algorithm would compute, had it been given the entire data.

In a series of experiments employing networks of up to 10,000 simulated sensors, we prove that our algorithm has good locality, incurs reasonable communication costs, and quickly converges to the correct answer whenever the input stabilizes. We further show that when faced with constant data updates, the vast majority of sensors continue to compute the optimal solution. Most importantly, the algorithm is extremely robust to sporadic changes in the data. So long as these do not change the global result, they are pruned locally by the network.

The rest of this paper is organized as follows. We first describe our notations and formally define the problem. Then, in Section 3, we give our version of the majority voting algorithm originally described in [20]. Section 4 describes the local k-facility location algorithm. Finally, in Section 5, we present some of the experimental results.

2   Notations, Assumptions, and Problem Definition

We assume a large number N of processors, which can communicate with one another by sending messages. We further assume that communication among neighboring processors is reliable and ordered. This assumption can be enforced using standard numbering, ordering and retransmission mechanisms. For brevity, we assume an undirected communication tree. As shown in [3], such a tree can be efficiently constructed and maintained using variations of Bellman-Ford algorithms [10]. Finally, we assume that failure is fail-stop and that the neighbors of a processor that is disconnected or reconnected for any reason are reported.

Given a database DB containing *input points* $\{p_1, p_2, \ldots, p_n\}$, a set M of m possible *locations*, and a cost function $d : DB \times M \to \mathcal{R}^+$, the task of a k-*facility location* algorithm is to find a set of *facilities* $C \subset M$ of size k, such that the cumulative distance of points from their nearest facility $\sum\limits_{p_i \in DB} \min\limits_{c \in C} d(p_i, c)$ is minimized.

To relate these definitions to the example given in the introduction, consider a database that includes a list of events that occurred in the last hour. Each event would have a heuristic estimate of its importance. Furthermore, each sensor would evaluate its hop distance from every relay and multiply this by the heuristic importance of each event to produce its cost. Given this input, a facility location algorithm would compute the best combination of relays such that the most important events need not travel far before they reach the nearest relay. The less important events, we assume, would be suppressed either in the sensor that produced them, or in-network by other sensors.

An *anytime* k-facility location algorithm is one which, at any given time during its operation, outputs a placement for the location such that the cost of this ad hoc output improves with time until the optimal solution is found. A *distributed* k-facility location algorithm would compute the same result even when the DB is partitioned into N mutually exclusive databases $\{DB^1, \ldots, DB^N\}$, each of which is deposited with a separate processor, and these are then allowed to communicate by passing messages to each other. A *local* k-facility location algorithm is a distributed algorithm whose performance does not depend on N but rather corresponds to the difficulty of the problem instance at hand.

The *hill-climbing* heuristic for k-facility location begins from an agreed upon placement of the facilities (henceforth, *configuration*). Then, it finds a single facility and a single empty location, such that by moving the facility to that free location the cost of the solution is reduced to the greatest possible degree. If such a step exists, the algorithm changes the configuration accordingly and iterates. If any configuration which can be produced by moving just one facility has a higher cost than the current configuration, the algorithm terminates and outputs the current configuration as the solution.

This paper presents a local anytime algorithm that computes the hill-climbing heuristic for k-facility location.

3   Local Majority Voting

Our k-facility location algorithm reduces the problem to a large number of majority votes. In this section, we briefly describe a variation of the local majority voting algorithm from [20], which we use as the main building block for the algorithm. The algorithm assumes that messages sent between neighbors are reliable and ordered, and that processor failure is reported to the processor's neighbors. These assumptions can easily be enforced using standard numbering, retransmission, ordering, and heart-beat mechanisms. The algorithm makes no assumptions on the timeliness of message transfer and failure detection.

Given a set of processors V, where each $u \in V$ contains a zero-one poll with c^u votes, s^u of which are one, and given the required majority $0 < \lambda < 1$, the objective of the algorithm is to decide whether $\sum_u s^u / \sum_u c^u \geq \lambda$. We call Δ the number of excess votes.

The following local algorithm decides whether $\Delta \geq 0$. Each processor $u \in V$ computes the number of excess votes in its own poll, $\delta^u = s^u - \lambda c^u$. It then stores the number of excess votes it reported to each neighbor v in δ^{uv} and the number of excess votes which have been reported to it by v in δ^{vu}. Processor u computes the total number of excess votes it knows of, as the sum of its own excess votes and those reported to it by the set G^u of its neighbors $\Delta^u = \delta^u + \sum_{v \in G^u} \delta^{vu}$. It also computes the number of excess votes it agreed on with every neighbor $v \in G^u$, $\Delta^{uv} = \delta^{uv} + \delta^{vu}$. When u chooses to inform v about a change in the number of excess votes it knows of, u sets δ^{uv} to $\Delta^u - \delta^{vu}$ – thus setting Δ^{uv} to Δ^u, and then sends δ^{uv} to v. When u receives a message from v containing some δ, it sets δ^{vu} to δ – thus updating both Δ^{uv} and Δ^u. Processor u outputs that the majority is of ones if $\Delta^u \geq 0$, and of zeros otherwise.

The crux of the local majority voting algorithm is in determining when u must send a message to a neighbor v. More precisely, the problem is to determine when sending a message can be avoided, despite the fact that the local knowledge has changed. In the algorithm presented here, there are two cases in which a processor u would send a message to a neighbor v: when u is initialized and when the condition $(\Delta^{uv} \geq 0 \wedge \Delta^{uv} > \Delta^u) \vee (\Delta^{uv} < 0 \wedge \Delta^{uv} < \Delta^u)$ evaluates true. Note that u must evaluate this condition upon receiving a message from a neighbor v (since this event updates Δ^u and the respective Δ^{uv}), when its input bit switches values, and when an edge connected to it fails (because Δ^u is then computed over a smaller set of edges and may change as a result). This means the algorithm is event driven and requires no synchronization.

We modify the local majority voting algorithm slightly in order to apply it to k-facility location. We add the ability to suspend and reactivate the vote using corresponding events. A processor whose voting has been suspended will continue to receive messages and modify the corresponding local variable, but will not send any messages. When the vote is activated, the processor will always check whether it is required to send a message as a result of the information ir received while in a suspended state.

4   Majority Based k-Facility Location

The local k-facility location algorithm which we now present is based upon three fundamental ideas: The first is to have every processor optimistically perform hill-climbing steps without waiting for a decision as to which is the globally optimal step. Having taken these steps, the processor continues to validate the agreement of the steps it took with the globally correct one. If there is no agreement, then these speculative steps are undone and better ones are chosen. The second idea is to choose the optimal step not by computing the cost of each step directly, but rather by voting on which pair of possible steps is more costly (i.e., more popular). The third idea is a pruning technique by which many of these votes can be avoided altogether; avoiding unnecessary votes is essential because, as we further explain below, computing votes among each pair of optional steps might be arbitrarily more complicated than finding the best next step.

4.1   Optimistic Computation of an Ad-Hoc Solution

Most parallel data mining algorithms use synchronization to validate that their outcome represents the global data $\bigcup_u DB^u$. We find this approach impractical for large-scale distributed systems — especially if one assumes that the data may change with time, and thus the global data can never be determined. Instead, when performing parallel hill-climbing, we let each processor proceed uphill whenever it computes the best step according to the data it currently possesses. Then, we use local majority voting (as we describe next) to make sure that processors which have taken erroneous steps will eventually be corrected. In the event that a processor is corrected, computations associated with configurations that were wrongly chosen are put on hold. These configurations are put aside in a designated cache in case additional data that accumulates will prove them correct after all.

We term the sequence of steps selected by processor u at a given point in time its *path* through the space of possible configurations and denote it $R^u = \langle C_1^u, C_2^u, \ldots, C_l^u \rangle$. C_1^u is always chosen to be the first k locations in M. C_l^u is the ad hoc solution C^u. u refrains from developing another configuration following a given C_l^u when no possible step can improve on the cost of the current configuration, or when two or more steps still compete on providing the best improvement.

Since the computation of all of the configurations along every processor's path is concurrent, messages sent by the algorithm contain a *context* – the configuration to which they relate. Since the computation is also optimistic, it may well happen that two processors u and v temporarily have different paths R^u and R^v. Whenever u receives a message in the context of some configuration $C \notin R^u$, this message is considered to be *out of context*. Rather than being accepted by u, it is stored in u's out-of-context message queue. Whenever a new configuration C enters R^u, u scans the out-of-context queue and accepts messages relating to C in the order by which they were received.

4.2 Locally Computing the Best Possible Step

For each configuration $C_a^u \in R^u$, processor u computes the best possible step as follows. First, it generates the set of possible configurations $Next[C_a^u]$, such that each member of $Next[C_a^u]$ is a configuration that replaces one of the members of C_a^u with a non-member location from $M \setminus C_a^u$. Next, for each $C \in \{C_a^u\} \cup Next[C_a^u]$ and each $p \in DB^u$, the cost incurred by p in C is computed such that $cost(p, C) = \min_{x \in C}\{d(p, x)\}$. Finally, for every $C_i, C_j \in Next[C_a^u]$, where $i < j$, processor u initiates a majority vote, $Majority_{C_a^u}^u \langle i, j \rangle$, which compares their relative costs and eventually computes $\Delta_{C_a^u}^u \langle i, j \rangle \geq 0$ if the global cost of C_i is higher than that of C_j (as we explain below). Correctness of the majority vote process guarantees that the best configuration $C_{i_{best}} \in \{C_a^u\} \cup Next[C_a^u]$ will eventually have negative $\Delta_{C_a^u}^u \langle i_{best}, j \rangle$ for all $j > i_{best}$, and positive $\Delta_{C_a^u}^u \langle j, i_{best} \rangle$ for all $j < i_{best}$. Hence, the algorithm will optimistically choose C_i as the next configuration whenever C_i has the maximal number of majority votes indicating it is the better one (even if some votes indicate otherwise).

To determine, by means of majority vote, which of two configurations has the lower cost, we set for every processor u, $\delta^u \langle i, j \rangle = \sum_{p \in DB^u} cost(p, C_i) - cost(p, C_j)$. This can be done for any $\delta^u \langle i, j \rangle \in [-x, x]$ by choosing, for example, $c = 2x$, $\lambda = 1/2$ and $s = x - \lambda$. Note that, as shown in [20], s^u and c^u can be set to arbitrary numbers and not just to zero or one. Further note that for every $C_i, C_j \sum_{p \in DB} cost(p, C_i) - \sum_{p \in DB} cost(p, C_j) = \sum_u \sum_{p \in DB^u} [cost(p, C_i) - cost(p, C_j)]$. Hence, if the vote comparing the cost of C_i to that of C_j determines that $\Delta^u \langle i, j \rangle \geq 0$, this proves the cost of C_i is larger than that of C_j.

Note that since every majority vote is performed using the local algorithm described in Section 3, the entire computation is also local. Eventual correctness of the result and the ability to handle changes in DB^u or G^u also follow immediately from the corresponding features of the majority voting algorithm.

4.3 Pruning the Set of Comparisons

The subsections above show how it is possible to reduce k-facility location to a set of majority votes. However, these reductions overshoot the objective of the algorithm. This is because while a k-facility location really only requires that the *best* possible configuration be calculated given a certain configuration, the reduction above actually computes a *full order* on the possible configurations. This is problematic because, for some inputs, computing a full order may be arbitrarily more difficult (and hence, less local and more costly) than computing only the best option.

To overcome this problem, the algorithm is augmented with a pruning technique that limits the progress of comparisons such that only a small number of them actually take place. Given a configuration C, processor u sets as its *best* possible configurations the ones with the maximal number of majority votes—indicating that these configurations are less costly. It sets as *contending* those possible configurations which are indicated to be less costly than one of the best configurations. Processor u keeps track of its best and its contending configurations and the best and contending configurations of its neighbors in G^u. For this purpose u reports, with every message it sends, which configurations it currently considers best or contending. u retains in an active state those majority votes that compare a configuration to either its own or its neighbors' best and contending configurations. u suspends the rest of the majority votes, meaning that it will not send messages relating to them even if it accepts messages or data changes. It can also be shown that this pruning technique does not affect the correctness of the algorithm.

5   Experiments

To evaluate the algorithm's performance we ran it on simulated networks of up to ten thousand processors using up to one thousand input points in each processor. The main conclusions are as follows:

- The algorithm is local. The number of messages per processor remains constant as the network size increases (see figure 2). Moreover, the number of interlocutors of each processor don't grow with network size.

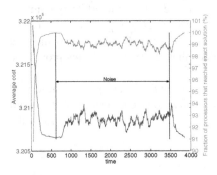

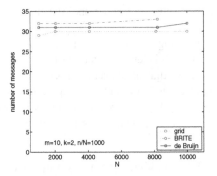

Fig. 1. Behavior in dynamic environment. The upper graph shows that more than 98% of nodes output the exact solution. The lower graph shows average solution cost. As the noise begins, the average cost deviates from the minimum but returns to it as the noise stops

Fig. 2. Messages per processor. The graph shows the number of messages each processor sends for 3 topology types and 5 different sizes. The number of messages stays constant as the network size increases

- The algorithm easily adapts to incremental data changes. In the dynamic data experiment, we swapped the databases of two random processors during a typical edge delay (we call those swaps *noise*). Throughout the experiment, not more than 2% of the processors deviated from the exact solution (see figure 1).
- The majority of processors converge rapidly. More than 90% of the processors converged to the exact solution after 4 edge delays. In addition, convergence time doesn't depend on network size.

References

1. Vijay Arya, Naveen Garg, Rohit Khandekar, Kamesh Munagala, and Vinayaka Pandit. Local search heuristic for k-median and facility location problems. In *STOC*, pages 21–29, 2001.
2. B. Awerbuch, A. Bar-Noy, N. Linial, and D. Peleg. Compact distributed data structures for adaptive network routing. *Proc. 21st ACM STOC*, May 1989.
3. Y. Birk, L. Liss, A. Schuster, and R. Wolff. A local algorithm for ad hoc majority voting via charge fusion. In *DISC*, 2004.
4. Moses Charikar and Sudipto Guha. Improved combinatorial algorithms for the facility location and k-median problems. In *FOCS*, pages 378–388, 1999.
5. Inderjit S. Dhillon and Dharmendra S. Modha. A data-clustering algorithm on distributed memory multiprocessors. In *Large-Scale Parallel Data Mining*, pages 245–260, 1999.
6. George Forman and Bin Zhang. Distributed data clustering can be efficient and exact. *SIGKDD Explor. Newsl.*, 2(2):34–38, 2000.
7. D. Foti, D. Lipari, C. Pizzuti, and D. Talia. Scalable Parallel Clustering for Data Mining on Multicomputers. In *IPDPS'00*, Cancun, Mexico, May 2000.
8. Guha and Khuller. Greedy strikes back: Improved facility location algorithms. In *SODA: ACM-SIAM)*, 1998.
9. P. Gupta and P. R. Kumar. The capacity of wireless networks. *IEEE Transactions on Information Theory*, 46(2):388 – 404, 2000.
10. J.M. Jaffe and F.H. Moss. A responsive routing algorithm for computer networks. *IEEE Transactions on Communications*, pages 1758–1762, July 1982.
11. K. Jain, M. Mahdian, and A. Saberi. A new greedy approach for facility location problems.
12. Kamal Jain and Vijay V. Vazirani. Primal-dual approximation algorithms for metric facility location and k-median problems. In *FOCS*, pages 2–13, 1999.
13. Jon Kleinberg, Christos Papadimitriou, and Prabhakar Raghavan. A microeconomic view of data mining. *Data Mining and Knowledge Discovery*, 1998.
14. Madhukar R. Korupolu, C. Greg Plaxton, and Rajmohan Rajaraman. Analysis of a local search heuristic for facility location problems. In *Proc. ACM-SIAM*, pages 1–10, 1998.
15. S. Kutten and B. Patt-Shamir. Time-adaptive self-stabilization. *Proc. PODC*, pages 149–158, August 1997.
16. S. Kutten and D. Peleg. Fault-local distributed mending. *Proc. PODC*, August 1995.
17. N. Linial. Locality in distributed graph algorithms. *SIAM J. Comp.*, 21:193–201, 1992.
18. M. Naor and L. Stockmeyer. What can be computed locally? *STOC*, pages 184–193, 1993.
19. M. Sviridenko. An improved approximation algorithm for the metric uncapacitated facility location problem, 2002.
20. R. Wolff and A. Schuster. Association rule mining in peer-to-peer systems. In *Proc. ICDM*, Melbourne, Florida, 2003.

jWebDust: A Java-Based Generic Application Environment for Wireless Sensor Networks[*]

Ioannis Chatzigiannakis[1], Georgios Mylonas[2], and Sotiris Nikoletseas[1]

[1] Research Academic Computer Technology Institute, P.O. Box 1122, 26110 Patras, Greece
{ichatz, nikole}@cti.gr
[2] Dept of Computer Engineering and Informatics, University of Patras, 26500, Patras, Greece
mylonasg@ceid.upatras.gr

Abstract. Wireless sensor networks can be very useful in applications that require the detection of crucial events, in physical environments subjected to critical conditions, and the propagation of data reporting their realization to a control center. In this paper we propose jWebDust, a *generic* and *modular* application environment for developing and managing applications that are based on wireless sensor networks. Our software architecture provides a range of services that allow to create *customized applications* with minimum implementation effort that are easy to administrate. We move beyond the "networking-centric" view of sensor network research and focus on how the end user (administrator, control center supervisor, etc.) will visualize and interact with the system.

We here present its open architecture, the most important design decisions, and discuss its distinct features and functionalities. jWebDust allows heterogeneous components to interoperate (real world sensor networks will rarely be homogeneous) and allows the integrated management and control of multiple such networks by also defining web-based mechanisms to visualize the network state, the results of queries, and a means to inject queries in the network. The architecture also illustrates how existing protocols for various services can interoperate in a bigger framework - such as the tree construction, query routing, etc.

1 Introduction

Wireless sensor networks are very large collections of small in size, low-power, low-cost sensor devices that collect and disseminate quite detailed information about the physical environment. Large numbers of sensor devices can be deployed in areas of interest (such as inaccessible terrains or disaster places) and use *self-organization and collaborative methods* to form a sensor network. The flexibility, fault tolerance, high sensing fidelity, low-cost and rapid deployment characteristics of sensor networks help to create many new and exciting application areas for remote sensing.

---

[*] This work has been partially supported by the IST / FET Programme of the European Union under contract numbers IST-2004-001907 (DELIS) and the Programme PYTHAGORAS under the European Social Fund (ESF) and Operational Program for Educational and Vocational Training II (EPEAEK II).

V. Prasanna et al. (Eds.): DCOSS 2005, LNCS 3560, pp. 376–386, 2005.

This wide range of applications is based on the use of various sensor types (i.e. thermal, visual, seismic, acoustic, radar, magnetic, etc.) in order to monitor a wide variety of conditions (e.g. temperature, object presence and movement, humidity, pressure, noise levels etc.) and report them to a (fixed or mobile) control center. Thus, sensor networks can be used for important applications, including (a) military (like forces and equipment monitoring, battlefield surveillance, targeting, nuclear, biological and chemical attack detection), (b) environmental applications (such as fire detection, flood detection, precision agriculture), (c) health applications (like telemonitoring of human physiological data) and (d) home applications (e.g. smart environments). For an excellent survey of wireless sensor networks see [1].

Another possible categorization of the applications for wireless sensor networks is based on the notification strategy of the authorities, i.e. the way that the authorities are updated on the monitoring state. For example, in a museum, it is important to report only when emergency situations arise, such as an incendiary fire. On the other hand, in habitat monitoring for instance, continuous monitoring of the physical environment is required so that information is gathered over a long period of time. Therefore, depending on the actual application, the following services are required [3]: (i) *Periodic Sensing* (the sensor devices constantly monitor the physical environment and periodically report their sensors' measurements to a control center), (ii) *Event driven* (to reduce energy consumption, sensor devices monitor the environment and send reports only when certain events are realized) and (iii) *Query based* (sensor devices respond to queries made by a supervising control center).

In the light of the above categories of applications and services, we present jWebDust, a software environment that allows the implementation of customized applications for wireless sensor network that can (i) provide a wide range of services, (ii) minimize the overall implementation effort and (iii) considerably reduce the needs for network administration. jWebDust is *modular* and *extendable* as the application implementor can use a selected set of features, modify some of them and provide new ones that best suit his needs; in this sense, jWebDust is able to deal with *several kinds of applications* (size, functionality, etc.).

jWebDust differentiates the system into two main groups: the networked sensor devices (from now on referred as *motes*) that operate using TinyOS and the rest of the network (e.g. control centers, database server, etc.) that is capable of executing Java code. Both system groups use an open architecture implementing the emerging component-based architecture. The standardized component interface and the exchange of data over broadly used protocols provide increased portability. This implies that the system can be used over different machine architectures as well as OS and server technologies.

We define and implement the *Mote Discovery Service* that reduces significantly the overall network administration needed in applications for wireless sensor networks. More specifically, this novel service keeps track of the motes that participate in the wireless sensor network and their technical characteristics (e.g. type of sensors attached to each device, available power, etc.). During the setup phase of the network, the motes report to the control center of the network and get registered in jWebDust's database without any further human interaction. In this sense, the time required by the administrator to register the devices that make up the network and their hardware characteristics

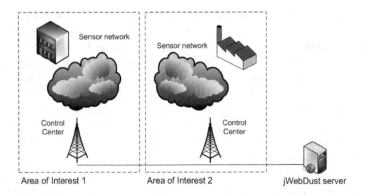

Fig. 1. Multiple wireless sensor networks form a single, unified, virtual sensor network

is greatly reduced, especially in the case where the network is comprised of heterogeneous devices [5,8]. Furthermore, in some cases, it is possible to redeploy additional motes in order to extend the lifetime of the network and/or increase the area of surveillance while the network is in operation [1,5]. In such cases, the discovery service improves the *scalability* of the applications that are based on jWebDust, since the need for network administration is unaffected as the size of the sensor networks increases.

A distinct feature of jWebDust is its ability to manage multiple wireless sensor networks, each with a different control center, under a common installation (e.g. see Fig. 1). This is done by introducing the notion of a *virtual sensor network* that somehow, *hides* the actual network topology and allows the user to control the motes as if they were deployed under a single, unified, sensor network. This abstraction significantly reduces the overhead of administering multiple networks. Furthermore, the idea of a *unified, virtual sensor network* allows the integration of *totally heterogeneous sensor networks*, i.e. not only regarding different kind of sensors attached to the motes of the network, but also different kind of CPU architectures and communication units.

In jWebDust, the information collected by the control center of each operating sensor network is stored in a single, central, relational database. Based on this database server, jWebDust provides a *web-based, user-friendly interface* that targets both to scientific as well as other less technically-trained personnel. This user interface is *customizable* and allows the designer to present the information in different ways and offers extendable statistics based on the needs of the application.

The component-based architecture that is used in designing jWebDust provides *high adaptability*. The software under development provides an open interface through which added functionality can be implemented and integrated at later stages, probably carried out by the final application implementor.

Although wireless sensor networks have attracted a lot of attention from researchers at all levels of the system hierarchy (from the physical layer up to the application layer), generic application environments that provide all the necessary tools and operations to allow the implementation of a wide range of applications are few [1]. To the best of our knowledge, there are only two such environments: TinyDB [11] and Mote-VIEW [9].

TinyDB is an application that allows multiple concurrent queries, event-based queries and time synchronization through an extensible framework that supports adding new sensor types and event types. The central idea of TinyDB is to provide an SQL-like interface to the programmer that makes the wireless sensor network *look like* a *DBMS*. A tree based routing scheme is used for multi-hop communication inside the network. In addition, TASK [10] (Tiny Application Sensor Kit) is built on top of TinyDB in order to further simplify application deployment and development. The kit includes a server that acts as a proxy for the sensor network on the internet, a relational database where readings from the sensor network are stored, and a frontend that help to choose, record and visualize motes' metadata. In contrast to the database schema used in TASK, our approach is more detailed and can be easily extended to the final application's functionality needs (e.g. adding new sensor types, mote types, etc.).

Mote-VIEW [9] is an application that provides tools to the user to visualize results from a sensor network, combined with a data logger that runs on the sensor network gateway. The data logger constantly listens to readings arriving from the network through a control center attached to the gateway and stores them in a relational database. The motes in the sensor network poll their sensors for readings at a sampling rate specified by the user and send them to the gateway using a multi-hop protocol. The user can check readings from the motes' sensors on the fly, see a visualization of the network's topology, produce graphs from selected motes' readings, and check their status.

2   The Architecture of jWebDust

jWebDust is designed on a component-based architecture with several and diverse design goals as a guide that emphasizes on autonomy, reliability, and availability. At a high-level, the components of jWebDust are organized, using the *N-tier* application model, as follows: (i) the *Sensor Tier* that consists of one or more wireless sensor networks deployed to areas of interest, (ii) the *Control Tier* that corresponds to the control centers where the wireless sensor networks report the realization of events, (iii) the *Data Tier* responsible for storing the information extracted from the wireless sensor network(s), (iv) the *Middle Tier* that is responsible for processing the data to generate statistics and other meaningful information and (v) the *Presentation Tier* that interfaces the information with the final user in an easy way based on the capabilities of the user's machine. The five tiers that make up jWebDust are shown in Fig. 2.

The Sensor Tier. Naturally, the sensor tier is the foundation of any application based on wireless sensor networks. The motes are usually scattered in the area of interest and form one or more sensor networks. Each of these scattered motes has the capability to collect data and route data back to the control center and the end users. Data are routed to the control center by a multi-hop architecture (described in greater details in Sec. 3) and then the control center communicates with the other tiers of the system possibly via internet.

The jWebDust firmware, executed by the motes, is based on the TinyOS operating system and implements a *mote discovery protocol* for reporting their sensing and processing characteristics to the control center and a *query dissemination protocol* for

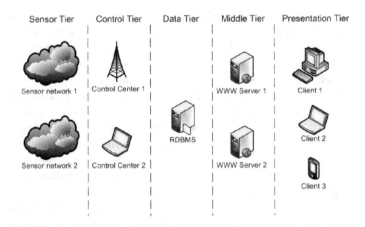

Fig. 2. N-tier Architecture

distributing queries in the sensor network and disseminating the data matching these queries back to the control center; these protocols are defined more formally in Sec. 3.

The Control Tier. The *Control Tier* consists of the control centers of each sensor network. Control centers are responsible for the gathering of all the readings coming from the sensor networks and the forwarding of the queries from the data tier to the motes; in other words, they *act as gateways* between the sensor tier and the data tier. On the hardware level, a typical control center consists of one mote connected to a desktop PC or laptop along with a network connection to the database server. The mote attached to the control tier is necessary to communicate with the sensor network. Alternatively, an embedded platform can be used (e.g. Stargate [9]).

One important feature of the control tier is the *ability to operate even when there is no connection to the data tier for sustained periods of time*. The importance of this feature is outlined by the fact that control centers themselves may have a wireless connection to the data tier and uninterrupted communication is not guaranteed. During such a disconnected period, any new queries made will not be forwarded to the sensor network (since the control center cannot be informed about these queries), while sensor readings received will not be sent to the data tier but will be stored locally. As soon as the control center establishes a connection with the data tier, all locally stored readings are forwarded to the data tier and the new queries are forwarded to the network.

Another jWebDust's feature is the support of multiple sensor networks in a way that they are seen as a *single virtual sensor network*. For each sensor network a unique ID is given to its respective control center. This sensor network ID helps to distinguish one mote from another, when they have they same mote ID but belong to different sensor networks, thus making it possible to manage different networks in a unified way.

The Data Tier. The components at the data tier are based on the relational database system and the functionality and methods provided are related to the services required by the *middle tier* and the *control tier*. Our database schema consists of ten tables that can be organized in three categories:

(i) *Mote related* tables. The information related to the hardware characteristics of the motes that comprise the wireless sensor networks are organized in this category with the use of five tables. The database scheme supports heterogeneous sensor networks, i.e. networks that consist of different kinds of motes (e.g. Telos, MicaZ motes etc.), which in turn may have different kinds of sensors attached to them (e.g. light, pressure, humidity, temperature etc.).

(ii) *Query related* tables. This group of tables holds information regarding the queries that are made on the network that may possibly address multiple sensors and/or multiple sensor types (for more information see Sec. 3).

(iii) *Sensor readings* table. All the information received from the sensor networks that match a specific query is stored in this table. Each record represents a reading coming from a specific mote in the network concerning a single sensor.

The Middle Tier. The Middle tier is comprised by all the components that make up the jWebDust logic and are responsible for delivering structures and data to the presentation tier. The components of the middle tier can be considered as applications that run on a server without a face (also referred to as *servlets*). These components have autotelic functionality that is executed independently in order to process certain data structures and produce specified output. In this extent, components contribute to a simpler distribution of workload and allow easy development of the presentation tier services. In jWebDust the components of the presentation tier and the middle tier can operate simultaneously and independently.

Components are formulated based on the data structures available and the operations required by each entity. For example, the management of queries is handled by a single component and the methods implemented perform actions such as `createQuery`, `deleteQuery`, `updateQuery`, etc. All these methods accept parameters given by the presentation tier. The logic of the creation of a new query (i.e. check if a similar query already exists or if the mote provides the sensors specified in the query) is standardized through a single component. The query component just described, along with other similar components that are part of the jWebDust system are also referred to as the executants. The executants provide a specified interface to the other components; thus if modifications to jWebDust logic are carried out, the implementation of this interface can be modified. As long as the interface remains unaffected, components that use the executants will not notice any change and will not require any further change.

The Presentation Tier. The presentation tier is basically the user interface; it is the layer most available to the end users, responsible for the collection of input and presentation of the information collected from the sensor network. The components that make up the presentation tier are based on different technologies. The terms thin client and rich client are used to describe the capabilities of the presentation tier given the available resources. We characterize as rich clients the components of the presentation tier that are rich of functionality. The support of thinner clients will ensure the interoperability of the system through the various available machine architectures. The end users are

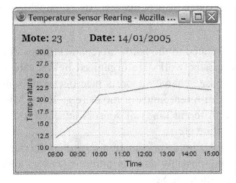 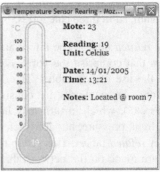

Fig. 3. GUI windows displaying sensor readings reported by the sensor network that match sample queries made by the user to specific motes, regarding different type of sensor devices

allowed to choose among the available client solution in order to maximize the available resources. The open architecture of jWebDust system allows new components to be introduced at later stages that will cope with these issues.

3   Sensor Network Services and Protocols

Mote Discovery Service. One of jWebDust's features is that every mote is able to register itself and its distinct features to the system, thus giving a clear view of the network and removing the need to initiate a mote discovery process periodically. The *mote discovery service* provides the user with a detailed view of the properties of each mote inside the (possibly heterogeneous) sensor network. Other existing applications do not provide such detailed information and thus do not cope well with the case of having motes with different sensors attached to them. As an example, consider the case where we want to have motes collecting temperature readings on the one side of the sensor network and on the other side motes with a different kind of sensor board collecting humidity readings. The use of different kinds of sensor boards could also be justified in terms of cost or because we want to take advantage of older equipment.

The operation of the discovery service relies on the correct programming of the motes; when installing the application firmware on the mote, the actual hardware characteristics are passed as parameters to the service. Each sensor type available to each mote and its mote type are represented by an integer, in correspondence to the integer IDs used in the data tier concerning the mote and sensor types (see Sec. 2). Based on this information, the discovery and registration of every mote in the network is achieved using a simple protocol.

Essentially, when a mote is powered on, and after its startup and initialization phase completes, the mote sends out to the control center a *short message* containing the mote's ID, type and a list of available sensors. By sending this message, the mote registers itself in jWebDust's data tier along with an amount of useful information when the control center receives the message. Note that after sending the message, the mote waits for an acknowledgement message from the control center. This is done to make sure that the registration message will reach the control center within a certain time period. If such an acknowledgment message is not received within the given time period (that can be adjusted depending on the expected network size and communication medium parameters), the mote sends another registration message. This process is repeated until an acknowledgement is properly received.

Time Synchronization Service. Time synchronization of the motes is a significant part of any sensor network as applications need some kind of collaboration between the motes in the network, which in turn means some kind of acceptable time synchronization. jWebDust uses a time synchronization scheme that follows an approach similar to TPSN [7]. Our method provides accurate time synchronization as it performs pairwise time synchronization between motes using the structure established by the communication protocol (see below). Every mote maintains a local clock that is synchronized with respect to a reference mote (e.g. the control center). In order to avoid collisions in the medium access level, at the beginning of the synchronization process, each mote decides to initiate time synchronization, individually, after a random period of time following the reception of a *time-sync* message.

Sensor Network Communication Protocol. Communication in wireless sensor networks presents many distinctive characteristics, which render the conventional and most ad-hoc routing approaches inadequate for use in these networks. Many routing schemes have been proposed specifically for sensor networks [4, 6], however only a few of them have been implemented in TinyOS. Routing protocols for TinyOS can be divided into two categories, based on the number of possible routing destinations.

The communication protocol used in jWebDust is based on the multi-hop tree-based routing protocol included in the TinyOS distribution, called MultiHopRouter, with an extension that aims towards reducing the energy dissipation of the motes. More specifically, we implement an extension that uses a mechanism for varying the transmission range [2] on top of MultiHopRouter. The implementation of such a mechanism is based on the ability of TinyOS to adjust the transmission range of the motes, e.g. in the MICA platform via the Potentiometer component.

The main idea is to periodically check whether a satisfactory number of nearby motes is active. This done by periodically broadcasting "hello" messages from the control center and flooding the network. If motes maintain network connectivity, this message will reach all the motes of the network, at a certain time point. By maintaining a flag in each mote, which should be up if any neighbors have sent "hello" messages over some period of time and down if no such message has been received, motes can decide on whether to modify their transmission range in order to overcome network connectivity issues or not. The concept of a varying transmission range protocol (VTRP) and a study of ways of modifying the transmission range are described in [2].

Sensor Query and Data Dissemination Protocol. The queries supported in jWebDust are categorized using two criteria; (i) the motes they are targeted to and (ii) the way information regarding the queries are reported to the control center. The first category includes *mote-specific* and *attribute-based* queries, while the second category includes *periodic sensing*, *event driven* and *query based* queries. These two categories are overlapping and the actual sensor queries are combinations of these categories. jWebDust gives the users the capability to send to the sensor network all these types of queries.

In the case of a *Mote-based* query, we are interested in targeting specific motes by using their IDs; the protocol will send one packet per each mote included in the query. On the other hand, *attribute-based* queries are not targeted to specific sensors but instead to the whole sensor network; the protocol will flood the network with the query and only those motes that match the specific attribute will respond to the query.

Periodic-update queries pose another time constraint along with the start and stop time constraints, regarding the time of reporting to the sink. They specify an interval time period between successive query reports. *Event-driven* queries instead of posing specific time constraints, use the notion of events in order to define the time to report back to the sink.

Examples of possible queries to a sensor network include "give me the temperature and humidity readings from motes where light reading is over 200 starting from 10:15 till 13:30" and "give me the light readings from mote 4 when temperature reaches 30°C". Thus, we have four types of possible queries, (i) attribute-based periodic update, (ii) attribute-based event-driven, (iii) mote-specific periodic update, and (iv) mote-specific event-driven queries.

The standard packet size in TinyOS is 34 bytes, leaving 29 bytes when using standard single hop communication and less when using multihop protocols. The space left suffices to send all types of queries. Timestamps occupy 5 bytes, while mote IDs occupy 2 bytes and sensor type IDs 1 byte. Events and attribute constraints can be expressed in the same way, occupying 4 bytes, 1 byte for the type of sensor ID, 1 byte for the relation used (equal, greater than, etc.) and 2 bytes for the value used in the constraint.

Because of the fact that a query can poll many sensors in the same mote and one packet might not be sufficient for sending the readings from all the requested sensors, more than one packets can be sent by motes answering a query back to the sink. The sequential messages will contain the readings that didn't fit in the previous ones.

Finally, the protocol supports the possibility of "cancelling" a query that is currently active. The user is able to dispatch a special cancel-query message that contains one or more IDs referring to the queries that are to be cancelled. Note that if the query that needs to be cancelled is *mote-based*, the message is sent directly to the mote involved; if it is an *attribute-based* the message needs to flood the whole network.

4   Concluding Remarks

We have presented the architecture and the design of the main components of jWebDust, and discussed several distinct features along with their implementation and functionality. The main strengths of jWebDust are:

1. Provides an environment to package and manage the plethora of lower level protocols with *minimum implementation and administration effort*.
2. Allows to implement *new functionality that can be easily integrated* with the rest of the architecture at all levels of the hierarchy (i.e. sensor level, control centers level, database level, user-interface level) to best suit the application needs.

3. Supports multiple/separate sensor networks (i.e. physically separated) with multiple/separate control centers and allows to handle them as a single *virtual* sensor network, even if more than one sensor share the same ID.
4. Handles sensor networks comprised of devices with *heterogeneous* characteristics. Real world sensor networks will rarely be homogeneous, motes-only deployments.
5. Supports Disconnected/Mobile Control Centers by taking special care of long disconnections of the control centers, and the sensor networks attached to these centers, from the upper parts of the architecture (i.e. database, web, etc.), so that information is not lost.
6. Many users can simultaneously query, monitor, and visualize the execution of the wireless sensor network through a Web-based user interface.
7. Offers an Extended Query Language that supports (i) multiple queries per sensor device, (ii) concurrent queries through out the network, (iii) monitoring of the active network queries, (iv) mechanisms to cancel a query.

We plan to continue the development of jWebDust by improving its features and introducing new ones. We plan to extend the Sensor Query and Data Dissemination Protocol by employing *data aggregation* techniques that can considerably reduce the communication overhead and improve the lifetime of the network. Also, we wish to consider investigate alternate architectures that can allow multiple control centers per sensor network; e.g. a JVM-equipped handheld that is carried by the administrator. To this end, we are considering the use of a many-to-many routing scheme within the sensor tier.

References

1. I.F. Akyildiz, W. Su, Y. Sankarasubramaniam, and E. Cayirci, *Wireless sensor networks: a survey*, Journal of Computer Networks **38** (2002), 393–422.
2. T. Antoniou, A. Boukerche, I. Chatzigiannakis, G. Mylonas, and S. Nikoletseas, *A new energy efficient and fault-tolerant protocol for data propagation in smart dust networks using varying transmission range*, 37th Annual Simulation Symposium (ANSS 2004), 2004, IEEE Press, pp. 43–52.
3. A. Boukerche, R.W.N. Pazzi, and R.B. Araujo, *A supporting protocol to periodic, event-driven and query-based application scenarios for critical conditions surveillance*, 1st International Workshop on Algorithmic Aspects of Wireless Sensor Networks (ALGOSENSORS 2004), Springer-Verlag, 2004, Lecture Notes in Computer Science, LNCS 3121, pp. 137–146.
4. I. Chatzigiannakis, T. Dimitriou, S. Nikoletseas, and P. Spirakis, *A probabilistic forwarding protocol for efficient data propagation in sensor networks*, 5th European Wireless Conference on Mobile and Wireless Systems beyond 3G (EW 2004), 2004, pp. 344–350.
5. I. Chatzigiannakis, A. Kinalis, and S. Nikoletseas, *Power conservation schemes for energy efficient data propagation in heterogeneous wireless sensor networks*, 38th Annual Simulation Symposium (ANSS 2005), 2005, IEEE Press.
6. I. Chatzigiannakis, S. Nikoletseas, and P. Spirakis, *Efficient and robust protocols for local detection and propagation in smart dust networks*, Journal of Mobile Networks and Applications **10** (2005), no. 1, 133–149, Special Issue on Algorithmic Solutions for Wireless, Mobile, Ad Hoc and Sensor Networks.

7. S. Ganeriwal, R. Kumar, and M. Srivastava, *Timingsync protocol for sensor networks*, 1st ACM International Conference On Embedded Networked Sensor Systems (SenSys 2003) (Los Angeles, CA, USA), 2003, pp. 138–146.

8. *Exploratory research: Heterogeneous sensor networks*, Intel Technology Journal: Research & Development at Intel (2004),
http://www.intel.com/research/exploratory/heterogeneous.htm.

9. *Mote-VIEW monitoring software, Crossbow Technology Inc.*,
http://www.xbow.com/Products/productsdetails.aspx?sid=88.

10. *Tiny application sensor kit (TASK), Intel Research, Berkeley*,
http://berkeley.intel-research.net/task/.

11. *TinyDB: A declarative database for sensor networks*,
http://telegraph.cs.berkeley.edu/tinydb/.

Networked Active Sensing of Structures

Krishna Chintalapudi[1], John Caffrey, Ramesh Govindan, Erik Johnson,
Bhaskar Krishnamachari, Sami Masri, and Gaurav Sukhatme

University of Southern California, Los Angeles, CA 90089, USA

Structural Health Monitoring (SHM) focuses on developing technologies and systems for detecting and locating damages in structures such as buildings, bridges, and aerospace structures. SHM techniques typically analyze changes in the structural response induced in a structure (from before and after possible damage) due to *ambient* (such as heavy winds or passing vehicles) or *forced* (shakers and impact hammers) excitation sources to detect and locate damages.

Untethered wireless sensor network-based structural sensing can significantly drive down cabling installation and maintenance costs while allowing flexible, dense, deployments. Use of wirelessly controlled actuators at various locations in the structure, capable of delivering deterministic excitations can lead to automated SHM sensor-actuator networks that allow for very low-duty cycle operations. We envision autonomous sensor-actuator networks that test structures by periodically exciting them at pre-determined locations and analyzing the structural responses. Such sensor-actuator networks promise to bring about a fundamental paradigm shift in SHM.

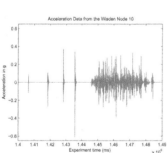

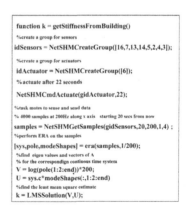

Fig. 1. Structural response collected by a Wisden node on a real structure

Fig. 2. A sample code in matlab for locating damage in a structure using NETSHM API

In our project, we have developed two software systems[1], Wisden and NET-SHM, for facilitating sensor network based SHM. Wisden enables continuous data acquisition over a self-configuring multi-hop wireless sensor network. NET-

---

[1] For our publications and software, please see http://net-shm.usc.edu/.

V. Prasanna et al. (Eds.): DCOSS 2005, LNCS 3560, pp. 387–388, 2005.

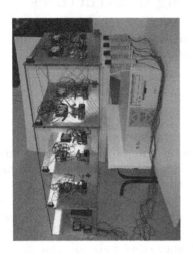

Fig. 3. A 4-story model building for forced excitation based testing using NETSHM.

Fig. 4. A 4-story model building for forced excitation based testing using NETSHM.

SHM provides a software platform for SHM engineers that allows them to program and implement and deploy various SHM algorithms in a high level language such as C/Matlab without having to understand the intricacies of the underlying sensor network.

NETSHM and Wisden distinguish themselves from other existing sensor network applications in terms of their need for high data-rates (in the range of a few Mbps), reliable data transfer (SHM algorithms are intolerant to losses), and tight time-synchronization among nodes (typically within a few 100 μsec). These requirements lead to very unique architectural and design choices and requirements. Figure 1 shows the structural responses collected by a Wisden node on a real structure. Figure 2 depicts a sample Matlab code in NETSHM that was used to locate damages in a 4-story building model (Figure 3).

We have also developed a realistic testbed for experimentation. Our testbed (Figure 4) is a full-scale realistic imitation of a 28' × 48' hospital ceiling. The ceiling is complete with real electric lights, fire sprinklers, drop ceiling installations and water pipes carrying water. It is designed to support 10,000 lb of weight. The entire ceiling can be subjected to uni-axial motion with a peak-to-peak stroke of 10 inches, using a 55,000 lb MTS hydraulic actuator having a ±5 inch stroke. The hydraulic pump delivers up to 40 GPM at 3000 PSI. The total weight of the moving portion of the test structure is approximately 12,000 lb.

We are currently constructing several robotic actuators that can be remotely commanded to move to various locations in the above seismic structure and impart local excitation. We are also investigating the use of other modalities, such as imaging, to help increase the fidelity of these algorithms.

Wireless Technologies for Condition-Based Maintenance (CBM) in Petroleum Plants

Kannan Srinivasan[1], Moïse Ndoh[1], Hong Nie[2], Congying (Helen) Xia[2], Kadambari Kaluri[2], and Diane Ingraham[2]

[1] National Research Council of Canada, Institute for Information Technology
Wireless Systems, Sydney, Nova Scotia, Canada
{Kannan.Srinivasan, Moise.Ndoh}@nrc-cnrc.gc.ca
http://iit-iti.nrc-cnrc.gc.ca
[2] Cape Breton University, Sydney, Nova Scotia, Canada
{Hong_Nie, Helen_Xia, Diane_Ingraham}@capebretonu.ca
http://www.capebretonu.ca

Abstract. Wireless devices and systems have been deployed in many communication sectors but have not yet been adequately adapted to harsh industrial environments. Cost effective, reliable and scalable wireless technologies have yet to be introduced in industrial plants, including petroleum plants. At this Stime, there are very few, if any, cost-effective Commercial-Off-The-Shelf (COTS) radios available that can be used to provide reliable data links under the very harsh conditions encountered in petroleum environments. This paper presents a strategic approach to research wireless system technologies and their application to the oil and gas industry, specifically for asset integrity and automated diagnostic CBM applications.

1 Introduction and Strategy

The failure of critical operational equipment in the petroleum industry is a major expense because downtimes mean a loss of revenue to the companies. Recent study [5] has shown that inaccurate predictions of lifetime of equipment costs about $1 trillion per year in replacing good equipment. To mediate this, companies may choose to employ Condition-Based Maintenance (CBM) strategies that maximize operational uptimes and minimize equipment repair and replacement. CBM requires timely, reliable, up to date, and cost effective data collection for reliable equipment failure prediction [8].

Currently the petroleum industry pays high labor charges for skilled technicians to run routes with data collectors on a regular basis as part of (CBM) procedures [5]. This labor intensive solution is not always timely and cost effective. An alternative solution involves hard wired data collectors. The costs of installing such dedicated solution is very expensive due to the cost of conventional (copper, fiber optic) wires. Some estimates are as high as $2000 per linear foot! Plants operating in rigorous environments with extremes of high and low temperatures, rain, snow, ice, and harsh

V. Prasanna et al. (Eds.): DCOSS 2005, LNCS 3560, pp. 389–390, 2005.

or corrosive chemicals mean even hardened cables must be replaced every 6 months and cost estimates in the order of $6 - $8 million are certainly possible depending on the project.

A reliable, robust and rugged wireless mesh networking architecture, apart from solving the cable issue, could provide near real time monitoring capability. This, integrated with existing CBM trend analysis software, can help predict equipment failure in a timely fashion. However, the electromagnetic environment in a petroleum plant is a challenge for implementing any wireless communication technology. We propose to design a power-efficient, robust wireless mesh network that involves careful modeling [7] of the wireless channel followed by suitable transceiver architecture implementation [6] [2], signal processing [1] [3] and cross layer protocol stack [4] designs.

References

1. Chiasserini. C. F, and Rao. R. R, "On the Concept of Distributed Digital Signal Processing in Wireless Sensor Networks.", IEEE MILCOM 2002, Anaheim, CA, USA, October 2002.
2. Guest Editors: Czylwik. A, Gershman. A, and Kaiser. T, "Special Issue on Advances in Smart Antennas," *EURASIP Journal on Applied Signal Processing*, vol. 2004, no. 9, Aug. 2004.
3. Lewis. F, Wireless Sensor Networks. Smart Environments: Technologies, Protocols, and Applications, John Wiley, New York (2004).
4. Madan. R, Cui. S, Lall. S, Goldsmith. A, Cross-Layer Design for Lifetime Maximization in Interference-Limited Wireless Sensor Networks, IEEE INFOCOM, March 2005.
5. McLean. C, Wolfe. D, Intelligent Wireless Condition-Based Maintenance, Sensors Magazine, Vol. 19 no. 6, pp. 1-12, June 2002.
6. Guest Editors: Poor. H. V , Barbarossa. S, Papadias. C, and Wang. X, Special Issue on MIMO Communications and Signal Processing, *EURASIP Journal on Applied Signal Processing*, vol. 2004, no. 5, May 2004.
7. Sarkar. T. K, Ji. Z, Kim. K, Medouri. A and Palma. M. S,"A Survey of Various Propagation Models for Mobile Communication ," IEEE Antennas Propagation Magazine, vol.45, no.3, pp. 51-82, June 2003.
8. Taylor. J. H, Sayda. A, "Intelligent Information, Monitoring, and Control Technology for Industrial Process Applications," FAIM 2005, Spain, July 2005.

SensorNet Operational Prototypes: Building Wide-Area Interoperable Sensor Networks – Extended Abstract*

Mallikarjun Shankar, Bryan L. Gorman, and Cyrus M. Smith

Oak Ridge National Laboratory, Oak Ridge, TN, USA
http://www.sensornet.gov

1 Overview

As sensor network deployments continue to expand, government agencies will benefit from the transmission and processing of sensor data aimed at enhancing public safety and utility services. Oak Ridge National Laboratory's SensorNet project is building a vendor-neutral interoperability framework based on emerging standards for plug-n-play access, control, and integration of online sensors, sensor-derived data repositories, and sensor-related processing capabilities [3]. Focussing on wide-area deployments and on a broad class of sensors and applications, the system prototypes communicate and analyze data from transducers (sensors[1] and actuators) using mechanisms that allow disparate entities to alert each other and share critical information. Deploying operational prototypes allows us to instantiate and field building block components early while collecting a broad range of user feedback in our incremental development.

2 Primary Components

Sensors and applications must modularly plug into the system so that it is easy to interconnect them without having to custom build an end-to-end channel over the wide-area. For the transducer interfaces, we adopt the methodology advocated by the IEEE 1451 working groups [1] currently developing standards for "Smart Transducers". A smart transducer includes sufficient descriptive information so that a control software component can automatically determine its operating parameters, decode the (electronic) data sheet, and issue commands to read and actuate the transducer.

For the applications interfaces, we build on the techniques enabling interoperability in the web-services arena and base our (application layer) communications protocols on these services. We take advantage of the framework of web-services primarily for lower frequency model interactions and user-directed control. We

---

* Supported by SensorNet®, Oak Ridge National Laboratory, managed by UT-Battelle, LLC, for the U. S. Department of Energy under Contract No. DE-AC05-00OR22725.
[1] Sensor and Transducer are terms used interchangeably in the community because of the legacy connotations of the former.

V. Prasanna et al. (Eds.): DCOSS 2005, LNCS 3560, pp. 391–392, 2005.

incorporate service directories and data dictionaries to make application services use consistent terminology so that they can interoperate. A majority of transducer data is geospatially referenced, and consequently we adopt specifications and standards developed by the Open Geospatial Consortium (OGC) [2] in defining our services. We use the notion of a data *feature* - a descriptive container for almost all data and sensor entities in SensorNet that aims to normalize diverse sets of information into a baseline canonical form.

3   Deployment Experiences: Some Lessons Learnt

In contrast to finer granularity deployments, ours allowed for greater disk storage, required remote connectedness (which we implemented through commercial wireless modems and satellite dishes), greater computing capability with a powered source, and a requirement to support a wide variety of sensors and applications. We found that while the emphasis on research for elemental sensors may involve their networking, power management, and information aggregation, the research challenges for intelligent/sophisticated sensors lie in integrating the sensor's specific behavior (e.g., placement considerations) with applications that task and share data from (hard and soft) sensors. Two broad observations follow:

– *Widespread Adoption and Data Services* We need a core set of standards or de facto procedures at several levels of any nationwide deployment hierarchy. While commercially available IP-based networks remain an effective resource to establish communication links, the sensor data-exchange protocols require enhanced capabilities that support flexible application integration. Unlike custom deployments, a large sensor network is likely to become part of a national infrastructure in much the same way as the Internet. This suggests that standards and task force bodies (that operate in spirit like the IETF) and community efforts such as RFC's can speed up the spread of ubiquitous sensor networks.
– *Management Plane* Management plane tasks are likely to be the most difficult (i.e., we usually know more about how to build the infrastructure than how to manage it). Although not necessarily foundational in terms of research, this will drive most of the commercial thrusts. We believe an important focus for research is language and system constructs needed for intelligent sensors. For example, a programming language or system abstraction for software plug-n-play in the context of sensor plug-n-play will ease rapid sensor networks deployment.

The functionality of the sensors in the sensor network and the applications that use them will ultimately dictate the rate at which sensor networks will proliferate in the real world – nonetheless, a standards-based approach, that generalizes across sensor categories and application protocols while fostering interoperability will best anticipate the needs of the future.

References

1. Kang Lee: Synopsis of IEEE 1451 Family, Sensors Expo, June 2004.
2. The Open Geospatial Consortium, http://www.opengeospatial.org
3. Bryan L. Gorman, Mallikarjun Shankar, and Cyrus M. Smith: Advancing Sensor Web Interoperability, *Sensors Magazine*, Vol. 22, No. 4, April 2005.

Project ExScal (Short Abstract)

Anish Arora, Rajiv Ramnath, Prasun Sinha, Emre Ertin, Sandip Bapat,
Vinayak Naik, Vinod Kulathumani, Hongwei Zhang, Mukundan Sridharan,
Santosh Kumar, Hui Cao, Nick Seddon, Chris Anderson[1], Ted Herman,
Chen Zhang, Nishank Trivedi[2], Mohamed Gouda, Young-ri Choi[3],
Mikhail Nesterenko, Romil Shah[4], Sandeep Kulkarni, Mahesh Aramugam,
Limin Wang[5], David Culler, Prabal Dutta, Cory Sharp, Gilman Tolle[6],
Mike Grimmer, Bill Ferriera[7], and Ken Parker[8]

[1] The Ohio State University, Columbus, USA
[2] University of Iowa, Iowa City, USA
[3] The University of Texas at Austin, USA
[4] Kent State University, Akron, USA
[5] Michigan State University, East Lansing, USA
[6] University of California at Berkeley, USA
[7] Crossbow, San Jose, USA
[8] Mitre Corporation, Washington DC, USA

Project ExScal (for Extreme Scale) fielded a 1000+ node wireless sensor network
and a 200+ node ad hoc network of 802.11 devices in a 1.3km by 300m remote
area in Florida during December 2004. In several respects, these networks are
likely the largest deployed networks of either type to date. We overview here
the key requirements of the project, describe briefly how they were met and
experimentally tested, and provide a pointer to our experimental results.

The Application. The ExScal concept of operation is to deploy a dense wireless
sensor network "tripwire" that detects, tracks, and classifies multiple, different
types of intruders in a long perimeter region. Application of this concept is
envisioned for protection of pipelines that are vulnerable to sabotage, borders
between nations that are prone to illegal crossing, and areas abutting critical
plants/thoroughfares that are vulnerable to terrorist threat.

The primary requirements for this application are:

1. *Low cost of covering a long perimeter over the mission lifetime.* This translates to selection of: sensing and communication modalities that have desirable range and enable low power operation, appropriate packaging, as well
   as node layouts that avoid nodal redundancy. Mission lifetime is 1-6 months.
2. *Accurate, timely, reliable, and robust operation.* This translates to low false
   alarm and loss omission rates in detection, tracking, and classification. Since
   the physical terrain is not assumed to be constrained, the network must deal
   with breaches anywhere along the long perimeter. Response must be quick,
   within a few seconds of intruder events over the mission lifetime. Quality of
   operation is required even if some nodes in a region are misplaced or their
   components fail, during deployment or operation.
3. *Low human effort.* This applies to all phases, including the placement of
   the nodes as well as in the operation, monitoring, maintenance, and reconfiguration of the network.

V. Prasanna et al. (Eds.): DCOSS 2005, LNCS 3560, pp. 393–394, 2005.

Architectural Principles. To meet these complex requirements, our design of ExScal relies on the following principles.

1. To contain cost, we design nodes that have competitive sensing and communication ranges and emplace them in a *planned topology*, specifically a *regular, hierarchical structure*, to efficiently cover the region. XSM (for Extreme Scale Mote) nodes have 30m+ reliable communications, passive infrared sensing with 15m/25m ranges for humans/vehicles, respectively, microphone sensing with 50m+ range for all-terrain vehicles, and magnetometer sensing with 8m ranges for vehicles. XSS (for Extreme Scale Stargate) nodes have 500m+ reliable 802.11b communications and GPS sensing with < 10m accuracy. Tier 1 of the hierarchy consists of a triangular grid of the XSMs, Tier 2 consists of rectangular grid of the XSSs, and Tier 3 consists of one master operator node. Application services exploit knowledge of this topology for efficiently utilizing system resources.

2. To meet the desired quality, we decompose ExScal into *multiple subapplications*: a Trusted Base program, a Deployment Application, a Localization Application, and a Perimeter Security Application; these execute in different phases of operation. Decomposition simplifies the design, allows us to configure and manage each subapplication separately (at run time if need be), and reduces operational resource requirements. Importantly, it frees us from using common services for all subapplications, instead we can use different services optimized for subapplication needs.

3. For cost-effective human manageability, ExScal operates by *command and control*: Tier 3 initiates, monitors, and regulates the operations of all XSSs; in turn, each XSS likewise monitors and manages with a section of (normally 20- 50) XSMs. The operator maintains/reconfigures ExScal effectively based on the feedback obtained from the network. *Autonomous* functions, specifically, configurable recoverability for nodes and their component, tolerance to several classes of faults (often by self-stabilization), and adaptation to certain classes of variable, non-uniform environments all support containment of human effort.

ExScal Experiments. 10000 XSMs and 300 XSSs were manufactured (these nodes are now commercially available). The footprint of the code we designed is ~200KB for an XSM and ~2MB for an XSS. We designed ExScal scenarios for node configuration at the factory, for field marking, for node deployment at site, for network configuration in the field, for ExScal operation, and for network teardown. Over a two week period, we collected data on field marking accuracy, deployment yield, localization accuracy, sensing performance and variability, environment data (especially wind data collected via microphones), communications and network management performance at each tier, and intruder traces. This data and other literature on the project is being made available at the ExScal website, http://www.cse.ohio-state.edu/~exscal .

Acknowledgement. Project ExScal was conceived in and funded by the DARPA NEST research program.

NetRad: Distributed, Collaborative and Adaptive Sensing of the Atmosphere Calibration and Initial Benchmarks

Michael Zink[1], David Westbrook[1], Eric Lyons[1], Kurt Hondl[2], Jim Kurose[1], Francesc Junyent[3], Luko Krnan[3], and V. Chandrasekar[4]

[1] Dept. Computer Science, University Massachusetts, Amherst MA 01003
{zink, kurose, westy}@cs.umass.edu
[2] National Severe Storms Laboratory, National Oceanic and Atmospheric Administration
Norman OK 73019
Kurt.Hondl@noaa.gov
[3] Dept. Electrical and Computer Engineering, University Massachusetts
Amherst MA 01003
{junyent, lkrnan}@ecs.umass.edu
[4] Dept. Electrical & Computer Engineering Colorado State University,
Fort Collins, CO 80523-1373
chandra@engr.colostate.edu

1 Introduction

We are currently building a NetRad prototype system to be deployed in southwestern Oklahoma, consisting of four mechanically scanned X-band radars atop small towers, and a central control site (later to be decentralized as the number of radars increases) known as the System Operations and Control Center (SOCC). The SOCC consists of a cluster of commodity processors and storage on which the Meteorological Command and Control (MC&C) components execute. NetRad radars are spaced approximately 30 km apart from each other and together scan an area of 80km x 80km and up to 3 km in height. In this paper, we overview the radar calibration process, as well as the initial benchmark execution times of the software modules we will demonstrate at DCOSS.

2 Radar Calibration

Before a radar (or any sensor) is deployed in the field, it must be calibrated to assess the measurement accuracy. The NetRad radar must perform very accurate measurements since one of the system's capabilities is to pinpoint tornadoes with spatial error no greater than 300m. Our NetRad radars are being calibrated with the aid of the well-calibrated CHILL weather radar from Colorado State University. Radar calibration involves three major steps (that we will illustrate in more detail in the DCOSS demonstration): *(i)* mechanical calibration, *(ii)* standalone calibration, *(iii)* calibration through comparison with the CHILL radar. The major task in the mechanical calibration step is to assure the antenna is pointed accurately in azimuth and elevation. In the standalone calibration step the radar scans non-meteorological

V. Prasanna et al. (Eds.): DCOSS 2005, LNCS 3560, pp. 395 – 396, 2005.

objects such as a launched sphere. In the final step, both radars (NetRad and CHILL) scan a meteorological event simultaneously and results are compared with each other. This step is performed to gain better insight in the accurate scanning of meteorological events. Should different results be obtained in the third step, steps *(i)* and *(ii)* are repeated. A result of this initial calibration routine will be a protocol for an automated calibration of the radar which can be performed in regular intervals during regular operation.

3   Initial Benchmarks for NetRad Control Loop

The Systems Operation Control Center (SOCC) is a centralized compute cluster on which the MC&C algorithms execute. There are five main components (as shown in Figure 1) *(i)* data ingest and storage, *(ii)* meteorological feature detection and multi-radar merging, *(iii)* feature repository, *(iv)* utility and task generation, and *(v)* optimization. The NetRad system is a "real-time" system in the sense that radars must be re-tasked by the MC&C every 30 seconds – the system "heartbeat" interval. Radars are retasked based on detected meteorological features and the projected future evolution of these features. In order to estimate execution times of the various software components, we performed a series of benchmarking experiments, using existing NEXRAD radar data as input to the NetRad MC&C components for the benchmarking. As a result of this experiment we obtained the following average processing times for each of the MC&C components:

MC&C Component	Average Processing Time (seconds)
Data ingest and storage	1
Feature detection and multi-radar merging	0.3
Feature repository	0.3
Utility and task generation	0.03
Optimization	0.03

This results in an overall average processing time of 1.66 seconds, which is well below 30 seconds showing us that the components are well suited for the NetRad system. Processing times depend on the number of meteorological feature present in the data, but for all data sets, runtimes were with 25% of the average value over all data sets.

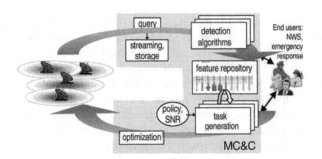

Fig. 1. MC&C components

Service-Oriented Computing in Sensor Networks

Jie Liu and Feng Zhao

Microsoft Research, One Microsoft Way, Redmond, WA 98052, USA
{liuj, zhao}@microsoft.com

Consider the following scenario in an urban business district. A traffic engineer needs to gather vehicle statistics at major intersections for possible improvement to road layout. Police officers want to check out suspicious vehicles as soon as they enter the area. Environmental protection workers monitor smog level and must warn people when the public health is at risk. For economic reasons, these users may all tap into a common pool of sensing sources — a network of networks that connect to different sensors such as vehicle sensors, cameras, and smog detectors scattered across the city. The large-scale networked sensor system is likely to be deployed by multiple venders over time, so that the cost of construction and maintenance may be shared by the different user groups. Such a system may consist of both mobile and stationary nodes with a wide of range of processing and communication capabilities. It must be integrated seamlessly into the Internet and must provide intuitive interfaces for remote user interactions. There will be multiple, concurrent users exercising different functionalities of the system for different purposes. The system will have to be self-monitoring and resource-aware, and has a certain level of autonomy to decide on the best use of available resources to fulfill multiple users' uncoordinated requests.

We have developed a service-oriented architecture, SOARNETS[1], that provides a web-service like interface to these systems, and the underlying service composition and scheduling techniques that allow multiple applications to co-exist. Sensor networks export utilities such as data collection in a region, data processing and aggregation, and communication. In SOARNETS, each service is a process that consumes input and produces output. For example, the sensors in the above scenario may provide a set of services, for example, human detection, vehicle detection, speed calculation, certain chemical element detections, picture taking, etc. The users can send their own queries independently. These queries are processed by composing a subset of these services on demand. This is done by wiring output of some services with input of other services as appropriate. When all queries are running, parts of the query processing services that are common across the applications can be shared.

We adopt the notion of services in sensor networks primarily for their interoperability, scalability, and retaskability. Services are open and self-descriptive. This allows systems from different venders to work together. Services are not necessarily tied to particular nodes. For example, a car detection service can be implemented using acoustic, magnetic, loop inductor, break beam, or image sensor. In a real system, there is typically a great deal of redundant information available to the users. Hiding this information detail behind the service interfaces makes interactions with

---

[1] SOARNETS stands for Service-Oriented Architecture for Networks of Sensors.

V. Prasanna et al. (Eds.): DCOSS 2005, LNCS 3560, pp. 397–398, 2005.

physical phenomena scalable to the number of sensor nodes. In a service-oriented architecture, services are reusable and often generic. Applications are composed after the services are deployed, thus the whole system is retaskable. The scarce resources can be shared among multiple concurrent tasks.

We have developed a prototype of SOARNETS architecture that consists of the following components: service planning, service embedding, and service scheduling. As part of the runtime system, a networked embedded system self-monitors its operating conditions, such as network connectivity, resource availability, as well as preconditions for registered services. This information is made available to the schedule time query planner. When a user issues a query, it is first decomposed by the planner into a set of services forming a *service composition graph* or workflow. The logical graph is then embedded onto the physical sensor nodes, resulting in an assignment of logical operations to distributed processors. For an optimal embedding of the graph, the assignment must take into account node locations, sensing modalities, network topology, service availability, and resource constraints. Since the system configuration may change during the course of executing a long-running query, the result may only serve as an initial embedding subject to run-time adaptation. The task assignment is then injected into the network in the form of a tasking description language – the microserver tasking markup language, or MSTML. This description is accepted by a service scheduler running on each node. In order to control memory consumption, not all services are preloaded. They are created only upon request. The service scheduler is also responsible for checking all other services running on the same node to determine whether part of a new task can be achieved using parts of other tasks. After this optimization step, the task is admitted to execute. Because some tasks are instantiated on-demand, service schedulers may negotiate with the planner to iteratively achieve a feasible and optimal service allocation. After the services are instantiated, the service execution engine executes and monitors the tasks across multiple nodes. When resources change in the network, some services may be migrated to other nodes. If the execution engine cannot determine alternate task composition locally, the schedule time planner may be invoked again. When tasks terminate, the execution engine is also responsible to clean up parts of the task that is not shared by other tasks.

We have built a proof-of-concept testbed to study the SOARNETS architecture. An initial prototype includes two microserver nodes, six sensor motes and a web camera deployed in our underground parking garage. Five Crossbow MicaZ motes are placed in a row, each attached to a break beam sensor. The break beam sensor emits an infrared beam which bounces back from a reflector on the opposite side of the lane. When a car drives by this section of the lane, it blocks the infrared beam one by one. By correlating the sequence of break beam signals, one can detect vehicles and estimate their lengths and speeds. A magnetometer is placed further down the road. When receiving a trigger, it can collect and transmit magnetometer readings to a microserver. Both the five break beam motes and the magnetometer motes communicate wirelessly with a microserver. A camera, connected to another microserver, is also placed near the magnetometer. When receiving a trigger, it can take a picture and send it to the microserver. The system has been used to answer user queries such as gathering vehicle arrival statistics or taking pictures of speeding cars or large trucks.

Wireless Sensors: Oyster Habitat Monitoring in the Bras d'Or Lakes

Diane Ingraham[1], Rod Beresford[1], Kadambari Kaluri[1],
Moise Ndoh[2], and Kannan Srinivasan[2]

[1] Cape Breton University, Box 5300 1250 Grand Lake Road,
Sydney, NS, Canada B1P 6L2
dingraham@mac.com, rod_beresford@capebretonu.ca,
kadambari_kaluri@capebretonu.ca
http://www.capebretonu.ca
[2] National Research Council of Canada, IIT - Wireless Systems,
Box 5300, 1250 Grand Lake Road, Sydney, NS, Canada B1P 6L2
{moise.ndoh, kannan.srinivasan}@nrc-cnrc.gc.ca
http://iit-iti.nrc-cnrc.gc.ca

Abstract. In 2002 a devastating oyster parasite (MSX) was found in the Bras d'Or Lakes of Cape Breton. Environmental parameters affecting the transmission and life cycle of MSX are not well understood. However field observations so far indicate temperature and salinity are critical. Wireless sensor technology can provide robust, reliable and near real time field data collection of such parameters. This paper addresses an implementation of wireless sensors for the timely and cost-effective monitoring and dissemination of bio-physical parameters of interest in the Bras d'Or Lakes.

1 Introduction

In 1957, the oyster population in Delaware Bay was devastated by a protozoan called MSX (multinucleated spherical X, now known as *Haplosporidium nelsoni*) [3]. Approximately 80% of the oyster population dies after initial exposure to this parasite, and after two years, mortality is closer to 95% [2]. By the mid-1990s, it had spread from Florida to Maine [4]. In 2002, it was discovered in the Bras d'Or Lakes. Prior to this, there were no known cases of MSX in Canada. Oysters from the Bras d'Or Lakes have historically been moved throughout Atlantic Canada with the activities of commercial aquaculture. The lifecycle of this parasite is not understood. Laboratory transmission experiments of MSX have been unsuccessful, and eradication of this parasite has not been possible [3]. Field evidence indicates that temperature and salinity are factors in the development and transmission of the parasite [3]. Temperature is one possible method for management of this parasite since oysters in aquacultural endeavours can be positioned in the water column where the temperature inhibits MSX.

The Bras d'Or Lakes, lying between severely glacially weathered cliffs rising to elevations of about 200m, form a brackish inland sea, open at both ends to the Atlantic Ocean. Bathymetric measurements indicate depths to 180m. The shallow (less than

V. Prasanna et al. (Eds.): DCOSS 2005, LNCS 3560, pp. 399–400, 2005.

10m) shoreline environment preferred by oysters is generally accessible and ideal for testing innovative biological and oceanographic monitoring solutions.

This monitoring is an imperative first step to understanding the life cycle of the MSX parasite and so a suitable, cost-effective, robust and reliable data collection technology is needed. Wireless sensor technology is one such technology which provides near real-time data collection capability using a mesh topology [1]. Wireless sensor technology also can make use of existing infrastructure such as the cellular network and the high-speed wireless (802.11/WiFi) that may be available nearby to share the monitored data over the Internet to the local communities including the First Nations [5] who are actively engaged in scientific investigations of the ecosystem of the Bras d'Or Lakes. Wireless sensor technology can be very robust and reliable through the use of mesh networking that will allow sensor nodes to effectively avoid failed nodes.

Although some commercial-off-the-shelf (COTS) wireless sensor nodes based on Zigbee standard [6] and proprietary technologies are available, it is not clear if any of them would be suitable for habitat monitoring of the Lakes. Hence our investigation will start with experimenting with these COTS radios in the Lakes followed by modeling of the wireless channel presented by the Lakes which will aid in proposing suitable wireless sensor technology. The investigation will also include proposing suitable protocol and hardware architectures that will satisfy the requirements of habitat monitoring in the Bras d'Or Lakes.

References

1. Akyildiz IF, Su W, Sankarasubramaniam Y, Cayirci E, Wireless sensor networks: a survey, *Computer Networks*, 38(4):393-422, March 2002.
2. Andrews JD and Wood JL (1967) Chesapeake Science, Vol. 8, Pg. 1-13.
3. Burreson EM, Stokes NA, and Friedman CS (2000) Journal of Aquatic Animal Health, Vol. 12 (1), Pg. 1-8.
4. Burreson EM (2003) Oyster Research and Restoration, US Coastal Waters: Strategies for the Future, The State of Oyster Disease, Abstract #7.
5. Unama'ki Institute of Natural Resources (2004) *Bras d'Or Lake Fisheries Resources and Pollution Source – a working atlas based on the Pitu'paq database* , Pg. 1-67.
6. Zigbee Alliance, http://www.zigbee.org.

Heavy Industry Applications of Sensornets

Philip Buonadonna, Jasmeet Chhabra, Lakshman Krishnamurthy,
and Nandakishore Kushalnagar

Intel Research
{phil.buonadonna, jasmeet.chhabra, lakshman.krishnamurthy,
nandakishore.kushalnagar}@intel.com

1 Introduction and Motivation

Sensors are a cornerstone of heavy industrial operations. Manufacturing plants and general engineering facilities, such as power plants or shipboard engine rooms, require a high degree of sensing to ensure product quality and/or efficient and safe operation. Wireless sensor networks are a natural fit to meeting the demands of scale, data access and cost. However, the nature and environment of industrial applications presents unique requirements for sensornets. To better understand these challenges, we conducted deployments in two settings: a semiconductor fabrication plant and an oil tanker. The context for the deployments was a condition based maintenance application which monitors machinery vibration to detect/preempt failures.

In this discussion, we focus on the oil tanker environment of the Loch Rannoch – a British Petroleum (BP) shuttle tanker operating in the North Sea (Fig 1). This environment was considered challenging for multiple reasons. The all steel construction of the engine room and the isolation of water-tight spaces lead to challenges for RF connectivity. Also, there was a great deal of vibration and temperature variation in the spaces for different operating conditions of the vessel. Finally, the ship's crew would have limited, if any, ability to diagnose and correct sensornet problems.

We conducted pre-deployment surveys of the site and folded the learnings from this survey into the architecture of the system. We then deployed a 26-node, 150 sensor point network using the Crossbow mica2 processor-radio platform with a custom

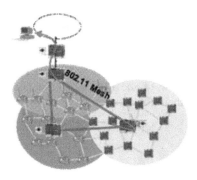

Fig. 1. Hierarchical network architecture

Fig. 2. The Loch Rannoch oil tanker (courtesy British Petroleum)

V. Prasanna et al. (Eds.): DCOSS 2005, LNCS 3560, pp. 401–402, 2005.
© Springer-Verlag Berlin Heidelberg 2005

data-capture board. and 3 XScale® based Stargate gateways. The network collected vibration data and routed it to Rockwell's Emonitor® analysis applications [1].

2 Planning

To better understand the challenges and requirements of the shipboard environment, we conducted an onboard site survey. The procedure involved deploying sensor nodes (without sensors) and gateways at the site and executing simple end-to-end data transfer applications. The results of the survey showed excellent, single-hop connectivity could be achieved throughout the engineering spaces, even across non-watertight bulkheads. 802.11b connectivity could be achieved when watertight hatchways were opened and we exploited this fact to implement disconnected operation across these boundaries. We attributed the connectivity results to the steel construction which likely tended to reflect rather than absorb or pass RF energy.

3 Architecture and Results

Fig. 2 shows the generic architecture we applied to the shipboard deployment. The design is hierarchical with 802.11 gateways providing a backbone for independent sensor node meshes. At the top of the architecture is the interface to the enterprise data applications – in this case the Emonitor® application. We incorporated caches at the gateways to permit operation across watertight boundaries. The gateways would upload data during opportunistic connectivity periods when access hatchways were opened.

The deployed sensornet operated for 4 months without major failures and with no required operator intervention. Some transient failures were observed that would cause certain nodes to not report data when expected. Approximately half the failures could be attributed to a localized hardware problem on one of the gateways or a ship wide power failure that shutdown the gateways. The cause of the other failures was unknown, but attributed to temporary RF interference. In each case of failure, the sensornet subsequently recovered and was able to resume normal operation.

Acknowledgements

The authors wish to acknowledge British Petroleum especially Harry Cassar, Peter Charrett, the CTO Office and the Masters and Crew of the M/V Loch Rannoch. Thanks also goes to Rockwell Automation, UK for their assistance.

References

1. Rockwell Automation: Emonitor Enshare. http://www.rockwell.com

Coordinated Static and Mobile Sensing for Environmental Monitoring

Richard Pon[1,2], Maxim A. Batalin[1,3], Victor Chen[1,2], Aman Kansal[1,2],
Duo Liu[1,2], Mohammad Rahimi[1,3],
Lisa Shirachi[1,2], Arun Somasundra[1,2], Yan Yu[1,4],
Mark Hansen[1,5], William J. Kaiser[1,2],
Mani Srivastava[1,2], Gaurav Sukhatme[1,3], and Deborah Estrin[1,4]

[1] Center for Embedded Networked Sensing, University of California,
Los Angeles, CA 90095
[2] Electrical Engineering Department, University of California, Los Angeles, CA 90095
[3] Computer Science Department, University of Southern California, Los Angeles, CA
[4] Computer Science Department, University of California, Los Angeles, CA 90095
[5] Statistics Department, University of California, Los Angeles, CA 90095

Abstract. Distributed embedded sensor networks are now being successfully deployed in environmental monitoring of natural phenomena as well as for applications in commerce and physical security. While substantial progress in sensor network performance has appeared, new challenges have also emerged as these systems have been deployed in the natural environment. First, in order to achieve minimum sensing fidelity performance, the rapid spatiotemporal variation of environmental phenomena requires impractical deployment densities. The presence of obstacles in the environment introduces sensing uncertainty and degrades the performance of sensor fusion systems in particular for the many new applications of image sensing. The physical obstacles encountered by sensing may be circumvented by a new mobile sensing method or Networked Infomechanical Systems (NIMS). NIMS integrates distributed, embedded sensing and computing systems with infrastructure-supported mobility. NIMS now includes coordinated mobility methods that exploits adaptive articulation of sensor perspective and location as well as management of sensor population to provide the greatest certainty in sensor fusion results. The architecture, applications, and implementation of NIMS will be discussed here. In addition, results of environmentally-adaptive sampling, and direct measurement of sensing uncertainty will be described.

1 Introduction

The first generation of networked embedded sensing systems have been successfully applied to distributed monitoring of environments. These first applications have stimulated rapid growth of applications based on an unprecedented capability for characterizing important environmental phenomena.[1] The primary challenges for operation of networked embedded sensing systems first appeared in the development

V. Prasanna et al. (Eds.): DCOSS 2005, LNCS 3560, pp. 403–405, 2005.

of scalable, low energy networking and cooperative detection. While progress has been made towards addressing these challenges, the deployment of first generation sensor networks has revealed a new class of problems associated with optimizing sensing fidelity and sustainability in complex environments. Specifically, the unpredictable evolution of events and the presence of physical obstacles to sensing introduces uncertainty in sensing and the results of sensor data fusion. Most importantly, the inevitable presence of unpredictable physical sensing obstacles is fundamental to environments of interest and creates a pervasive limitation that threatens to degrade the performance of distributed sensor systems.

2   Coordinated Static and Mobile Sensing

Perhaps the most important challenge for sensor networks is the development of capabilities for sensor network self-awareness where the sensor network itself is capable of determining its own sensing fidelity. Of course, with only a fixed sensor distribution and an accompanying unpredictable set of obstacles, this self-awareness may be unachievable since obstacles may not be identifiable within an environment by fixed sensors alone. Now, since it is the presence of physical obstacles that create uncertainty, then physical reconfiguration of sensors is required to circumvent obstacles. This introduces further challenge, of course, since such mobility must be autonomous and generally sustainable in the environment. However, a new generation of networked embedded systems incorporating controlled and precise mobility is now being explored. These Networked Infomechanical Systems (NIMS) directly address the fundamental objective of self-awareness by enabling motion of sensor node networks to circumvent obstacles, probe sensing fidelity, and optimize sensor and sample distribution.[2]

NIMS introduces a new networked embedded system capability that provides the ability to explore large volumes, adds new networking flexibility and functionality, and new logistics for support of distributed sensors, as well as the capability for self-awareness. This requires, in turn, the development of new methods for scalable and optimized coordination of mobility among nodes for many possible objectives. NIMS also introduces infrastructure-supported mobility to enable low energy transport and retain inherent low operating energy, rapid deployment characteristics, and environmental compatibility of distributed sensors.

3   Coordinated Static and Mobile Sensing Systems and Applications

The first investigations of natural environment phenomena with NIMS have revealed characteristics of fundamental field variables that were not visible with previous sensing methods based only on static sensing. An example is the spatiotemporal distribution of solar illumination, important in the understanding of fundamental ecosystem phenomena. The complexity and rapid evolution of light fields immediately demonstrated that fixed sensors or even simple actuated sensor scanning methods would be inadequate for high fidelity mapping. A series of new methods have been developed that merge static sensing (providing constant vigilance for temporal change in environmental phenomena) with mobile sensing, providing the ability to intensively sample phenomena.   These include adaptive sampling and task allocation methods that provide a means to efficiently manage the distribution of sampling points in a variable field to achieve a specified fidelity threshold.[3] Others include methods that exploit controlled mobility to benefit network performance.

NIMS systems include multiple embedded platform types. Horizontally actuated nodes operate with motion along a horizontally suspended cable. This embedded device, hosting adaptive sampling and other applications also controls the motion of an independently operating and vertically actuated sensor node.   This latter device carries microclimate monitoring devices. These nodes maintain network access to the static nodes also distributed in the environment with static nodes suspended by the cable itself or distributed at the surface. Imager systems carried by the horizontally actuated node include angular perspective actuation. New embedded software systems include embedded statistical computing tools for in-network processing in support of adaptive sampling. The runtime environment and software interfaces connecting application level software systems and sensor and actuator systems follows the Emstar architecture.[4]

NIMS system applications and this tool is being adopted by a community of researchers. First, at the James San Jacinto Mountain reserve,[5] NIMS systems are in use for microclimate and solar radiation mapping. Also, at this same location, a second NIMS system has been adopted for investigation of interaction between surface and subsurface (forest soil) environmental phenomena including measurements of gas transport. NIMS systems have also been deployed for measurement of water quality and contamination in the Los Angeles area watershed. NIMS systems are also under development for deployment in the Merced River of California for characterization of the influence of agricultural processes on river water quality. Finally, a sensing architecture has been designed for deployment in tropical rain forests for characterization of fundamental biological science phenomena as well as for investigation of the impact of fragmentation on forest ecosystems.

References

[1] D. Estrin, G.J. Pottie, M. Srivastava, "Instrumenting the world with wireless sensor networks," ICASSP 2001, 2001.

[2] W. Kaiser, G. Pottie, M. Srivastava, G. Sukhatme, J. Villasenor, D. Estrin, "Networked Infomechanical systems (NIMS) for Ambient Intelligence," Center for Embedded Networked Sensing Technical Report, No. 31, December 2003.

[3] M. Batalin, M. Rahimi, Y. Yu, D. Liu, A. Kansal, G. S. Sukhatme, W. J. Kaiser, M. Hansen, G. J. Pottie, M. B. Srivastava, and D. Estrin "Call and Response: Experiments in Sampling the Environment," Proceedings of SenSys 2004, pp. 25-38, 2004.

[4] L. Girod and J. Elson and A. Cerpa and T. Stathopoulos and N. Ramanathan and D. Estrin, "EmStar: a Software Environment for Developing and Deploying Wireless Sensor Networks," USENIX, 2004.

[5] http://www.jamesreserve.edu

Ayushman: A Wireless Sensor Network Based Health Monitoring Infrastructure and Testbed*

K. Venkatasubramanian, G. Deng, T. Mukherjee, J. Quintero,
V. Annamalai, and S.K.S. Gupta

Arizona State University, Tempe AZ 85281, USA
sandeep.gupta@asu.edu

With a rapidly aging population, automated health monitoring systems provide an effective means of reducing the resulting healthcare professional shortage. To this end, we at the IMPACT lab at Arizona State University are developing Ayushman, a sensor network based health monitoring infrastructure and testbed[1]. Ayushman provides a medical monitoring system that is dependable, energy-efficient, secure, and collects real-time health data in diverse scenarios, from home based monitoring to disaster relief. Further, Ayushman is designed to be a testbed which allows researchers to test their communication protocols and systems in a realistic environment. Fig.1 presents Ayushman's architecture,

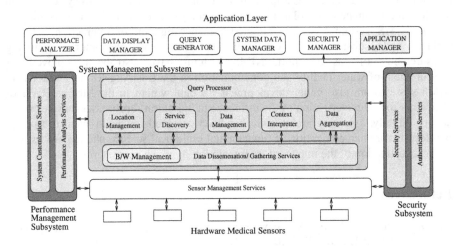

Fig. 1. Ayushman: Architecture

which is organized as a collection of services. Apart from the *application layer* and *sensor management service* (which provides an abstraction of the medical sensors), Ayushman has three main components: *the system management subsystem*, which provides health monitoring services, *the performance management*

---

* This work is supported by NSF grant ANI-0086020.
[1] Ayushman Project - http://shamir.eas.asu.edu/~mcn/Ayushman.html

V. Prasanna et al. (Eds.): DCOSS 2005, LNCS 3560, pp. 406–407, 2005.

subsystem, which monitors and controls the entire system for testing purposes and *the security sublayer,* which provides security and authentication services to all components. So far, we have implemented the query processor and data storage/gathering services, and we are currently working on resource management and reliability issues.

Studying Upper Bounds on Sensor Network Lifetime by Genetic Clustering

Min Qin and Roger Zimmermann

Computer Science Dept., University of Southern California, Los Angeles, CA 90089
mqin, rzimmerm @usc.edu

1 Introduction

In this paper we introduce the maximum lifetime sensor clustering (MLSC) protocol for calculating the upper bound on the lifetime of a sensor network. The new protocol can maximize the K-of-N network lifetime. Also, a novel genetic clustering algorithm is introduced to form optimal clusters. Simulation results show that our model can extend the sensor network lifetime to five times over that of existing approaches.

2 MLSC Protocol

In MLSC, sensors are organized into clusters and a linear programming model is introduced for calculating a cluster head rotation schedule. To reduce time and space complexity of the algorithm, we use a genetic algorithm to divide and conquer the original problem. Through a series of crossover and mutation functions, the offsprings of the initial solutions progress towards the optimal solution. Figure 1 shows the result of our algorithm compared to several other algorithms [1, 2].

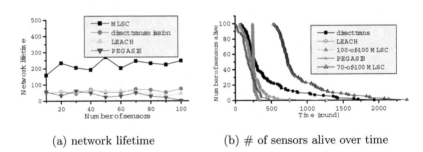

(a) network lifetime   (b) # of sensors alive over time

Fig. 1. Simulation results of the MLSC protocol

References

1. Heinzelman, W., Chandrakasan, A., Balakrishnan, H.: Energy-Efficient Communication Protocols for Wireless Microsensor Networks. Proc. of the Hawaiian Int'l Conf. on Systems Science, January 2000.
2. Lindsey, S., Raghavendra, C. S.: PEGASIS: Power Efficient GAthering in Sensor Information Systems. 2002 IEEE Aerospace Conference, March 2002, pp. 1-6.

V. Prasanna et al. (Eds.): DCOSS 2005, LNCS 3560, p. 408, 2005.
© Springer-Verlag Berlin Heidelberg 2005

Sensor Network Coverage Restoration

Nitin Kumar, Dimitrios Gunopulos, and Vana Kalogeraki

Department of Computer Science and Engineering,
University of California, Riverside, California, 92521
nkumar, dg, vana @cs.ucr.edu

Abstract. Wireless sensor networks are emerging as a new computational platform consisting of small, low-power and inexpensive nodes that integrate a modest amount of sensing, computation and wireless communication capabilities. These have found popular applications in a broad set of areas including environmental monitoring, habitat monitoring and disaster recovery. Typically sensor nodes are deployed over a geographical area for the purpose of detecting, tracking and monitoring events of interest. Reports produced upon the observation of specific events are then processed locally at the sensor nodes and transmitted over multiple hops to a centralized sink in order to reach an operations center or to be analyzed further. Since sensor nodes are deployed in a large land region, the objective is to achieve complete coverage of the region, that is, every location in the region lies in the observation field of at least one sensor node. However the initial placement of sensors may not achieve this goal for various reasons: the number of original sensors may have been too low, the original placement may have been random (for example, sensors deployed from the air) leaving parts of the region uncovered, or, some of the sensors have malfunctioned, leaving coverage holes.

In this paper we study the coverage restoration problem in sensor networks. The fundamental question is *"Given a two-dimensional area, a piece of land for example, and an initial set of sensors, how can we determine the number of sensor nodes required to completely cover the region"*. Essentially, the coverage restoration problem reflects how well an area is monitored by sensors. In abstract terms, our approach determines uncovered area in the sensor network field and proposes the deployment of nodes to completely cover the area. Our mechanism consists of two steps: (a) estimating the regions uncovered by sensors and (b) identifying the minimum number and location of sensors required to cover this region. The key idea of our technique is to make an efficient and yet very accurate representation of the uncovered area that uses techniques from discrepancy theory. By representing the uncovered area as a set of points, we can use efficient and simple algorithms for finding small sets of sensors to cover the uncovered areas. We partition the sensor network into cells, and run these algorithms locally. We formulate this problem as a disk covering problem where the goal is to cover a set of points on the plane by a set of disks. This problem is known to be NP-complete. However, there exist various approximate solutions that run in polynomial time and have a bounded error ratio. We use one of such proposed algorithm for our experimental purpose. The technique we propose is distributed, and minimizes the communication costs. We present a comprehensive set of experiments that demonstrate that our technique is highly effective in achieving a good coverage of a given sensor network area.

V. Prasanna et al. (Eds.): DCOSS 2005, LNCS 3560, p. 409, 2005.
© Springer-Verlag Berlin Heidelberg 2005

A Biologically-Inspired Data-Centric Communication Protocol for Sensor Networks

Naoki Wakamiya, Yoshitaka Ohtaki, Masayuki Murata, and Makoto Imase

Graduate School of Information Science and Technology, Osaka University
Suita, Osaka 565-0871, Japan
wakamiya, y-ohtaki, murata, imase @ist.osaka-u.ac.jp

1   Introduction

In sensor networks, each application has unique characteristics different from others, including the area of region of deployment, a type of object to monitor or obtain information of, the number of sensor nodes, the number of sources and sinks, and the frequency of needs for sensor data. In addition, those characteristics dynamically change in accordance with changes in conditions of surroundings, a topology of a sensor network, and user's requirements. Therefore, a communication protocol for sensor networks must be adaptive to diverse and dynamically changing characteristics of applications and networks.

In this paper, we propose a new communication protocol for robust and scalable sensor networks. Our protocol, ARCP (Ant-based Rendezvous Communication Protocol) is based on the foraging behavior of ants. Ants generated at sensor nodes which obtain or need sensor data wander around a sensor network while leaving pheromones. When an ant finds a trail of the other at a node, the node becomes a rendezvous point where ants meet and sensor data are passed from one to another. Since each ant moves on its own decisions in accordance with pheromones and no one ant has the dominant influence, ARCP is a fully distributed, robust, and scalable protocol.

2   Ant-Based Rendezvous Communication Protocol

In our ARCP, ants are emitted from both of sources and sinks. *Data provision ants* are generated at sources at the rate R_s. *Data gathering ants* are generated at sinks at the rate S_g. To receive sensor data at the desired rate, S_g is updated with the current data reception rate R_c and the desired reception rate R_t as $S_g^+ = S_g + \beta(R_t - R_c)$. Each of ant is categorized into forward and backward ant depending on its direction of migration. Forward ants wander around a sensor network. Forward data provision ants leave data provision pheromones which indicate the direction of sources. Forward data gathering ants leave data gathering pheromones which indicate the direction of sinks.

When an ant finds a trail, *i.e.*, pheromones, of the other type of ant, a rendezvous point (RP) is established at the node. An ant which established or finds an RP becomes a backward ant. By following pheromones, it goes back to

V. Prasanna et al. (Eds.): DCOSS 2005, LNCS 3560, pp. 410–411, 2005.
© Springer-Verlag Berlin Heidelberg 2005

its originating node while leaving rendezvous pheromones at nodes it traverses. Other forward ants from sinks or sources can easily reach the rendezvous point by following rendezvous pheromones. The probability that a forward ant chooses the next-hop node n is given as $P_r(n) = T_r(i, n)/\sum_{j \in N_i} T_r(i, j)$, where N_i is a set of neighbors of node i and $T_r(i, n)$ corresponds to the rendezvous pheromone value to node n. To find other trails or RPs, forward ants ignore rendezvous pheromones at the probability p and choose the next-hop node at random. Data provision ants put sensor data at a rendezvous point and data gathering ants take sensor data of interest from a rendezvous point. All pheromones are defined by data-oriented attributes. Thus, for example, rendezvous pheromones attract both forward ants as far as they carry or look for sensor data of the same attributes. Pheromones are accumulated by ants and evaporate as time passes.

Since there is no centralized control and ants behave independently of others, there occasionally appear many rendezvous points. In addition, since rendezvous points are accidentally generated, they are not located at preferable positions. We consider that an RP which is located at the center, from a viewpoint of hop count, TTL of ants, and emission rate of ants, of sources and sinks is appropriate. In our proposal, each ant pulls an RP in accordance with the distance from its origin to the RP. When a forward ant arrives an RP, it moves the RP to the node from which it came with probability $\gamma h/TTL$, where h corresponds to the number of hops that the ant experienced to the RP and γ is a constant ($0 < \gamma < 1$). TTL is a initial TTL value of the ant set by its origin. When two or more RPs are moved to a node, they are merged together.

3   Simulation and Evaluation

Figures 1 and 2 show simulation results to compare the adaptability of our protocol to the number of sinks and sources in a sensor network of 60 nodes with that of Directed Diffusion protocols. Although not shown in figures, in all cases, all sinks received sensor data at the expected rate. The reason that ARCP introduced much load as the number of sinks increased is that backward data gathering ants did not return to their originating sinks by being attracted by data gathering pheromones of other sinks. This is left as a future work.

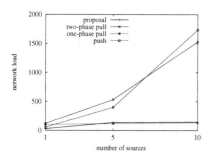

Fig. 1. Network load (one sink)

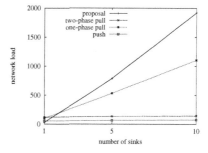

Fig. 2. Network load (one source)

RAGOBOT: A New Platform for Wireless Mobile Sensor Networks

Jonathan Friedman, David Lee, Ilias Tsigkogiannis, Sophia Wong, Dennis Chao, David Levin, William Kaiser, and Mani Srivastava

University of California, Los Angeles, Los Angeles CA 90024, USA

Abstract. We present Ragobot, a fully-featured validated platform for use in mobile sensor networks. Ragobot is a robot of small dimensions with features that surpass those provided by many other robots in this category. Ragobot hardware and software are implemented with modularity as one of the main considerations; therefore, these are easy to upgrade and customize according to the needs of each specific application. Moreover, we present Ragoworld, a controlled physical space for the development and evaluation of mobile sensor network algorithms.

1 RAGOBOT Project

To explore mobility in wireless sensor networks, we created Ragobot, a small-sized robot platform that provides a large number of features, some not supported even by larger-size robots. At only 13.3 cm long by 6cm wide, Ragobot is positioned at the intersection of size and capability. Ragobot is able to operate on terrain that is impassable by other platforms ([1], [2], [4], [3]) while being smaller than a Cotsbot and an Amigobot, and only consuming about as much power (4W at full power, 40mW at idle, and even less during sleep) as a Robomote. Significant emphasis was placed on the analog design of Ragobot allowing for simultaneous movement and sensor operation without noticeable degradation of sensor performance as occurs in other platforms ([2], [3]). For more information visit http://www.ragobot.com.

References

1. ActiveMedia, Inc., www.amigobot.com
2. Sarah Bergbreiter, Kristofer S. J. Pister, "CotsBots: An Off-the-Shelf Platform for Distributed Robotics" IROS 2003, Las Vegas, NV, October 27-31, 2003.
3. Karthik Dantu, Mohammad H. Rahimi, Hardik Shah, Sandeep Babel, Amit Dhariwal, and Gaurav Sukhatme, "Robomote: Enabling mobility in sensor networks" International Conference on Information Processing in Sensor Networks, 2005.
4. Luis Navarro-Serment, Robert Grabowski, Chris Paredis, Pradeep K. Khosla, "Modularity in Small Distributed Robots", Proceedings of the SPIE conference on Sensor Fusion and Decentralized Control in Robotic Systems II, September, 1999.

V. Prasanna et al. (Eds.): DCOSS 2005, LNCS 3560, p. 412, 2005.

Energy Conservation Stratedgy for Sensor Networks

Hyo Jong Lee

Chonbuk National University,
Jeonju, Jeonbuk Prov., Korea

Abstract. A simple but efficient power conserving algorithm for the sensor networks is presented in this paper. The goal is to predict sleep time for the sensor networks based on statistics of data transfer. The proposed sensors in this paper can increase the sensor's life time from 247% to 600% within the range of 1% tolerable data error.

1 Power Conservation for Quality Controlled Sensor Networks

The only possible way to save power consumption from sensors is to predict how long the current data will stay within a specified error tolerance. If the probability that the sensed value stays is high, the sensor should go to *sleeping* mode; otherwise, it should stay in *active* mode. The principle of data sample is described in the following section before we explain the algorithm.

The picewise constant approximation was developed during the Quality-Aware Sensor Architecture project at UC Irvine. The method explores the characteristics of inertial data to archive. It was designed to work as batch-mode for archival data. Thus, it cannot be used for real-time data compression. Poor Man's Compression(PMC) was used to reduce the size of a time series data. It requires $O(1)$ of computational space and time. It was proved that the maximum error of time series is bound to 2ϵ. The compression algorithm can be implemented using either the midrange or mean of the sensor readings. The difference of two algorithms is not significant, although the PMC*midrange* is slightly better than the PMC *mean one*.

In the real time PMC mean value is adjusted with ϵ according to data change direction. When a new value is out of the error range and its value is greater than the mean value, a new mean value will be adjusted by adding ϵ to move upwardly. If a new value is smaller than the mean value, a new mean value will be adjusted by subtracting ϵ to lower the current mean value.

The new modified piecewise constant approximation shows significant power conservation compared to the regular PMC. The life span of sensors can be increased from 247% to 600% with 1% of error tolerance. The rate of power conservation becomes higher with greater error tolerance.

V. Prasanna et al. (Eds.): DCOSS 2005, LNCS 3560, p. 413, 2005.

Meteorological Phenomena Measurement System Based on Embedded System and Wireless Network

Kyung-Bae Chang[1], Il-Joo Shim[1], Seung-Woo Shin[1], and Gwi-Tae Park[1]

[1] ISRL, Korea University, 1, 5ga Anam-dong Sungbuk-Gu,
136-713 Seoul, South Korea
{lslove, ijshim, ezshini, gtpark}@korea.ac.kr
http://control.korea.ac.kr

Abstract. This paper aims at the reliable gathering of meteorological data of all the regions, by the use of wireless networks, in order to prevent meteorological disasters. For simple installation in areas like mountains, embedded system technology, which shows high performances and small sizes, to process the meteorological data.

1 Introduction and Conclusion

Because the countries 70% is consisted with mountain areas, base-station constructions for the gathering of meteorological data of specific regional (mountains, valleys, islands) with a wired transmission system is difficult and expensive. Also, this interferes with the construction of infra for the prevention of disasters resulting from recent meteorological changes. In order to solve these problems, this paper proposes a method, which uses a highly efficient embedded system to consist AWS. And also, by the world's first use of CDMA 1x EVDO services, we propose a method of gathering meteorological data by SMS service using domestic wireless networks, which leads the wireless internet market.

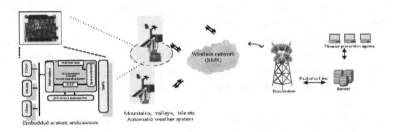

Fig. 1. Automatic Weather Observation system

By using the existing wireless networks, system construction on far regions, like mountains or islands, can be a lot easier and economically profitable. Research involving the integration of the wireless network of meteorological data using Wibro, which is a high speed and high capacity portable internet network that will be constructed hereafter, is being planned.

V. Prasanna et al. (Eds.): DCOSS 2005, LNCS 3560, pp. 414–414, 2005.

An Architecture Model for Supporting Power Saving Services for Mobile Ad-Hoc Wireless Networks[*]

Nam-Soo Kim[1], Beongku An[2] and Do-Hyeon Kim[3]

[1] School of Computer and Communication Engineering,
Cheongju University, Cheongju, Chungbook, 360-764, South Korea
nskim@cju.ac.kr
[2] Department of Electrical, Electronic & Computer Engineering,
Hongik University Jochiwon, Chungnam, 339-701, South Korea
beongku@wow.hongik.ac.kr
[3] Department of Computer & Telecommunication Engineering,
Cheju National University, Cheju, 690-756, South Korea
kimdh@cheju.ac.kr

Abstract. In this paper, we propose an architecture model for supporting power saving services in mobile ad-hoc wireless networks. The main feature of the proposed power saving architecture is the combination of the following two ideas such as the direction guided transmission for the information transmission using a directional antenna and the receive antenna diversity with maximal ratio combining (MRC) technique. The numerical results show the transmission power can be reduced to 22.47*(N-1) dB, where N mobile nodes within the direction guided transmission area have directional transmit antennas with the half-power beamwidth (HPBW) of radian and two branch receiving diversity in mobile ad-hoc wireless networks.

1 Introduction

Mobile ad-hoc wireless networks are self-organized among mobile nodes which function as both host and router with wireless interface. These mobile nodes that have the problems of power supply using battery have the functions of data transmission as intermediate nodes because the radio distance is limited. That is the battery capacity of mobile node limits the network survival and routing. In this paper, we propose an architecture model for supporting power saving services using the combination of the direction guided transmission protocol (DGTP) and receiver diversity technology in mobile ad-hoc wireless networks[1][2].

2 The Proposed Architecture Model

For supporting the power saving services in mobile ad-hoc wireless networks[2], we propose the combination architecture model of the direction guided trans-

[*] This work was supported by the Korea Research Foundation for research program of regional universities.

V. Prasanna et al. (Eds.): DCOSS 2005, LNCS 3560, pp. 415–416, 2005.

mission based on the known geographical information of the target location and the receive antenna diversity[1] with MRC. The proposed architecture model is shown in fig. 1(a). The proposed architecture model consists of two power saving structures as follows. The first structure is the power saving scheme using receive antenna diversity while the second structure is the power saving scheme using direction guided transmission protocol(DGTP).

3   Numerical Results and Conclusion

For the numerical analysis, we assume the horizontal transmit angle from a mobile node is $2\pi/3$ [rad]. The gain from a pair of mobile nodes is 4.77 dB, and the total transmit gain in mobile ad-hoc wireless networks with N mobile nodes within the direction guided area becomes 4.77 *(N-1) dB. Fig. 1(b)shows the SNR gain as function of number of receive antennas. In conclusion, the proposed combination architecture model of the direction guided transmission protocol with the receive antenna diversity can drastically reduce the transmit power in mobile ad-hoc wireless net-works[2] and can also prolong the network liability.

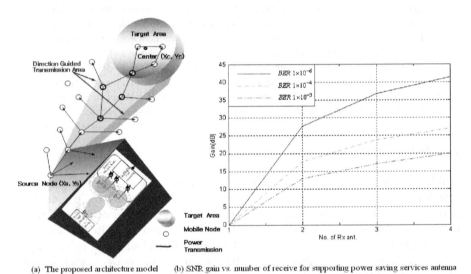

(a) The proposed architecture model   (b) SNR gain vs. number of receive for supporting power saving services antenna

Fig. 1. The proposed architecture model and numerical results

References

1. W. Stutzman, G. Thiele, Antenna theory and design, John Wiley & Sons, 1998. 2.
2. Beongku An, Symeon Papavassiliou, "An Entropy-Based Model for Supporting and Evaluating Stability in Mobile Ad-hoc wireless Wireless Networks," IEEE Communications Letters, vol. 6, no. 3, August 2002.

Distributed Recovery Units for Demodulation in Wireless Sensor Networks

Mostafa Borhani and Vafa Sedghi

Sharif University of Technology, Tehran Polytechnic University of Technology
borhani@ee.sharif.edu
vafa_sedghi@yahoo.com

Abstract. In this paper, the distributed recovery units for FM and PPM demodulation in wireless sensor networks is proposed and is showed that, the method introduces very little amount of distortion to recovered signal. With simulations, we have examined several different parameters. Besides, two different filtering schemes are simulated. Effect of iteration coefficient (λ) on stability and convergence rate is investigated for different sampling and filtering methods. Next, we have studied the noise effect in demodulation with this method. Simulations results are included to verify convergence.

1 Introduction

Demodulation techniques is one the most important topics in the wireless networks. By using the distributed techniques we can remove vast computational complexity of demodulation blocks.

In this work, we concentrate on distributed recovery units for FM and PPM demodulation in wireless sensor networks and employ techniques to demodulate signals. We have simulated this technique and examined the performances of this method for various parameters.

2 Recovery Unit

In [1-2] a method called Iterative method is proposed to reduce the distortion. A block diagram of this method is shown in figure 1. In this figure, the G operator is all that is FM demodulator or PPM demodulator. The number of blocks in this figure is equal to the number of blocks. In reference [2] it's shown that when number of iterations approach infinity, the distortion of signal reaches zero.

In FM/PPM demodulation, G=PS, that S means sampling and P low pass filter therefore i[th] iteration output signal equals:

$$x_i(t) = \lambda x_0(t) + (1 - \lambda G)x_{i-1}(t) \tag{1}$$

Filtering may be done using FFT or FIR methods. Besides, the constant λ can affect rate of improvement of signal and it can make the system unstable, if not selected properly. All of these effects are discussed in the following sections.

V. Prasanna et al. (Eds.): DCOSS 2005, LNCS 3560, pp. 417–419, 2005.

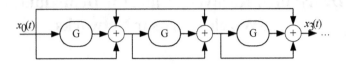

Fig. 1. Block diagram of iterative method

3 Simulation Results

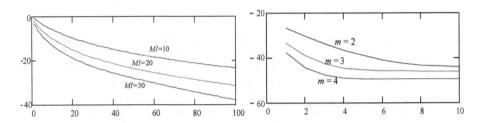

Fig. 2. MSE as a function of number of blocks for different sampling rate for PPM(left) /FM (right)demodulator

3.1 The Lambda effect

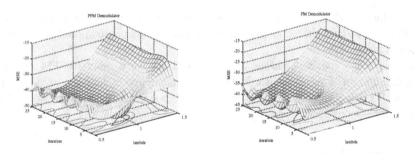

Fig. 3. The lambda effect for iterative method in PPM(left)/FM(right) demodulator

3.2. The Noise effect

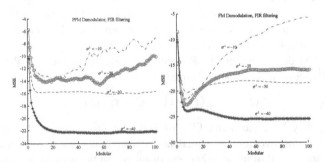

Fig. 4. The noise effect on iterative method for PPM(left)/FM(right) demodulator

4   Conclusions

In this paper, we examined iterative method for demodulation of FM and PPM signals. Our simulations proved that, the method is strongly capable to reconstruct the message signal. One can see that, the amount of distortion decreases as the number of blocks increases in FM and PPM demodulators. Two schemes for filtering are considered and the effect of choosing lambda is studied in both FM and PPM demodulators.

We see that iterative method is very sensitive to input noise and its performance degrades a great amount when the noise is presented at the input.

Acknowledgement. With Great thanks from Dr. F.A. Marvasti.

References

[1] F. Marvasti, A Unified Approach to Zero-Crossing and Nonuniform Sampling of Single and Multidimensional Signals and Systems, Nonuniform Publication, 1987.

[2] F. Marvasti, *"The reconstruction of a signal from the FM zero-crossing"*, Trans. of IECE of Japan, vol. E68, no. 10, Oct. 1985.

System Integration Using Embedded Web Server and Wireless Communication

Kyung-Bae Chang[1], Il-Joo Shim[1], Tae-Kook Kim[1], and Gwi-Tae Park[1]

[1] ISRL, Korea University, 1,
5ga Anam-dong Sungbuk-Gu 136-713 Seoul,
South Korea
{lslove, ijshim, prince, gtpark}@korea.ac.kr
http://control.korea.ac.kr

Abstract. As the buildings become intelligent with the development of infor-mation communication, many facilities have appeared. Usually, each facility performed independent functions, but with the development of the internet, the facilities integrate with each other and perform various new abilities. At pre-sent, in order to offer linked service by integrating, PC servers are being used. However, the use of a PC server is expensive, consumes a lot of electricity and needs a lot of room. Therefore, this paper proposes the Embedded Web Server, which can substitute the existing PC server and control devices through the Web. Also, we propose integration through wireless communication, which can solve problems like complicating wiring and space restriction.

1 Introduction and Conclusion

With the development of information communication techniques and internet ser-vices, buildings are becoming more intelligent. For example, the control and monitor-ing of the building's air-conditioning/heating devices, illumination and security de-vices can be done remotely.

At present, a PC server is used to integrate facilities and control through the inter-net. These PC servers are expensive, consume a lot of power and needs a lot of space. To solve these problems, we propose the small sized Embedded Web Server.

The embedded web server provides various interfaces like Ethernet, RS232, USB, IrDa, PCMCI and CF for the PC's functions and system integration. Because it per-forms one specific function, to integrate the many facility's systems, a low costing, small sized, low power embedded web server can be designed.

This paper proved that by the use of the proposed Embedded Web server and wire-less communication, the building's system can be integrated, controlled and monitored.

V. Prasanna et al. (Eds.): DCOSS 2005, LNCS 3560, p. 420, 2005.
© Springer-Verlag Berlin Heidelberg 2005

Author Index

Lecture Notes in Computer Science

For information about Vols. 1–3452

please contact your bookseller or Springer

Vol. 3504: A.F. Frangi, P.I. Radeva, A. Santos, M. Hernandez (Eds.), Functional Imaging and Modeling of the Heart. XV, 489 pages. 2005.

Vol. 3503: S.E. Nikoletseas (Ed.), Experimental and Efficient Algorithms. XV, 624 pages. 2005.

Vol. 3502: F. Khendek, R. Dssouli (Eds.), Testing of Communicating Systems. X, 381 pages. 2005.

Vol. 3501: B. Kégl, G. Lapalme (Eds.), Advances in Artificial Intelligence. XV, 458 pages. 2005. (Subseries LNAI).

Vol. 3500: S. Miyano, J. Mesirov, S. Kasif, S. Istrail, P. Pevzner, M. Waterman (Eds.), Research in Computational Molecular Biology. XVII, 632 pages. 2005. (Subseries LNBI).

Vol. 3499: A. Pelc, M. Raynal (Eds.), Structural Information and Communication Complexity. X, 323 pages. 2005.

Vol. 3498: J. Wang, X. Liao, Z. Yi (Eds.), Advances in Neural Networks – ISNN 2005, Part III. L, 1077 pages. 2005.

Vol. 3497: J. Wang, X. Liao, Z. Yi (Eds.), Advances in Neural Networks – ISNN 2005, Part II. L, 947 pages. 2005.

Vol. 3496: J. Wang, X. Liao, Z. Yi (Eds.), Advances in Neural Networks – ISNN 2005, Part II. L, 1055 pages. 2005.

Vol. 3495: P. Kantor, G. Muresan, F. Roberts, D.D. Zeng, F.-Y. Wang, H. Chen, R.C. Merkle (Eds.), Intelligence and Security Informatics. XVIII, 674 pages. 2005.

Vol. 3494: R. Cramer (Ed.), Advances in Cryptology – EUROCRYPT 2005. XIV, 576 pages. 2005.

Vol. 3493: N. Fuhr, M. Lalmas, S. Malik, Z. Szlávik (Eds.), Advances in XML Information Retrieval. XI, 438 pages. 2005.

Vol. 3492: P. Blache, E. Stabler, J. Busquets, R. Moot (Eds.), Logical Aspects of Computational Linguistics. X, 363 pages. 2005. (Subseries LNAI).

Vol. 3489: G.T. Heineman, I. Crnkovic, H.W. Schmidt, J.A. Stafford, C. Szyperski, K. Wallnau (Eds.), Component-Based Software Engineering. XI, 358 pages. 2005.

Vol. 3488: M.-S. Hacid, N.V. Murray, Z.W. Raś, S. Tsumoto (Eds.), Foundations of Intelligent Systems. XIII, 700 pages. 2005. (Subseries LNAI).

Vol. 3486: T. Helleseth, D. Sarwate, H.-Y. Song, K. Yang (Eds.), Sequences and Their Applications - SETA 2004. XII, 451 pages. 2005.

Vol. 3483: O. Gervasi, M.L. Gavrilova, V. Kumar, A. Laganà, H.P. Lee, Y. Mun, D. Taniar, C.J.K. Tan (Eds.), Computational Science and Its Applications – ICCSA 2005, Part IV. XXVII, 1362 pages. 2005.

Vol. 3482: O. Gervasi, M.L. Gavrilova, V. Kumar, A. Laganà, H.P. Lee, Y. Mun, D. Taniar, C.J.K. Tan (Eds.), Computational Science and Its Applications – ICCSA 2005, Part III. LXVI, 1340 pages. 2005.

Vol. 3481: O. Gervasi, M.L. Gavrilova, V. Kumar, A. Laganà, H.P. Lee, Y. Mun, D. Taniar, C.J.K. Tan (Eds.), Computational Science and Its Applications – ICCSA 2005, Part II. LXIV, 1316 pages. 2005.

Vol. 3480: O. Gervasi, M.L. Gavrilova, V. Kumar, A. Laganà, H.P. Lee, Y. Mun, D. Taniar, C.J.K. Tan (Eds.), Computational Science and Its Applications – ICCSA 2005, Part I. LXV, 1234 pages. 2005.

Vol. 3479: T. Strang, C. Linnhoff-Popien (Eds.), Location- and Context-Awareness. XII, 378 pages. 2005.

Vol. 3478: C. Jermann, A. Neumaier, D. Sam (Eds.), Global Optimization and Constraint Satisfaction. XIII, 193 pages. 2005.

Vol. 3477: P. Herrmann, V. Issarny, S. Shiu (Eds.), Trust Management. XII, 426 pages. 2005.

Vol. 3476: J. Leite, A. Omicini, P. Torroni, P. Yolum (Eds.), Declarative Agent Languages and Technologies. XII, 289 pages. 2005.

Vol. 3475: N. Guelfi (Ed.), Rapid Integration of Software Engineering Techniques. X, 145 pages. 2005.

Vol. 3474: C. Grelck, F. Huch, G.J. Michaelson, P. Trinder (Eds.), Implementation and Application of Functional Languages. X, 227 pages. 2005.

Vol. 3468: H.W. Gellersen, R. Want, A. Schmidt (Eds.), Pervasive Computing. XIII, 347 pages. 2005.

Vol. 3467: J. Giesl (Ed.), Term Rewriting and Applications. XIII, 517 pages. 2005.

Vol. 3466: S. Leue, T.J. Systä (Eds.), Scenarios: Models, Transformations and Tools. XII, 279 pages. 2005.

Vol. 3465: M. Bernardo, A. Bogliolo (Eds.), Formal Methods for Mobile Computing. VII, 271 pages. 2005.

Vol. 3464: S.A. Brueckner, G.D.M. Serugendo, A. Karageorgos, R. Nagpal (Eds.), Engineering Self-Organising Systems. XIII, 299 pages. 2005. (Subseries LNAI).

Vol. 3463: M. Dal Cin, M. Kaâniche, A. Pataricza (Eds.), Dependable Computing - EDCC 2005. XVI, 472 pages. 2005.

Vol. 3462: R. Boutaba, K.C. Almeroth, R. Puigjaner, S. Shen, J.P. Black (Eds.), NETWORKING 2005. XXX, 1483 pages. 2005.

Vol. 3461: P. Urzyczyn (Ed.), Typed Lambda Calculi and Applications. XI, 433 pages. 2005.

Vol. 3460: Ö. Babaoglu, M. Jelasity, A. Montresor, C. Fetzer, S. Leonardi, A. van Moorsel, M. van Steen (Eds.), Self-star Properties in Complex Information Systems. IX, 447 pages. 2005.

Vol. 3459: R. Kimmel, N.A. Sochen, J. Weickert (Eds.), Scale Space and PDE Methods in Computer Vision. XI, 634 pages. 2005.

Vol. 3458: P. Herrero, M.S. Pérez, V. Robles (Eds.), Scientific Applications of Grid Computing. X, 208 pages. 2005.

Vol. 3456: H. Rust, Operational Semantics for Timed Systems. XII, 223 pages. 2005.

Vol. 3455: H. Treharne, S. King, M. Henson, S. Schneider (Eds.), ZB 2005: Formal Specification and Development in Z and B. XV, 493 pages. 2005.

Vol. 3454: J.-M. Jacquet, G.P. Picco (Eds.), Coordination Models and Languages. X, 299 pages. 2005.

Vol. 3453: L. Zhou, B.C. Ooi, X. Meng (Eds.), Database Systems for Advanced Applications. XXVII, 929 pages. 2005.